T0132618

POLITICAL ESSAY ON
THE KINGDOM OF NEW SPAIN

VOLUME 2

ALEXANDER
VON HUMBOLDT
IN ENGLISH

A series edited by Vera M. Kutzinski and Ottmar Ette

POLITICAL ESSAY ON THE KINGDOM OF NEW SPAIN

VOLUME 2

A Critical Edition

ALEXANDER VON HUMBOLDT

Edited with an Introduction
by Vera M. Kutzinski and Ottmar Ette

Translated by J. Ryan Poynter, Kenneth Berri, and Vera M. Kutzinski

With Annotations by Giorleny D. Altamirano Rayo,
Tobias Kraft, and Vera M. Kutzinski

THE UNIVERSITY OF CHICAGO PRESS CHICAGO AND LONDON

The University of Chicago Press, Chicago 60637
The University of Chicago Press, Ltd., London

For more information, contact the University of Chicago Press, 1427 E. 60th St., Chicago, IL 60637.
Published 2019
Printed in the United States of America

28 27 26 25 24 23 22 21 20 19   1 2 3 4 5

ISBN-13: 978-0-226-65155-2   (cloth)
ISBN-13: 978-0-226-65169-9   (e-book)
DOI: https://doi.org/10.7208/chicago/9780226651699.001.0001

Library of Congress Cataloging-in-Publication Data
Names: Humboldt, Alexander von, 1769–1859, author. | Kutzinski, Vera M., 1956– editor, translator, writer of introduction, writer of added commentary. | Ette, Ottmar, editor, writer of introduction. | Poynter, J. Ryan, translator. | Berri, Kenneth, translator. | Altamirano Rayo, Giorleny D., writer of added commentary. | Kraft, Tobias, 1978– writer of added commentary. | Humboldt, Alexander von, 1769–1859. Works. English.
Title: Political essay on the Kingdom of New Spain : a critical edition / Alexander von Humboldt ; edited with an introduction by Vera M. Kutzinski and Ottmar Ette ; translated by J. Ryan Poynter, Kenneth Berri, and Vera M. Kutzinski ; with annotations by Giorleny Altamirano Rayo, Tobias Kraft, and Vera M. Kutzinski.
Other titles: Essai politique sur le royaume de la Nouvelle-Espagne. English (Kutzinski and Ette)
Description: Chicago ; London : The University of Chicago Press, 2019. | Includes bibliographical references and index.
Identifiers: LCCN 2019005351 | ISBN 9780226651385 (volume 1 ; cloth : alk. paper) | ISBN 9780226651415 (volume 1 ; e-book) | ISBN 9780226651552 (volume 2 ; cloth : alk. paper) | ISBN 9780226651699 (volume 2 ; e-book)
Subjects: LCSH: Mexico—Description and travel. | Mexico—Geography. | Mexico—Population. | Mexico—Statistics. | Mines and mineral resources—Mexico. | Mexico—Social conditions—To 1810. | Agriculture—Mexico. | Mexico—Politics and government—1540–1810.
Classification: LCC F1211 .H92613 2019 | DDC 917.2—dc23
LC record available at https://lccn.loc.gov/2019005351

♾ This paper meets the requirements of ANSI/NISO Z39.48-1992 (Permanence of Paper).

Contents

VOLUME 3

X Plants That Furnish Raw Materials for Manufacturing and for Trade—Raising Livestock—Fishing—Agricultural Yield Estimated According to the Amount of the Tithes. *3*

XI State of the Mining of New Spain—Gold and Steel Production—Average Wealth in Ore—Amount of Mercury Consumed Annually in the Amalgamation Process—Amount of Precious Metals That Have Flowed from the One Continent to the Other Since the Conquest of Mexico. *61*

General Tableau of the Mines of New Spain 70

I. Intendancy of Guanajuato 70

II. Intendancy of Zacatecas 70

III. Intendancy of San Luis Potosí 71

IV. Intendancy of Mexico City 71

V. Intendancy of Guadalajara 72

VI. Intendancy of Durango 73

VII. Intendancy of Sonora 73

VIII. Intendancy of Valladolid 74

IX. Intendancy of Oaxaca 75

X. Intendancy of Puebla 75

XI. Intendancy of Veracruz 75

XII. Old California 75

Output of the Mining District of Guanajuato 100

Comparative Tableau of the Mines of the Americas and Europe 117

VOLUME 4

BOOK V. The State of Manufacturing and Trade in New Spain *289*

XII The Manufacturing Industry—Cottons—Woolens—Cigars—Soda and Soap—Gunpowder—Coins—The Exchange of Products—Trade in the Interior—Roads—External Trade through Veracruz and Acapulco—Obstacles to This Trade—Yellow Fever. *291*

Tableau I: Veracruz Trade Balances in 1802 *322*

A. *Spain's Imports in Mexico, in Agricultural Products and Products of Domestic Industry* *322*

B. *Spanish Imports in Mexico, in Agricultural Products and Products from Foreign Industry* *324*

C. *America's Imports (Spanish Colonies) from Mexico* *325*

D. *Mexican Exports to Spain* *326*

E. *Mexican Exports to Other Parts of Spanish America* *327*

Results: Veracruz Trade Balance in 1802 *328*

Observations *328*

Tableau II: Trade Balance of Veracruz in 1803 *330*

A. *Imports of National Agricultural Products from Spain to Mexico* *330*

B. *Imports of Products of National Industry from Spain to Mexico* *331*

C. *Imports of Agricultural Products and Products of Foreign Industries from Spain to Mexico* *332*

D. *Imports from America (Spanish Colonies) to Mexico* *333*

E. *Exports from Mexico to Spain* *334*

F. *Exports from Mexico to Other Parts of Spanish America* *335*

Results: Veracruz Trade Balances in 1803 *336*

Observations *336*

Veracruz Trade in 1804 *338*

From the Port of Alvarado *341*

From the Port of Veracruz *341*

Total Trade Balance of Veracruz, from 1796 to 1820 *342*

Tableau I: Gross Yield from the National Revenue of New Spain *372*

Tableau II *373*

A. *Value of Precious Metals Sent on Behalf of the Crown from Veracruz to Spain* *373*

B. *Amount of Piasters Sent from Veracruz to Cádiz and to the Antilles on Behalf of the Crown* *373*

C. *Export of Precious Metals from Veracruz to Havana, Puerto Rico, and Louisiana, as Much on Behalf of the Crown (as Situados) as by Individuals* *374*

Results *374*

Tableau III: Amount of Piasters Exported from Veracruz to Spain and the Spanish Colonies, as Much on Behalf of the Crown as by Individuals *375*

Imports *375*

Exports *375*

Imports and Exports from the Spanish Colonies of the New Continent *386*

Meteorological and Nosographic Table of Veracruz (Latitude 19°11′52″) on the Centigrade Thermometer *403*

The State of Hospitals in Veracruz in 1806 *419*

Average Temperature in Veracruz (Centigrade Thermometer) *421*

BOOK VI. State Revenue—Military Defense *425*

XIII The Current Revenue of the Kingdom of New Spain—Its Gradual Increase since the Beginning of the Eighteenth Century—Sources of Public Revenue. *427*

Comparative Tableau of the Revenue of New Spain *433*

XIV Collection Costs—Public Expenses—Situados—The Net Product Deposited in the Royal Treasury in Madrid—The State of the Military—National Defense—Final Summary. *434*

Budget of the Public Revenue of New Spain for 1803 *437*

The Finances of the Spanish Monarchy in 1804 *444*

Comparative Tableau for 1804 *446*

I. *General Tableau of the Army in 1804* *447*

II. *Detailed Tableau Representing the Divisions of the Line Troops* *448*

III. *Detailed Tableau Representing the Divisions of the Military* *449*

Notes *465*

Note A *465*

NOTE A *BIS* *466*

Note B *466*

A. *Singuilucan* *467*

B. *Dolores* *469*

Note C *470*

Census of the Population of Mexico City, Compiled in September 1820 *471*

Census of the Population of Mexico City in 1790 *472*

I. *Religious Orders (Male)* *472*

II. *Religious Orders (Female)* *474*

III. *Lay Persons* *476*

IV. *Castes* *477*

V. *Schools for Men* *478*

VI. *Schools for Girls* *479*

VII. *Hospitals* *480*

VIII. *Prisons* *482*

IX. *Inhabitants of Mexico City by Occupation* *483*

X. *Summary* *483*

Note D *484*

Note E *484*

Note F *484*

Supplement *489*

Astronomical Positions *489*

Measurements of Elevation *491*

Mining Output *492*

Excerpt from the Will of Hernán Cortés *495*

Geographical and Physical Atlas of the Kingdom of New Spain *507*

Index of Names *513*
Subject Index *545*
Toponym Index *579*

VOLUME 3

CHAPTER X

III.1

Plants That Furnish Raw Materials for Manufacturing and for Trade—Raising Livestock—Fishing—Agricultural Yield Estimated according to the Amount of the Tithes.

Although Mexican agriculture, like that of all countries that provide for the needs of their own population, is oriented primarily around foodstuffs, New Spain is nevertheless rich in strictly colonial commodities, that is, in products that furnish raw materials for trade and the manufacturing industry of Europe. In this respect, this vast kingdom combines the advantages of New England with those of the Antilles. It became competitive with the latter islands after the civil war in Saint-Domingue, when the devastation of the French sugar refineries made it profitable to produce colonial commodities on the American continent. It is worth noting that the growth of III.2 this production in Mexico has surpassed that of cereals. In these climes, the same area of land, for example an arpent of 6,400 square meters, produces eighty to one hundred francs' worth of wheat, 250 francs' worth of cotton, and 450 francs' worth of sugar.[1] Given this enormous difference in the value of the harvests, it is perhaps unsurprising that the Mexican colonist prefers the colonial commodities to barley and European wheat. But this predilection will not succeed in disturbing the balance that exists to this day between the different branches of agriculture, because, fortunately, a large part of New Spain has a climate that is cold rather than temperate and is not, therefore, suitable for growing sugar, coffee, cocoa, indigo, or cotton.

1. This is the estimate that colonists in 1804 saw as the most exact for Louisiana in the lands neighboring the city of New Orleans. There, one *acre* yields ten *bushels* of wheat, 258 lbs. of cotton, and 1,000 pounds of sugar. This is the average yield, but one can easily imagine the extent to which local circumstances must impact these results.

Sugarcane cultivation has experienced such rapid growth in recent years that exports of sugar from the port of Veracruz are currently over half a million arrobas (6,250,000 kg), which, at three piasters per arroba in 1803, equals seven and a half million francs. We have already observed above that the ancient Mexica were only familiar with bee-honey syrup, III.3 *metl* (agave) syrup, and corn syrup. Sugarcane, which has been grown in the East Indies, in China,[1] and in the South Sea islands since early antiquity, was introduced by the Spanish to the island of Saint-Domingue, whence it passed to the island of Cuba and thence to New Spain. Pierre d'Atienza planted the first sugarcanes around the year 1520,[2] on the outskirts of the city of Concepción de la Vega. ▼ Gonzalo de Velosa built the first cylinders; and, by 1535, on the island of Saint-Domingue, there were over thirty sugar refineries, several of which were staffed by a hundred black slaves [esclaves nègres] and which had incurred between ten and twelve thousand ducats in construction costs. It is worth noting that among the first sugarmills (*trapiches*) the Spanish built at the beginning of the sixteenth century, there were already some that were powered not by horses but by hydraulic wheels, although these same water mills (*trapiches* or *molinos de agua*) were recently introduced as a foreign invention by refugees from Cap Français.

III.4 In 1553, sugar was in such great abundance in Mexico that it was exported from Veracruz and Acapulco to Spain and Peru.[3] The latter exports ceased long ago, because Peru today produces more sugar than it needs for local consumption. Since the population of New Spain is concentrated in the interior of the country, fewer sugar refineries are found along the coasts,

1. I am even inclined to believe that the method we use to produce sugar came to us from eastern Asia. In Chinese paintings depicting the arts and crafts I saw in Lima, I recognized the cylinders placed on the field and set in motion by a rotary cutter, the boiler crews, and the purgeries, just as one sees in the Antilles islands.

2. Not in 1506, as is generally thought. Oviedo, who came to the Americas in 1513, clearly states that he witnessed the establishment of the first sugar refineries in Saint-Domingue. *Historia natural de Indias*, book IV, chap. 8.

3. "In addition to gold and silver, Mexico also produces a great deal of sugar and cochineal (two very precious goods), feathers and cotton. Very few Spanish ships return without cargo; this is not the case in Peru which nevertheless has the unfounded reputation of being richer than Mexico. This latter region has also maintained a larger number of inhabitants. It is a beautiful and very populous land, where nothing would be lacking if only the rains were more frequent. New Spain sends horses, beef, and sugar to Peru." This remarkable passage from López de Gómara, who so deftly portrays the condition of the Spanish colonies in the mid-sixteenth century, is found only in the edition of the *Conquista de México* published in Medina del Campo, 1553, folio 139. It is missing from the French translation printed in Paris in 1587, p. 191.

where the abundant rains might actually favor sugarcane cultivation, than on the slope of the Cordilleras and in the highest parts of the central plateau. The main plantations are in the intendancy of Veracruz, near the cities of Orizaba and Córdoba; in the intendancy of Puebla, near Huautla de las Amilpas, at the foot of Popocatépetl; in the intendancy of Mexico City, to the west of the Nevado de Toluca, and to the south of Cuernavaca, in the San Gabriel plains; in the intendancy of Guanajuato, near Celaya, III.5 Salvatierra, and Penjamo, and in the Santiago valley; in the intendancies of Valladolid and Guadalajara, to the southwest of Patzcuaro and Tecolotlan. Although the average temperature that best suits sugarcane is twenty-four or twenty-five degrees Centigrade, this plant can still be cultivated successfully in places where the average heat over the course of a year does not exceed nineteen or twenty degrees. But, since the heat drops by one degree on the Centigrade thermometer[1] with every two hundred meters of elevation, the mean temperature on the steep slopes of the mountains in the tropics is generally found to be twenty degrees at an elevation of 1,200 meters above sea level. On plateaus with a large surface area, the sun's glare raises the heat level so high that the mean temperature of Mexico City is 17.0° rather than 14.7°; that of Quito is 14.4° rather than 13.2°. These data suggest that, on the central plateau of Mexico, the maximum height at which sugarcane can grow vigorously, without suffering from winter frosts, is not 1,200 meters but 1,400 to 1,500 meters. With favorable exposures, especially in the valleys sheltered from the north winds by mountains, the upper growth line for sugar cultivation can even surpass 2,000 meters. Indeed, while the San III.6 Gabriel plains, which have several magnificent sugar refineries, are at an elevation of only 980 meters, the outskirts of Celaya, Salvatierra, Irapuato, and Santiago, on the other hand, have an absolute elevation of over 1,800 meters. People have assured me that the sugarcane plantations of Río Verde, located to the north of Guanajuato, at 22° latitude, are at an elevation of 2,200 meters in a narrow valley surrounded by high Cordilleras, and so hot that the inhabitants suffer from intermittent fevers. While examining Cortés's will,[2] I discovered that in that great man's time, there were sugar

1. See my "Mémoire sur les réfractions," in *Recueil d'observations astronomiques*, vol. I, p. 107, and *Lignes isothermes*, pp. 125–31.

2. "It is my wish that someone examine whether, in my *estados*, land was taken from the indigenous peoples to be planted with vines; I also wish for investigations to be carried out on the plot of land that I have, in recent years, given to my servant, Bernardino del Castillo, to establish a sugar refinery near Coyoacan." (*Testamento que otorgó al Excellentissimo Señor Don Hernán Cortés made in Seville, August 28, 1548 [1547],* art. 48.)

refineries near Coyohuacan in the Valley of Mexico. This curious fact proves that this valley—as several other phenomena suggest—is colder today than it was at the beginning of the conquest because, at that time, a large number of trees dampened the effect of the north winds that blow with great fury today. Those who are accustomed to seeing the sugarcane plantations in the Antilles islands will be similarly surprised to learn that, in the kingdom of New Granada, the largest quantities of sugar are harvested not

III.7 in the plains, on the banks of the Magdalena River, but on the slope of the Cordilleras, around the Guaduas valley on the road from Honda to Santa Fe, in terrain that, according to my barometric measurements, is between 1,100 and 1,600 meters above sea level.

Fortunately, the increase in sugar production in Mexico has not been accompanied by a similar increase in the number of blacks [nègres] brought into the country. Although in the intendancy of Puebla, near Cuautla de las Amilpas, there are plantations (*haciendas de caña*) with an annual production of over twenty or thirty thousand arrobas[1] (228,000 to 342,000 kilograms), almost all Mexican sugar is crafted by Indians, and, therefore, by free labor. One can easily foresee that the lesser Antilles islands, despite their location, which is favorable to trade, will not be able to sustain competition from the continental colonies much longer, if the latter continue to devote themselves with the same ardor to growing sugar, coffee, and cotton. In the physical world, just as in the moral world, everything eventually reverts to the order prescribed by nature, and if some islands with little surface area, and whose population has been exterminated, have traded their prod-

III.8 ucts more actively than the neighboring continent, it is only because the inhabitants of Cumaná, Caracas, New Granada, and Mexico began quite late to take profit from the immense advantages that nature has bestowed upon them. Having emerged from centuries of lethargy and no longer burdened by the constraints that flawed policies placed on the development of agriculture, the Spanish colonies on the continent will wrest control over the different branches of trade from the Antilles islands. This change, for which the events of Saint-Domingue have paved the way, will have a beneficial influence on the reduction in the trade in Africans [la traite des nègres]. Suffering humanity shall owe to the natural progress of things what it might

1. This is a very significant yield: on the island of Cuba, it is only found on one plantation, that of the ▼ Marquis *de Arcos*, called *Río Blanco*, between Jaruco and Matanzas, which yields 40,000 arrobas of sugar yearly. Fewer than eight plantations have produced 35,000 arrobas ten years in a row.

have reason to expect from the wisdom of European governments. The colonists of Havana, well aware of their true interests, are therefore closely monitoring the development of sugar cultivation in Mexico and Guatemala, and coffee cultivation in Caracas. They have long feared competition from the continent, especially since the lack of fuel and the extremely high cost of food, slaves, metal utensils, and the livestock necessary for a sugar refining have significantly decreased the net revenue of the plantations.

In addition to the advantage of its population, New Spain has another very important one, namely, an enormous amount of amassed capital among the mine owners and in the hands of the merchants who have retired from trade. To understand how important this advantage is, one must recall that on the island of Cuba, the establishment of a large sugar refinery, which, through the work of three hundred blacks [nègres], produces 400,000 kilograms of sugar annually, requires an advance of two million livres tournois, and that it generates between 300,000 and 350,000 pounds of revenue. The Mexican colonist can select the climate best suited to sugarcane cultivation from along the coasts and valleys of various depths; he fears the effect of frost less than does the Louisiana colonist. But the extraordinary lay of the land in New Spain places huge constraints on the shipment of sugar from Veracruz. The plantations that exist today are, for the most part, quite far from the coast facing Europe. Since the country still lacks canals or roads suitable for carts, mule freight increases the price of sugar at Veracruz by one piaster per arroba, or eight sous per kilogram. These constraints will be greatly reduced by the road that is currently being built from Mexico City to Veracruz, via Orizaba and Jalapa, along the eastern slope of the Cordilleras. The development of colonial agriculture will likely contribute to populating the coastline of New Spain, which has remained uncultivated and deserted for centuries. III.9

[Due to the difficulty of internal communications, sugar from the outskirts of Cuernavaca and Valladolid de Michoacan has never been profitably exported from the port of Veracruz, except during the period when, because of the destruction of Saint-Domingue, prices rose to forty-eight and fifty-six piasters per crate (half a *tercio*, or sixteen arrobas, or 184 kilograms). Mexican sugar exports from Veracruz are as follows [in the table on the following page]: III.10

Exports decrease as prices drop. In Veracruz, these prices were three piasters per arroba from 1802 to 1804; two and a half piasters from 1810 to 1812. Today (1825), one arroba is worth only one and six-tenths piasters. For exports of Mexican sugar to become decoupled from the very high prices in Europe, one would have to move the sugar refineries from the

1802	439,132 arrobas, valued at	1,476,435	piasters
1803	490,292	1,514,882	
1804	381,509	1,097,505	
1810	121,050	272,362	
1811	101,016	251,040	
1812	12,230	30,575	

interior to the coasts, between Alvarado and Tabasco. In the republic of Centro-America, the banks of the Ulúa[1] may one day compete with the most fertile districts of the island of Cuba.]

In Mexico, the level of sweetness of the *vezú*, or the juice squeezed from the sugarcane, depends on whether the plant grows in the plain or on an elevated plateau. The same difference applies among cane grown in Malaga, in the Canary Islands, or in Havana. The elevation of the soil has the same effects on plant life everywhere, as does the difference in geographic latitude. Climate also has an influence on the proportion of the amounts of liquid sugar to crystallizable sugar contained within the cane juice; sometimes, the *vezú* has a very sweet flavor but only crystallizes with great difficulty. The chemical composition of the *vezú* is not always the same, and ▾ Mr. Proust's fine experiments have put into relief the phenomena that occur within the workshops of America, many of which are a source of despair for sugar refiners.

III.11

The calculations that I made on the island of Cuba suggested that a hectare of land produces on average twelve cubic meters of *vezú*, from which is extracted at most 10 to 12 percent, or 1,500 kilograms of raw sugar, through processes commonly used to this day, in which a great deal of sugary matter is broken down by heat. In Havana and in the hot and fertile regions of New Spain, one *caballería* of land—which contains eighteen square *cordeles* (with twenty-four *varas*), or 130,118 square meters—yields 2,000 arrobas, or 23,000 kg, annually. The average yield is nevertheless only 1,500 arrobas, which makes 1,320 kilograms of sugar per hectare. In Saint-Domingue, the yield of a *plot* of land of 3,403 toises, or 12,900 square meters, is assessed at 4,000 pounds, which also equals 1,550 kg per hectare. In general, the fertility of the soil in the equinoctial Americas is such that all of the sugar consumed in France, which I place (in 1804) at twenty

1. *Redactor general de Guatemala*, 1825, p. 25.

million kg,[1] could be produced on a single seven-square-league plot, an area that does not even equal one-thirtieth of the smallest department of France. III.12

In plots of land suitable for watering, and in which plants with tuberous roots, such as potatoes and yams, have preceded sugarcane cultivation, the annual yield rises to three or four thousand *arrobas* per *caballería*, or 2,660 to 3,540 kg of raw sugar per hectare. But if a crate of sixteen *arrobas* is valued at twenty-four piasters, which is the price in Havana (in 1824), one finds that, according to these figures, one hectare of irrigated land can produce 2,020 livres tournois' worth of sugar, while the same hectare would yield only 260 pounds of wheat, assuming both a harvest that is ten times larger and a price of sixteen livres tournois for one hundred kilograms of wheat. In comparing these two types of farming, one must not forget that the advantages that sugarcane presents are severely undermined by the enormous advances required for setting up a sugar refinery.

Most of the sugar that New Spain produces is consumed in the country itself. It is likely that, despite the large number of Indians, this consumption amounts to over twenty-four million kilograms, for sugar consumption in the island of Cuba is probably 60,000 crates of sixteen *arrobas*, or 184 III.13 kilograms, in 1825. Those who have not witnessed firsthand the enormous amount of sugar consumed in Spanish America, even in the least well-off families, will be surprised to learn that, in 1825, the overall demand for sugar in France was only four times larger than on the island of Cuba, whose free population is still under 450,000 inhabitants.

I would have liked to place the sugar exports of New Spain next to those of the Antilles in a single table, but it was impossible for me to include all the data for the year 1803 alone. I was unable to obtain solid information on the total output of the British islands' sugar refineries, which has increased prodigiously. In 1803, the island of Cuba exported 158,000 *cajas* from the port of Havana and 30,000 *cajas*, including smuggled goods, from the port of Trinidad and from Santiago de Cuba.

[Since that time, exports from Havana have increased so much that, between 1815 and 1819, they equaled 206,360 crates in an average year; 215,593 crates in 1820; 236,669 crates in 1821; 300,211 crates in 1823;

1. In 1788, France extracted from her colonies a total of 872,867 quintals of raw sugar, 768,566 of clayed sugar, and 242,074 of white refined sugar. Of this amount, according to Mr. Peuchet, only 434,000 quintals of refined sugar was consumed in the kingdom itself. The tables published by the ministry of ▼ Mr. Chaptal show that sugar imports in France rose to 515,100 quintals in year IX [1800–1].

and 245,329 crates in 1824 (a relatively unproductive year). If one adds[1] internal consumption to these exports from Havana; the amounts of sugar III.14 that move from Matanzas, Trinidad, Santiago de Cuba, and Baracoa; and the effects of customs fraud, then it is undeniable that the total production on the island of Cuba today is between 400,000 and 450,000 crates in years of average fertility, that is, between 73,500,000 and 82,700,000 kilograms.[2] Sugar production in the French part of Saint-Domingue was 80,360,000 kilograms in 1788 (in 1799, it was barely twenty and a half million). Exports from Saint-Domingue to France amounted to 70,315,147 kilograms in 1788;[3] from all the French colonies in the Antilles in the same year (that is, in these establishments' most prosperous period), they amounted to 2,286,943 kilograms. More recently, between 1817 and 1821, France received 35,545,400 kilograms of sugar from her colonies in an average year; in these last few years, this amount has even surpassed forty million kilograms. Exports from all the British colonies of the Antilles, which amounted to only 128,569,728 kg of sugar in 1802, rose to 3,583,660 cwt, or 182,014,091 kg,[4] in 1823 (including British Guiana), not III.15 counting 4,833,844 gallons of rum. Jamaica alone provided the metropole with 1,417,758 cwt (72,007,928 kg) of sugar and 2,951,110 gallons of rum. Sugar exports from Jamaica today are therefore relatively similar to those from Saint-Domingue in 1788 and one-fifth greater than those of the island of Cuba in years of average productivity.][5]

According to the remarkable account that Mr. Beckford has included in his *Indian Recreations*, published in Calcutta, sugarcane is cultivated

1. Average exports from Havana over the past eight years: 237,000 crates; from the other ports: 70,000; smuggling: at least one-quarter or 77,000; internal consumption: 60,000; total production: 444,000 crates; total exports, legal and illegal: 384,000 crates of sugar.

2. For measurement conversion: 100 Spanish pounds, or one quintal = 45.97 kg; one arroba = twenty-five Spanish pounds = 11.49 kg; therefore, one caja of sugar, or sixteen arrobas = 400 Spanish pounds = 183.904 kg. On the other hand, 1 cwt = 112 English pounds = 50.796 kg, assuming that one French pound = 489/91,000 kg.

3. That is: 822,628 quintals raw sugar, 566,285 clayed sugar, 49,090 white refined sugar. Peuchet, *Stat[istique] de la France*, p. 407.

4. [▼ Tooke,] *Statist[ical] Illustrations on the British Empire*, 1825, p. 54.

5. According to customs records, in the especially productive year of 1823, the two ports of Havana and Matanzas together exported 15,139,188 piasters' worth of local goods, including 5,254,680 arrobas of sugar (327,855 crates) and 979,864 of coffee. Nevertheless, an estimate of exports of 360,000 crates of sugar, or 66,240,000 kg, in 1823 for the entire island (the ports of Havana, Matanzas, Trinidad, Santiago de Cuba, and Baracoa) would not be excessively high.

mainly in Bengal, in the districts of Peddapore and Benares, in the Godavari Delta, and on the banks of the Elyseram River. They irrigate the plantations there, as is the custom in many parts of Mexico and in the Güines valley to the southeast of Havana. To prevent the soil from becoming exhausted, they alternate between growing leguminous plants and sugarcane, which is generally three meters tall and four centimeters thick. In Bengal, one acre (4,044 square meters) yields 2,300 kilograms of sugar, which equals 5,750 kilograms per hectare. The soil is therefore twice as productive as that of the Antilles, while the cost incurred by the free Indian is almost three times less than the price of one day's work by the black slave [nègre esclave] on the island of Cuba. In Bengal, six pounds of cane juice yield one pound of crystallized sugar, whereas in Jamaica, eight pounds are necessary to produce the same amount of sugar. If one considers *vezú* to be a salt-bearing liquid, one finds—if these estimates are credible—that in Bengal, this liquid contains 16 percent sugary matter, and in Jamaica 12 percent. East Indies sugar is, therefore, at such a low price that the grower sells it at four and four-fifths *rupees* per quintal, or twenty-six centimes per kilogram, which is more or less one-third of the value of this commodity on the Havana market. Although sugarcane cultivation is spreading in Bengal with stunning speed, total production is still far below what is widely believed. Mr. Beckford estimates that the harvest (not total exports) of Bengal in 1802 was one-quarter that of Jamaica at most.

III.16

One of the most interesting problems of political economy is how to determine the consumption of commodities that, in the current state of European civilization, are the main objects of colonial industry. One can obtain more or less exact results for the *lower and upper limits*, in two different ways: (1) by discussing the exports from the countries that provide the largest amounts of these commodities, which are, with regard to sugar, the Antilles, Brazil, the Guyanas, Île-de-France [Mauritius], Bourbon [Réunion], and the East Indies; (2) by examining European imports of colonial commodities, and by comparing their annual consumption to the population, wealth, and national habits of each country. Whenever there is a single source for a product, such as tea, studies of this kind are easy and quite definitive; but the difficulty increases in the tropical regions, which all produce a more or less considerable amount of sugar, coffee, and indigo. In this case, to establish an *upper limit* for the *minimum* consumption, one must begin by focusing on large quantities. If one knows that, according to customs records, the British, Spanish, and French Antilles

III.17

export 269 million kilograms of sugar, then it is of little consequence that the Dutch and Danish Antilles produce eighteen to twenty-two million. If Brazil, Demerary, Berbice, and Essequibo export 155 million kilograms of sugar, then any uncertainty about production in Suriname and Cayenne, which together provide less than twelve million kilograms, has very little impact on the estimate for overall consumption in Europe.

I have already discussed elsewhere (*Relation historique*, [vol. II, p. 122]) III.18 the problem whose solution will be discussed in this note; relying on less numerous and less accurate materials, I had thought then that European sugar consumption in the year 1818 amounted to only 450 million pounds. Even for that time, this sum would seem to err on the low side, perhaps by a fifth or even by a quarter; but one must not forget that, from 1818 to 1823, the price of sugar from the Americas dropped 38 percent, and that there is an inverse relationship between consumption and price (*Table of Prices* in Tooke, *Appendix to Part IV*, 1824, p. 53, and *Statist[ical] Illustr[ations] of the British Empire*, 1825, p. 56). In France, for example, consumption increased by over 40 percent from 1788 to 1825: in 1788, it was twenty-one million kilograms; in 1818, thirty-four million; and in 1825, over fifty million. It is because of the very high rate of growth in colonial trade and European prosperity that it is important to set a numerical figure for the state of affairs at a given time. Works of this kind provide points of comparison whose importance will be keenly felt by persons in a future century, who, like Mr. Tooke, will wish to trace the progressive development of the industrial system of the two worlds. We shall begin with a brief overview of sugar production, or, rather, the amounts exported legally to the ports of Europe and the United States. [See the table on the following two pages.]

I have meticulously indicated the sources from which I have taken the items in the general table; without indications of the documents consulted, III.22 studies of this kind are of little value. The reader must be equipped to examine the partial data. Today, the only uncertainty that remains is about small amounts (for example, exports from Puerto Rico, Curaçao, and St. Thomas) or about irregular sugar production in Brazil. If one estimates these variations and all the areas of uncertainty that remain to be thirty-five million kilograms, the total sum of exports would still vary by no more than one-fourteenth.

In comparing the populations of the island of Cuba, Great Britain, the United States, and France to the amounts of raw sugar consumed annually in these different countries, one finds quite a remarkable downward pro-

	MILLIONS OF KILOGRAMS	
ANTILLEAN ARCHIPELAGO	287	III.19
BRITISH ANTILLES	165	
We have estimated above (volume III, page 410) the average exports from Jamaica to the ports of Great Britain and Ireland (exports that must not be confused with production) from 1816 to 1824 to have been 1,597,000 cwt, or 81,127,000 kilograms. Those of the rest of the British Antilles amounted to 1,634,000 cwt, or 83,007,000 kilograms; for a total 3,231,000 cwt, or over 164 million kilograms. If one considers only the last five years (1820–24), the same official figures reveal, in an ordinary year, for Jamaica, 1,573,000 cwt, or 79,908,000 kilograms; for the other British Antilles, 1,564,000 cwt, or 79,451,000 kilograms; for a total 159,359,000 kilograms. Depending on whether one takes the averages since 1816 or 1820, the difference is only four and a half million kilograms, or 88,500 cwt, an amount that is a great deal less than the variations experienced by sugar exports from Jamaica to Europe from year to year. Listing the British Antilles according to the amount of sugar that they currently trade, one arrives at the following order: Jamaica, St. Vincent and Barbados (with almost equal production), Grenada, Antigua, Trinidad, Tobago, St. Kitts, St. Lucia, Dominica, Nevis, Montserrat, and Tortola.		III.20
SPANISH ANTILLES	62	
Only recorded amounts are included in this table: with contraband, Cuban exports alone are over seventy million kilograms.		
FRENCH ANTILLES	42	
The slave population of the French and Spanish Antilles is in the same proportion to their sugar exports, which proves how fertile the soil of the island of Cuba is, for nearly a third of the slaves of this island live in the large cities. (*Personal Narrative*, quarto edition, volume III, page 391, 423.)		
DUTCH, DANISH, AND SWEDISH ANTILLES	18	

(*continued*)

	MILLIONS OF KILOGRAMS
BRAZIL In 1816, according to studies by ▼ Baron Delessert, exports were even 5,200,000 kilograms higher; but, as we have already recalled above, in years of serious drought, exports drop to ninety-one million kilograms.	125
III.21 BRITISH, DUTCH, AND FRENCH GUIANAS If one considers only the last five years (1820–24), exports from Demerara, Essequibo, and Berbice (or British Guiana), equaled 30,937,000 kg. One sees that cultivation in this part of Guiana increases at the same time as that of the British Antilles tends to decrease somewhat. From 1816 to 1824, the average for British Guiana was 525,000 cwt, or twenty-six and a half million kg, which suggests an annual export growth of four and a half million kg, or one-eighth, while production in the British Antilles also decreased by four and a half million kilograms, or one thirty-fifth (according to the comparison of the averages from 1816 to 1824 and from 1814 to 1824).	40
LOUISIANA	13
EAST INDIES, ÎLE-DE-FRANCE, AND BOURBON Île-de-France [Mauritius], twelve million kg; the East Indies, at most ten million kg; Bourbon, eight million kg. The exports to the United States are combined with the exports to Europe, as in the rest of this table. If the East Indies were one day to replace the British Antilles, their exports would have to be sixteen times greater.	30
TOTAL	495

gression, depending on the degree of wealth and especially depending on national habits. [See table on the following page.]

III.23 Deducting the thirty-eight million kilograms of sugar for consumption in the United States and British Canada from the 495 million kilograms of raw sugar traded annually in Europe and the United States leaves 457 million kilograms of sugar (of which seven-eighths are raw and one-eighth clayed) for annual imports to Europe. This is the *lowest* possible *upper limit*, for the data for these calculations are all taken from customs records, without adding the amount of smuggled goods.

COUNTRY	ANNUAL RAW SUGAR CONSUMPTION IN KG	FREE POPULATION	ANNUAL SUGAR CONSUMPTION PER HEAD
Island of Cuba	11 million	450,000	24 2/5 kg
Great Britain[1]	142 million	14,500,000	9 4/5 kg
United States of America	36 million	9,400,000	3 4/5 kg
France	52 million	30,600,000	1 4/5 kg

III.22

[1] In Great Britain, partial consumption of sugar from India increased in

1808	to	23,526 cwt
1809		9,313
1810		42,145
1820		90,625
1821		121,859
1822		124,009

Dividing the amount of raw sugar consumed in Europe by the number of inhabitants (208 and a half million), one arrives at two and one-fifth kilograms per person, but this result is only a sterile mathematical abstraction that leads to reflections as useless as the attempts to distribute the population contained within the cultivated regions of the United States or Russia across a total area of 174,000 or 616,000 square nautical leagues. Europe has fifty-one percent or one hundred and six million, inhabitants in the British Empire, the Netherlands, France, Germany (in the strict sense), Switzerland, and Italy who consume a prodigious amount of sugar, and thirty-five percent, or seventy-three million, spread across Russia, Poland, Bohemia, Moravia, and Hungary, countries in which the extreme poverty of the bulk of the inhabitants makes for extremely low consumption. In terms of luxury or the artificial needs of society, these are the extremes on the scale. To give a sense of the wealth of the population of Germany, I shall recall here that nearly forty-five million kilograms of sugar were imported in 1821 in the port of Hamburg alone, while in 1824, imports were 44,800 crates (or 29,120,000 kg) from Brazil; 23,800 crates (or 4,379,000 kg) from Havana, and 10,600 barrels (or 8,480,000 kg) from London; for a total of 41,979,000 kilograms. In 1825, 31,920 crates (or 20,748,000 kg) were imported from Brazil; 42,255 crates (or 7,774,900 kg), from Havana, and 20,506 barrels (or 16,404,800 kg) from England; for a total of

III.24

44,927,000 kilograms. In 1825, imports into Hamburg were, therefore, only one-quarter lower than imports into the whole of France. The port of Bremen imported nearly five million kilograms; in the same year, the port of Antwerp imported 10,758,000 kilograms. In southern Germany, where sugar consumption is also very high, the complications posed by both shipping and smuggling make statistical research very difficult. How, for example, can one accept ▾ Mr. Memminger's claim that in the kingdom of Württemberg, a land of great prosperity, 1,446,000 inhabitants could consume only 980,000 kilograms of sugar per year?

III.25 Deducting the 204 and a half million kilograms of raw sugar consumed in France and in the three United Kingdoms from the 457 million kilograms imported into Europe, and allowing for two kilograms per person (a very plausible estimate) for the combined population of seventy-six million in the Netherlands, Germany (strictly speaking), Switzerland, Italy, the Iberian Peninsula, Denmark, and Sweden, leaves nearly one hundred and a half million kilograms for Asia Minor, the Barbary Coasts, the western governments of Siberia, and the parts of Europe inhabited by peoples of the Slavic, Hungarian, and Turkish races. Yet, the populations of Morocco, Algiers, Tunis, and Tripoli are quite significant: combined, they total twenty-four million. Asia Minor has over four million inhabitants. If one includes only the population of the coastline, dotted with large trading cities, one can reasonably assume exports of ten million kilograms of raw sugar to the coasts of Africa, Asia Minor, and Syria. Based on these figures, it must be concluded that the eighty million inhabitants of the Slavic, Magyar, and Turkish parts of Europe (Russia, Poland, Bohemia, Moravia, Hungary, and Turkey) consume one and thirteen-hundredths kilograms per person. This result is indeed surprising, if one compares the current state of civilization in these countries to that of France. One would expect their consumption to be much lower; nevertheless, far from being exaggerated, the estimates for sugar exports from the Americas and the East Indies to Europe and the United III.26 States are likely lower than actual exports. If customs fraud makes consumption in Great Britain and France (two countries that have served as examples in the preceding arguments) more significant than is assumed, and if one can accept that the French and the English consume even more than one and four-fifths and nine and four-fifths kilograms per person, one must not forget that the same source of error affects export estimates for the Americas and the East Indies. In 1810, when Great Britain consumed almost 177 and a half million kilograms, the rate was twelve and one-fifth kilograms per person. It would be desirable for a writer accustomed to precision in numerical research and who could draw from good sources to discuss, in a separate work, the important issues of the European sugar, coffee, tea, and cocoa consumption in a given period. This work would require several years to carry out, for some documents are not printed and can only be obtained through an assiduous correspondence with the major trading houses of Europe. I was unable to delve into this research to the fullest possible extent. The time is approaching when colonial commodities will be largely the product not of colonies but of independent countries, not of islands but of the great continents of the Americas and Asia. The history of trade among peoples is lacking in numerical data relating to the state of society as a whole, and this knowl- III.27 edge gap can only be filled at a later point, when, with great revolutions threatening the industrial world, someone has the courage to collect the materials that are currently scattered and to subject them to an exacting critical eye.

Cotton is one of the plants whose cultivation among the Aztec people is as ancient as that of the century plant, corn, and quinoa. The highest quality cotton grows on the western coasts, from Acapulco to Colima and to the port of Gualán, especially to the south of the Jorullo volcano, between the villages of Petatlan, Teipa, and Atoyaque. Since people there are not yet familiar with any ▾ machines that separate cotton fibers from the seed, the high cost of shipping places serious constraints on this branch of Mexican agriculture. One *arroba* of cotton (*algodón con pepa*), which fetches eight francs at Teipa, costs fifteen at Valladolid, because it is hauled there on muleback. The section of the eastern coast that stretches from the mouths of the Guasacualco and Alvarado rivers to Panuco could provide an enormous amount of cotton for the Veracruz trade; but this coastline is nearly uninhabited, and the lack of labor makes supplies prohibitively expensive for the establishment of agriculture. New Spain provides Europe with only 25,000 arrobas (or 312,000 kilograms) of cotton annually. This amount, although not particularly significant in itself, is still already six times higher than what the United States exported of its own production in 1791 (based on information that I owe to the kindness of Mr. Albert Gallatin, former III.28 treasury secretary in Washington). But the speed with which industry increases among a free and wisely governed people is so great that, according to a note provided to me by this very statesman, the ports of the United States exported in

	POUNDS OF FOREIGN COTTON, AND	POUNDS OF DOMESTIC COTTON
1797	2,400,000	1,200,000
1800	3,660,000	14,120,000
1802	3,400,000	24,100,000
1803	13,493,544	37,712,079

These data suggest[1] that cotton production grew 377 times in twelve years. Comparing Mexico's physical position to that of the United States, it is clear that these two countries alone could one day produce all the cotton linters that Europe uses in its factories. In a paper published a few years ago, the enlightened merchants who make up the chamber of commerce in Paris maintained that cotton imports to Europe total thirty million kilograms of cotton linters per year; from 1802 to 1806, Great Britain alone received nearly thirty-two million kilograms in an average year; in 1825, over one hundred million.

III.29

Flax and *hemp* could one day be profitably cultivated anywhere that the climate does not allow for cotton crops, as in the *Provincias internas* and even the equinoctial region, on plateaus where the average temperature is

1. Cotton exports from the various ports of the United States, which totaled 100,000 pounds in 1790, and 1,300,000 pounds in 1795, amounted to:

1815	to	82,998,747 pounds	1821	to	124,893,406 pounds
1816	to	81,947,116	1822	to	144,675,095
1817	to	85,649,328	1823	to	173,723,270
1818	to	92,470,178	1824	to	142,369,663
1819	to	87,997,045	1825	to	166,784,629
1820	to	127,860,152			

From 1802 to 1806, Great Britain received from the United States 104,000 bales of cotton in an average year; and from 1802 to 1823, 359,300 bales in an average year ([Powell,] *Statistical Illust[rations]*, 1825, p. 58). In these same periods, cotton imports from Brazil into the ports of Great Britain only doubled: in 1802, 74,720 bales were imported; in 1823, 148,070 bales. British cotton imports totaled 77,393,000 English pounds in 1802; 180,233,795 pounds in 1823; and 222,576,000 pounds in 1825. Egypt, whose exports still amounted to very little in 1823, already furnishes Great Britain with half of its total consumption. In 1825, imports from the United States to all British ports amounted to 425,100 bales, while Egypt's were 103,400 bales (▼ Nicholson, Suppl[ement] to *Lon[don] New Price-Current*; 1825, p. 17).

below fourteen degrees on the Centigrade thermometer. The Abbot Clavijero maintains that flax grows wild in the intendancy of Valladolid and in New Mexico, but I very much doubt that this claim is based on the precise observation of a traveling botanist. Whatever the case may be, it is quite certain that, to this day, very little hemp and flax is cultivated in Mexico. Spain had several progressive ministers who sought to promote these two branches of colonial industry. This favor, however, was only ever fleeting, and the Council of the Indies, whose influence is as long-lasting as that of any institutional body with longstanding principles, was accused of having sought the opposition of the metropole to the cultivation of hemp, flax, grapes, olive trees, and mulberry trees. Blind to its real interests, the government preferred to see the Mexican people wearing cotton cloths purchased in Manila and in Canton, or sent to Cádiz on English ships, to protecting cotton manufacturing in New Spain. One may hope that the mountainous part of the Sonora, in the intendancy of Durango and in New Mexico, might one day compete in flax production with Galicia and Asturias. As for hemp, it would be important to introduce not the European species but, rather, the one that is cultivated in China (perhaps a variety of *Cannabis indica*), the stalk of which reaches a height of five to six meters. It can be presumed, in any case, that hemp and flax cultivation will only spread with great difficulty in the region of Mexico where the cotton tree can be successfully cultivated. The retting of the hemp demands greater care and labor than does the separation of the cotton fiber from seed, and in a country with a small workforce and an attitude of general sloth, the people prefer to cultivate things that can be quickly and easily used.

III.30

III.31

[It does not seem fair to blame the Spanish government for having opposed flax cultivation; I hasten to correct this involuntary error, using the information that Mr. José Cia, nephew of the worthy viceroy, Mr. de Azanza, was kind enough to send me. By an order issued on June 13, 1545, to the viceroys and governors of the Indies, Emperor Charles V commanded that they "hagan sembrar y beneficiar en las Yndias lino y cáñamo y procuren que los Yndios se apliquen a esta granjería y entiendan en hilar y tesar lino" [sow and cultivate flax and hemp in the Indies and make sure that the Indios apply themselves to this farming and know how to dress and spin flax] (*Recopilación de leyes*, tit. 18, book 4, chap. 20). Under the reign of Charles III, at a time when the prices of foreign hemp and flax at Cádiz were very high, the minister reiterated Charles V's encouragement through the decree of January 12, 1777. It was explicitly declared that flax cultivation was not prohibited in the Americas and that the *Alcades mayores* were to protect it. The government even

contracted a few private individuals to bring European colonists and to teach the Indians how to prepare hemp and flax. Plots of land that had belonged to the Jesuits were ceded to these individuals. ▼ Don Luis Parilla, *Director de las*
III.32 *Temporalidades*, was placed at the head of the establishment at Chalco; a few suits made of native flax were manufactured in Mexico City and sent to Madrid and to San Blas in 1783; but having spent less than 122,000 piasters, the court abandoned Parilla's project. The European farmers returned to Europe at the end of 1786; no more purchases were made for the *cuenta de la Real Hacienda*, but the Indians were still permitted to devote themselves to hemp and flax cultivation. The viceroys, Count de Revillagigedo and Marquis de Branciforte, urged the bishops and priests to promote this branch of industry. The court gave new orders in 1792 and 1795, but the ease of obtaining cotton cloth, even in the cold region of Mexico, made these praiseworthy efforts useless. To prove, moreover, that flax and hemp cultivation were never actually prohibited, one has only to cite article forty-three of the *Reglamento del libre comercio* (October 12, 1778), according to which flax or hemp, *if it comes from Spanish America*, is exempt from all import duties.]

Coffee farming began on the island of Cuba and in the Spanish colonies on the continent only after the destruction of Saint-Domingue.[1] In 1804, the is-
III.33 land of Cuba already produced 12,000 quintals, while the province of Caracas produced nearly 5,000. New Spain has sugar refineries in greater number and of greater size than does the Tierra Firme, but coffee production is more or less non-existent there, although there is no doubt whatsoever that this cultivation would succeed perfectly well in the temperate regions, especially at the same height as the cities of Jalapa and Chilpanzingo. The custom of drinking coffee[2] is still so rare in Mexico that the country as a whole consumes only four to five

1. In 1783, the French part of Saint-Domingue produced only 445,734 quintals of coffee, but, five years later, it produced 762,865. Nevertheless, the price in 1783 was fifty francs per quintal, and in 1788, it was ninety-four francs, which demonstrates the extent to which coffee use spread throughout Europe, despite the increase in price. According to Mr. Raynal, Yemen produces 130,000 quintals annually; Mr. Page cites 150,000 quintals, nearly all of which are exported to Turkey, Persia, and the Indies. L'Île-de-France [Mauritius] and l'Île de Bourbon together produce 45,000 quintals. (Based on the information that I have attempted to obtain, it seems that the whole of Europe consumed around sixty-eight million kilograms of coffee in 1818. A coffee tree on good land yields half a kilogram of coffee, and 3,500 trees fit on one hectare of land.)

2. This use has increased considerably since 1802, and coffee exports from Havana to Mexico have become quite significant. In 1823, according to customs records, the island of Cuba exported 979,864 *arrobas* from the ports of Havana and Matanzas alone, excluding contraband. On the other hand, coffee imports in France in 1818 totaled 6,796,000 kilograms, and coffee consumption in England was barely two million kilograms.

hundred quintals annually, while consumption in France, whose population is barely five times larger than that of New Spain, amounted (in 1803) to almost 5,880,000 kg; from 1820 to 1823, it was 8,197,900 kg in an average year.

III.34 Cacao (*cacari* or *cacava quahuitl*) farming was already quite widespread in Mexico in Montezuma's time, and it was there that the Spaniards became acquainted with this valuable tree, which they then transplanted to the Canaries and to the Philippines. The Mexicans prepared a drink called *chocolatl*, in which a bit of corn flour, vanilla (*tlilxochitl*), and the fruit of a species of capiscum (*mecaxochitl*) were blended with the cocoa (*cacahuatl*[1]). They even knew how to mold chocolate into bars, and not only this art but also the instruments they used to grind the cocoa and the word *chocolatl* passed from Mexico to Europe. It is all the more stunning, then, to note that cacao farming is almost entirely neglected today. One finds barely any of these trees on the outskirts of Colima or on the banks of the Guasacualco. The cacao plantations in the province of Tabasco are relatively small, and Mexico takes all the cocoa necessary for its own consumption from the kingdom of Guatemala, from Maracaibo, from Caracas, and from III.35 Guayaquil. Yearly consumption appears to be 30,000 *fanegas*, at a weight of fifty kilograms each. The ▾ Abbot Hervás claims that Spain consumes 90,000 *fanegas*.[2] This estimate, which appears a bit too low to me, suggests that Spain consumes only a third of the cocoa imported into Europe annually. But the research that I conducted on site from 1799 to 1803 led me to conclude that annual cocoa exports were as follows:

In the provinces of Venezuela and Maracaibo	145,000 fanegas
In the province of New Andalusia (Cumaná)	18,000
In the province of New Barcelona	5,000
In the kingdom of Quito, from the port of Guayaquil	60,000

1. Hernández, [*Quatro libros*,] book II, chap. 15; book III, chap. 46; book V, chap. 13. In Hernández's time, one distinguished between four varieties of cocoa, known as *quauhcahuatl*, *xochicucahuatl*, and *tlalcacahuatl*. The latter variety had very small grains. The tree that produced it was probably similar to the cacao tree that we have found growing wild on the banks of the Orinoco to the east of the mouth of the Yao. The cacao tree that has been cultivated for centuries has a bigger, sweeter, and oilier seed. The *Theobroma bicolor* should not be confused with the *T. cacao*, the drawing of which I have included in our *Plantes équinoxiales* (vol. I, plate XXX *a* and *b*, p. 104), and which is exclusive to the province of Chocó.

2. [Hervás y Panduro,] *Idea del Universo*, vol. V, p. 174. In 1818, I estimated cocoa consumption in Europe to be 23,000,000 pounds (See *Relat[ion] hist[orique]*, vol. II, p. 120–22).

If one values a *fanega* at only forty piasters, then these eleven and a half million kilograms of cocoa total 45,600,000 livres tournois, in peacetime. In the Spanish colonies, chocolate is not considered a luxury item but, rather, a good of the highest necessity. It is indeed a healthy and highly nutritious food and of great help to travelers. The chocolate manufactured in Mexico is of superior quality, because the Veracruz and Acapulco trade III.36 introduces into New Spain the famous Soconusco (*Xoconochco*) cocoa from the coasts of Guatemala; that of Gualán, from the gulf of Honduras, near Omoa; that of Oritucu, near San Sebastián, in the province of Caracas; that of *Capiriqual*, from the province of Nueva Barcelona; as well as that of *Esmeralda*, from the kingdom of Quito.

In the time of the Aztec kings, cocoa beans served as currency at the main market of Tlatelolco, as shells did in the Maldives Islands. For chocolate, they used the Soconusco cocoa cultivated at the eastern edge of the Mexican empire and the small beans called *tlalcacahuatl*. The lower quality varieties were reserved for use as currency. "Knowing," wrote Cortés in his first letter to the emperor Charles V, "that there was an abundance of gold in the province of Malinaltebeque, I urged Lord Montezuma to establish a farm there for Your Majesty. He put so much zeal into this endeavor that in less than two months they had already sown sixty fanegas of corn and ten of broad beans. They had also planted two thousand *cacap* (cacao trees), which produce a fruit, similar to the almond, that is sold after being ground. This bean is so highly valued that it is used as currency across the whole country, and they spend it not only in the markets but everywhere III.37 else."[1] Even today, cocoa is used as a billon in Mexico City. Since the smallest currency denomination in the Spanish colonies is a half-real (*un medio*), equal to two soles, the people find it convenient to use cocoa as coins. Six grains represent one sol.

The use of *vanilla* passed from the Aztecs to the Spanish. As we have observed above, chocolate was flavored with several spices, among which the vanilla pod had pride of place. Today, the Spanish only trade this valuable product in order to sell it to the other peoples of Europe. Spanish chocolate does not contain any vanilla, and even in Mexico City, there is a widespread belief that this flavor is harmful to one's health, especially to persons with a highly sensitive nervous system. One hears serious talk about vanilla causing nervous conditions (*la baynilla dapasmo*). Only a few

1. Lorenzana, [*Historia de Nueva-España*,] p. 91, § 26. Clavijero, [*Storia antica del Messico*,] I, p. 4; II, p. 219; IV, p. 207.

years ago, the same thing was said about the use of coffee in Caracas, which has nevertheless begun to spread among the indigenous peoples.

When one considers the extremely high price that vanilla constantly obtains in Europe, one is surprised by the carelessness of the inhabitants of Spanish America, who neglect the cultivation of a plant that nature spontaneously produces in the tropics wherever there is heat, shade, and a great deal of humidity. All of the vanilla that Europe consumes comes from Mexico, and exclusively from Veracruz. It is harvested in an area of land only a few square leagues in size. There is no doubt, however, that the coast of Caracas and even Havana could one day engage in a very significant trade in vanilla. We have found, in the course of our plant-collecting trips, highly aromatic and extraordinarily large vanilla pods in the mountains of Caripe next to Paria; in the beautiful valley of Bordones near Cumaná; on the outskirts of Turbaco near Cartagena de Indias; in the province of Jaén on the banks of the Amazon River; and in Guiana, at the foot of the granitic boulders that form the great cataracts of the Orinoco. Some inhabitants of Jalapa who trade in the exquisite Mexican vanilla from Misantla were struck by the excellent quality of the vanilla that Mr. Bonpland brought back from the Orinoco and that we gathered in the groves that surround the *Raudal de Maypures*. On the island of Cuba, one finds vanilla plants (*Epidendrum vanilla*) on the coasts of Bahía Honda and at Mariel. The vanilla of Saint-Domingue has very long fruit of relatively little aroma; while encouraging vegetation, high humidity is adverse to the development of aromatic spices. Traveling botanists should not assess the quality of vanilla on the basis of the scent that this liana emits in the forests of the Americas; the odor largely comes from the flower, which, in the deep and humid valleys of the Andes, is often four to five centimeters long. III.38

The author of the *Histoire philosophique des deux Indes*[1] complains about the relative lack of information that he was able to obtain about vanilla cultivation in Mexico. He did not even know the names of the districts where it is produced. I was able, on site, to record more detailed and more precise information. In Jalapa and Veracruz, I consulted with individuals who have traded in vanilla from Misantla, Colipa, and Papantla for thirty years. What follows are the fruits of my research into the current state of this interesting branch of national industry. III.39

1. Raynal, [*Histoire*,] vol. II, p. 68, § 16; ▼ Thiéry de Menonville, *De la culture du Nopal*, p. 142. A small amount of vanilla is also grown in Jamaica, in the parishes of St. Anne and St. Mary. Browne, [*Civil and natural history of Jamaica*,] p. 326.

All of the vanilla that Mexico sends to Europe comes from the two intendancies of Veracruz and Oaxaca. This plant is especially prevalent on the eastern slope of the Cordillera of Anahuac, between 19° and 20° latitude. The indigenous peoples, having realized quite early on how difficult harvesting was, despite this abundance, because of the vast stretch of land that one had to traverse each year, promoted the species by gathering together a large number of plants in a narrower space. This technique did not entail a great deal of maintenance: it sufficed to clear the soil a bit and to plant two cuttings of Epidendrum at the foot of a tree or, rather, to attach III.40 segments cut from the stalk to the trunk of a Liquidambar, an Ocotea, or a Piper arborescens.

The cuttings are generally four to five decimeters long. They are attached with lianas to the trees on which the new stalk is meant to climb. Each cutting bears fruit in the third year. Over a period of thirty to forty years, each stalk produces up to fifty pods, especially if the growth of the vanilla plant is not stunted by the proximity of other lianas that might smother it. In Mexico, the *baynilla cimarona*, or wild vanilla, which was not planted by humans, and which grows in a plot of land covered by shrubs and other climbing plants, bears fruit that is both extremely dry and in scant quantity.

In the intendancy of Veracruz, the districts famous for the vanilla trade are the *Subdelegación de Misantla*, with the Indian villages of Misantla, Colipa, Yecuatla (near Sierra de Chiconquiato), and Nautla, all of which were formerly part of the *Alcaldía mayor de la Antigüa*; the *Jurisdicción de Papantla*, those of Santiago and San Andrés *Tuxtla*. Misantla is thirty leagues northwest of Veracruz and twelve leagues from the sea coasts. It is a charming spot, untroubled by the scourge of the *mosquitos* and *jején* [gnats] that are so numerous in the port of Nautla, on the banks of the Río de Quilate, and in Colipa. If the Misantla River, the mouth of which is near Barra de Palmas, were made navigable, this district would soon reach a high level of prosperity.

III.41 The indigenous peoples of Misantla gather vanilla in the mountains and the forests of Quilate. The plant flowers in February and March. The harvest is bad if, during this time, the north winds are frequent and accompanied by heavy rains. Whenever the humidity is too high, the flower drops without bearing fruit. An extreme drought is also harmful to the growth of the pod. In addition, because of the milk that the green fruit contains, the latter is not attacked by any insects. Cutting begins in March and April, once the *subdelegate* has published a ban announcing that the harvest is open to the Indians; it lasts through the end of June. The indigenous peoples, who remain in the forests of Quilate for eight days in a row, sell the fresh, yellow vanilla to the *gente de razón*, who are whites, mestizos [métis], and mulattos [mulâtres]. Only the latter know the *beneficio de la baynilla*, that is, the method for drying it with care, preserving its silvery sheen, and bundling it for shipment to Europe. The yellow fruits are spread out on cloths and placed in the sun for a few hours. When they are sufficiently warm, they are wrapped in woolen sheets to be sweated: the vanilla then blackens and is finally dried through exposure to the sun's heat from morning to evening.

The method of vanilla preparation in Colipa is far superior to the *beneficio* widely used in Misantla. People claim that when the packets are unwrapped in Cádiz, they find barely 6 percent of waste in the vanilla from Colipa, while in the vanilla from Misantla, there are twice as many rotten or spoiled pods. The latter variety is more difficult to dry, because its fruit is larger and more aqueous than that of Colipa, which, harvested in savannas rather than on mountains, is called *baynilla de acaguales*. When the rainy weather prevents the inhabitants of Misantla and Colipa from exposing the vanilla to the rays of the sun until it has turned a blackish color and is covered in silvery streaks (*manchas plateadas*), they are forced to turn to the use of an artificial heat source. Using small reed tubes, they make a frame, hung by ropes and covered by a woolen cloth, on which they spread out the pods. Fire is placed underneath, but at a significant distance. They dry the pods by gently moving the frame and by gradually heating up the reeds and the cloth. A great deal of care and experience is necessary to succeed in drying the vanilla through this method, which is called *beneficio de poscoyol*. Whenever artificial heat is used, the losses are generally very high.

III.42

In Misantla, the vanilla fruit is gathered into packets called *mazos*. One *mazo* contains fifty pods; a thousand (*millar*) are therefore distributed across twenty *mazos*. Although all the vanilla that is traded appears to be the product of a single species of Epidendrum (*tlilxochitl*), the harvested fruit is nevertheless divided into four different classes. The characteristics of the soil, the humidity of the air, and the heat of the sun have an extremely powerful impact on the size of the pods and on the amount of oily and aromatic matter that they contain. These four classes of vanilla are, from highest to lowest quality: *baynilla fina* (within which there is a distinction between *grande fina* and *chica fina*, or *mancuerna*), *zacate*, *rezacate*, and *basura*. Each class is easily recognizable in Spain by the way in which the packets are tied. The *grande fina* is usually twenty-two centimeters long, and each *mazo* of it weighs ten and a half ounces in Misantla, and nine to ten in Colipa. The *chica fina* is five centimeters shorter than the *grande*

III.43

fina and can be bought for half the price. The *zacate* is a very long but extremely thin and aqueous vanilla pod. The *basura*, one packet of which contains one hundred pods, is only used to fill the bottom of the crates that are shipped to Cádiz. The lowest-quality vanilla from Misantla is called *baynilla cimarona* (wild), or *baynilla palo*. It is very thin and nearly devoid of juice. A sixth variety, the *baynilla pompona*, bears very large and beautiful fruit. It has been shipped to Europe on several occasions and, through the Genoa merchants, to the Levant; but since its odor is different from the vanilla called *grande fina*, it has not yet found a market there.

III.44 Based on what we have just explained about vanilla, one can see that the quality of the product is similar to that of cinchona, which depends not only on the species of cinchona used but also on the ground elevation, the exposure of the tree, the time of the harvest, and the care with which the bark was dried. Both the vanilla and cinchona trade are controlled by a few individuals referred to as *habilitadores*, because they advance money to the *cosecheros*, that is, to the Indians who perform the harvest and who thereby make themselves dependent upon the entrepreneurs. It is the latter who make nearly all the profit from this branch of Mexican industry. There is such little competition among buyers in Misantla and Colipa that one needs to be highly experienced in order not to be cheated while purchasing prepared vanilla. A single spotted (*manchada*) pod can lead to the loss of an entire case during the crossing from America to Europe. Specific names (*mojo negro*, *mojo blanco*, *garro*) are used to refer to the flaws found either on the pod or on the petiole (*garganta*). It is for this reason that prudent buyers examine the packets included in a single shipment several times.

Over the last twelve years, the *habilitadores* have paid an average price of twenty-five to thirty piasters for a thousand pods of first-class vanilla, ten piasters for a thousand pods of *zacate*, and four piasters for the same III.45 amount of *rezacate*. In 1803, the price of *grande fina* was fifty piasters, and that of *zacate* was fifteen piasters. Far from paying the Indians in cash, the buyers offer in exchange—and at a very high price—brandy, cocoa, wine, and, especially, cotton cloth made in Puebla. It is from this exchange that the monopolists derive a large part of their profit.

The district of *Papantla*, which was formerly an *Alcaldía mayor*, is located eighteen leagues north of Misantla. It produces very little vanilla, which is also poorly dried, though quite aromatic. The Indians of Papantla, like those of Nautla, are suspected of sneaking into the forests of Quilate to gather the fruit of the Epidendrum planted by the indigenous peoples of Misantla. The village of *Teutila* in the intendancy of Oaxaca is famous for the high quality of the vanilla that grows in the neighboring forests. It appears that this variety was the first to be brought to Spain, in the sixteenth century, for even today the *baynilla de Teutila* is still regarded in Cádiz as preferable to all others. Indeed, they dry it with great care, pricking it with pins and hanging it from century plant threads, but it weighs about one-ninth less than vanilla from Misantla. I do not know how much vanilla is harvested in the province of Honduras and exported each year from the tiny port of Trujillo, but it appears to be a relatively insignificant amount.

In especially productive years, the Quilate forests produce eight hundred thousand vanilla pods. In very rainy years, a bad harvest yields only two hundred thousand. The product is estimated to be, on average: III.46

in Misantla and Colipa	700,000
in Papantla	100,000
in Teutila	110,000

The value of these 910,000 pods is between thirty and forty piasters in Veracruz. To this should be added the products of the Santiago and San Andrés Tuxtla harvests, but I do not have sufficient information about them. It often occurs that one year's harvest is not shipped in its entirety to Europe, but one portion is reserved and added to the following year's harvest. In 1802, 1,793 thousand vanilla pods left the port of Veracruz. It is surprising to learn that total European consumption does not exceed this amount.

The same eastern slope of the Cordillera on which vanilla is harvested also produces sarsaparilla (*zarza*), of which nearly 250,000 kilograms were exported from Veracruz in 1803,[1] and jalap (*purga de Jalapa*), which is the root, not of the Mirabilis jalapa, the M. longiflora, or the M. dichotoma, but, rather, of the *Convolvulus jalapa*. This bindweed grows at an absolute height of 1,300 to 1,400 meters, across the entire mountain chain that stretches from the volcano of Orizaba to the Cofre de Perote. We did not find it on our plant-collecting trips around the city of Jalapa, but the Indians who live in the neighboring villages brought us some lovely root samples, which they had gathered near the Banderilla to the east of San Miguel el Soldado. This precious tonic is harvested in the *Subdelegación de Jalapa* III.47

1. The sarsaparilla that is traded comes from several species of Smilax. See the description of the eleven new species that we brought back, in Mr. Willdenow's *Species*, vol. IV, part 1, p. 773, and in our *Nov[a] Gen[era]*. vol. I, p. 270.

around the villages of Santiago, Tlachi, Tihuacan de los Reyes, Tlacolula, Xicochimalco, Tatatila, Ixhuacan, and Ayahualulco; in the *Jurisdicción de San Juan de los Llanos* near San Pedro Chilchotla and Quimixtlan; and in the *Partidos* of the cities of *Córdoba*, *Orizaba*, and *San Andrés Tuxtla*. True *Purga de Jalapa* exclusively favors a temperate, almost cold climate, in shady valleys and on mountain slopes. I was, therefore, surprised to learn, after my return to Europe, that Thiéry de Menonville,[1] a learned traveler who has exhibited tremendous devotion to the good of his country, had claimed to have encountered jalap in great abundance in the arid, sandy III.48 land surrounding the port of Veracruz, thus in a climate that is both excessively hot and located at sea level.

Raynal[2] suggests that Europe consumes 7,500 quintals of jalap each year; this estimate appears to be over 200 percent too high, for, according to the precise figures that I was able to obtain in Veracruz, only 2,921 quintals of jalap were exported from that port in 1802, and 2,281 in 1803.

During our stay in New Spain, we did not see any of the bindweed that is claimed to provide the *Michoacan root* (the *tacuache* of the Tarasca Indians, the *tlalantlacuitlapilli* of the Aztecs). We did not even hear about it during our travels in the former kingdom of Michoacan, which is part of the intendancy of Valladolid. The Abbot Clavijero[3] tells of a doctor of the ▾ last king of Tzintzuntzan who taught the missionary monks that had followed Cortés's expedition how to identify this remedy. Does there, in fact, exist a root that is exported from Veracruz under the name of *mechoacan*, or does this remedy, which is identical to Marcgrave's *jeticucu*,[4] come to us from the coasts of Brazil? It even appears that true jalap was formerly called *mi-* III.49 *choacan*, and that, through one of those misunderstandings so common in the history of drugs, this name subsequently passed to the root of another plant.

Mexican *tobacco* farming could one day become an indispensable branch of agriculture, if the tobacco could be freely traded. But ever since the introduction of the monopoly, that is, since the establishment of the

1. Thiéry [de Menonville, *Voyage à Guaxaca*,] p. 59. This jalap from Veracruz also seems identical to the one that Mr. Michaux has found in Florida. See the paper by Mr. Desfontaines on the Convolvulus jalapa ["Mémoire sur le Jalap"] in *Annales du Muséum [d'histoire naturelle]*, vol. II, p. 120.

2. [Raynal,] *Hist[oire] philos[ophique]*, vol. II, p. 68.

3. [Clavijero,] *Storia antica di Messico*, vol. II, p. 212.

4. Linnaeus, *Mat[eria] medica*, 1749, [vol. I], p. 28. ▾ Murray, *Apparatus medicaminum*, vol. I, p. 62.

royal plantation (*el estanco real de Tabaco*) by the *visitador* Don José de Gálvez, in 1764, one needs special permission to plant tobacco, and the grower is required to sell it to the plantation at the price that the latter arbitrarily sets, based on the quality of the produce. In addition, tobacco cultivation is restricted to the areas surrounding the cities of Orizaba and Córdoba, and to the partidos of Huatusco and Songolica, located in the intendancy of Veracruz. Officials bearing the title of *Guardas de Tabaco* [tobacco guards] travel across the country to pull up tobacco planted outside of the aforementioned districts and to fine those farmers who grow enough for their own consumption. It was thought that limiting cultivation to an area of land of four or five square leagues would reduce smuggling. Before the establishment of the *plantation*, the intendancy of Guadalajara, especially the partidos of Autlan, Ezatlan, Ahuxcatlan, Tepic, Santixpac, and Acaponeta, were famous for the abundance and excellent quality of the tobacco that they produced. These lands, once happy and flourishing, have experienced a decrease in population since the plantations were transferred to the eastern slope of the Cordillera. III.50

The Spanish became acquainted with tobacco in the Antilles. The word, adopted by all peoples of Europe, comes from the language of Haiti, or Saint-Domingue, for the Mexica called the plant *yetl* and the Peruvians called it *sayri*.[1] In Mexico and Peru, the indigenous peoples smoked and snuffed tobacco in powder form. At Montezuma's court, the nobles used tobacco smoke as a narcotic, not only for their siestas after dinner but also in order to sleep in the morning, immediately after the morning meal, as is still the custom in several parts of equinoctial America. They rolled the dry leaves of the *yetl* into *cigars* and inserted them in tubes made of silver, wood, or reed. They often mixed in the resin of the *Liquidambar styraciflua* with III.51

1. Hernández, [*Quatro libros*,] book V, chap. 51, p. 173. Clavijero, [*Storia antica di Messico*,] vol. II, p. 227. Garcilaso [de la Vega El Inca, *Comentarios reales*], book II, chap. 25. The Mexica already recommended tobacco as a cure for toothaches, head colds, and colic. The Caribs used chewed tobacco leaves as an antidote. In our travels on the Orinoco, we have seen chewed tobacco applied successfully to bites from poisonous grass snakes. After the famous *Bejuco del Guaco*, the knowledge of which we owe to Mr. Mutis, tobacco is probably the most powerful antidote in the Americas. Tobacco cultivation spread so quickly that by 1559 it was already being sown in Portugal, and at the beginning of the seventeenth century, they planted it in the East Indies. Beckmann, [*Beyträge zur*] *Geschichte der Erfindungen*, book 3, p. 366; Crawfurd, *Hist[ory] of the Ind[ian] Archip[elago]*, vol. I, p. 409. The Haitian word *tabacco* did not refer to the plant but, rather, to the tube through which they smoked it. *Rel[ation] hist[orique]*, vol. II, p. 622. With the exception of green N. undulata, the genus Nicotiana is entirely Mexican. Brown, *Botany of Congo*, p. 53.

other aromatic material. In order to breathe in the tobacco smoke more easily, they held the tube in one hand and blocked their nostrils with the other. Many people even enjoyed inhaling the smoke through the nose. Although *Picietl* (*Nicotiana rustica*) was widely cultivated in ancient Anahuac, it nevertheless seems that only wealthy people used tobacco; we see today that this custom is entirely unknown among the racially pure Indians because nearly all of them descend from the lowest class of the Aztec people.[1]

In Veracruz, the amount of tobacco harvested in the districts of Orizaba and Córdoba is estimated at eight to ten thousand tercios (eight *arrobas*), which equals 1,600,000 to 2,000,000 pounds; but this estimate seems a bit too low. The King pays the grower two and a half reales per pound of tobacco; that is, twenty-one sols per kilogram. Later in this work, we shall see, based on information that I have taken from official documents, that Mexican plantations typically sell over thirty-eight million francs' worth of tobacco for smoking and in powder form on the domestic market, and that they return to the King a net profit of over twenty million livres tournois. This must seem an extremely high level of consumption for New Spain, III.52 especially since, from a population of 5,800,000 souls, one must subtract two and a half million indigenous residents who do not smoke. In Mexico, moreover, plantations are of much greater importance to the tax office than in Peru, because the number of whites is more significant in Mexico and because the custom of smoking cigars is more widespread there, even among women and children of a young age. In France, where, according to studies by ▾ Mr. Fabre de l'Aude, there are eight million people who use tobacco, total consumption is over forty million pounds; but the value of imports of foreign tobacco amounted to only 14,142,000 livres tournois in 1787.[2]

Rather than exporting its domestic tobacco, New Spain takes almost 56,000 pounds of it from Havana. The humiliation to which the planters have been subjected, combined with the preference given to coffee cultivation, have nevertheless resulted in a severe decrease in production on Cuban tobacco farms. Today, that island provides barely 150,000 arrobas, while prior to 1794, the harvest in good years was estimated to be 315,000 arrobas (7,875,000 pounds' worth[3]), of which 160,000 arrobas

1. See above vol. I, p. 338 [p. 226 in this edition].

2. Peuchet, [*Statistique élémentaire*,] pp. 315 and 409.

3. Raynal, [*Histoire*] (vol. III, p. 268) estimated the harvest at only 4,675,000 pounds' worth. Prior to 1775, Virginia produced over 55,000 *hogsheads* of tobacco, or thirty-five

were consumed on the island itself, and 128,000 were shipped to Spain. III.53 This branch of colonial industry is of key importance, even in its current state of monopoly and constraint. The *Renta de Tabaco* [tobacco tax] that the Peninsula imposes provides net revenue of six million piasters, revenue that is largely obtained through the sale of tobacco from the island of Cuba, shipped to Seville. The storehouses of Seville sometimes hold eighteen to nineteen million pounds' worth of powdered tobacco alone, the value of which rises to the exorbitant sum of two hundred million livres tournois.

Indigo cultivation, very widespread in the kingdom of Guatemala and in the province of Caracas, is almost completely neglected in Mexico. The plantations found along the western coasts do not produce enough for even the few native cotton cloth factories. Every year, indigo is imported from the kingdom of Guatemala, where the plantations' total output had a value of twelve million livres tournois in 1803. This dye, on which Mr. Beckmann has conducted scholarly studies, was known to the Greeks and the Romans by the name of *Indicum*. The word *anil*, which passed into the Spanish language, comes from the Arab word *nir*, or *nil*. Hernández, in speaking of Mexican indigo, calls it *anir*. In Dioscorides's time, the Greeks took their indigo from Gedrosia; and in the thirteenth century, ▼ Marco Polo painstakingly described its preparation in Hindustan. Raynal erroneously claims that the Europeans introduced the cultivation of this precious plant to the III.54 Americas. Several species of indigofera are unique to the New Continent.[1] Ferdinand Columbus, in the biography of his father, cited indigo among the products of the island of Haiti. Hernández described the method that the indigenous peoples of Mexico used to separate the powder from the sap of the plant, a method that differs from the one that we use today. The small, fire-dried indigo cakes were called *mohuitli*, or *tlellohuilli*. The plant was also referred to by the name *xiuhquilipitzahuac*. Hernández[2] proposed to the court that indigo cultivation be introduced into southern Spain. I do not know whether his advice was heeded, but it is certain that indigo was quite common in Malta until around the end of the seventeenth century. The species of *indigofera* from which they extract indigo in the colonies

---

million pounds, per year. Jefferson, [*Notes on the State of Virginia*,] p. 323. From 1822 to 1823, in an average year, Cuban production equaled 400,000 arrobas.

1. Humboldt, Bonpland, and Kunth, *Nov[a] Gen[era] et Spec[ies]*, vol. VI, p. 454; and *Rel[ation] hist[orique]*, vol. II, p. 506.

2. Hernández, [*Quatro libros*,] book IV, chap. 12, p. 108. Clavijero, [*Storia antica del Messico*,] II, 189. Beckmann, [*Beyträge zur Geschichte der Erfindungen*,] vol. IV, 474–532. ▼ Berthollet, *Élémens de l'art de la teinture*, vol. II, p. 37.

today are Indigofera tinctoria, I. anil., I. disperma and I. argentea, as the most ancient hieroglyphic paintings of the Mexica reveal. Even thirty years after the conquest, the Spanish, who had not yet found materials for making ink, wrote with indigo, a fact established by the documents preserved in the archives of the Duke of Monteleone, who was the last offspring of the Cortés

III.55 Family. In Santa Fé, people still write using the juice expressed from the fruit of the uvilla (Cestrum mutisii), and there is a court order enjoining the viceroys to use only this blue uvilla [ink], because it was acknowledged to be more permanent than the best ink from Europe.

[The following table presents exports of indigo from Guatemala in nine consecutive years; the results have been taken from customs records (see *Redactor general de Guatemala*, 1825, n. 5121).

YEAR	POUNDS OF INDIGO	VALUE IN PIASTERS
1794	592,262	641,393
1795	1,108,789	1,066,786
1796	1,184,201	1,369,881
1797	159,665	211,650
1798	151,317	141,859
1799	533,637	469,592
1800	450,606	398,096
1801	331,897	332,063
1802	1,479,641	1,921,356

This fluctuation in exports was the effect of both the stagnation in trade and the lack of European ships. Indigo cultivation in San Salvador is done almost entirely by free workers. Ever since the price of indigo began to rise again in Europe, its cultivation in Guatemala has also increased again. According to notes that I owe to the good offices of ▼ Mr. García Granados, current indigo exports (1825) from Guatemala equal 12,000 tercios (fifty-

III.56 nine Spanish pounds, or six arrobas) per year. In Guatemala, the price of one pound is nine reales de plata, or one and one-eighth piasters, such that the total value of exports of 1,800,000 pounds amounts to over two million piasters in the republic of *Centro-América*. Between 1821 and 1824, in an average year the ports of Great Britain received 5,077,878 English pounds of indigo from the Americas and the East Indies; in 1825, it was over 7,539,500.]

Having carefully examined the crops that are the mainstays of agriculture and trade in Mexico, it remains for us to take a quick glance at products from the animal domain. Although the most sought-after of these products, cochineal, originated in New Spain, it is nevertheless certain that the most important products to the welfare of the inhabitants were introduced there from the Old Continent. The Mexica had not made any attempts at domesticating either species of wild oxen (Bos americanus and Bos moschatus) that roam the plains bordering the North River in herds. They were completely unfamiliar with the llama from the Cordillera of the Andes, which does not live above the southern hemisphere. They did not know how to make profitable use of the wild ewes of California,[1] nor of the goats in the mountains of Monterey. Among the numerous varieties of dog[2] that are native to Mexico, only one, the *techichi*, was used for meat. The need for domestic animals was probably not as evident before the conquest, at a time when each family cultivated only a small parcel of land, and when many people consumed almost exclusively plants. Nevertheless, the lack of these animals forced a sizable class of inhabitants, that of the Tlamama, to perform the role of beasts of burden and to spend their lives on the main roads. They carried large leather bags (*petlacalli* in Nahuatl, *petacas* in Spanish) that contained goods weighing between thirty and forty kilograms. III.57

Since the middle of the sixteenth century, the most useful animals from the Old Continent—oxen, horses, ewes, and hogs—have multiplied to a surprising extent in all parts of New Spain, especially in the vast plains within the borders of the *Provincias internas*. It would be unnecessary to refute[3] here ▾ Mr. de Buffon's baseless claims regarding the supposed degeneration of the domesticated animals introduced to the New Continent. These ideas spread easily, not only because they flattered the vanity of the Europeans but also because they were linked to brilliant theories about the primitive state of our planet. Now that the facts are being examined with greater care, physicists are able to acknowledge the harmony that exists in III.58

1. On the wild ewes and goats in the mountains of old and new California, see above, vol. II, p. 263 [p. 442 in this edition].

2. See my *Tableau de la nature*, vol. I, pp. 127–24. The Comanches, a tribe from the northern provinces, use Mexica dogs to transport tents, as do several peoples of Siberia. See above, vol. II, p. 230 [p. 424 in this edition]. The Peruvians of Sausa [Xauxa] and Huanca ate their dogs (*runalco*), and the Aztecs sold at market the meat of the mute dogs, *techichi*, which they neutered in order to fatten them up. Lorenzana, [*Historia de Nueva-España*,] p. 103. Compare with *Rel[ation] hist[orique]*, vol. II, pp. 624, 692, 708.

3. This refutation can be found in the excellent work by Mr. Jefferson, [*Notes on the State of Virginia*], pp. 109–66. See also Clavijero, [*Storia antica del Messico*,] vol. IV, p. 105–60.

the very places where that eloquent natural historian thought to find only contrasts.

Horned animals live in great abundance along the eastern coasts of Mexico, especially at the mouths of the Alvarado, Guasacualco, and Panuco Rivers, where numerous herds find perpetually green pastures. Both the capital of Mexico and the large neighboring cities nevertheless obtain their meat supplies from the intendancy of Durango. Like most of the peoples of Asia east of the Ganges,[1] the indigenous peoples have almost no taste for milk, butter, or cheese. Cheese is highly sought-after by the mixed-race castes and forms quite a significant branch of domestic trade. In the statistical table that the intendant of Guadalajara drew up in 1802, and which I have often had occasion to cite, the yearly value of curried leather III.59 is estimated at 419,000 piasters, that of tallow [suet] and soap at 549,000 piasters. The city of Puebla alone produces 200,000 arrobas of soap each year, as well as 82,000 cowhides. But exports of these two items from the port of Veracruz have thus far been relatively insignificant. In 1803, they amounted to barely 140,000 piasters. It even appears that, in the sixteenth century, before domestic consumption had risen along with the increase in the whites' numbers and luxuries, New Spain shipped a greater amount of leather to Europe than it does today. Father Acosta[2] reports that a fleet that reached Seville in 1587 carried 63,340 [350] Mexican hides. Horses from the northern provinces, especially those of New Mexico, are as famous for their excellent qualities, as are the horses of Chile. Both kinds descend, it is claimed, from the Arab race. They roam in wild herds on the savannas of the *Provincias internas*. Exports of these horses to Natchez and New Orleans have grown more significant with each passing year. There are several families in Mexico whose holdings in their *hatos de ganado* [livestock ranches] range from thirty to forty thousand total cattle and horses. The number of mules would be even higher if so many of them did not die on the main roads from the fatigue that overwhelms them after journeys lasting III.60 several months. Veracruz trade alone is thought to require nearly 70,000

1. For example, in southeastern Asia, the Chinese and the inhabitants of Cochinchina. The latter never milk their cows, although milk from the tropics and from the hottest parts of the Earth is quite excellent. ▼ Macartney, *Voyage*, vol. II, p. 153, and vol. IV, p. 59. Even the Greeks and the Romans only learned how to make butter from their contacts with the Scythians, the Thracians, and the peoples of Germanic race. Beckmann, [*Beyträge zur Geschichte der Erfindungen*,] book III, p. 289.

2. [José de Acosta, *Historia natural y moral de las Indias*,] book IV, chap. 3[3], [p. 276].

mules per year. Over five thousand of them are used as luxury items in the carriage teams[1] of Mexico City.

The raising of *sheep* was once utterly neglected in New Spain, as it was in all of the Spanish colonies in the Americas. It is likely that the first woolly animals, introduced in the sixteenth century, were not of the voyager *merinos* breed, and that they were certainly not of the Leonese, Segovian, or Sorian breed. Since that time, no one has undertaken to improve the breed. In the part of Mexico located outside the tropics, it would be easy to introduce the herding system referred to in Spain as the *mesta*, a system in which the ewes change climates with each passing season and are always in harmony with the latter. There is no need to worry that these herd migrations might cause long-term damage to Mexican agriculture. Today, the wool that is considered the most beautiful comes from the intendancy of Valladolid.

III.61 It is worth noting that both the common *pig*[2] and the chickens that are found in all the islands of the South Sea were unknown to the Mexica. The Peccaries (*Sus tajassu*) that one often encounters in the huts of the indigenous peoples of South America could have easily been domesticated, but this animal lives only in the plains region. Of the two most common pig breeds in Mexico today, one was introduced from Europe and the other from the Philippines. Their numbers have grown considerably on the central plateau where the Toluca valley boasts a very lucrative ham trade.

Before the conquest, *poultry* was quite scarce among the indigenous peoples of the New Continent. Raising these birds requires particular care in recently cleared lands, where the forests are teeming with carnivorous quadrupeds of all sorts. Furthermore, a tropics-dweller feels the need for domestic animals less acutely than does the temperate zone resident because the fertility of the soil relieves him from having to plow a large area of land, and because the lakes and rivers are covered with innumerable birds that III.62

1. Havana has 2,500 barouches, called *volantes*, which require more than 3,000 mules. In 1802, it was estimated that there were 35,000 horses in Paris.

2. ▼ Pedro de Cieza [de León] and Garcilaso de la Vega [El Inca] recorded in their works the names of the colonists who first raised European domestic animals in the Americas. They report that, in mid-sixteenth-century Peru, two pigs cost 8,000 livres tournois; a camel cost 35,000; a donkey, 7,000; a cow, 1,200; and a sheep, 200. Cieza [de León], *Cronica del Perú* (Antwerp, 1554), p. 65. Garcilaso, [*Comentarios reales*,] vol. I, p. 328. In addition to the scarcity of items for sale, these exorbitant prices signal the abundance of precious metals. ▼ General Belalcázar, who had purchased a sow for 4,000 francs in Buga, could not resist the temptation of eating it at a feast. Such was the luxury that prevailed within the army of the *Conquistadores*.

are easy to catch and that provide abundant food. European travelers are surprised to note that the savages of South America go through enormous trouble to tame monkeys, *manaviri* (Ursus caudivolvula), and squirrels, while they do not seek to domesticate the large number of useful animals that live in the surrounding forests. Nevertheless, before the Spaniards' arrival, the most civilized peoples of the New Continent had already raised in their poultry yards several gallinaceans [gallinaceous birds], such as Hoccos (Crax nigra, C. globicera, and C. pauxi), turkeys (Meleagris gallopavo), several species of pheasants, ducks, and water chickens, yacous or guans (Penelope, *pava de monte*), and Aras (Psittaci macrouri), which are regarded as a delicacy when they are young. At that time, the cock, which originated in the East Indies and is common in the Sandwich Islands, was completely unknown in the Americas. This fact, which is of great importance to theories about the migration of the peoples of the Malay race, was first contested in Spain at the end of the sixteenth century. Learned etymologists argued that the Peruvians must have had chickens before the discovery of the New World, because the Inca language [Quechua] has a particular word, *gualpa*, which refers to the cock. They did not know that *gualpa*, or *huallpa*, is a

III.63

contraction of *Atahualpa*, and that the natives of Cuzco had derisively given to the cocks brought by the Spanish the name of a prince who was hated for his cruel treatment of the ▾ Huescar family, because they perceived—which seems quite strange to the European ear—a certain resemblance between the song of this bird and the name of Atahualpa. This anecdote, recorded in *Garcilaso*'s work (vol. I, p. 331), was also told to me in 1802 in Cajamarca, where I encountered, in the ▾ *Astorpilco* family, the descendents of the last Inca of Peru; these wretched Indians dwell within the ruins of Atahualpa's palace. Garcilaso reports that the Indians imitated the cock's cry by pronouncing *four-syllable* words in a rhythmic manner. The supporters of Huescar had composed a number of burlesque songs to poke fun at Atahualpa and three of his generals, named Quilliscacha, Chalchuchima, and Ruminavi. If one consults languages in the same way one consults historical monuments, one must distinguish very carefully between what is truly ancient and what has been naturalized through use. The Peruvian word *micitu*, meaning cat, is just as modern as the word *huallpa*. The Peruvians formed *micitu* from the root *miz* because, having witnessed the Spaniards use this word when calling their cats, they thought that *miz* was the name of the animal itself.

It is a very curious physiological phenomenon that, on the plateau of the city of Cusco, which is higher and colder than that of Mexico City, chickens only started to become acclimated and propagate after a thirty-year period. III.64 Until that time, all the chickens died as soon as they hatched from the egg. Today, the various chicken breeds—especially those from Mozambique, which have black flesh—have become common in both hemispheres wherever the peoples of the Old Continent ventured. Several tribes of primitive Indians who live near the European settlements were able to obtain some of these. While we were in Tomependa on the banks of the Amazon River, we saw a few Xibaro Indian families who have settled in Tutumbero, in a nearly inaccessible spot between the Yariquisa and the Patorumi cataracts; it was in these Indians' huts that we had seen chickens when we first visited them several years ago.

New Spain provided Europe with the largest and most useful of all domestic gallinaceous birds: the turkey (*Totolin* or *Huexolotl*), which was once found wild on the ridge of the Cordilleras from the Isthmus of Panama to New England. Cortés recounts that several thousand of these birds, which he calls chickens (*gallinas*), were being fed in the poultry yards of Montezuma's castles. From Mexico, the Spanish brought turkeys to Peru, to Terra Firma (*Castilla del oro*), and to the Antilles islands, where Oviedo wrote about them in 1515. Hernández already astutely observed that the wild turkeys of Mexico were much bigger than the domestic turkeys. Today, the former are only found in the northern provinces. They have withdrawn III.65 to the north as the population has grown and as the forests have unavoidably shrunk. We have learned from Mr. Michaux, a scholarly traveler to whom we owe a very interesting description of the lands located to the west of the Allegheny Mountains,[1] that the wild turkey of Kentucky sometimes weighs up to forty pounds, an enormous weight for a bird capable of flying very fast, especially when pursued. When the English landed in Virginia in 1584, turkeys had already existed in Spain, Italy, and England for fifty years.[2] It was not, therefore, from the United States that this bird first came to Europe, as many naturalists have mistakenly suggested.

Guinea fowl (*Numida meleagris*), which the ancients so astutely called *aves guttatæ*, are quite rare in Mexico, although they have become feral on the island of Cuba. As for the musk duck (Anas moschata), which the Germans call Turkish duck, and which is now commonly found in our poultry yards, Europe owes this as well to the New Continent. We found it living wild on the banks of the Magdalena River, where the male grows to an

1. Michaux, *Voyage* [*à l'ouest des monts Alléghanys*,] p. 190.
2. Beckmann, [*Beyträge zur Geschichte der Erfindungen*,] vol. III, pp. 238–70.

III.66 extraordinarily large size. The ancient Mexica had domestic ducks, which they plucked yearly, the feathers being an important trade item. These ducks appear to have mixed with the species introduced from Europe. The goose is the only one of our poultry-yard birds that is found virtually nowhere in the Spanish colonies of the New Continent.

Both mulberry tree cultivation and the raising of silkworms had been introduced by Cortés a few years after the siege of Tenochtitlan. On the ridge of the Cordilleras, there exist two mulberry trees native to the equinoctial regions, *Morus celtidifola* and *M. corylifolia*, which we found growing wild in the kingdom of Quito near the villages of Pifo and Puembo. The leaves of these mulberry trees are not as tough as those of the red mulberry tree (M. rubra) of the United States, and the silkworms eat it just as they do the leaves of the white mulberry of China. This latter tree, which, according to ▼ Olivier de Serres, was not planted in France until the reign of Charles VIII, around the year 1494, was already quite common in Mexico toward the middle of the sixteenth century. At that time, they harvested a very large amount of silk in the intendancy of *Puebla*, on the outskirts of *Panuco*,[1] and in the province of *Oaxaca*, where a few *Mixteca* villages still bear the name *Tepexe de la Seda* (silk) and *San Francisco de la Seda*. The III.67 policies of the Council of the Indies, which are consistently detrimental to Mexican manufacturing; increasing trade with China; and the efforts of the Company of the Philippines to sell Asian silks to Mexicans appear to be the main factors that gradually destroyed this branch of colonial industry. Only a few years ago, a private individual in Querétaro suggested that the government establish large mulberry tree plantations in one of the most beautiful valleys of Mexico, *La Cañada* of the baths of San Pedro, inhabited by over three thousand Indians. Less work is involved in raising silkworms than in raising cochineal, and the indigenous peoples' character makes them particularly well-suited to any task that demands tremendous patience and meticulous care. *La Cañada*, which is two leagues northeast of Querétaro, enjoys a continually mild and temperate climate. Today, only avocado trees (*Laurus persea*) are cultivated there, and the viceroys, who, afraid of hurting what are referred to in the colonies as the privileges of the metropole, were unwilling to allow avocado farming to be replaced by mulberry tree cultivation.

New Spain has several species of indigenous caterpillars, which spin silk similar to that of the *Bombyx mori* of China, but which have not yet

1. [Garcilaso de la Vega El Inca,] *La Florida del Inca*. Madrid, 1723, vol. I, p. 258.

been sufficiently examined by entomologists. These are insects that produce Mixteca silk, which was already an item of trade in Montezuma's time. Handkerchiefs made of this Mexican silk are still manufactured today in the intendancy of Oaxaca. We purchased a few of these on the road from Acapulco to Chilpanzingo. The cloth is rough to the touch, like certain silks from India that are also the product of insects very different from the silkworm of our mulberry trees.

III.68

In the province of Michoacan and in the mountains of Santa Rosa, to the north of Guanajuato, one sees oval-shaped sacs resembling the nests of Troupials and Caciques hanging from different species of trees, especially from the branches of the *Arbutus madroño*. These sacs, called *capullos de madroño*, are the work of a large number of caterpillars belonging to the genus Bombyx de Fabricius, insects that live in large groups and spin silk. Each *capullo* [cocoon] is eighteen to twenty centimeters long and ten wide. They are of a bright white color and are formed by removable layers. The inner layers are the thinnest and are extraordinarily transparent. The material from which these large pockets are formed resembles Chinese paper. Its tissue is so dense that one can barely recognize the threads that are pasted crosswise on top of one another. I found a large number of these *capullos de madroño* while coming down from the Cofre de Perote to Las Vigas, at an absolute elevation of 3,200 meters. One can write on the inner layers of these cocoons without having to treat them. It is true natural paper, which the ancient Mexica knew how to use profitably, by pasting several layers together to form a white, glossy cardboard. We had living Bombyx madroño caterpillars from Santa Rosa delivered to us by post in Mexico City. They are of a blackish olive color and are covered in down. They are twenty-five to twenty-eight millimeters long. We did not witness their metamorphosis, but we did take note that, despite the beauty and the extraordinary shine of this *madroño* silk, it will be nearly impossible to put it to good use because of the difficulties involved in unwinding it. Since several caterpillars work together, their threads cross one another and interlace. I thought it necessary to enter into these details because a number of individuals, whose zeal eclipses their knowledge, recently focused the attention of the French government on the indigenous silk of Mexico.

III.69

Wax is an item of huge importance to a country where religious worship is marked by a great deal of splendor. An enormous amount of it is used in the church festivals not only in the capital but also in the smallest Indian villages. Beehives are extremely productive in the Yucatán peninsula, especially on the outskirts of the port of Campeche, from which 582 *arrobas* of wax were shipped to Veracruz in 1803. Between six and seven hundred beehives are collected in a *colmenar* [beehive]. This Yucatán wax comes from a variety of bees native to the New Continent and reputed to be III.70 missing a stinger, probably because their weapon is quite ineffective and somewhat negligible in size. It is this feature that led people in the Spanish colonies to give the name of *little angels* (*angelitos*) to the bees that ▾ Mr. Illiger, Mr. Jurine, and Mr. Latreille described, using the names Melipona and Trigona. I do not know whether the Campeche bee is different from the *Melipona fasciata* that Mr. Bonpland encountered on the eastern slope of the Cordilleras.[1] What is certain is that the wax from American apiaries is more difficult to whiten than the wax from European domestic bees. New Spain takes nearly 25,000 *arrobas* of wax from Havana every year, and the price of these imports amounts to over two million livres tournois. Only a very small portion of the wax from the island of Cuba comes from the wild *Trigones* that live in the trunks of the *Cedrela odorata*; the majority of it is produced by a variety of bee native to northern Europe (*Apis mellifica*), the farming of which has become quite widespread since 1772. In 1803, the island of Cuba exported 42,670 *arrobas* of wax, including smuggling. At that time, the price of one *arroba* was between twenty and twenty-one piasters; but the average price at peacetime is only fifteen piasters, or seventy-five livres tournois. In the Americas, the presence of neighboring sugar refiner- III.71 ies is very harmful to the bees. Extremely greedy for honey, these insects drown themselves in the cane juice, which puts them in a state of immobility and drunkenness whenever they drink it in excess.

The raising of *cochineals* (*grana*, *nochitztli*) dates back to the earliest times in New Spain. It likely began before the incursion of the Toltec peoples. Cochineals were more common in the time of the dynasty of Aztec kings than they are today. *Nopalries* existed not only in Mixtecapan (Mixteca) and in the province of Huaxyacac (Oaxaca), but also in the intendancy of Puebla, on the outskirts of Cholula and Huejotzingo. Both the humiliations to which the indigenous peoples were subjected at the start of the conquest and the low price at which the Encomenderos forced the farmers to sell them their cochineals led to this branch of Indian industry's falling into neglect everywhere, except in the intendancy of Oaxaca. As little as forty years ago, the Yucatán peninsula still had significant nopalries. In a single night, all the nopals on which the cochineals lived were cut down.

1. See the insects that we collected over the course of our expedition and that Mr. Latreille described in our *Recueil d'observations de zoologie et d'anatomie comparée*, p. 251.

The Indians claim that the government resorted to this violent measure to raise the price that the inhabitants of Mixteca paid for a commodity over which it wanted to reserve exclusive rights. The whites, on the other hand, claim that the indigenous peoples, angry and displeased about the price that the merchants had established for cochineals, conspired to destroy both the insects and the nopals.

The number of cochineals that the intendancy of Oaxaca ships to Europe can be estimated—in a regular year, and including all three types (*grana*, *granilla*, and *polvos de grana*)—at 4,000 *zurrones*, or 32,000 *arrobas*, which, if one arroba is equal to seventy-five piastras fuertes, totals 2,400,000 piasters, or twelve million livres tournois. The following amounts were exported from Veracruz: III.72

In	1802,	46,964 arrobas, or	3,368,557 piasters' worth
	1803,	29,610	2,238,673

But since one part of a year's harvest is often included with the next year's, one must not determine the progress of farming by exports alone. It appears that *nopalries* are growing very slowly in Mixteca. In the intendancy of Guadalajara, barely eight hundred arrobas of cochineal are harvested each year. Raynal[1] estimates total exports from New Spain at 4,000 quintals, an estimate that is too low by half. The East Indies have also begun to contribute to the cochineal trade, but only negligible amounts. ▾ Captain Nelson took the insect from Rio de Janeiro in 1795. Nopalries were established on the outskirts of Calcutta, Chittagong, and Madras. They have met with great difficulty there in obtaining the right species of cactus for the insect's food. We do not know whether this Brazilian cochineal, transported to Asia, is of the mealy species of Oaxaca or the cottony cochineal (*grana silvestre*). III.73

I shall not repeat here what Thiéry de Menonville and other naturalists after him have published about the farming of nopal and about the raising of the valuable insect that it feeds. Mr. Thiéry conducted his research with as much shrewdness as he displayed in courage while carrying out his projects. His observations on the cochineal introduced to Saint-Domingue are certainly very precise, but since he did not know the local language and was afraid of arousing suspicion by demonstrating excessive curiosity, he was

1. Vol. III, p. 72 [p. II.41 in this edition].

only able to gather somewhat incomplete information on Mexican nopalries during his stay in the intendancy of Oaxaca. I had occasion to observe wild cochineals in the kingdom of New Granada, in Quito, in Peru, and in Mexico; I was not to have the good fortune to see cultivated cochineals, but having consulted individuals who lived for a long time in the mountains of *Mixteca*, and having at my disposal excerpts from several manuscripts that the ▼ Count of Tepa, during his stay in Mexico City, had commissioned from a number of alcaldes [mayors] and clerics from the diocese of Oaxaca, I trust that I can provide some useful information about an insect that has become an object of main importance to European manufacturing.

Is the *mealy* cochineal (*grana fina*) in Mixteca unambiguously distinct III.74 from the *cottony* or wild cochineal (*grana silvestre*), or is the latter the original form of the former, which would thus be the product of degeneration incurred by its being raised and cared for by humans? It is just as difficult to resolve this dilemma as it is to determine whether the domestic ewe descends from the mouflon, the dog from the wolf, and the ox from the aurochs. Everything regarding the origin of species and the theory of a variety that has become constant, or of a type that perpetuates itself, belongs to problems of zoonomy, on which it is wise not to pronounce affirmatively.

The cultivated cochineal differs from the wild cochineal not only in size but also because it is mealy and covered in white powder, while the wild variety is wrapped in a thick cotton which makes its segments difficult to distinguish from one another. The metamorphoses of both insects are, however, identical. In the parts of South America where people have devoted themselves to raising wild cochineals for centuries, they have not succeeded in making them lose their down. It is true that in Saint-Domingue, they were persuaded that in the nopalries established by Mr. Thiéry [de Menonville], the insects in the care of human ingenuity increased in size, and that there was a noticeable change in the thickness of their cottony exterior. But Mr. Latreille, a learned entomologist who is inclined to regard the wild cochineal as a different species from the cultivated cochineal, believes that the insects' down only appeared to have thinned, and that this must III.75 be attributed to the thickening of their bodies. Since the segments on the back of the female are more expanded, the hairs that cover this part must have appeared further apart from one another and, for that reason, scarcer. A few people who have stayed on the outskirts of the city of Oaxaca for an extended period have assured me that one occasionally sees among the tiny newborn *coccus* a number of them covered in quite long hair. One might be tempted to regard this as proof that even after nature has deviated from an original type, it nevertheless returns to it from time to time; it is in this manner that ▾ Mr. Duchêne's Fragaria monophylla seed consistently produces a few normal strawberry plants with split leaves. But one must not forget that when the cultivated cochineal emerges from its mother's body, it has a wrinkled back covered in twelve silk threads that are often very long but that disappear with adulthood. People who have not closely compared the *nymphs* of the cultivated cochineal with that of the wild cochineal are naturally struck by the presence of these hairs. The cultivated cochineal looks powdery ten days after its birth, after it has shed its coat, fringed with tiny silk threads. The wild cochineal, on the contrary, becomes hairier as it grows older. Its down thickens, and the insect resembles a small white speck during the period before the two sexes mate.

In the Oaxaca nopalries, one occasionally observes the winged male of the cultivated cochineal mating with a wild cochineal female. This has been cited as proof of the sameness of the species. Yet, in Europe, we commonly see *ladybugs* mate despite being fundamentally different from one another in shape, size, and color. If two species are closely related, we should not be surprised to see them mating. III.76

Are the cultivated cochineal and the plant on which they are raised both found in their wild state in Mexico? Mr. Thiéry thought it possible to respond negatively to this question. This naturalist seems to suggest that both the insect and the nopal of the Oaxaca plantations were imperceptibly modified in shape through their extensive cultivation. This supposition, however, appears to me to be as spurious as the one that maintains that wheat, corn, and the banana tree should be regarded as degenerated plants, and, to cite an example taken from the animal realm, the llama—which is unknown in its wild state—should be regarded as a variety of the vicuña of the High Andes. The Coccus cactus has innumerable insect and bird enemies. Wherever the cottony cochineal propagates itself without assistance, it is only found in relatively small numbers. Yet, one can easily conceive that the mealy cochineal must have been even scarcer in its native land because it is more delicate and because, not being covered in down, it is more sensitive to the cold and the humidity of the air. While debating the question of whether the cultivated cochineal can propagate without human assistance, ▾ Ruiz de Montoya,[1] the subdelegate of the province of Oaxaca, cites in his paper the quite remarkable fact that "at a distance of III.77

1. ["Informe sobre el cultivo de la grana en la jurisdicción de Nejapa,"] *Gazeta de literatura de México*, 1794 [1770], p. 228.

seven leagues from the village of Nejapa, there is a place where, owing to its particularly favorable circumstances, the loveliest *grana fina* can be gathered on wild, extremely tall and spiny nopals, without one ever having to go through the trouble cleaning the plants or renewing the *nymphs* of the cochineal." Furthermore, one must not be surprised that even in a country where this insect is indigenous, it might completely cease being found in its wild state as soon as people began seeking it out and raising it in nopalries. It is likely that the Toltec, before undertaking such a difficult type of cultivation, might have gathered cultivated cochineal on the nopals that grew wild on the mountainsides of Oaxaca. If they had harvested the females before the latter had lain their eggs, the species would have soon become extinct, and it was in order to guard against this progressive extinction, and to prevent the cottony and mealy cochineals from mixing on the same cactus—the former would have taken all the food away from the latter—that the Toltecs established nopalries.

The plants on which either species of cochineal propagates are fundamentally different from one another. This well-established fact is among the III.78 many indications of an original and specific difference between the *grana fina* and the *grana silvestre*. If the mealy cochineal were simply a variety of the cottony cochineal, is it likely that the former would die on the same cacti that botanists call Cactus opuntia, C. tuna, and C. ficus indica that serve as food for the cottony cochineal. Mr. Thiéry, in the work[1] that we have often had occasion to cite, claims that in Saint-Domingue, in the Cul-de-Sac Plain, cottony (or wild) cochineals do not appear on the Cactus tuna but on the C. pereskia, which he classes among the *segmented [jointed]* prickly pears. I am afraid that this botanist may have confused a variety of opuntia with the real C. pereskia, which forms a tree with wide, oily leaves, on which I have never encountered a single cochineal. I also regard it as very dubious that the plant which Linné called Cactus coccinillifer, and which we cultivate in Europe, is the nopal on which the Indians of Oaxaca raise mealy cochineals. Mr. De Candolle,[2] who has shed a great deal of light on this matter, seems to share my opinion; he cites, as being identical with the cochineal prickly pear, Thiéry de Menonville's *nopal sylvestre*, which is entirely different from the nopal of the plantations. Indeed, Linné had given the name Cactus coccinellifer to the prickly pear, which several European

1. [Thiery de Menonville, *De la culture du Nopal*,] pp. 275–82.
2. *[Histoire naturelle des] plantes grasses* by ▼ Mr. Redouté and Mr. De Candolle, installment 24.

botanical gardens had received at the same time as the cottony cochineal. III.79 It is a purple-flowered species (▼ Plukenet's *Ficus indica vermiculos proforens*) that grows wild in Jamaica, Cuba, and nearly everywhere in the Spanish continental colonies. I showed this cactus to some very knowledgeable people who had carefully examined the Oaxaca nopalries; they uniformly assured me that it is fundamentally different from the *nopal of the plantations*, and that the latter, as Mr. Thiéry also notes, does not occur in a wild state. Moreover, the Abbot Clavijero,[1] who lived in Mixteca for five years, expressly states that the fruit of the nopal on which cultivated cochineals are propagated is small, not very tasty, and white, whereas the fruit of the Cactus coccinillifer Lin. is red. The famous ▼ Ulloa suggests in his works that true nopal has no needles. But he seems to have confused this plant with a prickly pear that we often encountered in the gardens (*conucos*) of the Indians of Mexico and Peru, and which the creoles call *Tuna de Castilla* because of its gigantic size, its excellent fruit, and its beautiful segments, which are of a bluish green and devoid of needles. This latter nopal, the most elegant of all the Opuntia, is indeed suitable for feeding the mealy cochineal, especially shortly after birth; it is, however, very rare in the Oaxaca nopalries. If, according to the theory held by a few distinguished botanists, III.80 the *Tuna* (or *Nopal*) *de Castilla* is merely a variety of the ordinary Cactus opuntia that has resulted from cultivation, then it is surprising to note that the prickly pears cultivated for centuries in our botanical gardens and those of the nopalries of New Spain have not also lost the needles with which their plump segments are armed.

The Indians of the intendancy of Oaxaca do not all follow the same method for raising cochineals that Thiéry de Menonville witnessed in operation during the course of his brief journey through San Juan del Ré, San Antonio, and Quicatlan. Those of the districts of Sola and Zimatlan[2] set up their nopalries on the mountain slopes or in the ravines located two to three leagues from their villages. They plant the nopals after cutting down and burning the trees that cover the terrain. If they continue to clear the ground twice a year, then the young plants will be ready to feed the cochineals as early as the third year. To that end, in April or May, a nopalry owner purchases branches or segments of *Tunas de Castilla* covered in tiny newborn cochineals (*semilla*). These rootless cuttings, separated from the trunks, retain their juice for several months. They sell for around three francs per

1. [Clavijero, *Storia antica del Messico*,] p. 115.
2. ▼ *Informe de Don Francisco Ibañez de Corvera* [manuscript].

III.81 hundred in the Oaxaca market. The Indians store the *nymphs* of the cochineal in caves or inside their huts for a twenty-day period, after which they expose the young coccus to the fresh air. They hang the segments on which the insects are placed in a shed covered with a straw roof. Cochineal grows so rapidly that, by August or September, one finds females already pregnant before the young have hatched. These mother cochineals are placed in *nests* made from a species of Tillandsia [Bromeliad] called *Paxtle*. It is in these nests that they are carried two or three leagues away from the villages and distributed among the nopalries, where the *nymphs* are deposited on the young plants. The mother cochineals lay their eggs over a thirteen- to fourteen-day period. If the plantation is in a low-lying area, one can expect the first harvest in less than four months. In colder, less temperate climates, one sees that the color of the cochineal is equally beautiful, but that the harvest occurs much later. In the plains, the mother cochineals grow to an even larger size, but the cochineals also find more predators there among the innumerable amount of insects (*xicaritas*, *perritos*, *aradores*, *agujas*, *armadillos*, *culebritas*), lizards, rats, and birds that devour them. A tremendous effort is required to clean the nopal segments; the Indian women use the tail of a squirrel or a stag for this purpose. They crouch down for hours on end next to a single plant, and, despite the exorbitant price of the cochineal, one III.82 might question the profitability of cultivating it in countries where people know how to put the time and labor of men [and women] to good use. In Sola, where very cold rain and sometimes even hail falls in January, the indigenous peoples protect the young cochineals by covering the nopals with mats made of rushes. The price of *grana fina* nymphs, which normally cost only five francs a pound, rises to around eighteen or twenty francs there.

In several districts of the province of Oaxaca, there are three cochineal harvests per year, the first of which—the one that yields the nymphs—is not very profitable, because the mother retains only a very small amount of dye if she dies naturally after giving birth. This first harvest provides the *grana de pastle*, or *nest cochineals*, called thus because one finds the mothers, after they have lain their eggs, in the very nests that were hung from the nopals. Near the city of Oaxaca, the cochineals are *seeded* in August; in the district of Chontale, this procedure takes place only in October; on colder plateaus, in November and December.

The cottony or wild cochineal, which is being introduced into the nopalries—and the male of which, according to Mr. Alzate's observation, is nearly as large as the mealy or cultivated cochineal male—causes serious damage to the nopals. The Indians, therefore, kill the wild variety cochineal whenever they find it, despite the fact that the dye it produces is very bold and quite beautiful. It appears that not only the fruit but also the green segments of several species of cactus could be used to dye cotton purple or red, and that the dye of the cochineal is not exclusively the product of the *animalization* of the vegetable juices within the insect's body. III.83

In a good year in Nejapa, it is expected that a pound of mealy cochineal *nymphs* placed on the nopals in October will yield twelve pounds of mother cochineal by January, while leaving a sufficient number of *nymphs* on the plant; that is, delaying the harvest until the mothers have given birth to half of their young. These new *nymphs* produce another thirty-six pounds by May. In Zimatlan and other Mixteca and Xicayan villages, people harvest barely three to four times the number of *cochineals* that they *seed*. If the south wind, which is harmful to the insect's growth, has not blown for a long time, and if the cochineal is not mixed with *tlasole*, that is, the remains of the winged males, it loses only two-thirds of its weight when dried in the sun.

Both species of cochineal (cultivated and wild) appear to contain more dye in temperate climes, especially in regions where the average air temperature is between eighteen and twenty degrees Centigrade. Cultivated cochineals can withstand very cold temperatures; in the province of Oaxaca, they are still farmed on plateaus where the thermometer nearly always hovers between ten and twelve degrees Centigrade. As for wild cochineals, we encountered them in large number in the most contrasting climes, in the mountains of Riobamba, at an absolute elevation of 2,900 meters, and in the plains of the province of Jaén de Bracamoros, under the blazing sun, between the villages of Tomependa and Chamaya. III.84

Around the city of Oaxaca, especially near Ocotlan, there are plantations (*haciendas*) that contain between 50,000 and 60,000 nopals, planted in rows like century plants, or *magueys de pulque*. The bulk of the cochineal that enters into trade is nevertheless produced by small nopalries owned by destitute Indians. Generally, the nopal is not allowed to grow higher than twelve decimeters, making it easier to remove the insects that prey on the cochineal. Preferred are cactus varieties that have more spines and down, because these defenses protect the cochineal from flying insects, and care is taken to cut off the flowers and the fruit to prevent these flying insects from depositing their eggs there.

The Indians who raise cochineals and who are called *Nopaleros*, especially those who live around the city of Oaxaca, follow an extremely old and quite extraordinary practice, that of *relocating the cochineals*. In this part of the Torrid Zone, it rains from May to October in the plains and

III.85 valleys while in the nearby mountain chain, called the *Sierra de Istepeje*, frequent rains occur only from December to April. Instead of keeping the insects inside their huts during the rainy season, the Indians place the leaf-covered mother cochineals in layers, in baskets made from extremely supple lianas. The Indians carry these baskets (*canastos*) on their backs with great haste into the Istepeje Mountains above the village of Santa Catalina, nine leagues from Oaxaca. The mother cochineals give birth to their young over the course of the journey. When the *canastos* are opened, they are filled with young *cocci that are* then distributed among the nopals of the Sierra. The cocci remain there until October, when the rains cease in the lowland regions. At this time, the Indians return to the mountains to retrieve the cochineals and to return them to the Oaxaca nopalries. It is in this manner that the Mexica relocate insects to spare them the nefarious effects of humidity, just as the Spanish relocate merinos to avoid the cold.

At harvest time, the Indians kill the mother cochineals in a wooden dish called *chicalpetl*, either by throwing them in boiling water, stacking them in layers exposed to the sun, or placing them on mats in the same circular ovens (*temazcalli*) used for the steam and hot-air baths that we have described III.86 above.[1] The latter method, which is the least commonly used, preserves the whitish powder that covers the insect's body and that raises its price in Veracruz and in Cádiz. The buyers prefer the white cochineal, because it is less likely to be fraudulently mixed with bits of gum, wood, corn, and red earth. In Mexico, there are very old laws (from the years 1592 and 1594) that strive to prevent cochineal fraud. Since 1760, there has even been a push to establish in the city of Oaxaca a jury of *Veadores*, who examine the bags (*zurrones*) before they are shipped out of the province. All cochineal sold must have the *grain* removed, so that the Indians cannot introduce foreign substances into the agglutinated masses called *bodoques*. None of these means have, however, succeeded in putting an end to fraud. The fraud perpetrated in Mexico by the *Tiangueros*, or *Zanganos* (*falsificadores*), pales by comparison with what this merchandise is subjected to in the ports of the peninsula and in the rest of Europe.

To complete the portrait of the animal products of New Spain, we must III.87 still take a quick glance at *pearl* fishing and *whaling*. It is likely that these

1. See vol. II, p. 283 [p. 452 in this edition]. Mr. Alzate, who has provided a good image of the temazcalli (["Memoria en que se trata del insecto grana ó cochinilla,"] *Gazetas de Literatura* de México, vol. III, p. 252), claims that the heat of the steam in which the Mexican Indian bathes is normally sixty-six degrees Centigrade.

two branches of fishing will one day become subjects of great importance for a country that has a coastal area of over 1,700 nautical leagues. Long before the discovery of the Americas, pearls were prized by the indigenous peoples. ▾ Hernando de Soto found them in great number in Florida, especially in the provinces of Ichiaca and Confachiqui where they decorated the princes' tombs.[1] Among the presents that Montezuma gave Cortés prior to his entrance into Mexico City (and which the latter sent to Emperor Charles V) were necklaces garnished with rubies, emeralds, and pearls.[2] We do not know whether the Aztec kings received a portion of the latter by trading with the barbarous and nomadic peoples who frequented the Gulf of California. What is more certain is that they commissioned pearl fishing on the coasts stretching from Colima, the northern border of their empire, to the province of Xoconochco (or Soconusco), especially near Tototepec, between Acapulco and the gulf of Tehuantepec, and in Cuitlatecapan. The Inca of Peru attached a great value to pearls, but Manco Capac's laws forbade the Peruvians from pursuing the diving profession, calling it of little III.88 use to the state and dangerous for those who devoted themselves to it.[3]

The areas that, since the discovery of the New World, have provided the Spanish with the most pearls are the following: the sound between the islands of Cubagua and Coche and the coast of Cumaná, the mouth of the Río de la Hacha; the Gulf of Panama near the *Islas de las Perlas*; and the eastern coasts of California. In 1587, 316 kilograms of pearls were exported to Seville, among them five kilograms[4] of the highest quality reserved for King Philip II. The pearl fisheries of Cubagua and Río de la Hacha were very productive, but not for long. Since the beginning of the seventeenth century, especially since the voyages of Yturbi and [Bernal de] Piñadero, pearls from California have become competitive in trade with those of the Gulf of Panama. At that time, the most skilled divers were sent to the coasts of the Sea of Cortés; pearl fishing nevertheless fell by the wayside again, and although there was an attempt to revive it at the time of Gálvez's expedition, this effort was rendered somewhat fruitless by the factors that we have examined above[5] in our description of California. It was not until 1803 that a Spanish monk, residing in Mexico City, focused the government's attention III.89

1. [Garcilaso de la Vega El Inca,] *La Florida del Inca*. Madrid, 1723, pp. 129, 135, and 140.
2. Gómara, *Conquista de México*. Medina del Campo, 1553, folio 25.
3. Garcilaso, [*Comentarios reales*,] book VIII, chap. 23.
4. [José de] Acosta, [*Historia natural y moral de las Indias*,] book IV, chap. [1]5, [p. 235].
5. See above vol. II, p. 265 [p. 443 in this edition].

once more upon the pearls of the Ceralvo coast of California. Since the divers (*buzos*) waste a great deal of time returning to the surface of the water to breathe, and since they wear themselves out needlessly by diving repeatedly to the bottom of the sea, this monk proposed fishing for pearls using a diving bell that would serve as a reservoir of atmospheric air and under which the diver would take shelter each time he needed to breathe. Outfitted with a mask and a flexible tube, he could walk on the bottom of the Ocean while breathing the oxygen provided by the bell at the other end of the tube. During my stay in New Spain, I witnessed a series of very interesting experiments designed to test this theory and conducted in a small pond near the castle of Chapultepec. It was probably the first time that a diving bell was built at an elevation of 2,300 meters—that is, at a height equivalent to that of the Simplon Pass. I do not know whether the experiments conducted in the Valley of Mexico have been repeated in the Gulf of California, or if pearl fishing has been revived after a hiatus of over thirty years, for, until now, almost all the pearls that the Spanish colonies provide to Europe have come from the Gulf of Panama.

Among the pelagic shells of New Spain, I must cite here the *Murex* [sea III.90 snails] from the coast of Tehuantepec in the province of Oaxaca, whose coat transudes a crimson colored dye, and the famous *Monterey shell* [abalone], which resembles the most beautiful Haliotis [ear-shells] of New Zealand. The Monterey shell is found on the coasts of New California, especially between the ports of Monterey and San Francisco. As we have observed[1] above, it is used in the fur trade with the inhabitants of Nootka. As for the gastropod of Tehuantepec, the Indian women collect its crimson liquid by walking along the shore and rubbing the coat of the *Murex* with de-seeded cotton.

The western coasts of Mexico, especially the part of the Great Ocean located between the Gulf of Bayona [Gulf of Bayonne], the three Islas Marías, and Cabo San Lucas, are abundant in *cachalot* [sperm whales], the hunting of which has become one of the prime objects of mercantile speculation for the British and the inhabitants of the United States because of the extremely high price of *spermaceti* (adipocere). The Spanish-Mexicans witness the arrival at their coasts of *sperm whalers,* who are forced to sail for over 5,000 nautical leagues, and who are quite improperly called *balleneros* (*whalers*); they are by no means tempted to take part in the hunt for these great cetacean mammals [baleen whales]. ▾ Mr. Schneider, as good

1. Marchand, *Voyage*, vol. II, p. 600, 641.

a physicist as he is a Hellenist, Mr. Lacépède, and Mr. Fleurieu have given very precise accounts of whaling and sperm whaling in both hemispheres. I shall record here the more up-to-date information I was able to gather during my stay on the South Sea coasts. III.91

Were it not for sperm whaling and for the Nootka trade in sea otter furs, the Anglo-Americans and the nations of Europe would rarely frequent the Great Ocean. Despite the strict economy of fishing expeditions, those that go beyond Cape Horn are too costly for baleen whales (*black whales*) to be the sole objective. The expenses that these voyages incur can only be compensated for by the high price that either need or luxury attach to the goods upon return. Yet, of all the oily liquids that enter into trade, only very few are dearer than spermaceti, the peculiar substance contained within the enormous cavities in the sperm whale's head. A single one of these gigantic cetaceans provides up to 125 English barrels[1] (at thirty-two and a half gallons apiece) of *spermaceti*. A container loaded with eight of these barrels, or 1,024 Paris pints, was sold in London for seventy to eighty pounds sterling before the Treaty of Amiens and for ninety-five to one hundred pounds during the war.

It was not Cook's third expedition, to the northwest coasts of the New Continent, but James Collnet's travels to the Galapagos Islands that revealed to the Europeans and the Anglo-Americans the abundance of sperm whales in the Great Ocean north of the equator. Until 1788, whalers had only frequented the coasts of Chile and Peru. At that time, only twelve to fifteen ships rounded Cape Horn each year to hunt sperm whale, whereas, during the time that I spent in the South Sea, there were over sixty ships under the English flag. III.92

The *Physeter macrocephalus* lives not only in the Arctic seas between the coasts of Greenland and the Davis Strait; it is found not only in the Atlantic Ocean between the Newfoundland Bank and the Azores Islands where the Anglo-Americans sometimes hunt it; this cetacean is also present south of the equator, on the coasts of Brazil and Guinea. It appears that, in the course of its periodic voyages, this whale veers closer to the African continent than to that of the Americas for only baleen whales are caught on the outskirts of Rio de Janeiro and Bahia. Sperm whaling has nevertheless sharply declined on the coasts of Guinea, since the navigators have started

1. A barrel contains 1.48 hectoliters, or around 178 and seven-eighths Paris pints (Adam Smith, *An Inquiry into the Nature and Causes of the Wealth of Nations*, translation by Mr. Garnier, vol. V, p. 451).

rounding Cape Horn with less apprehension, and since they have become more focused upon the wealth of cetaceans in the Great Ocean. Physeters III.93 can be found in quite large packs in the Mozambique Channel and south of the Cape of Good Hope, but the animal is generally smaller there, and the sea's constant swells and rough waters makes the *harpooners'* maneuvering extremely difficult.

The Great Ocean combines all the circumstances that make sperm whaling both easy and lucrative. Richer in mollusks, fish, porpoises, tortoises, and seals of all species, it offers more food to the *spouting* cetaceans than the Atlantic Ocean. The latter are therefore present in greater numbers; they are fatter and larger. The calm that reigns for a great part of the year in the equinoctial part of the South Sea greatly facilitates the pursuit of sperm whales and baleen whales. The former do not stray far from the coasts of Chile, Peru, and Mexico, which are steep (*acantiladas*) and lapped by extremely deep waters. As a general rule, the sperm whale avoids the shoals, whereas the baleen whale seeks them out. For this reason, the latter cetacean is very common on the low-lying coasts of Brazil, while the former is abundant near the coast of Guinea, which are higher and accessible everywhere for the largest ships. Generally, the geological makeup of the two continents is such that the western coasts of the Americas and Africa resemble one another, while the eastern and western coasts of the New Con- III.94 tinent offer the most remarkable contrast in terms of their elevation above the bottom of the neighboring Ocean.

Most of the British and Anglo-American ships that sail into the Great Ocean have the dual objective of sperm whaling and engaging in illegal trade with the Spanish colonies. They round Cape Horn after attempting to leave smuggled goods at the mouth of the Río de la Plata, or at the *presidio* of the Malvinas Islands. They start hunting sperm whales near the small, deserted islands of Mocha and Santa María to the south of Concepción in Chile. On Mocha, there are wild horses that the inhabitants of the nearby coast introduced there and that occasionally serve as food for the sailors. The island of Santa María has very beautiful and abundant springs. There, one finds feral pigs and a variety of turnip that is very large and very nutritious, and that is believed to be native to these climes. After a month-long stay in these parts and after having engaged in smuggling on Chiloé Island, the whalers (*balleneros*) usually sail along the coasts of Chile and Peru to Cabo Blanco located at 4°18′ southern latitude. The sperm whale is very common throughout these parts, up to a distance of fifteen to twenty leagues from the continent. Before Captain Collnet's expedition, whaling did not extend beyond Cabo Blanco, or around the equator; but over the past fifteen to twenty years, the *balleneros* have pushed farther to the north all the way past Cabo Corrientes on the Mexican coastline of the intendancy of Guadalajara. It is around the Galapagos archipelago, where it is very III.95 dangerous to make landfall, that the cetaceans are the most common and gigantic in size. In the spring, all the macrocephalic sperm whales from the coasts of Mexico, Peru, and the Gulf of Panama gather near the Galapagos to mate. During this period, Mr. Collnet saw young sperm whales that were two meters long. Farther north from the *Islas Marías*, in the Gulf of California, one finds no more physeters, only baleen whales.

Whalers can easily distinguish between sperm whales and baleen whales from afar by the way in which the former spurt water from their blowholes. Sperm whales can remain underwater longer than baleen whales; when they surface, their respiration is interrupted more often; they let water sit for less time in the membranous pockets located above their nostrils. Their jets of water are more frequent, aimed more to the front, and higher than those of the other *spouters*. The female sperm whale is four to five times smaller than the male. Her head provides only twenty-five English *barrels* of *adipocere*, whereas the male's head yields between one hundred and one hundred and twenty-five. The females (*cow whales*) travel together in large groups escorted by two or three males (*bull whales*) who swim in circles around III.96 their herd. Very young females, who yield only twelve to sixteen barrels of adipocerous matter, and which the English whalers call *school-whales*, swim so close to one another that they rise up halfway from the water. It is almost unnecessary to observe here that the adipocere, which is not part of the animal's brain, is found not only in all known species of sperm whale (*Catadontes Lac.*), but also in all physalia and physeters. The spermaceti extracted from the head cavities of the sperm whale—cavities that must not be confused with that of the cranium—is only one-third of the thick, adipocerous oil yielded by the rest of the body. The *spermaceti* from the head is of the highest quality; it is used in candle production, while that of the body and the tail is only used—in England—to give *sheen* to linens.

To be profitable, sperm whaling must be conducted with the strictest economy. Ships of 180 to 300 tons are used. The crew consists of only sixteen to twenty-four individuals, including the captain and the first officer, who are themselves obliged to cast the harpoon, like ordinary seamen. In London, the cost of outfitting a 180-ton ship lined in copper and stocked for a two-year campaign is estimated at 7,000 pounds sterling. Each whaling ship in the South Sea has two rowboats; each of those must be outfitted with four seamen, a ship's apprentice, a helmsman, a 130-fathom-long rope, III.97 three spears, five harpoons, an axe, and a lantern to make it visible from afar at night. The ship owner gives the seamen only food and a very modest sum of money in the form of an advance. Their pay depends on the outcome of the hunt; since the entire crew takes part in it, each individual has a right to the profit. The captain receives one sixteenth of the entire profit, the first officer one twenty-fifth, the second officer one thirty-fifth, the petty officer one sixtieth, and the deck hand one eighty-fifth. The hunt is seen as successful if a two-hundred-ton ship returns to port carrying eight hundred *barrels* of spermaceti. Over the last few years, the sperm whale, ceaselessly hounded, has become more timid and harder to catch; but, in order to encourage navigation on the South Sea, the British government grants advances to any ship that sets out to hunt sperm whales. These advances are from 300 to 800 pounds sterling, depending on the tonnage of the ship. The Anglo-Americans conduct whaling with even greater economy than do the British.

Ancient Spanish laws forbid whaling ships, like other foreign ships, from entering into the ports of the Americas, except in cases of distress or lack of water or supplies. The Galapagos Islands, where the whalers occasionally disembark their sick, have springs, but these springs are very poor and very unreliable. Cocos Island (5°35′ northern latitude) is very rich in III.98 water, but when one heads north from the Galapagos, this small, isolated island is hard to find because of the force and irregularity of the currents. The whalers have strong reasons for preferring to take on water from the coast. They look for pretexts for entering into the ports of Coquimbo, Piseo, Tumbez, Payta, Guayaquil, Realejo, Sonsonate, and San Blas. Only a few days—often only a few hours—are necessary for the crew of whaling ships to forge ties with the inhabitants, to sell British goods to them, and to take on loads of copper, vicuña, cinchona, sugar, and cocoa. This trade in smuggled goods is carried out by people who do not speak the same language, often through signs, and with a good faith exceedingly rare among the civilized peoples of Europe.

It would be unnecessary to enumerate here the advantages that the inhabitants of the Spanish colonies would have over the British and the Americans of the United States if they decided to take part in sperm whaling. From Guayaquil and Panama, it takes only ten to twelve days to reach the area where this cetacean abounds. The voyage from San Blas to the *Islas Marías* takes barely thirty-six hours. If they devoted themselves to sperm whaling, the Spanish-Mexicans would have 4,000 fewer leagues to travel than the Anglo-Americans do; they would have supplies at a lower cost; they would find ports everywhere in which they would be welcomed as friends, and which would provide them with new provisions. It is true that spermaceti is still not sought-after on the continent of Spanish America. The clergy persists in confusing adipocire with tallow, and the American bishops have declared that the candles that shine on the altars must be made only of beeswax. In Lima, however, people have begun to sidestep the bishops' vigilance by mixing spermaceti with wax. A number of merchants who have purchased British catches have had large quantities of spermaceti, and adipocere, used in the church festivals, has become a new and very lucrative branch of trade. III.99

It is not the lack of manpower that might prevent the inhabitants of Mexico from devoting themselves to sperm whaling. Only two hundred men would be needed to outfit ten whaling ships and to collect nearly a thousand tons of spermaceti per year. In time, this substance could become an export item almost as important as cacao from Guayaquil and copper from Coquimbo. In the current state of affairs in the Spanish colonies, the sloth of the inhabitants forms an obstacle to carrying out such plans. How to find seamen who might willingly embrace such a tough profession, such a miserable life as that of the sperm whalers? How to find them in a country where, according to the ideas of the lower classes, bananas, salted meat, a hammock, and a guitar are all that is needed to be happy? The potential for material gain is too weak a stimulus in a region where beneficent nature offers humans a thousand ways to procure a comfortable, peaceful existence without ever having to leave one's country or having to fight against the monsters of the Ocean. III.100

The Spanish government has long had misgivings about sperm whaling, which draws the British and the Anglo-Americans[1] to the coasts of Peru and Mexico. Before this practice was established, the inhabitants of the western coasts of the Americas had never seen any other flag flying on these seas than the Spanish flag. Political factors might have prompted the metropole to spare no expense to encourage local sperm whaling, perhaps less with an eye to direct profits than to excluding competition from foreigners, and to prevent them from forging ties with the indigenous peoples. Privileges granted to a company that was based in Europe, and that existed in

1. According to the official information that I owe to Mr. Gallatin, Secretary of the Treasury in Washington, in 1800, 1801, and 1802, there were eighteen to twenty whaling ships (between 2,800 and 3,200 tons) from the United States. A third of these ships departed from the port of Nantucket. In 1805, 1,146 *barrels* of spermaceti were brought into this port.

name only, could not give primary impetus to the Mexicans and the Peruvians. Outfitting for sperm whaling should be done in the Americas—in Guayaquil, Panama, or San Blas. On these coasts, there are always a certain number of British sailors who have abandoned their whaling ships, either out of discontent or to seek their fortune in the Spanish colonies. The first III.101 expeditions could be organized by mixing these seamen, who are well-experienced in sperm whaling, with the *zambos* of the Americas, who are daring enough to fight crocodiles in hand-to-hand combat.

We have just examined in this chapter the true natural wealth of Mexico, for the products of the land are, in effect, the only basis for long-lasting prosperity. It is reassuring to see that the work of humans, for the past half-century, has been directed more toward this fertile and inexhaustible source than toward mining, the riches of which have no impact on public affluence and change only the nominal value of the annual output of the land. The land tax that the clergy collects under the name of the tithe measures the level of this output; it gives a precise indication of the development of the agricultural industry, provided that one compares two different periods between which the price of commodities has not changed significantly. Here is the table of the value of these tithes,[1] taking as an example two series of years, from 1771 to 1780, and from 1780 to 1789.

III.102

NAME OF DIOCESE	PERIOD	VALUE OF TITHES	PERIOD	VALUE OF TITHES
		In Piasters		In Piasters
Mexico City	1771–1780	4,132,630	1781–1790	7,082,870
Puebla de los Ángeles	1770–1779	2,965,601	1780–1789	3,508,884
Valladolid de Michoacan	1770–1779	2,710,200	1780–1789	3,239,400
Oaxaca	1771–1780	715,974	1781–1790	863,237
Guadalajara	1771–1780	1,889,724	1781–1790	2,579,108
Durango	1770–1779	943,028	1780–1789	1,080,313

1. I have borrowed this table from a handwritten paper by ▼ Mr. Maniao, based on official documents and bearing the title of *Estado de la Renta de Real Hacienda de Nueva España en un año común del quinqueñio de 1784 hasta 1789*. The numbers contained within this table are slightly different from those published by Mr. Pinkerton ([*Modern Geography*,] vol. III, p. 234), following ▼ Estala's work, which I have thus far been unable to obtain.

This table suggests that tithes in these six dioceses of New Spain amounted

---

From 1771 to 1779, to 13,357,157 piastras fuertes
From 1779 to 1789, to 18,353,821 piastras fuertes

The total increase over the latter ten years was, therefore, five million piasters, or two-fifths of the total output. These same data also reveal that agricultural development in the intendancies of Mexico City, Guadalajara, Puebla, and Valladolid outstripped that of the province of Oaxaca and Nueva Vizcaya. The tithes nearly doubled in the archdiocese of Mexico City, for those that were collected during the ten years prior to 1780 were in a ratio of ten to seventeen to those collected ten years later. In the intendancy of Durango, or Nueva Vizcaya, this increase was at a ratio of only ten to eleven. III.103

The famous author of *An Inquiry into the Nature and Causes of the Wealth of Nations*[1] estimated the output of the land in Great Britain on the basis of property tax revenues. In the Political Tableau of New Spain that I presented to the court of Madrid in 1803, I attempted a similar estimate, based on the value of the tithes paid to the clergy. This work suggested that the annual output of the land in Mexico is at least twenty-four million piasters. The results upon which I focused while drafting this first table were very shrewdly discussed in a paper that the municipal body of the city of Valladolid de Michoacan presented to the king in October of 1805 on the occasion of an ordinance delivered on the property of the clergy. According to this paper, a copy of which I have before me, to these twenty-four million piasters should be added three million for the production of cochineal, vanilla, jalap, Tabasco pepper, and sarsparilla, which are not subject to tithes, and one million for sugar and indigo, on which, in lieu of whole tithes, a tax to the clergy of only 4 percent is levied. If one adopts these data, one finds that total agricultural production amounts to twenty-nine million piasters

1. Adam Smith, [*Recherches sur la nature.*] Translation by Mr. Garnier, vol. IV, p. 246.

III.104 per year, or over 145 million francs, which, if one converts them to a *natural measure* and takes as a base the current price of wheat in Mexico, which is fifteen francs per ten myriagrams,[1] is equivalent to *ninety-six million myriagrams of* wheat. The total amount of precious metals mined yearly in the Kingdom of New Spain represents barely *seventy-four million myriagrams of wheat*, which demonstrates the interesting fact that the value of the gold and silver of the mines in Mexico is almost a quarter less than the value of agricultural output.

Despite the constraints that hinder it from all sides, the cultivation of the land has in recent years made advances that are all the more significant in that immense amounts of capital have been invested in land by families that had become rich either through the Veracruz or Acapulco trade or through mining. The Mexican clergy owns barely two to three million piasters' worth of real estate (*bienes raíces*); but the capital that the monasteries, chapters, confraternities, hospices, and hospitals have invested in land amounts to forty-four and a half million piasters, or over 222 million livres tournois. Here, based on an official document,[2] is a table showing this III.105 capital, which is referred to as the *Capitales de capellanías y obras de la jurisdicción ordinaria*:

Archbishopric of Mexico City	9,000,000 piasters
Bishopric of Puebla	6,500,000
Bishopric of Valladolid (very exact estimate)	4,500,000
Bishopric of Guadalajara	3,000,000
Bishoprics of Durango, Monterey, and Sonora	1,000,000
Bishoprics of Oaxaca and of Mérida	2,000,000
▾*Obras pias* of regular clergy	2,500,000
Fond dotal of the churches and communities of monks and nuns	16,000,000
	44,500,000

This enormous sum, which is in the hands of landowners (*haciendados*) and mortgaged on real estate, was almost deducted from Mexican agriculture in 1804. The ministry in Spain, out of ideas for avoiding a situation of

1. See above, vol. II, p. 449 [p. 547 in this edition].
2. *Representación de los vecinos de Valladolid al Excelentísimo Señor Virey, en fecha del 24 octubre del año 1805* (manuscript).

national insolvency brought about by the overabundance of paper money (*vales*), attempted a very risky operation. A royal decree, delivered on December 26, 1804, ordered not only the sale of property belonging to the Mexican clergy but also the collection of all capital belonging to the monks, which was to be sent to Spain and deposited in a sinking fund of royal bonds (*caja de consolidación de vales reales*). Instead of complaining about III.106 this decree and explaining to the sovereign how detrimental it would be, if carried out, to agriculture and to the overall wellbeing of the inhabitants, the council of finance (which is presided over by the viceroy, and which bears the title of *Junta superior de Real Hacienda*) launched into making collections. The resistance on the part of the landowners was so fierce that, from May 1805 to June 1806, the sinking fund collected only the modest sum of 1,200,000 piasters. One can only hope that administrators who are more enlightened as to the true interests of the state have since brought to an end an operation whose effects would have been felt in the future.

Reading the excellent work *on the land laws* that was presented to the Council of Castile in 1795,[1] one recognizes that, despite the difference in climate and other local circumstances, Mexican agriculture is impeded by the same political factors that halt industrial development in the Peninsula. The vices of feudal government have passed in their entirety from one hemisphere to the other, and in Mexico, the effects of abuses were all the more insidious because the vast distance made it more difficult for the supreme authority to address wrongs and mobilize its resources. In New Spain, as in old Spain, the land is primarily in the hands of a few powerful III.107 families who have gradually taken over all private property. In the Americas, as in Europe, large districts are condemned to serving as cattle pasture and, therefore, to perpetual unproductiveness. As for the clergy and its influence on society, the circumstances on the two continents are not the same. The clergy is much smaller in number in Spanish America than in the Peninsula. The missionary monks have contributed to spreading agricultural development among the barbarous peoples. The introduction of the *majorats* and the reduction of the Indians to a state of mindlessness and extreme poverty pose a greater threat to industrial development there than does the monks' mortmain.

The ancient legislature of Castile forbids monasteries from being the sole proprietors of real estate, and although this sage law has often been

1. Mr. de Laborde has provided a translation of this paper by ▼ Mr. Jovellanos in the fourth volume of his *Itinéraire descriptif de l'Espagne*, pp. 103–294.

broken, the clergy has been unable to acquire especially large plots of land in a country where devotion does not hold the same sway over the minds of the people as it does in Spain, Portugal, and Italy. Since the suppression of the Jesuit order, very few plots of land have belonged to the Mexican clergy. Its true wealth, as we have just noted, consists in tithes and in the capital invested in the small-scale farmers' plantations. This capital is put to good use, which increases the productive capacity of the local labor force.

III.108 One may, however, be surprised to note that the large number of monasteries founded since the sixteenth century throughout Spanish America have all been concentrated within the cities. Had they been scattered throughout the countryside and placed on the ridge of the Cordilleras, they could have had the same beneficial influence on agriculture as is evident in the north of Europe, on the banks of the Rhine and in the chain of the Alps. Those who have studied history know that the monks in the time of Phillip II no longer bore any resemblance to those of the ninth century. The luxury of the cities and the climate of the Indies are contrary to the austere customs and the orderly spirit that characterized the first monastic institutions, and while traversing the mountainous deserts of Mexico, one regrets not finding there, as one does in Europe and Asia, those solitary sanctuaries where religious hospitality comes to the aid of travelers.

CHAPTER XI

III.109

State of the Mining of New Spain—Gold and Steel Production—Average Wealth in Ore—Amount of Mercury Consumed Annually in the Amalgamation Process—Amount of Precious Metals That Have Flowed from the One Continent to the Other Since the Conquest of Mexico

Now that we have examined Mexican agriculture as the primary source of national wealth and the inhabitants' prosperity, it remains for us to draw a portrait of the mineral products that have been the object of mining in New Spain for the last two and a half centuries. This overview, extraordinarily impressive to those who base their calculations only upon the nominal value of things, is less so if one takes into account the intrinsic value of the metals that are mined, their relative utility, and the direct impact that they have on the manufacturing industry. The mountains of the New Continent, like those of the Old, hold iron, copper, lead, and a large number of other mineral substances indispensable to agriculture and artisanry. If human labor in the Americas has been almost exclusively directed toward III.110 extracting gold and silver, this is because the considerations that motivate the members of a society are very different from those that should guide society as a whole. Everywhere that the soil is able to sustain both indigo and corn production, the former is always given preference over the latter, although it is in the common interest to privilege plants that serve as food for humans over those that provide goods for exchange with the outside world. Similarly, on the ridge of the Cordilleras, iron or lead mines, however rich they may be, remain abandoned because the colonists' attention is entirely focused on gold and silver veins, even when the *outcroppings* of gold and silver present only weak indications of wealth. Such is the lure of these precious metals—which, by general agreement, have become the representative signs of livelihood and labor—whose exploitation in the central part of Mexico has given life to all the other branches of agriculture and the manufacturing industry.[1]

The Mexican people are clearly capable of procuring, through foreign III.111 trade, all the things that their country does not provide them. But in the midst of great wealth in gold and silver, the need for the materials that are the most indispensable to the mechanical arts makes itself felt whenever trade with the metropole or other parts of Europe and Asia is interrupted, and whenever war blocks maritime communications. Twenty-five to thirty million piasters are sometimes amassed in Mexico City, while both factories and the exploitation of mines are hindered by the lack of steel, iron, and mercury. A few years before my arrival in New Spain, the price of iron had risen from twenty francs per quintal to 240 francs, that of steel from ninety to 1,300 francs. In these periods when foreign trade grinds to a complete halt, Mexican industry undergoes a momentary revival. It is only then that people begin manufacturing steel, using iron and mercury ore from the mountains of the Americas; at this time, the nation, enlightened as to its own interests, feels that true wealth consists in the abundance of consumable goods, of *things*, and not in the amassing of a *sign* that represents them. During the second-to-last war between Spain and Britain, an attempt was made to work the iron mines of Tecalitan near Colima, in the intendancy of Guadalajara. The *Tribunal de Minería* spent over 150,000 francs to extract the mercury from the veins of San Juan de la Chica. But the effects of such laudable efforts were somewhat short-lived. The Treaty of Amiens brought III.112 to an end the ventures that seemed to give the miners' labor a purpose that was more conducive to public prosperity. Maritime communications had barely been restored when the people once more began preferring to buy iron, steel, and mercury in European markets.

As the population of Mexico increases, and as its inhabitants, less dependent on Europe, begin to focus their attention on the great variety of useful products that the land holds, the mining system will change in appearance. An enlightened administration will encourage works aimed at extracting mineral substances that have an *intrinsic value*. Private citizens will no longer sacrifice their own interests and those of the res publica to deep-seated prejudices; they will feel that mining coal, iron, or lead can

1. The influence that the mines exert on the development of the Mexican population is very skillfully explained in the important work that Don Fausto Delhuyar has just published in Madrid under the title *Memoria sobre el influjo de la minería en la agricultura, industria, población y civilisación de la Nueva-España*, 1825, pp. 8, 25, 67, 78, and 125.

become as profitable an undertaking as mining a vein of silver. In the current state of affairs in Mexico, precious metals are the sole object of the colonists' industry, and when, later in this chapter, we use the word mine—*Real*, *Real de minas*—this, in the absence of explicit statements to the contrary, should be taken to refer to a gold or silver mine. Having spent a great deal of time since my early youth studying the art of mining, and having directed myself for several years the underground works in a part of Germany that contains a wide variety of ore, it was natural for me to be very keenly interested in ex- III.113 amining carefully the state of the mines and factories of New Spain. I have had occasion to visit the famous mines of Tasco, Pachuca, and Guanajuato. I stayed for over a month in Guanajuato, the veins of which are wealthier than everything that has been discovered in other parts of the world, and I was able to compare the different kinds of mining works in Mexico with those that I had observed the previous year in the mines of Peru. Since the large number of materials that I have gathered about these subjects can be usefully employed only in conjunction with a geological description of the country, I must reserve this analysis for the *Personal Narrative* of my journey into the interior of the New Continent. Without entering into detailed and purely technical discussions, I shall thus restrict myself here to examining those things that can lead to general conclusions.

What is the geographic location of the mines that provide the vast amount of silver that the Veracruz trade sends to Europe annually? Was this amount of silver produced by a large number of small, scattered mining operations, or can it be considered as having been furnished almost in its entirety by three or four metalliferous veins of extraordinary wealth and *power*? What is the total amount of precious metals mined yearly in Mexico? What is the relationship between this amount and the total production of all of the mines of Spanish America? What is the average esti- III.114 mated wealth, in ounces per quintal, of Mexico's silver ore? What is the ratio between the amount of ore that is smelted and the amount of ore from which the gold and silver are extracted through amalgamation? What influence does the price of mercury have on the development of mining, and how much mercury is regarded as being lost in the Mexican amalgamation process? Is it possible to know the exact amount of precious metals that have passed from the Kingdom of New Spain to Europe and Asia since the conquest of Tenochtitlan? Considering both the current state of mining and the geological composition of the country, is it likely that the annual production of Mexico's mines might increase, or should one assume, as have several famous writers, that silver exports from the Americas have already reached their *peak*? These are some of the general questions whose solutions shall occupy us in this work; they are connected to the most fundamental problems of political economy.

Long before the arrival of the Spanish, the indigenous peoples of Mexico, like those of Peru, were acquainted with the use of several metals. It seems certain that they were not content to use only those they found in their raw state on the surface of the ground, especially in riverbeds and in the ravines formed by torrents; rather, the indigenous peoples knew how to III.115 hollow out galleries and dig access and ventilation tunnels, and that they had instruments capable of chiseling into the rock. In the personal narrative of his expedition, Cortés tells us that gold, silver, copper, lead, and tin were sold at the main market of Tenochtitlan. The inhabitants of Zapoteca and Mixtecapan,[1] two provinces that are today part of the intendancy of Oaxaca, used the washing method to remove gold from the alluvial soil. These peoples paid their tributes in two different ways: either by gathering particles or grains of raw gold in leather bags or in small baskets woven from very thin rushes, or by melting down the metal into bars. These bars, similar to those that are still used in trade today, appear in ancient Mexica paintings. In Montezuma's time, the indigenous peoples were already working the argentiferous [silver] veins of Tlachco (Tasco) in the province of Cohuixco, as well as the ones that stretch across the Sumpango mountains.[2]

Gold and silver vases were manufactured in all the large cities of Anahuac, although the Americans prized silver much less than the peoples of the Old Continent did. During their first stay in Tenochtitlan, the Spanish III.116 could not admire enough the skill of the Mexica goldsmiths, among whom those of Azcapotzalco and Cholula were regarded as the most famous. When, seduced by his extreme gullibility, Montezuma saw the fulfillment of ▾ Quetzalcoatl's mysterious prophecy[3] in the arrival of the white, bearded men and forced the Aztec nobility to pay homage to the king of Spain, the total value of the precious metals offered to Cortés was estimated at 162,000 *pesos de oro*. "In addition to the large amount of gold and silver,"

1. Especially the inhabitants of the ancient cities of Huaxyacac (Oaxaca), Cojolapan, and Atlacuechahuayan.

2. Clavijero, [*Storia antica del Messico*,] 43, *II*, 125, 165, *IV*, 204. Mr. Delhuyar seems to doubt the existence of these ancient *underground* works (*[Memoria sobre] el influjo de la minería*, 1825, p. 27).

3. See my work entitled *Vues des Cordillères et monumens des peuples indigènes de l'Amérique*, p. 30.

said the Conquistador, in his first letter to Emperor Charles V,[1] "I was presented with such precious pieces of gold work and jewelry that, unwilling to let them be melted down, I set aside over one hundred thousand ducats' worth to offer them to your Imperial Highness. These items were of the greatest beauty, and I doubt that any other prince in the world has ever possessed anything like them. So that Your Highness might not think that I am merely telling tales here, I must add that King Montezuma had commissioned replicas in gold and silver, in gemstones and in bird feathers, of all the products of the land and the sea that were known to him, all to such a high degree of perfection that they gave the impression of seeing the objects themselves. Although a large portion of those that he had given me were intended for Your Highness, I commissioned the indigenous peoples to execute several other pieces of gold work based on some drawings with which I furnished them, such as images of saints, crucifixes, medallions, and necklaces. Since the *quinto*, the duty on silver paid to Your Highness, amounted to over one hundred mark, I ordered the native goldsmiths to convert them into plates of various sizes, spoons, cups, and other drinking vessels. All of these works were rendered with the greatest precision." Reading this passage, one has the impression of hearing the account of a European ambassador sent to China or Japan. It would nevertheless be difficult to accuse the Spanish general of exaggerating, especially if one considers that Emperor Charles V was able to judge with his own eyes the degree of perfection or imperfection of the items that were sent to him. The art of smelting had also advanced significantly among the Muisca in the kingdom of New Granada, the Peruvians, and the inhabitants of Quito. For several centuries, the people of the latter country have preserved precious pieces of ancient American gold work in the royal treasury (*en cajas reales*). It was only in recent years, because of an economic system that one might call barbarous, that they melted down these works, which proved that several peoples of the New Continent had attained a much higher degree of civilization than what is normally attributed to them.

III.117

III.118

Before the conquest, the Aztec peoples extracted *lead* and *tin* from the veins of Tlachco (Tasco, to the north of Chilpanzingo) and Ixmiquilpan; *cinnabar*, which served as a coloring agent for painters, came from the

1. Lorenzana, [*Historia de Nueva-España*,] p. 99. The spoils of gold that the Spanish took following the taking of Tenochtitlan was estimated at only 130,000 *castellanos de oro* (L.c., p. 301).

mines of Chilapan. Of all the metals, copper was the most commonly used in the mechanical arts. To a certain extent, it was substituted for iron and steel. All their tools—weapons, axes, chisels—were made from the copper extracted from the mountains of Zacatollan and Cohuixco. Everywhere on the globe, the use of copper appears to have preceded that of iron, and the abundance of copper in its raw state in the northernmost parts of the Americas may have contributed to the extraordinary predilection that the Mexica peoples, who emerged from these very regions, have demonstrated in constantly using it. Nature provided the Mexica[1] with enormous rock slabs composed of iron alloyed with nickel and cobalt; these slabs, which are found scattered across the surface of the ground, are fibrous, malleable, and so tough that it takes a great deal of effort to pry a few fragments from them with the help of one of our steel tools. The iron that most mineralo-

III.119 gists have regarded as both *raw* and *telluric*[2] because it cannot be identified as *meteoric* in origin is most often mixed with lead and copper; it is extremely scarce in all parts of the globe and is largely produced either by volcano fire or by the burning of coal deposits.[3] One must not, therefore, be surprised that at the beginning of civilization, the Americans, like most other peoples, focused their attention on copper rather than on iron. But why is it that these very Americans, who treated a wide variety of ore with fire,[4] were not led to discover iron by the mixture of combustible substances with the red and yellow *ochre*[5] that is extremely common in several parts of Mexico? If, on the other hand—as I am inclined to believe—they were acquainted with iron, how is it that they failed to appreciate it for its true value? Such considerations seem to indicate that the civilization of the

III.120 Aztec peoples did not date back very far. We know that in Homer's time, the use of copper still prevailed over that of iron, although the latter had been known for a long time.

1. See above, vol. II, p. 235 [p. 427 in this edition].

2. *Raw iron* from Gross Kaminsdorf (Saxony), Grenoble, Miedziana Góra (Galicia), and the Sholey Mountains (New York).

3. *Volcanic iron* from Graveneire (Auvergne). *Pseudo-volcanic iron* from La Bouche (department of Allier).

4. According to information that I gathered near Riobamba, among the Indians of the village of Licán, the ancient inhabitants of Quito succeeded in smelting silver ore by stratifying it with charcoal and blowing on the fire through long bamboo stalks. A large number of Indians were placed in a circle around the hole that contained the ore, so that the air currents were blown from several stalks at once.

5. Yellow ochre, called *tecozahuitl*, was used for painting, as was cinnabar. Ochre was one of the items that appeared on the list of tributes of Malinaltepec.

Several distinguished scholars, who were nonetheless unfamiliar with chemical research, have claimed that the Mexica and the Peruvians had a particular secret for tempering copper and for *converting it into steel.* It is known that axes and other Mexica tools were nearly as sharp as our own steel instruments, but they owed their extreme hardness both to the combination of the copper with tin and to this tempering process. What the first historians of the conquest called *hard or sharp copper* resembled the Greeks' χαλκός [chalkos] as well as the Romans' *æs.* Mexica and Peruvian sculptors created large works in diorite, basaltic porphyry, and other very hard rocks. Jewelers cut and drilled holes in emeralds and jade stones, using both a metal tool and a siliceous powder.[1] From Lima I brought back an ancient Peruvian chisel, in which ▼ Mr. Vauquelin found 94 percent copper and 6 percent tin. This alloy had been so well forged that, as its molecules had bonded, its specific weight had become 8.815, whereas, according to Mr. Briche's experiments,[2] chemists obtain this maximum density only by alloying six parts tin with one hundred parts copper. It appears that the Greeks used both tin and iron to harden copper. According to Mr. Vauquelin's analysis, even a Gallic axe that ▼ Mr. Dupont de Nemours found in France and that cuts wood like a steel axe without breaking or yielding, contains 87 percent copper, 3 percent iron, and 9 percent tin. ▼ Mr. D'Arcet's ingenious experiments have proven that, whereas the addition of tin generally detracts from the ductility of copper, the alloy of copper with tin becomes very malleable when, after getting it red-hot, one plunges it into cold water. It is likely that both the Mexica and the Gauls used this method for tempering bronze, which has long been known to several Asian peoples.

III.121

Tin being one of the least widespread metals on the globe, one should be surprised that the practice of hardening copper through the addition of tin can be found on both continents.[3] There is only one type of ore, sulfurated tin (*zinnkies* [stannite or pyrite])—which has to this day only been found in Wheal Rock in Cornwall—that contains both copper and tin in nearly equal parts. ▼ Mr. Haüy regards this as tin that has not been mineralized by sulfur, but simply *associated* with pyritic copper. We do not know whether the Mexica peoples mined the veins that contained both copper

III.122

1. *Vues des Cordillères et monumens des peuples de l'Amérique*, vol. II, pp. 85, 146, and 338 (octavo edition).

2. [Briche, "Observations sur la fonte des pièces de canon,"] *Journal des mines* [6 (XXXV),] p. 881.

3. It was through this custom, so widespread across both worlds, that ▼ Kirwan attempted to prove, *a priori*, the existence of raw tin.

and oxidized tin ore, or whether the latter metal, which one encounters in the protruding and fibrous shape of wood-tin in the alluvial soil of the intendancy of Guanajuato, was added to pure copper in a constant proportion. Be that as it may, it is certain that the lack of iron was less keenly felt among peoples who knew how to alloy different metals in a manner that was so advantageous to the mechanical arts. The Mexica's sharp tools were made either of copper or of obsidian (*iztli*). This latter substance was even the object of large mining operations, traces of which are still recognizable in the innumerable wells hollowed into the *Mountain of Knives* (*Cerro de las Navajas*) near the Indian village of Atotonilco el Grande.

In addition to the bags of cocoa, each of which contained three *xiquipilli* III.123 (or 24,000 grains)[1] and the *patolquachtli*, small bales of cotton sheets, the ancient Mexica used a few metals as currency, that is, as signs representing things. In the main market of Tenochtitlan, all kinds of goods could be purchased in exchange for gold powder stored in aquatic birds' quills. These III.124 quills were required to be transparent, so that one could identify the size

1. The Mexica had three numerical signs (numerals) for groups of twenty units, the square of twenty (four hundred), and twenty to the third power (eight thousand) (*Vues des Cordillères*, vol. I, pp. 345 and 369; vol. II, pp. 123, 183, and 234). Like the Romans and the Egyptians, they normally used juxtaposition (collating) of the various group signs to express eighty, 800, or 24,000; however, in counting their historical cycles, which began in the year 1091 of our era, according to the chronology of the major *bindings* of the years (groups of four times thirteen, or fifty-two), the Mexica added *exponents* or *indicators* to the hieroglyphs for the bindings (*xiuhmolpilli*). This is an incontrovertible trace of the *graphic Arithmetic* artifice commonly used among the Chinese, the Malabarians, and the ancient peoples of Iran who spoke the *pelvi* language; it is also the writing artifice, *gobar*, used by the Arabs. To avoid the puerile *juxtaposition* of the Romans and the Egyptians, peoples in East Asia add above the numerical hieroglyphs that indicate groups of ten, one hundred, and one thousand an *exponent*, or, rather, an *indicator*, that indicates the number of times by which the group should be multiplied. I believe to have proven, in a paper that I read before the *Académie des Inscriptions et Belles-Lettres* in 1819 (["Séance du lundi 20 septembre,"] *Annales de chimie et de physique*, vol. XII, p. 93) that the *positioning method* attributed to the Hindus originated not only in the *system of indicators* added to the hieroglyphs for the groups, but also in the use of the *tangible Arithmetic* techniques (small cords, quippus, rosaries, abacuses, sampans). Following the Hindu method, incorrectly called the Arab method, the hieroglyphs for the groups are omitted and only *indicators* are preserved and arranged such that a zero (sign of *emptiness*) marks the place where there is nothing (οὐδὲν [ouden], *surya*). I would go so far as draw the attention of those mathematicians who study the history of numerical signs to this way of envisaging *position value*, an invention whose felicitous influence was felt quite late in Europe. As ▼ Mr. Colebrooke was kind enough to remind me in a recent letter, Sanskrit exhibits traces of *position* even in figurative language. *Surya*, because of the famous twelve suns in Asian Mythology, means twelve; *manu*, because of the myth of fourteen *Manus*, means fourteen; thus, to express the year 1214, the word *surya-manu* is used.

of the grains of gold. In several provinces, they used as common currency pieces of copper to which they had given the shape of a Roman T. Cortés[1] reported that, having attempted to have cannons melted down in Mexico, and having sent emissaries to locate tin and silver mines, he learned that on the outskirts of Tachco (Tlachco or Tasco), the indigenous peoples used for their exchanges smelted pieces of tin as thin as the smallest coins of Spain.

Such are the inaccurate ideas that we have inherited from the first historians about the use that the indigenous peoples made of gold, silver, copper, tin, lead, and of mercury mines. I thought it necessary to enter into such detail, not only to cast some light on the ancient culture of these lands, but also to demonstrate that in the years immediately after the destruction of Tenochtitlan, the European settlers merely followed the indications that the indigenous population provided to them.

III.125

In its current state, the Kingdom of New Spain has nearly 500 sites (*reales y realitos*) famous for the mines found in their vicinity. Over two-thirds of these sites are identified on the general map of the country placed at the opening of my Mexican atlas. Both the names of the mines and their grouping can be found there, but because I was obliged to use somewhat inaccurate materials, the relative deposits of the mines and the distances between them need to be rectified. It is likely that 500 *reales* include nearly 3,000 mines (*minas*), if one uses this word to refer to the network of interconnected *underground works* through which one or more metallic *deposits* are exploited. These mines are divided into thirty-seven districts, or administrative divisions, which fall under the responsibility of an equal number of Mining Councils, called *Diputaciones de minería*. We shall bring together in a single table the names of these *Diputaciones* and of the *Reales de minas* found in the twelve intendancies of New Spain. The material used in this work was taken in part from an unpublished paper that the director of the High Mine Council, Don Fausto Delhuyar, drew up for the viceroy, the Count of Revillagigedo.

1. In his final letter to Charles V, this great man bemoans the fact that, after the taking of the capital, he was left without artillery and without arms. "Nothing," he writes, "invigorates human genius more (*no hay cosa que mas los ingenios de los hombres aviva*) than the sense of danger. Seeing myself in the situation of losing all that had cost us so much strain to win, I had to find the means for making cannons using the materials found in the country." I record here the remarkable passage in which Cortés writes about the use of tin as currency: "Topé entre los naturales de una provincial que se dice Tachco ciertas *piecezuelas de estaño* a manera de moneda muy delgada y procediendo en mi pesquisa hallé en que la dicha provincial y aun en otras se trataba *por moneda*" (Lorenzana, [*Historia de Nueva-España*,] p. 379, § XVII).

III.126

General Tableau of the Mines of New Spain

I. Intendancy of Guanajuato

From 20°55′ To 21°30′ Northern Latitude, and from 102°30′ to 103°45′ Western Longitude

Diputaciones de minería, or administrative divisions

1. Guanajuato

Reales, or sites surrounded by mines: Guanajuato. Villalpando. Monte de San Nicolás. Santa Rosa. Obejera. Santa Ana. San Antonio de las Minas. Comanja. Capulin. Comanjilla. Gigante. San Luis de la Paz. San Rafael de los Lobos. Durazno. San Juan de la Chica. Rincón de Centeno. San Pedro de los Pozos. Palmar de Vega. San Miguel el Grane. San Felipe.

II. Intendancy of Zacatecas

From 22°20′ to 24°33′ Northern Latitude, and from 103°12′ to 105°9′ Western Longitude

Diputaciones de minería, or administrative divisions

2. Zacatecas
3. Sombrerete
4. Fresnillo (III.127)
5. Sierra de Pinos

Reales, or sites surrounded by mines: Zacatecas. Guadalupe de Veta Grande. San Juan Bautista de Panuco. La Blanca. Sombrerete. Madroño. San Pantaleón de la Noria. Fresnillo. San Demetrio de los Plateros. Cerro de Santiago. Sierra de Pinos. La Sauceda. Cerro de Santiago. Mazapil.

III. Intendancy of San Luis Potosí

From 22°1′ to 27°11′ Northern Latitude, and from 100°35′ to 103°20′ Western Longitude

Diputaciones de minería, or administrative divisions

6. Catorce
7. San Luis Potosí
8. Charcas
9. Ojo Caliente
10. San Nicolás de Croix [San Nicolás Tolentino]

Reales, or sites surrounded by mines: La Purísima Concepción de Álamos de Catorce. Matehuala. Cerro de Potosí. San Martin Bernalejo. Sierra Negra. Tule. San Martin. Santa María de las Charcas. Ramos. Ojo Caliente. Cerro de San Pedro. Matanzillas. San Carlos de Vallecillo. San Antonio de la Iguana. Santiago de las Sabinas. Monterrey. Jesús de Río Blanco. Las Salinas. Boca de Leones. San Nicolás de Croix. Borbón. San José Tamaulipan. III.128 Nuestra Señora de Guadalupe de Sihue. La Purísima Concepción de Revillagigedo. El Venado. L. Tapona. Guadalcázar.

IV. Intendancy of Mexico City

From 18°10′ to 21°30′ Northern Latitude, and from 100°12′ to 103°25′ Western Longitude

Diputaciones de minería, or administrative divisions

11. Pachuca
12. El Doctor
13. Zimapan
14. Tasco
15. Zacualpan
16. Sultepec
17. Temascaltepec

Reales, or sites surrounded by mines: Pachuca. Real del Monte. Moran. Atotonilco el Chico. Atotonilco el Grande. Zimapan. Lomo del Toro. Xacala. San José del Oro. Verdozas. Capula. Santa Rosa. El Potosí. Las Plomosas. El Doctor. Las Alpujarras. El Pinal, or Los Amotes. Huascazoluya. San Miguel del Río Blanco. Las Aguas. Maconi. San Cristóbal. Cardonal. Xacala. Juchitlán el Grande. San José del Obraje Viejo. Cerro Blanco. Cerro del Sotolar. San Francisco Xichu. Jesus María de la Targea. Coronilla, or La Purísima III.129 Concepción de Tetela del Río. Tepantitlan. San Vicente. Tasco. Tehuilotepec. Coscallan. Haucingo. Huautla. Sochipala. Tetlilco. San Esteban. Real del Limón. San Jerónimo. Temascaltepec. Real de Ariba. La Albarrada. Ixtapa. Ocotepec. Chalchitepeque. Zacualpan. Tecicapan. Chontalpa. Santa Cruz de Azulaques. Sultepec.

Juluapa. Papaloapa. Los Ocotes. Capulatengo. Alcozauca. Totomitxtlahuaca.

V. Intendancy of Guadalajara

From 19°0′ to 23°12′ Northern Latitude, and from 103°30′ to 108°0′ Western Longitude

Diputaciones de minería, or administrative divisions

18. Bolaños
19. Asientos de Ibarra
20. Hostotipaquillo

Reales, or sites surrounded by mines: Bolaños. Jal[a]pa. San José de Guichichila. Santa María de Guadalupe, or de la Yesca. Asientos de Ibarra. San Nicolás de los Angeles. La Ballena. Talpan. Hostotipaquillo. Copala. Guajacatan. Amaxac. Limón. Tepanteria. Iocotan. Tecomatan. Ahuacatancillo. Guilotitan. Platanarito. Santo Domingo. Iuchipila. Mezquital. Xalpa. San José Tepostitlan. Guachinango. San Nicolás del III.130 Rojo. Amatlan. Natividad. San Joaquin. Santíssima Trinidad de Pozole. Tule. Motage. Frontal. Los Aillones. Ezatlan. Posesión. La Serranilla. Aquitapilco. Eliso. Chimaltitan. Santa Fe. San Rafael. San Pedro Analco. Santa Cruz de los Flores.

VI. Intendancy of Durango

FROM 23°55′ TO 29°5′ NORTHERN LATITUDE, AND FROM 104°40′ TO 110°0′ WESTERN LONGITUDE

Diputaciones de minería, or administrative divisions

21. CHIHUAHUA
22. PARRAL
23. GUARISAMEY
24. COSIGUIRIACHI
25. BATOPILAS

Reales, or sites surrounded by mines: San Pedro de Batopilas. Uruachi. Cajurichi. Nuestra Señora de Loreto. San Joaquín de los Arrieros. El Oro de Topago. San Juan Nepomuceno. Nuestra Señora del Monserrate del Zapote. Uriquillo. San Agustín. Nuestra Señora del Monserrate de Urique. Guarisamey. San Vicente. Guadalupe. Gavilanes. San Antonio de las Ventanas. San Dimas. San José de Tayoltita. Cosiguiriachi. Río de San Pedro. Chihuahua el viejo. San Juan de la Cieneguilla. Maguarichi. Cajurichi. San José del Parral.[1] Indehe. Los Sauces. Nuestra Señora de la Merced del Oro. Real III.131 de Todos Santos. San Francisco del Oro. Santa Barbara. San Pedro. Huejoquilla. Los Peñoles. La Cadena. Cuencamé. San Nicolás de Yerbabuena. La Concepción. Santa María de las Nieves. Chalchihuites. Santa Catalina. S. Miguel del Mezquital. Nuestra Señora de los Dolores del Orito. San Juan del Río. San Lucas. Panuco. Avinito. San Francisco de la Silla. Texamen. Nuestra Señora de Guadalupe de Texame. San Miguel de Coneto. Sianori. Canelas. Las Mesas. Sabatinipa, *or* Matabacas. Topia. San Rafael de las Flores. El Alacrán. La Lagartija. San Ramón. Santiago de Mapimi.

VII. Intendancy of Sonora

FROM 23°15′ TO 31°20′ NORTHERN LATITUDE, AND FROM 107°45′ TO 113°20′ WESTERN LONGITUDE

Diputaciones de minería, or administrative divisions

1. On a few proofs of my general map of New Spain, the name *Parral* is confused with that of the village of Valle San Bartolomé. The exact location of Parral—which is also found on my itinerary map, plate 7 of the *Mexican atlas*—is marked here by the symbol used to indicate the administrative center of a provincial mining council.

26. Álamos
27. Copala
28. Cosala
29. San Francisco Javier de la Huerta
30. Guadalupe de la Puerta
III.132 31. Santísima Trinidad de Peña Blanca
32. San Francisco Javier de Alisos

Reales, or sites surrounded by mines: San José de Copala. Real del Rosario. Plomosas. Santa Rosa, *or* Las Adjuntas. Apomas. San Nicolás de Panuco. Santa Rita. Trancito. Charcas. Limón. Santa Rosa de las Lagunas. Tocuistita. Corpus. Reyes. Cosala. Palo Blanco. El Cajón. Santiago de los Caballeros. San Antonio de Alisos. San Roque. Tabahueto. Norotal. Los Molinos. Surutato. Los Carcamos. San Juan Nepomucens. Bacatopa. Loreto. Tenoriba. Aguacaliente. Monserrate. Sivirijoa. Baroyeca. Yecorato. Zataque. Cerro Colorado. Los Álamos. Guadalupe. Río Chico. La Concepción de Haigamé. Santísima Trinidad. La Ventana, *or* Guadalupe. Saracachi. San Antonio de la Huerta. San Francisco Javier. Hostimuri. Quisuani. El Aguage. Higane. San José de Gracia. El Gabilán. El Populo. San Antonio. Todos Santos. El Carizal. Nacatabori. Racuach. S. Ildefonso de Cieneguilla. San Lorenzo. Nacumini. Cupisonora. Tetuachi. Basochuca. Nacosari. Bacamuchi. Cucurpe. Motepore.

VIII. Intendancy of Valladolid

From 18°25′ to 19°50′ Northern Latitude, and from 102°15′ to 104°50′ Western Longitude

Diputaciones de minería, or administrative divisions

33. Angangueo
III.133 34. Inguaran
35. Zitaquaro
36. Tlalpujahua

Reales, or sites surrounded by mines: Arigangueo. El Oro. Tlapaxahua. San Augustin de Ozumatlan. Zitaquaro. Istapa. Los Santos Reyes. Santa Rita de Chirangangeo. El Zapote. Chachiltepec. Sanchiqueo. La Joya. Paquaro. Xerecuaro. Curucupaseo Sinda. Inguaran. San Juan Guetamo Ario. Santa Clara. Alvadeliste. San Nicolás Apupato. Río del Oro. Axuchitlan. Santa María del Carmen del Sombrero. Favor. Chichindaro.

IX. Intendancy of Oaxaca

From 16°35′ to 17°55′ Northern Latitude, and from 98°15′ to 100°0′ Western Longitude

Diputaciones de minería, or administrative divisions

37. Oaxaca

Reales, or sites surrounded by mines: Zolaga. Talea. Hueplotitlan. La Aurora de Ixtepexi. Villalta. Ixtlan. Betolatia. Huitepeque. Río de San Antonio. Totomistla. San Pedro Nesicho. Santa Catalina Lachateo. San Miguel Amatlán. Santa María Iavecia. San Mateo Capulalpa. San Miguel de las Peras.

X. Intendancy of Puebla

III.134

From 18°15′ to 20°25′ Northern Latitude, and from 99°45′ to 100°50′ Western Longitude

Scattered mines: La Cañada. Tulancingo. Toltecanila. San Miguel Tenango. Zautla. Barrancas. Alatlanquetepec. Temetzla. Ixtacmaztitlan.

XI. Intendancy of Veracruz

From 20°0′ to 21°15′ Northern Latitude, and from 99°0′ to 101°5′ Western Longitude

Scattered mines: Zomelahuacan. Giliapa. San Antonio de Xacala.

XII. Vieja California

Mine: Real de Santa Ana

Those who have studied the geological composition of a large mining area know that it is nearly impossible to arrive at any general conclusions from observations of a wide variety of layers and metalliferous veins. A physicist can detect the relative age of the various *formations*; he manages to decipher the laws that govern the stratification of the rocks, the similarity of the layers, often even the angle that the latter form with the horizon or with the meridian of the location. But how might one discern the tandem workings of the secret causes that have determined the distribution of metals below the earth's surface, the thickness, orientation,

III.135

and incline of veins, the nature of their *mass*, and their particular structure? How might one reach general conclusions by observing a multitude of minor phenomena that have been modified by local disturbances and appear to be the product of a set of chemical affinities, the effects of which are confined to a very small area? The difficulties multiply when, as in the mountains of Mexico, the *veins*, *layers*, and *tiers* (*Stockwerke*) are found scattered amidst an endless number of rocks of very different admixtures and of very different formations. If one had an exact description of the four or five thousand veins that are currently being mined in New Spain, or that have been mined over the past two centuries, one could probably identify similarities in the mass and the structure of these veins that would indicate that they originated in the same period. One would find that these *masses* (*Gangausfüllungen*) are partly identical to those found in the veins of Saxony and Hungary, on which Mr. Werner, the century's leading mineralogist, has shed so much light. But we still have far to go to understand the metalliferous mountains of Mexico, and, despite the large number of observations that I was able to make myself while traveling around the country, across a total length III.136 of over four hundred leagues, I shall not attempt to sketch here a general portrait of the Mexican mines from a geological perspective. Rather, I shall limit myself to identifying the rock types that constitute the bulk of New Spain's riches.

In the current state of the country, the veins are the focus of the largest mines. It is quite rare to find ore distributed into *layers* or *tiers* there. The Mexican veins are mostly found in *primary* and *transition* rocks (*Ur-* and *Übergangsgebirge*), less commonly in mountains of secondary formation (*Flözgebirge*), which only occupy vast areas of terrain north of the tropic of Cancer, to the east of the Río del Norte, in the Mississippi Basin, and to the west of New Mexico, in the plains washed by the Zaguananas and San Buenaventura Rivers and abundant in muriatic salts.

One observes that in the Old Continent, the crests of the high mountain chains are most commonly composed of *granite*, *gneiss*, and *micaceous schist*. These rocks are rarely visible on the ridge of the Cordilleras of the Americas, especially in the central section between 18° and 22° northern latitude. Enormously thick layers of trachyte, amphibolic porphyry, diorite (*Grünstein*), amygdaloid, basalt, and other trappean formations cover the rocks that we are accustomed to regarding as primary, thereby concealing III.137 the latter from geologists. The coastlines of Acapulco are formed of granite. Moving toward the plateau of Mexico, one sees this granite pierce through the porphyry for the last time between Sumpango and Sopilote;[1] farther east, in the province of Oaxaca, the granite and the gneiss surge upward into large plateaus traversed by auriferous veins. It is my understanding that tin, which many geologists believe to be the oldest metal on the globe—along with titanium, scheelite, and molybdenum—has not yet been detected in the granite of Mexico. The fibrous tin (*wood-tin*) of Gigante belongs to alluvial formations, and the tin veins of the Sierra de Guanajuato are located in mountains of porphyry. In the Comanja mines, there is a mass of *syenite*, which appears to be of quite ancient formation and which contains an argentiferous vein. The Guanajuato vein, the richest in all the Americas, runs through a mass of *schist* (*Thonschiefer*) that often verges on *talcose schist* (*Talkschiefer*), and that seems to be an example of *transition schist*. The *serpentine* of Zimapán is devoid of metal.

For the most part, the *porphyry* rocks of Mexico can be considered eminently rich in gold and silver. Determining their *relative age* is one of the most difficult geological problems to resolve. What characterizes them all III.138 is the ubiquity of amphibole and the almost complete absence of quartz, so common in the primary porphyry rocks of Europe, especially in those that form layers in the gneiss. *Common feldspar* is rarely found in Mexican porphyry; it is peculiar only to the most ancient formations—those of Pachuca, Real del Monte, and Moran—whose veins yield twice as much silver as the whole of Saxony. Most often, only *vitreous feldspar* can be found in the porphyry of Spanish America. The rock that is traversed by the rich auriferous vein of Villalpando, near Guanajuato, is porphyry with a base verging on *Klingstein* (phonolite), in which amphibole is extremely scarce.[2] Several formations in New Spain exhibit great similarity to the problematic rocks of Hungary which ▾ Mr. de Born had described, using the somewhat vague term *saxum metalliferum*, and on which ▾ Mr. Beudant has recently shed new light.[3] The Real de Zimapán veins, the most revealing in terms of III.139

1. For fuller details on the granite of the western ridge of the Mexican Cordilleras (granite that was perhaps formed later than the *gneiss*, and that contains syenitic beds connected to *basanite* veins), see my *Essai géognostique sur le gisement des roches dans les deux hémisphères*. 1823, p. 78.

2. The three groups of *porphyritic formations* in New Spain are characterized in my *Essai géognostique*, pp. 170–88. They are the group of Mojouera, Sumpango, and Achichintla, on the road from Acapulco to Mexico City, pp. 173–79; the central plateau group, of Pachuca, Moran, and Puebla, pp. 179–85; finally, the group of Guanajuato and Santa Rosa, pp. 185–87.

3. [Beudant,] *Voyage minéralogique et géologique* [,] *en Hongrie*, vol. II, p. 594; vol. III, p. 528.

the theory of ore *deposits*, run through masses of porphyry with a *grünstein* base, which appear to be examples of trappean rocks of new formation. These same veins in the Zimapán district provide oryctological collections with a wide variety of interesting minerals, such as fibrous zeolite, stilbite, grammatite [tremolite], pycnite, native sulfur, fluorspar, barite, suberiform asbestos, green garnets, carbonate and lead chromate, orpiment, chrysoprase, and a new type of opal of the rarest beauty, which I introduced to Europe, and which ▼ Mr. Karsten and Mr. Klaproth have described by the name *Feuer-Opal* [fire opal].

The *transition rocks* containing silver ore include the transition limestone (*Übergangskalkstein*) of the Real del Cardonal, Xacala, and Lomo de Toro, north of Zimapán. In the last of these three sites, it is not veins that are mined but, rather, *tiers* of galena [blue lead], a few nests of which have, over a short period of time, yielded over 124,008 quintals of lead, according to Mr. Sonneschmidt's observation. In Mexico, the alternating layers of *Grauwake* [*greywacke*] and *Grauwakenschiefer* [*greywacke schist*] are no less rich in metals than in many parts of Germany. It is within this rock, the formation of which immediately preceded that of the secondary rocks, that several of the Zacatecas veins appear to be located.

III.140 As learned geologists explore more of northern Mexico, it will be determined that the metallic riches of Mexico are not restricted to primary formations and to transition rocks, but that they also extend to *secondary formations*. I do not know whether the lead that is mined in the eastern part of the intendancy of San Luis Potosí exists in veins or in layers; but it seems likely that the silver veins of the Real de Catorce, like those of El Doctor and Xaschi, near Zimapán, mostly run through *alpine limestone* (*Alpenkalkstein*). This rock lies on top of a pudding stone of siliceous cement, which can be regarded as the oldest of the secondary formations. The alpine limestone and the *Jurassic limestone* (*Jurakalkstein*) contain the famous silver of Tasco and of Tehuilotepec, in the intendancy of Mexico City, and it is in these calcareous rocks that the numerous veins that have been the focus of very old mining operations in this area have demonstrated the greatest riches. They are more *barren* in the *strata* of primary schist (*Urthonschiefer*), which, as can be observed in the Cerro de San Ignacio, serves as a base for the secondary formations.

This overview of the *metalliferous deposits* (*Erzführende Lagerstätten*) suggests that veins are present within a wide variety of rocks in the Cordilleras of Mexico, and that the rocks that currently produce nearly all of the III.141 silver exported annually from Veracruz are *transition schist*, *Grauwacke*, and *alpine limestone*, run through by the *primary veins* of Guanajuato, Zacatecas, and Catorce. It is also within a *primary* (*or transition?*) *schist*, upon which rests a layer of porphyry embedded with garnets, that the riches of *Potosí*, in the kingdom of Buenos Aires, lie. In Peru, by contrast, it is in *alpine limestone* that one finds the mines of Hualgayoc (or Chota) and Yauricocha (or Pasco), which together produce twice as much silver every year as all the mines of Germany. The more one studies the geological composition of the globe on a large scale, the more one learns that there is hardly any rock that has not been found to be eminently metalliferous in certain areas. The wealth of the veins is most often independent of the nature of the layers through which these veins run.

In the most famous mines of Europe, one observes that the underground works are focused either on a large number of relatively unproductive veins, as in the primary mountains of Saxony, or on a very small number of extraordinarily rich *lodes*, as at Clausthal, in the Harz Mountains, and near Schemnitz in Hungary. The Cordilleras of Mexico present numerous examples of these two types of mining operations; nevertheless, the mining districts with the most constant and most significant wealth—those of Guanajuato, Zacatecas, and the Real del Monte—each have only a single primary vein (*veta madre* [motherlode]). In Freiberg, the vein called *Halsbrükner Spath*, the thickness of which is two meters, and which has been surveyed over a length of 6,200 meters, is considered a remarkable phenomenon. The *Veta madre* of Guanajuato, from which over six million marks of silver have been extracted in the last ten years, has a thickness of forty to forty-five meters. It is mined from Santa Isabela and San Bruno to Buenavista, over a length of more than 12,700 meters.

III.142

In the Old Continent, the veins of Freiberg and Clausthal, which run through *gneiss* and *grauwacke* [*greywacke*] mountains, rise to the *surface* on plateaus with an elevation of only 400 meters and 580 meters above sea level, respectively. This elevation may be regarded as the mean [average] height of the richest mines of Germany. In the New Continent, nature deposited metallic riches right on ridge of the Cordilleras, sometimes in locations not far from the limit of the perpetual snows. The most famous mining operations of Mexico are found at absolute heights of 1,800 meters to 3,000 meters. In the Andes, the mining districts of Potosí, Oruro, Pasco, and Hualgayoc are located in an area whose elevation surpasses that of the highest peaks of the Pyrenees. Near the small city of Micuipampa, whose main square, according to my measurements, is at an elevation of 3,618 meters above sea level, a *mass* of silver ore known by the name of the *Cerro*

III.143

de Hualgayoc provided a tremendously rich output at an absolute height of 4,100 meters.

We have explained elsewhere[1] how advantageous it is for Mexican mining that the most important metalliferous deposits are located in the midsection of the country, where the climate is not unfavorable to agriculture and to the growth of plants. The large city of Guanajuato is located in a ravine whose base is less than two hundred meters below the level of the lakes contained within the valley of Tenochtitlan. We do not know the absolute heights of Zacatecas and of the Real de Catorce. These two sites are located on plateaus that appear higher than the ground at Guanajuato. Nevertheless, the temperate climate of these Mexican cities, which are surrounded by the richest mines in the world, contrasts with the extremely cold and unpleasant climates of Micuipampa, Pasco, Huancavelica, and other Peruvian cities.

When one compares the amount of silver delivered to the mint annually from a small-sized district (for example, Freiberg in Saxony) to the large number of mines that are in operation, one notices, even from the most cursory review, that this silver is provided by only a small portion of the underground works, and that nine-tenths of the mines have practically no impact on the total amount of ore extracted from the earth. Similarly, in Mexico, only a very small number of mines produce the 2,500,000 marks of silver shipped annually to Europe and to Asia from the ports of Veracruz and Acapulco. The three districts that we have mentioned on several occasions, namely, Guanajuato, Zacatecas, and Catorce, provide over one-half of this sum. A single vein, that of Guanajuato, *produces nearly one-quarter of all Mexican silver and one-sixth of the total output of the Americas as a whole.*

III.144

In the general table presented below, the principal mines are mixed in with those from which only a very small amount of metal is extracted. The disproportion between these two categories is so great that over nineteen-twentieths of Mexican mines fall under the latter category, the total output of which probably does not equal two hundred thousand marks. Similarly, in Saxony, the mines surrounding the city of Freiberg produce nearly fifty thousand marks of silver annually, while the rest of the *Erzgebirge* produces only seven to eight thousand marks. Here is a list of the richest mining districts of New Spain (in 1804), in order of the amount of silver that was extracted from them in that period:

1. See above, chap. III, vol. I, p. 276 [p. 186 in this edition]; and chap. IX, vol. II, p. 375 [p. 508 in this edition].

GUANAJUATO, in the intendancy of the same name,
CATORCE, in the intendancy of San Luis Potosí, III.145
ZACATECAS, in the intendancy of the same name,
REAL DEL MONTE, in the intendancy of Mexico City,
BOLAÑOS, in the intendancy of Guadalajara,
GUARISAMEY, in the intendancy of Durango,
SOMBRERETE, in the intendancy of Zacatecas,
TASCO, in the intendancy of Mexico City,

Batopilas, in the intendancy of Durango,
Zimapán, in the intendancy of Mexico City,
Fresnillo, in the intendancy of Zacatecas,
Ramos, in the intendancy of San Luis Potosí,
Parral, in the intendancy of Durango.

The precise materials that might enable one to retrace the history of mining in New Mexico are completely lacking. It appears certain that, of all the veins, those of Tasco, Sultepeque, Tlapujahua, and Pachuca were first worked by the Spanish. It was near Tasco, west of Tehuilotepec, in the *Cerro de la Compaña*, that Cortés bored a gallery of efflux through the mica schist upon which the alpine limestone is superposed, as we have indicated above. This gallery, called *el Socabón del Rey*, was begun in such large dimensions that it can be traversed on horseback over a length of more than ninety meters. It has just been completed, by virtue of the patriotic zeal of a miner from Tasco, Don Vicente de Ansa, who succeeded in cutting the primary vein up to a distance of 530 meters from the *mouth* of the gallery. III.146 Mining in Zacatecas has closely followed that of the *ore deposits* of Tasco and Pachuca. The San Bernabé vein was first mined as early as 1548, therefore twenty-eight years after the death of Montezuma, a circumstance that must seem even more remarkable since the city of Zacatecas is at a distance of over one hundred leagues, as the crow flies, from the valley of Tenochtitlan. It is claimed that a number of muleteers, en route from Mexico City to Zacatecas, discovered the silver ore of the Guanajuato district. In this district, the San Bernabé mine, near the basaltic hill of *Cubilete*, exhibits the oldest underground works. The primary vein of Guanajuato (*la veta madre*) was discovered later, while the shafts of Mellado and Rayas were being dug. The first of these shafts was begun on April 15, 1558, and the second was begun on April 16. The Comanjas mines are probably even older than those of Guanajuato. Given that the total output of the mines of

Mexico until the beginning of the eighteenth century was only six hundred thousand marks of gold and silver per year, one can conclude that, in the sixteenth century, relatively little attention and energy were devoted to ore extraction. The veins of Tasco, Tlalpujahua, Sultepeque, Moran, Pachuca, and Real del Monte, as well as those of Sombrerete, Bolaños, Batopilas, and Rosario, occasionally produced tremendous riches; but their output was less regular than that of the mines of Guanajuato, Zacatecas, and Catorce.

III.147

The silver extracted in the thirty-seven mining districts into which the Kingdom of New Spain is divided is deposited into the *provincial treasury coffers*, established in the intendancies' administrative centers. It is from the receipts of these *cajas reales* that one can determine the amount of silver produced in the various parts of the country. Here is a table of eleven provincial treasuries:

Between 1785 and 1789, the following amounts were deposited into the following *cajas reales*:

Guanajuato	2,469,000 marks of silver
San Luis Potosí (Catorce, Charcas, San Luis Potosí)	1,515,000
Zacatecas (Zacatecas, Fresnillo, Sierra de Pinos)	1,205,000
Mexico City (Tasco, Zacualpa, Sultepeque)	1,055,000
Durango (Chihuahua, Parral, Guarisamey, Cosiguiriachi)	922,000
Rosario (Rosario, Cosala, Copala, Alamos)	668,000
Guadalajara (Hostotipaquillo, Asientos de Ibarra)	509,000
Pachuca (Real del Monte, Moran)	455,000
Bolaños	364,000
Sombrerete	320,000
Zimapán (Zimapán, Doctor)	248,000
Five-year sum:	9,730,000 marks of silver

III.148

The part of the Mexican mountains that produces the greatest amount of silver today lies between parallels twenty-one and twenty-four and a half. The famous mines of Guanajuato are at a distance of only thirty leagues from those of San Luis Potosí, as the crow flies. San Luis Potosí is thirty-four leagues from Zacatecas, Zacatecas is thirty-one leagues from Catorce, and Catorce is seventy-four leagues from Durango. It is quite remarkable that the metallic riches of New Spain and Peru are found at a nearly equal distance from the equator in both hemispheres.

Within the vast expanse that separates the *ore deposits* of Potosí and La Paz from those of Mexico, there are no other mines that place a large amount of precious metals into circulation than those of Pasco and Chota. Traveling north from the Cerro de Hualgayoc, one finds only the gold washes of Chocó, those of the province of Antioquia, and the recently discovered silver veins of Vega de Supía. The same goes for the Cordillera of the Andes as for all the mountains of Europe in which metals are unevenly distributed. The province of Quito and the eastern part of the kingdom of New Granada, from 3° southern latitude to 7° northern latitude, the isthmus of Panama and the mountains of Guatemala present, over a length of 600 leagues, vast expanses of terrain in which no veins have been mined successfully thus far. It would, however, be inaccurate to suggest that these lands, most of which have been devastated by volcanoes, are entirely lacking in gold and silver ore. Numerous *metalliferous deposits* may have been concealed there by the superposition of strata of tra- III.149 chyte, basalt, amygdaloid, syenite porphyry, and other rocks that geologists include under the umbrella category of *trappean formation*.

As for the Mexican mines in particular, they can be considered as forming eight *groups* (*Erz-refiere*), which are almost all located either on the ridge or on the eastern slope of the Cordillera of Anahuac. The first of these groups is the one with the largest output; it includes the contiguous districts of Guanajuato, San Luis Potosí, Chalcas, Catorce, Zacatecas, Asientos de Ibarra, Fresnillo, and Sombrerete. The second encompasses the mines located to west of the city of Durango, as well as those of the province of Sinaloa; for the mining operations of Guarisamey, Copala, Cosala, and Rosario are similar enough to one another to be included in a single geological division. The third group, the northernmost of New Spain, is that of Parral, which covers the mines of Chihuahua and Cosiquiriachi. It extends from 27° to 29° latitude. To the north-northeast of Mexico City lies the *fourth* group, that of the Real del Monte (or Pachuca), and the *fifth* group, that of Zimapán (or Doctor). Bolaños (in the intendancy of Guadalajara), Tasco, and Oaxaca are, respectively, the central points of the sixth, seventh, and eighth groups of New Spain mines. This overview should III.150 suffice to demonstrate that this kingdom, like the Old Continent, contains vast expanses of land that appear to be utterly devoid of metalliferous veins. To this day, no significant mining operation has been undertaken in the intendancy of Puebla, that of Veracruz, the plains of secondary formation located on the left bank of the Río del Norte, or New Mexico.

III.151

Principal Mines of Mexico divided into eight groups	Surface Area of the land	Places that may be regarded as central points for each group (in sq. lg.)	Output of each group in marks of silver
First group (*central group*) from 21°0′ to 24°10′ northern latitude and from 102°30′ to 105°15′ western longitude.	1,900	Guanajuato Catorce Zacatecas	1,300,000
Second group (*Durango and Sonora group*) from 23°0′ to 24°45′ northern latitude and from 106°30′ to 109°50′ western longitude.	2,800	Guarisamey (Durango) Rosario (Copola)	400,000
Third group (*Chihuahua group*) from 26°50′ to 29°10′ of northern latitude and from 106°45′ to 108°50′ of western longitude.	3,100	Cosiquiriachi Parral Batopilas	uncertain
Fourth group (*Viscaína group*) from 20°5′ to 20°15′ of northern latitude and from 100°45′ to 100°52′ western longitude.	25	Real del Monte (Pachuca)	120,000
Fifth group (Zimapán group) from 20°40′ to 21°30′ northern latitude and from 100°30′ to 102°0′ western longitude.	750	Zimapán	60,000
Sixth group (*New Galicia group*) from 21°5′ to 22°30′ northern latitude and from 105°0′ to 106°30′ western longitude.	1,050	Bolaños	230,000
Seventh group (*Tasco group*) from 18°10′ to 19°20′ northern latitude and from 101°30′ to 102°45′ western longitude.	1,200	Temascaltepec Tasco Zacualpan	260,000
Eighth group (*Oaxaca group*) from 16°40′ to 18°0′ northern latitude and from 98°15′ to 99°50′ western longitude.	1,400	Oaxaca Villalta	uncertain
Average output of New Spain's mines, including the mines of the northern part of Nueva Biscay and those of Oaxaca: over 2,500,000 marks of silver.			

Rather than showing the relative wealth or the uneven distribution of metals from a geographical perspective, the following table shows the amount of silver that, in the current state of the mines, is extracted in the various parts of the Kingdom of New Spain. The mines have been arranged in the order that we have explained above, with both the name of the administrative center that is the central point of the group and the surface area of the region in which the various mining operations are found. A few groups can naturally be divided into several districts, which form an equal number of subdivisions or particular systems.

We shall later compare the output of the silver mines of Mexico with that of the various mines of Europe. For the moment, however, it suffices to observe that the two and a half million marks of silver exported annually from Veracruz *are equivalent to two-thirds of the silver that is annually extracted worldwide.* Together, the eight groups into which we have divided the mines of New Spain occupy a surface area of twelve thousand square leagues, or one-tenth of the total area of the Kingdom. If one focuses on the tremendous wealth of a very small number of mining operations, for example the Valenciana mine and the Reyas mine in Guanajuato, or the primary veins (*vetas madres*) of Catorce, Zacatecas, and Real del Monte, one can easily see that over 1,400,000 marks of silver are produced in an area of land that is smaller than the mining district of Freiberg. III.152

While the amount of *silver* that is extracted annually from the mines of Mexico is *ten times* larger than what is produced by all the mines of Europe, *gold*, on the other hand, is not much more abundant in New Spain than it is in Hungary and Transylvania. The latter two countries place into circulation nearly 5,200 marks, while the gold delivered to the mint of Mexico City amounts, in an average year, to only 7,000 marks. It can be expected that, at peacetime, when the lack of mercury does not slow down the amalgamation process, the annual output of New Spain is: III.153

In *silver*,	22
In *gold*,	1
	23 million piasters

For the most part, Mexican gold comes from alluvial formations, from which it is extracted through washes. These formations are common in the province of Sonora, which, as we have observed above,[1] can be considered

1. Chap. VIII, vol. II, p. 240 [p. 430 in this edition].

the Chocó of North America. A great deal of gold, spread throughout the sands that fill the base of the Río Hiaqui valley, to the east of the Tarahumara missions, has already been collected. Farther north, in the Pimeria Alta, at 31° latitude, grains (*nuggets*) of native gold with a weight of five to six pounds have been found. The extraction of gold in these deserted regions is hindered by the forays of the savage Indians, the exorbitant cost of supplies, and the lack of water, necessary for the washes.

Another portion of Mexican gold is extracted from the veins that traverse the primary rock mountains. Gold veins are most common in the province of Oaxaca, either in the gneiss or in the mica schist (*Glimmerschieffer*). The latter rock is especially rich in gold in the famous mines of

III.154

the Río San Antonio. These veins, the *gangue* of which is milky quartz, are over a half-meter thick, but their wealth is very unevenly distributed; they are often quite *narrow*, and the extraction of gold is, in general, somewhat insignificant. The same metal is present, either in pure form or mixed with silver ore, in most of the veins that are mined in Mexico. There is barely a single silver mine that is not also auriferous. One often recognizes native gold crystallized into either octahedra or strips, or into a *knitted* shape, in silver ore from the mines of Villalpando and Rayas, near Guanajuato, in those of Sombrero (intendancy of Valladolid), and Guarisamey, to the west of Mezquital, in the province of Guadalajara. The gold of Mezquital is regarded as the purest; that is, as the gold that is least alloyed with silver, iron, and copper. In Villalpando, in the Santa Cruz mine, which I visited in September of 1803, the primary vein is *shot through* with a large number of small *rotten veins* (*hilos del desposorio*) that are extraordinarily rich. The *clayey silt* that fills these *networks* contains such a large amount of gold, disseminated in extremely fine particles, that the miners are forced, whenever they emerge from the mine, nearly naked, to bathe themselves in large tanks, to prevent them from carrying off the auriferous clay that becomes attached to their bodies. The silver ore of Villalpando normally contains only two ounces of gold per *load* (*carga*,

III.155

of twelve arrobas); but its wealth often amounts to eight to ten ounces per *load*, or one and seven-tenths of an ounce per quintal. It is useful to recall here that, in the Harz Mountains, the pyrite of Rammelsberg contains only one twenty-nine millionth part of gold, which is nevertheless extracted profitably.[1]

1. ▾ Brongniart, *Minéralogie*, vol. II, p. 345.

According to the registers of the provincial treasury,[1] the mining district of Guanajuato produced:

PERIOD	GOLD MARKS	SILVER MARKS	GOLD CONTAINED IN SILVER
1766–1775	9,044	3,422,414	0,0026
1776–1785	13,254	5,281,214	0,0025
1786–1795	7,376	5,609,356	0,0013
1796–1803	13,356	4,410,553	0,0029
In 38 years	43,030	18,723,537	0,0023

This table suggests that the silver extracted from the Guanajuato vein contains between one- and three-thousandths of its own weight in gold.

It has been claimed, falsely, that platinum exists in the auriferous sands of the Sonora. To date, this metal has not yet been discovered north of the isthmus of Panama, on the continent of North America, although the existence of rhodium, identified by Mr. del Río in a few Mexican gold bars, makes it likely that platinum and palladium are in close proximity. Platinum in grain form is only found in two places in the known world, in Chocó, one of the provinces of New Granada, and near the South Sea coasts, in the III.156 province of Barbacoas, between 2° and 6° northern latitude. It is peculiar to alluvial formations, which are smaller than a single department of France. The *lavaderos* (*lavages*) that today produce platinum are all located to the south of the *threshold*[2] separating the sources of the Río Atrato from those III.157

1. *Estado de la Tesorería principal de Real Hacienda de Guanajuato, del 23 de diciembre de 1803* (manuscript).

2. I owe a number of these ideas about platinum to the kind communications of Mr. ▼ Joaquin Acosta, a highly educated young Colombian officer who resided for a long time in Chocó and who recently came (1825) to Paris to continue his studies here. I learned from this traveler that the small canal between the Río San Juan and the Quebrada de la Raspadura, which small boats were able to traverse in times of prodigious rains (see above, vol. I, p. 235 [p. 427 in this edition]), and about which I first read in a very interesting official document (*Relación del estado del Nuevo Reyno de Granada que hace el Arzobispo de Córdoba a su sucesor el Estimado Fray don Francisco Gil y Lemos, 1789*), was long ago filled in by alluvial deposits, and that the only memory that the *Arastradero de San Pablo* has preserved is of a channel (*zanja*) by which a monk enabled the fish of the Quebrada de la Raspadura and of the Río Quibdo, a tributary of the Atrato across the isthmus of San Pablo, to pass into the Río de San Juan. ▼ Don Ignacio Cavero, who, in his role as the secretary general of

of the Río San Juan; among the wealthiest of these, we shall cite the *lavaderos* of Santa Lucia and Tado, which belong to Doña Petronila Castro, and which produce two-thirds platinum and one-third gold; the *lavaderos* of Santa Rosa (belonging to Don José Antonio Zaa), of Viroviro, of Condoto, and of Tajuato (belonging to ▾ Don José María Mosquera), of Santa Barbara, of Yro, etc. In those parts, the price of this metal in grain form is between four and five piasters, while in Paris it normally sells for between 130 and 150 francs and sells today (1826) for 240 francs. It is absolutely untrue that platinum was ever found near Cartagena, near Santa Fe de Bogotá, on the island of Puerto Rico, on that of Barbados, and in Peru,[1] although

III.158 these various locations are cited in the most respected and most widespread works: perhaps chemical analysis will one day confirm for us that platinum

---

the viceroyalty, drafted the *Statistical Report* of the Viceroy-Archbishop, and whom I met in 1801 in Cartagena de Indias, when he was general administrator of customs, provides in this report a meticulous description of the small streams (Quiadocita, Platinita, and Quiado) that the monk had redirected to flow into the canal; he distinguishes these from the streams by which one might enlarge this body of water to transport from one sea to the other not only small dinghies but also large dugout canoes and *lanchas*. I felt it necessary to record information here about the state of affairs in 1789. Don José Ignacio Pombo, a merchant keenly interested in improving his country, insists, in his *Notícias sobre las Quinas oficiñales* (p. 63), published in 1814, on the ease with which a navigable canal 2,000 leagues long could be dug in the perfectly flat terrain (*corto espacio de tierra llana*) between the Río Quibdo and the Río San Juan, to open up trade contact between Peru and Cartagena de Indias. The existence of an old channel through which the fish of the Quebrada de la Raspadura passed into the Río de San Juan is sufficient proof that this isthmus of Chocó is no more of an obstacle to canalization than the isthmus of Javita, through which I traveled from the Tuamini, a tributary of both the Atabapo and the Orinoco, to the Pimichin, a tributary of both the Río Negro and the Amazon. I doubt, however, that the small canal of Chocó, such as it existed in the time of the Viceroy-Archbishop in 1789, was ever useful for domestic trade, and I was very likely mistaken when, following the information that had been given to me in Cartagena de Indias and in Popayan, I said (*Relat[ion] hist[orique]*, vol. III, p. 128) that large amounts of cocoa had crossed the Raspadura canal.

1. Haüy, *Minéralogie*, vol. III, p. 370, and also ▾ Phillips, [*An Elementary Introduction to the Knowledge of*] *Mineral[ogy]*, 1823, p. 325. In a paper included in the *Annales de ciencias naturales*, published by ▾ Abbot Cavanilles, one reads that platinum is found in *Chocó*, in *Barbados* (Barbacoas?), and in Cartagena, a seaport located 130 leagues away from the gold washes of Tado. Over eighteen years ago, however, Mr. Berthollet produced a very exact account of the sites that produce platinum (["Rapport fait au Bureau de Consultation, sur les moyens proposés par M. Jeanety pour travailler le Platine,"] *Annales de Chimie*, July 1792). I brought back to Europe, and deposited in the mineralogy exhibition room in Berlin, a *nugget* of platinum of extraordinary size. It weighs 1,088 and eight-tenths grains. According to Mr. Tralles, its specific weight is 18,947. Karsten, *Miner[alogische] Tabellen*, 1808, p. 96. In the *lavaderos* of Condoto, however, a *nugget* of platinum has since been found that is two inches four and a half lines in diameter, and weighs one pound, nine ounces, in English measurements. This *nugget*, the largest known, is preserved in the Museum of Madrid.

exists in some silver ore of Mexico, as it is assumed to exist, according to Mr. Vauquelin's analyses, in the *Fahlerz* (gray copper) of Guadalcanal, in Spain.

[Since the publication of the first edition of the *Political Essay*, platinum has been discovered in Brazil, and in the Ural Mountains in Siberia. In the first of these two countries, the destruction of layers of chloritic quartz and ferruginous breccia appears to have produced a wash formation in which one finds not only gold but also platinum, palladium, and diamonds in Correjo das Lagens; platinum and diamonds only in Rio Abaeté; gold and diamonds only in Tejuco.[1] Just as one encounters palladium grains mixed with platinum grains in Brazil, the Ural Mountains exhibit grains of iridium and osmium in their platiniferous formations.] III.159

The silver produced by the veins of Mexico is extracted from a wide variety of ore, whose composition makes them similar to those of the *metalliferous deposits* of Saxony, the Harz Mountains, and Hungary. Travelers should not expect to find a complete collection of these ores in the School of Mines in Mexico City. Since the mining operations are all in the hands of private individuals, and since the Mexican government still exerts only a weak influence over the administration of the mines, it was not possible for the faculty to collect complete information about the *structure* of the veins, the *layers*, and the *tiers* of ore. In Mexico City, as in Madrid, public collections exhibit the rarest ore from Siberia and Scotland, while one seeks in vain anything that might shed new light on the mineralogical geography of the country. One must hope that the cabinet of the School of Mines will benefit from that great institution's students being sent to the most distant provinces from the capital to convince the mine owners that it is in their interest to facilitate the means of instruction. Without an individual knowledge of the localities, without an in-depth study of the minerals that compose the *rock masses* of the veins, or the *content* of the *tiers* and the *layers*, any changes that are proposed in order to perfect the amalgamation process will amount to no more than fanciful projects. III.160

In Peru, most of the silver extracted from the ground comes from *pacos*, mud-colored ore, which Mr. Klaproth[2] was kind enough to analyze, at my request, and which consist of an intimate mixture of nearly imperceptible particles of native silver and brown iron oxide. In Mexico, on the other hand, the bulk of the silver that is placed into circulation each year

1. See my *Essai géognostique sur le gisement des roches*. 1823, p. 92.

2. Klaproth, *Beiträge zur chemischen Kentniss der Mineral-Körper*, vol. 4, p. 4.

comes from the category of ore that the Saxon miner calls *dürre Erze*, or meager ores,[1] especially from *silver sulfide* (or vitreous—*Glaserz*—silver), from *gray arsenious copper* (*Fahlerz*) and *antimonious copper* (*Grau-* or *Schwarz-Giltigerz*), from *muriated silver* (*Hornerz*), from *prismatic black silver* (*Sprödglaserz*), and from *red silver* (*Rothgiltigerz*). We are not including native silver among these ores, because it is not found in large enough quantities to constitute a significant portion of the total output of the mines of New Spain.

III.161 Silver sulfide and prismatic black silver are very common in the Guanajuato and Zacatecas veins, as well as in the Veta Vizcaína, in Real del Monte. The silver extracted from the Zacatecas ore exhibits the remarkable property of not containing any gold. The richest *Fahlerz* comes from the Sierra de Pinos and from the Ramos mines. The *Fahlerz* from the latter mines is accompanied by *Glaserz*, by hepatic copper pyrite (*bunt Kupferz*), by brown blende (zinc sulfide), and by vitreous copper (*Kupferglas*), which is mined exclusively for its silver content, not for its copper content. *Graugiltigerz*, or gray antinomous copper, described by Mr. Karsten, is found in Tasco, and in the Rayas mine, southeast of Valenciana. Muriated silver, which is so scarce in the veins of Europe, is, on the contrary, very abundant in the mines of Catorce, Fresnillo, and the Cerro de San Pedro, near the city of San Luis Potosí. The muriated silver of Fresnillo is often of an olive-green color bordering on leek-green. Magnificent samples of this very color have been found in the Vallorecas mines, which are located in the district of Los Alamos, in the intendancy of Sonora. In the Catorce veins, muriated silver is accompanied by lead molybdate (*Gelb-Bleierz*) and lead phosphate (*Grün-Bleierz*). According to Mr. Klaproth's most recent analyses, it seems III.162 that the muriated silver[2] of the Americas is a pure mixture of silver and muriatic acid, while the *Hornerz* of Europe contains iron oxide, alumina, and, especially, a bit of sulfuric acid. Red silver mines are among the primary riches of Sombrerete, Cosala, and Zolaga, near Villalta, in the province of

1. See the very informative work by ▼ Mr. D'Aubuisson, which bears the title *Description des mines de Freiberg* [*Des mines de Freiberg en Saxe et de leur exploitation*]. Throughout this chap., for items related to the art of mining and to ore deposits, I have used the terminology developed by ▼ Mr. Brochant, Mr. D'Aubuisson, and Mr. Brongniart.

2. Today, mineralogists distinguish between four types of muriated silver: common, earthy, conchoidal, and streaked. The latter two types, which are extremely beautiful, have been described by Mr. Karsten. They are among the minerals that I brought back from Peru (Karsten, in the *Magazin der Berliner Gesellschaft Naturforschender Freunde*, vol. 1, p. 156. Klaproth, *Beiträge*, vol. 4, p. 10).

Oaxaca. It was from the red silver ore of the Veta Uegra mine,[1] near Sombrerete, that over 700,000 thousand marks of silver were extracted over the course of five to six months. It is claimed that the *rising-terrace operation* that produced such an enormous amount of metal, the largest that a vein has ever produced from a single point of its *mass*, was less than thirty meters long. Mines of true *white silver* (*Weissgiltigerz*) are very rare in Mexico. The *grayish white* variety of it, very rich in lead, can nevertheless be found in the intendancy of Sonora, in the Cosala veins, where it is accompanied by argentiferous galena, red silver, brown blende, quartz, and baryta sulfate. This latter substance, extremely uncommon among the *gangues* of Mexico, is also present at Real del Doctor, near Baranca de las Tinajas, and at Sombrerete, especially in the mine called La Campechana. Fluor-spar has not yet been found outside the veins of Lomo del Toro, near Zimapán, at Bolaños, and at Guadalcázar, near Catorce. It is always either meadow-green or violet-blue.

III.163

In some parts of New Spain, the miners' labors are focused on a mixture of brown iron oxide and native silver, dispersed in the form of tiny particles imperceptible to the naked eye. This ochreous mixture, which is called *paco* in Peru, and which we have had occasion to discuss above, is the target of a significant operation in the Angangueo mines, in the intendancy of Valladolid, as well as at Yxtepexi, in the province of Oaxaca. The ore of Angangueo, known as *colorados*, have an earthy appearance. *Near the surface*, the brown iron oxide is blended with native silver, silver sulfide, and prismatic black silver (*Sprödglaserz*), all three in a state of decomposition. At great depths, the Angangueo vein offers no more than galena and iron pyrite, relatively scarce in silver. The blackish *pacos* of the Aurora mine of Yxtepexi, which must not be confused with the *negrillos* of Peru, owe their wealth more to the *glaserz* than to the imperceptible *filaments* of *branched* native silver. This vein is quite uneven in its output, barren at certain points and abundant at others. The *colorados* of Catorce, especially those of the Concepción mine, are brick-red in color and mixed with muriate of silver. In general, one sees that in Mexico, as in Peru, these oxidized iron masses occur exclusively in the section of the veins nearer to the earth's surface. To a geologist's eye, the *pacos* of Peru bear a very striking resemblance to the earthy masses that the miners of Europe call the iron *cap* of the veins (*eiserne Hut*).

III.164

1. See above, chap. VIII, vol. II, p. 240 [p. 430 in this edition].

Native silver, much less abundant in the Americas than is generally assumed, has been found in large masses, occasionally at a weight of over two hundred kilograms, in the veins of Batopilas, located in Nueva Vizcaya. These mines, barely worked today, are among the northernmost of New Spain. Nature presents the same ore there as in the Kongsberg vein, in Norway. Those of Batopilas contain spindly, dendritic, and knitted silver, running through layers of carbonate of lime [calcite]. Moreover, the native silver in the veins of Mexico is consistently accompanied by *Glaserz*, just as in the European mountain veins. One frequently encounters these two minerals together in the extremely wealthy mines of Sombrerete, Madroño, Ramos, Zacatecas, Tlapujahua, and Sierra de Pinos. From time to time, one can also identify small branches or cylindrical filaments of native silver in the famous Guanajuato vein, but these masses have never been so large as

III.165 those formerly extracted from the *Encino* mine, near Pachuca and Tasco, where the native silver is occasionally enveloped in sheets of selenite. At Sierra de Pinos, near Zacatecas, this latter metal is consistently accompanied by radial blue copper (*strahlige Kupferlazur*), crystallized into tiny four-faced prisms.

A significant portion of the silver produced annually in Europe is derived from the *argentiferous lead sulfide* (*silberhaltiger Bleiglanz*) that is found either in the veins that traverse the *primary and transition mountains* or in particular *strata* (*Erzflöze*) in rocks of *secondary formation*. In the Kingdom of New Spain, most veins also exhibit a small amount of argentiferous galena; but there are only very few mines in which lead ore form the primary target of operations. Among these can be cited only the mines of the districts of Zimapán, Parral, and San Nicolás de Croix. I have observed that, in Guanajuato, as in several other mines of Mexico[1] as well as the whole of Saxony, galena contains more silver when it occurs in smaller grains.

III.166 A substantial amount of silver is produced by the smelting of martial pyrites (*gemeine Schwefel Kiese*), of which New Spain offers varieties that are sometimes even richer than *Glaserz*. At Real del Monte, on the Vizcaína vein near the San Pedro shaft, martial pyrites were found which contained up to three marks of silver per quintal.

1. The galena from the new Talpan mine, in the Cerro de las Vigas, lying within the district of Hostotipaquillo, is among the very richest in silver. This galena, which sometimes verges on *compact* and *antimonial lead sulfide* (*Bleischweif*), is accompanied by several coppery pyrites and by carbonate of lime [calcite].

At Sombrerete, the great abundance of pyrites dispersed throughout the red silver mine seriously hinders the amalgamation procedure.

We have just identified the ore from which Mexican silver comes. It remains for us to examine the *average wealth* of these ore, by considering them all mixed together. It is a widely held belief in Europe that large masses of native silver are extremely common in Mexico and in Peru, and that the mines where mineralized silver is reserved for either amalgamation or smelting generally contain more ounces or marks of silver per quintal than the *poor ore* of Saxony and Hungary. Possessed of this same belief, I was doubly surprised, upon my arrival in the Cordilleras, to find that the number of *poor mines* far exceeds that of the mines that we in Europe designate as *wealthy*. A traveler who visited Mexico's famous *Valenciana* mine after having examined the *metalliferous deposits* of Clausthal, Freiberg, and Schemnitz, would have difficulty understanding how a vein that, for a large part of its *depth*, contains silver sulfide dispersed throughout its *gangue* in nearly imperceptible fragments could regularly produce 30,000 marks per month, that is, an amount of silver equal to half that produced by all the mines of Saxony over the course of one year. III.167

It is an established fact that enormously heavy blocks of *native silver* (*papas de plata*) have been extracted from the Batopilas mines in Mexico and from the Guantahajo mines in Peru. But if one studies closely the history of the principal mines of Europe, one finds that the Kongsberg veins in Norway; those of Scheeberg in Saxony; and the famous *ore masses* of Schlangenberg in Siberia, have revealed masses of much greater size. In general, it is not by the size of the blocks that one can determine the wealth of the mines in different countries. France as a whole produces only 8,000 marks of silver per year, yet there are veins (those of Sainte-Marie-aux-Mines) from which thirty-kilogram masses of native silver have been extracted.

It appears that, in all climes, when the veins were formed, silver was distributed unequally: sometimes concentrated in a single point, sometimes dispersed throughout the *gangue* and alloyed with other metals. Sometimes, in the midst of the poorest ore, one encounters large masses of native silver, a phenomenon that appears to derive from a particular set of chemical affinities whose mode of action and laws are unknown to us. Rather than being concealed within galena or within relatively unargentiferous pyrites, rather than being spread throughout the *mass of the vein*, across a very large area, the silver is concentrated in a single block. The wealth of a particular point can thus be considered the primary cause for the poverty of the surrounding ore, and one can understand, on the basis of this overview, why the richest sections of a vein are separated from one another by stretches III.168

of *gangue* that are nearly devoid of metals. In Mexico, as in Hungary, large masses of native silver and of Glaserz appear only *in nodules*. The composite rocks exhibit the same phenomena as do the vein *masses*. By carefully examining the structure of the granite, one discovers the effects of a particular attraction among the mica, amphibole, and feldspar crystals, a large number of which are accumulated in a single point, while the surrounding parts are almost devoid of them.

Although the New Continent has not, to this point, exhibited blocks of native silver as large as those of the Old Continent, this metal is nevertheless found in a state of absolute purity in greater abundance in Peru and Mexico than anywhere else on the globe. In stating this opinion, I am not consider- III.169 ing the native silver that manifests itself in the form of strips, branches, or cylindrical filaments in the mines of Guantahajo, Potosí, and Hualgayoc, or in those of Batopilas, Zacatecas, and Ramos; I am, rather, basing this statement on the enormous abundance of ore called *pacos* and *colorados*, in which the silver is not *mineralized* but rather dispersed in fragments so tiny that they can only be detected by means of a microscope.

The studies that have been conducted by the director general of the mines of Mexico, Don Fausto Delhuyar, and by several members of the High Council of Mines suggest that by gathering all the silver ore that are extracted each year, one would find that their combined *average wealth* would be 0.0018 to 0.0025 silver; that is, in miners' parlance, *one quintal of ore* (of one hundred pounds, or 1,600 ounces) *contains three to four ounces of silver*. This important result has been confirmed by the account of an inhabitant of Zacatecas, who has directed large metallurgical operations in several mining districts of New Spain and who has just published a very interesting work on American amalgamation. Mr. Garcés,[1] whom we have III.170 already had occasion to cite above, explicitly states that "the vast majority of Mexican ore are so poor that the three million marks of silver that the kingdom produces in good years are extracted from ten million quintals of ore, which is treated partly by smelting and partly by the amalgamation procedure." According to these figures, the average [mean] wealth would amount to only two and two-fifths ounces per quintal, a result that contrasts sharply with the claim of an otherwise estimable traveler,[2] who reports that

1. *Nueva teórica [y práctica] del beneficio de los metales [de oro y plata,]* by *José Garcés y Eguía, Perito facultativo de minas y Primario de beneficios de la minería de Zacatecas* (Mexico City, 1802), pp. 121 and 125.

2. ▼ The Jesuit Och. (Murr's *Nachrichten vom spanischen Amerika*. vol. I, p. 236).

the veins of New Spain are of such extraordinary wealth that the indigenous peoples neglect to mine them when the ore contains less than a third of their weight in silver, or seventy marks per quintal! Since the most erroneous ideas about the *content* of American ore have been spread throughout Europe, I shall provide more detailed information about the mining districts of Guanajuato, Tasco, and Pachuca, which I have visited.

In Guanajuato, the mine belonging to the Count of Valenciana produced, from January 1, 1787, to June 1, 1791, a total of 1,737,052 marks of silver, which were extracted from 84,368 *montones* of ore. In the table[1] presenting the general state of the mine, a *montón* is estimated at thirty-two quintals, or nine and four-hundredths *cargas*, from which it results that the average wealth of the ore twenty years ago was five and one-tenth ounces of silver per quintal. Performing the calculation on the output of the single year 1791 yields nine and three-tenths ounces per quintal. At that time, when the mine was in its most flourishing state, of the total amount of ore, there were: III.171

5/1,000	of	rich ore (*polvillos* and *xabones*) containing per quintal	22 marks	3 ounces
28/1,000	of	rich ore (*apolvillado*)	9	3
152/1,000	of	rich ore (*blanco bueno*)	3	1
815/2,000	of	poor ore (*ganzas, tierras ordinarias*, etc.)		3

The ratio of rich ore to poor ore was around three to fourteen. In 1791, the ore that contained only three ounces per quintal produced (at all times, we are referring only to the Valenciana mine) over 200,000 marks of silver, while enough rich ore existed (between three and twenty-two marks per III.172

1. *Estado de la mina Valenciana remitido por el mano del Ex[celentísi]mo Señor Virey de Nueva España al Secretario de Estado don Antonio Valdés* (manuscript). I have reproduced the figures displayed on this table drawn by the administrator of the Valenciana, Don José Quixano. Furthermore, one *montón* (pulverized mass of ore) is calculated as thirty-five quintals at Guanajuato; thirty quintals at Real del Monte, Pachuca, Sultepeque, and Tasco; twenty quintals at Zacatecas and Sombrerete; eighteen quintals at Fresnillo; and fifteen quintals at Bolaños. At Guanajuato, one *carga* is generally valued at fourteen *arrobas*, such that ten *cargas* equal one montón there (Garcés [y Eguía, *Nueva teórica*,] p. 92). Since the *content of the montón* is used to determine the wealth of the ore, it is extremely important to have an exact knowledge of this measurement when performing metallurgical calculations.

quintal) to yield over 400,000 marks. Today the *average wealth* of the entire Guanajuato vein can be estimated at four ounces of silver per quintal of ore. The vein's southwestern section, which traverses the Rayas mine, nevertheless exhibits ore whose *content* regularly amounts to over three marks.

In the mining district of Pachuca, the output of the Vizcaína vein is divided on the *sorting tables* into three classes, whose respective wealth in 1803 ranged from four to twenty marks for each *montón* of thirty quintals. The first-class ores, which are the richest, contain between eighteen and twenty marks; those of the second class, between seven and ten marks. The poorest mines, which form the third class, are estimated at only four marks of silver per *montón*. This suggests that in the triage, *good* is from four and eight-tenths to five and three-tenths ounces of silver per quintal, *average* is from one and eight-tenths to two and seven-tenths ounces, and the *lowest* is one and three-fiftieths ounces. In the mining district of Tasco, the ore of Tehuilotepec contain twenty-five marks of silver per *tarea* of four *montones*, or one hundred quintals; those of Guautla yield forty-five. Their average wealth is thus between two and three and six-tenths ounces of silver per quintal of ore.

It is not, therefore, as it has too long been believed, by virtue of the intrinsic value of the ore, but rather through the great abundance in which they are found within the earth and the facility of conducting operations that the mines of the Americas are distinct[1] from those of Europe. The three mining districts that we have just mentioned produce over one million marks of silver per year alone, and looking at the combined data, we cannot doubt that the average *content* of Mexican ore amounts, as we reported above, to three to four ounces of silver per quintal. This also suggests that these ore are somewhat wealthier than those of Freiberg, but that they contain much less silver than the ore of Annaberg, of Johann-Georgenstadt, of Marienberg, and of other districts of the *Obergebirge*, in Saxony. From 1789 to 1799, in a regular year, 156,752 quintals were extracted,[2] which yielded 48,952 marks of silver, such that the *average*

III.173

1. The silver ore of Peru do not appear, in general, to be wealthier than those of Mexico. The value of their content is calculated not by *montón* but by *cajón* (crate), which contains twenty-four *cargas*, where each *carga*, in turn, contains ten *arrobas*, or two and a half quintals. In Potosí, the *average wealth* of the ore is fifty-three hundredths of an ounce per quintal; in the Pasco mines, it is one and three-fiftieths of an ounce per quintal.

2. D'Aubuisson, [*Des mines de Freiberg en Saxe*,] vol. 2, p. 128. (The comparison of ore *content* from the two continents presupposes that the concentrations are equal; but given that

content was two and thirty-nine eightieths ounces per quintal of ore. In the *metalliferous deposits* of the *Obergebirge*, on the other hand, the average wealth has amounted to ten, and in especially felicitous periods, even fifteen ounces per quintal. III.174

We have just given an overview of the rocks that contain the principal mines of New Spain; we have just examined at which points, at which latitudes, and at which heights above sea level nature has collected the greatest metallic riches; we have indicated which ore yield the immense amount of silver that flows annually from one continent to the other. It remains for us to provide a few details about the most significant mining operations. We shall limit ourselves to three of these *mine groups* that we have described above, that is, to the central group and to those of Tasco and Vizcaína. Those who are familiar with the state of mining operations in Europe will be struck by the contrast between the great mines of Mexico—for example, those of Valenciana, Rayas, and Tereros—and the mines in Saxony, the Harz, and Hungary that are considered to be very wealthy. If it were possible to relocate the latter mines among the operations of Guanajuato, Catorce, or Real del Monte, both their wealth and the sheer volume of their output would appear, to the inhabitants of the Americas, just as unremarkable as the height of the Pyrenees in comparison to that of the Cordilleras.

The *central group* of the New Spain mines, the wealthiest swath of land on the globe in terms of silver, is located on the same parallel as Bengal, at a latitude where the equinoctial zone verges on the temperate zone. This group encompasses the three mining districts of Guanajuato, Catorce, and Zacatecas, the first of which covers an area of 220 square leagues, the second covers 750, and the third covers 730, if one calculates the surface area using the position of the isolated mines (*realitos*) that are located farthest from the seat of the administrative division. III.175

The *Guanajuato district*, the southernmost of the group, is as remarkable for its natural wealth as for the gigantic works that humans have carried out in the bosom of the mountains. In order to have more exact sense of the position of these mines, we invite the reader to recall what we have stated above[1] in our description of each of the provinces, and to glance at

the *concentration* in Mexico is less perfect, the numerical results that the ore *content* yields are less favorable than they will be once the preparation of the *schlich* is perfected.)

1. Chap. VIII, vol. II, p. 160 [p. 388 in this edition]. I drew up a geological map of the environs of the city of Guanajuato, which appeared in the *Personal Narrative* of my travels in the equinoctial regions of the Americas. This is merely a simple sketch based in part on the

the physical tableau of the central plateau shown on the fourteenth plate of the Mexican atlas attached to the present work.

III.176 In the center of the intendancy of Guanajuato, on the ridge of the Cordillera of Anahuac, rises a group of porphyritic peaks known as the Sierra de Santa Rosa. This group of mountains, partly arid and partly covered in arbutus trees and evergreen oaks, is surrounded by fertile plains that are plowed with care. To the north of the Sierra, the *Llanos* de San Felipe stretch as far as the eye can see; to the south, the plains of Iraputo and of Salamanca offer the delightful spectacle of a rich and populous land. The *Cerro de los Llanitos* and the *Puerto de Santa Rosa* are the highest peaks of this mountain group. Their absolute height is between 2,800 and 2,900 meters; but since the neighboring plains that form part of the great *central plateau* of Mexico are at an elevation of 1,800 meters above sea level, to the eyes of the traveler accustomed to the imposing appearance of the Cordilleras these porphyritic summits seem no more than relatively insignificant hills. The famous Guanajuato vein, which alone has produced an amount of silver equivalent to fourteen hundred million francs since the end of the sixteenth century, runs through the southern slope of the Sierra de Santa Rosa.

Traveling from Salamanca to Burras and Temascatio, one notices a wall of mountains that border the plains as they extend from the southeast to the northwest. The crest of the vein follows this very direction. When one is at III.177 the foot of the Sierra, after having passed by the Jalapita farm, one comes across a narrow ravine, dangerous to cross when the water level peaks, the *Cañada de Marfil*, which leads to the city of Guanajuato. As we have observed above, the population of this city is over seventy thousand souls. In such a wild place, it is surprising to see such large, beautiful buildings in the midst of wretched Indian huts. The home of▼ Colonel Don Diego Rul, one of the owners of the Valenciana mine, would not be out of place in the loveliest streets of Paris or Naples. Its façade displays Ionic columns; its design is both simple and remarkable for its great stylistic purity; the construction of this building, which is nearly uninhabited, cost over eight hundred thousand francs, a considerable sum in a country where the price of both a day's work and materials is quite modest.

The name of Guanajuato was relatively unknown in France, England, and Germany when I undertook my journey, yet the wealth of the mines of this district is certainly greater than that of the *metalliferous deposit* of

---

use of *perpendicular bases*, which were barometrically measured. See above, vol. I, p. 22 [p. 23 in this edition], and my *Recueil d'observations astronomiques*, vol. I, p. 372.

Potosí. Over the last 230 years, the Potosí deposit, discovered in 1545 by the ▼ Indian Diego Hualca, has provided, according to a still-unpublished account,[1] 788,257,512 piastras fuertes or, counting eight and a half piasters per mark, a total of 92,736,294 marks of silver; that is: III.178

From 1556 to 1578	49,011,285 piasters	or	5,766,033 silver marks
From 1579 to 1736	611,399,451	or	71,929,347
From 1737 to 1789	127,847,776	or	15,040,914
	788,258,776 piasters	or	92,736,294 silver marks

In these three periods, the following amounts were thus extracted from the Cerro de Potosí in an average year:

From 1556 to 1578	262,092 silver marks equivalent	to 2,227,782 piasters
From 1579 to 1736	458,148	to 3,994,258
From 1737 to 1789	289,248	to 2,458,606

The output of the Guanajuato vein is almost double that of the Cerro de Potosí. Currently, Guanajuato is the lode that alone provides all the silver of the Guanajuato district mines: *between 500,000 and 600,000 marks of silver and between 1,500 and 1,600 marks of gold* are currently extracted from this vein in a typical year.

In [the table below, I represent] the annual amounts of gold and silver that the Guanajuato mines provided from 1766 to 1803. The metals that are removed from the ore through amalgamation have been distinguished here from those that are obtained through smelting. One mark of gold contains fifty *castellanos*, which are equivalent to 400 *tomines*, or 4,800 *granos*. This table, which is based on official documents,[2] suggests that in a thirty-eight- III.180

1. ▼ *Extrait du livre de compte de la trésorerie royale de Potosí, fait sur les lieux par M. Frédéric Mothès.* (*Razón de los Reales derechos que se han cobrado en las Cajas Reales de la planta que producido el Cerro de Potosí.*) This unpublished paper, which is in my possession, provides the output of Potosí, year by year, from 1558 to 1789. The records of the treasury do not contain any information about the years prior to 1556, although two miners from Porco, Juan de Villaroel and Diego Centeno, had first mined this vein in 1545.

2. *Razón de los Castellanos de oro de ley 22 quilates y marcos de plata, de 12 dineros de los beneficios de azogue y fuego, manifestados en la tesorería principal de Real Hacienda de Guanajuato, desde 1º de enero 1766 hasta 31 de diciembre 1803* (manuscript). One mark of

III.179 Output of the Mining District of Guanajuato

PERIOD	GOLD					
	Removed through Amalgamation			Removed through Smelting		
	castell.	tom	gran	castell.	tom	gran
1766	702	3	9	35,542	4	0
1767	552	0	0	46,325	4	10
1768	0	0	0	40,130	0	0
1769	0	0	0	31,543	0	0
1770	5,361	6	8	46,945	0	0
1771	7,938	3	8	47,980	0	3
1772	7,759	2	2	50,917	3	8
1773	5,135	4	0	35,662	0	0
1774	1,985	5	9	30,835	5	1
1775	6,235	4	8	50,671	7	0
1776	22,527	4	0	81,642	4	4
1777	21,673	6	3	74,481	3	3
1778	23,034	6	8	50,100	6	3
1779	31,115	2	3	50,686	3	5
1780	25,044	0	0	29,123	4	1
1781	30,790	2	6	27,781	0	1
1782	24,645	2	10	15,975	7	8
1783	32,887	3	4	20,830	0	7
1784	28,332	4	10	25,194	3	1
1785	26,823	2	4	20,012	0	5
1786	25,217	0	5	12,275	5	4
1787	21,820	0	2	13,124	5	4
1788	13,160	7	4	10,374	2	9
1789	16,451	5	4	16,927	0	10
1790	21,219	2	2	13,135	4	9
1791	25,654	6	7	23,407	5	0
1792	16,855	3	1	8,434	5	0
1793	28,257	2	10	16,360	1	4
1794	23,090	1	0	7,084	2	1

SILVER					
Removed through Amalgamation		Removed through Smelting			
marks	ounces	marks	ounces	tom	gran
207,412	5	86,407	1	0	0
185,439	2	77,847	3	0	0
194,579	4	87,906	0	1	8
194,628	2	106,444	3	3	11
233,235	6	123,782	0	6	0
299,016	1	120,845	2	5	11
287,160	7	96,412	0	7	0
267,621	7	136,799	4	4	1
243,601	4	98,957	0	3	2
277,589	7	96,727	7	5	5
434,175	7	164,756	1	7	1
452,226	4	169,921	0	1	1
431,850	5	93,152	5	0	5
418,215	2	118,200	5	0	9
338,470	4	138,821	1	1	2
403,772	7	162,184	0	7	0
309,734	1	148,302	4	1	2
430,957	5	113,145	3	2	1
386,861	7	100,319	3	2	1
365,308	2	100,836	5	3	1
316,332	5	96,300	7	6	4
365,038	3	103,223	3	0	3
403,894	3	93,657	1	5	7
487,321	6	137,120	2	4	7
463,807	6	131,318	0	4	8
623,921	5	143,685	5	7	3
541,735	6	93,711	6	4	1
440,581	4	76,035	3	1	8
443,366	3	81,206	3	3	4

(*continued*)

PERIOD	GOLD					
	Removed through Amalgamation			Removed through Smelting		
	castell.	tom	gran	castell.	tom	gran
1795	31,518	1	0	24,441	5	7
1796	43,538	5	6	10,505	7	7
1797	34,454	0	0	13,962	6	3
1798	92,074	6	9	34,393	7	5
1799	67,332	1	4	31,316	6	7
1800	71,791	2	4	24,833	6	9
1801	49,305	0	8	31,579	5	6
1802	46,459	0	4	40,401	1	2
1803	59,772	1	1	17,100	2	8

year period, the Guanajuato mining district produced 165 million piasters' worth of gold and silver, and that, from 1786 to 1803, the output in a regular year was 556,000 marks of silver, which are equivalent to 4,727,000 piasters. Together, all of the veins of Hungary and Transylvania yield only 85,000 marks of silver.

Taking the averages of four periods—three of which cover five years and one of which covers eight—one obtains the following results:

PERIODS	VALUE OF TOTAL GOLD AND SILVER OUTPUT FROM THE GUANAJUATO MINES	SILVER IN AN AVERAGE YEAR	VALUE OF GOLD AND SILVER IN AN AVERAGE YEAR
1766–1775	130,320,503 piasters	342,241 marks	3,032,050 piasters
1776–1785	46,692,863	528,121	4,669,286
1786–1795	48,682,662	560,936	4,868,266
1796–1803	39,306,117	551,319	4,913,265

silver has been valued at eight and a half piasters, and one mark of gold at 136 piasters (one piaster at 5 liv. 5 s).

SILVER					
Removed through Amalgamation		Removed through Smelting			
marks	ounces	marks	ounces	tom	gran
462,444	5	104,652	6	1	0
404,639	2	84,486	7	6	6
592,512	1	114,540	2	6	10
521,888	4	104,048	5	3	3
406,286	5	93,679	4	2	5
397,119	4	109,557	0	7	2
221,590	1	118,860	1	7	0
319,719	0	177,460	1	4	0
659,992	7	84,172	4	7	0

What is the nature of the *metalliferous deposit* that yielded these immense riches, and that can be considered the Potosí of the northern hemisphere? What is the bed of the rock that traverses the Guanajuato vein? These questions are too important for me not to draw up here a geological portrait of such a remarkable land. III.181

The oldest rock known in the Guanajuato district is *Thonschiefer* (clayey schist), which probably rests on the granitic rocks of Zacatecas and the Peñon Blanco.[1] It is either ash gray or grayish-black in color and often shot through[2] with a multitude of small veins of quartz that, at great depths, verges on *Talkschiefer* (talcose schist) and *schistous chlorite*. Should this *Thonschiefer* be regarded as primary formation rock, or do the thin-sheeted layers it contains—which are saturated with carbon—not make it resemble *transition Thonschiefer*? These layers (*hojas de libro*) are most often found[3] close to the surface; they are occasionally visible[4] at great depths. III.182 While digging the main shaft (*tiro general*) of Valenciana, they discovered

1. Sonneschmidt, *[Mineralogische] Beschreibung der [vorzüglichsten]Bergwerks-Reviere von Mexico,* pp. 194 and 292.

2. In the *quebrada* of San Roquito, which is connected to the Acabuca ravine.

3. In the Valenciana mine.

4. In the Mellado, Animas, and Rayas mines.

alternating beds of *syenite*, amphibolic schist (*Hornblendschiefer*), and true *serpentine* that form *sub-layers* in the *Thonschiefer*. This extraordinary phenomenon of syenite alternating with serpentine is also present on the island of Cuba near Regla and Guanabacoa,[1] where serpentine abounds in *glimmering diallage* (*Schillerspath*). The same *Thonschiefer* of Guanajuato, which is found at the *bottom* of the Valenciana mine, reappears on the surface, eight hundred meters above, on the ridge of the Sierra de Santa Rosa; I doubt that it has been found at higher elevations. These strata are oriented very regularly at eight to nine hours on the miner's compass;[2] they are at an III.183 inclination of 45° to 50° to the southwest. This is the direction most of the oldest rocks of Mexico follow.

Upon the *Thonschiefer* rest two very different formations: one, of porphyry, at considerable heights, to the east of the Marfil valley and to the northwest of Valenciana; the other, of *sandstone*, in the ravines and on low-lying plateaus.

The *porphyry* forms gigantic rocky masses that have the strangest appearance from afar, often resembling ruined walls and bastions. These rocks, cut steep and at an elevation of three to four hundred meters above the surrounding plains, bear the name of *bufas* in this country. Enormous balls with concentric layers repose on isolated boulders. These porphyry rocks lend a wild character to the outskirts of the city of Guanajuato, which is likely to surprise the European traveler who imagines that nature has de-

1. *Relation historique* (quarto edition), vol. III, p. 368.

2. Or from the southeast to the northwest. Since 1791, I have been struck by the great law of the parallelism of layers, which one encounters in immense stretches of terrain, and which can be regarded as one of the most interesting geological phenomena. In my writings, I have ceaselessly called travelers' attention to a subject on which it would be easy to collect a large number of observations in a very short time. See my *Expériences sur l'irritation de la fibre musculaire et nerveuse*, vol. I, p. 8; [Savaresi,] *Lettre de M. Savaresi à M. de Fourcroy, en date du 5 pluviose an VI* [1791]; the *Tableau géologique de l'Amérique méridionale* (*Journal de physique*, 1800); and *Géographie des plantes*, p. 117. The direction of the high mountain chains appears to exert the strongest influence on the direction of the layers, even at considerable distances from the central crest. This influence manifests itself in the Pyrenees, in Mexico, and, especially, in the High Alps. (I recently gathered in my *Essai géognostique sur le gisement des roches*, pp. 54–60, everything related to parallelism or, rather, to the *loxodromism* of the layers in both hemispheres.) See also ▼ Ebel, *Über den Bau der Erde im Alpengebirge*, vol. I, p. 220, vol. II, p. 201–15, and p. 357; Steininger, *Über die erloschenen Vulkane*, p. 3; Eschwege, *Geognost[ische] Gemälde von Brasilien*, p. 6; Riepl, *Darstellung der Eisenerz Gebilde*, p. 55; Franklin, *[Narrative of a] Journey to [the Shores of] the Polar Sea*, p. 534.

posited great metallic riches only within round-topped mountains and in places where the terrain has a gentle, uniform appearance. This porphyry makes up the largest part of the *Sierra de Santa Rosa*; it generally has a greenish tint and varies according to the nature of its base, its texture, and the crystals it contains. The oldest layers appear to be those with a base of compact feldspar, which has long been mistakenly referred to as *Hornstein-Porphyr.*[1] The newer layers, on the contrary, display vitreous feldspar, embedded within a rock mass that verges on jade felsite at some points, at others on Werner's *phonolith*, or *Klingstein*. The latter is most similar to the *Porphyrschiefer* (porphyritic schist) of *Mittelgebirge* in Bohemia. One would be tempted to place them among the rocks of *trappean formation*, if these very layers did not, at Villalpando, contain the richest gold mines. What all of the porphyry rocks of the Guanajuato district have in common is that amphibole is as scarce in them as are quartz and mica. The *direction* and *inclination* of their layers correspond to those of the *Thonschiefer*.

III.184

On the southern slope of the Sierra, at generally lower elevations than the porphyry that lies within the plains of Burras and Cuevas, particularly between Marfil, Guanajuato, and Valenciana, the *Thonschiefer* is covered in *sandstone* of old secondary formation. This sandstone is a breccia with a clayey cement blended with iron oxide, in which are embedded some *jagged fragments* of quartz, Lydian stone, syenite, porphyry, and flaky *Hornstein*. Layers containing fragments six to eight centimeters thick occasionally (near Cuevas) alternate with other layers in which grains of quartz are agglutinated by ochreous cement. At other points (in the Marfil ravine and the Salgado road), the cement becomes so abundant that the embedded pieces disappear entirely, and one encounters beds of yellowish-brown schistous clay eight to nine meters thick alternating with large-pebble breccia. This sandstone formation resembles more the *red* or *coaly sandstone* of Europe than the fragmentary rocks of transition formation,[2] which, in Switzerland,

III.185

1. Porphyry with a base of hornstein [hornstone] or neopetre [secondary petrosilex].

2. The sandstone of transition formation peculiar to both transition schist and transition limestone is the *old red sandstone* (of Mitcheldean in Hertfordshire) of the British geognosts. The red or coaly sandstone (*Totes Liegende*) of the German and French geognosts is the *new red conglomerate* (of Exeter and Teignmouth) of the British, positioned between the *magnesian limestone* (*Zechstein*, *Alpenkalk*) and the transition rocks. In Britain and Belgium, the coaly terrain (*coal measures*) is more closely linked to the transition rocks than to the secondary formations. In northern Germany, on the contrary, the vast majority of the coal occurs in the red sandstone (*Totes Liegende*) and quartziferous porphyry terrain. Through the elimination of a limestone formation positioned between red sandstone (*new*

rises to an absolute elevation of over one thousand meters in Oltenhorn and III.186 the Diablerets; it displays little regularity in the direction of its layers. Their inclination is generally the opposite of that of the *Thonschiefer strata*. Near Guanajuato, the sandstone formation backs onto the porphyry of the Bufa; but near Villalpando, porphyry serves as a base for the sandstone that appears *on the surface* at an absolute elevation of 2,600 meters.

Breccia, within which are embedded fragments of primary and transition rock, should not be confused with another sandstone that might almost be referred to as a *feldspathic agglomerate*, and that appears to be of more recent formation. This agglomerate (*lozero*), from which the most beautiful cut stones are extracted, is reddish-white, sometimes apple-green; like slabbed sandstone (*Leuben* or *Waldplattenstein* of Suhl), it is divided into very thin sheets; it contains grains of quartz, small fragments of *Thonschiefer*, and large numbers of feldspar crystals, some broken, others intact. In the *lozero* of Guanajuato, these various substances held together by an argilloferruginous cement (Cañada de Serena and almost the entire mountain of the same name), as they are in the porphyritic-looking rock of Suhl. It is likely that the destruction of the porphyry had the strongest influence on the formation of the feldspathic sandstone of Guanajuato. The most seasoned mineralogist would be inclined to take it, at first glance, for either clay-based porphyry or porphyritic breccia. Around Valenciana, the *lozero* III.187 forms masses two hundred toises thick; their elevation surpasses that of the mountains formed by intermediary porphyry. Near Villalpando, two-foot-thick layers of feldspathic agglomerate with very small grains alternate twenty-eight times with blackish-brown schistous clay. I saw either this agglomerate (or *lozero*) resting on red sandstone everywhere, and on the southwestern slope of the Cerro de Serena, descending toward the Rayas mine, it seemed even quite obvious to me that the *lozero* forms a layer within the coarse conglomerate of Marfil. I doubt, therefore, that this remarkable formation could belong to the *pumiceous trachytic conglomerates*, which is where Mr. Beudant appears to place it on the basis of its similarity to some rocks of Hungary. The clayey cement often becomes so abundant that the embedded pieces are barely visible and the rock mass verges on compact argillite (*Thonstein*). In this condition, the *lozero* provides the beautiful cut

---

red conglomerate) and the multicolored sandstone (*new red sandstone* or *red marl*), a single mass of secondary sandstone occasionally exhibits (for example, in southern Germany and in some parts of New Granada) both red sandstone (*Totes Liegende*) and multicolored sandstone (*bunter Sandstein*).

stone of Querétaro (the Caretas and Guimilpa quarries), which is highly sought-after for construction. I have seen columns made of it that were fourteen feet high and two and a half feet in diameter, flesh-colored, or brick- or peach-flower-red. These lovely colors turn gray upon contact with the air, most likely due to the atmosphere's effect on the dendriform manganese in the rock fissures. The fracture in the columns from Querétaro is even, like that of the lithographic stone of the Jura. It is with difficulty that one finds within this argillite a few miniscule fragments of Thonschiefer, quartz, feldspar, and mica. It is probable that the unbroken crystals of the *lozero* (or feldspathic sandstone) did not develop within the rock mass itself, but, rather, that they are there by accident. I shall merely recall here that the red sandstone of Europe and its porphyry are also sometimes characterized by a *localized suppression* of crystals and embedded fragments. The *lozero* appears to me to be a superposed sandstone formation, perhaps even subordinate to the red or coaly sandstone; and if the Old Continent does not offer us an identical rock, we nevertheless see the embryonic form of this kind of pseudo-porphyritic structure in the sandstone beds bearing feldspar crystals, either broken or intact, that are occasionally embedded within the large red sandstone formation of the Munsfeld and the Thüringerwald.[1] These phenomena all seem to confirm the theory of a great geologist,[2] according to which red sandstone is produced by the upheaval of the porphyry itself.

III.188

The *sandstone* formations of Guanajuato serve as a base for other secondary formations, which exhibit the greatest similarity to the secondary rocks of central Europe in their *depositing*, that is, in the *order of their superposition*. In the plains of Temascatio (at *lo de Sierra*), one sees a compact limestone (*dichter Kalkstein*), often filled with bubble-like cavities lined with calcareous spar and manganese ore, either earth-colored or glowing. In a few spots, this limestone—the *even*, almost *conchoidal fracture* of which makes it resemble *Jurassic formation*—is covered, by beds of fibrous *gypsum* mixed with hardened clay.

III.189

We have just detailed the numerous rocks that rest on the *Thonschiefer* of Guanajuato and that are secondary formations of sandstone, calcareous rock, and gypsum on the one hand and formations of porphyry, syenite, serpentine, and amphibolic schist on the other. The ravine of Marfil, which leads from the plains of Burras to the city of Guanajuato, separates, so to

1. ▼ Freiesleben, *Über das Kuphenchiefergebirge*, vol. IV, pp. 82, 85, 95, and 194.

2. Mr. Léopold de Buch (see *Geogn[ostische] Briefe [an Herrn Alexander von Humboldt] über das südl[iche] Tyrol*, 1824, p. 75.)

speak, the porphyritic region from the region in which *syenite* and *Grünstein* [greenstone] predominate. To the east of the ravine rise very steep porphyry mountains; their rifts form the most fantastic shapes: to the west, one comes across a terrain with a slightly undulating surface covered in basaltic cones.

From the Esperanza mine, located to the northwest of Guanajuato, to the village of Comangillas, famous for its thermal springs, over an area of more than twenty square leagues, the *Thonschiefer* of Guanajuato serves as a base for syenite layers that alternate with *transition Grünstein* (*diabase* or

III.190

diorite). These layers are generally only four to five decimeters thick; they are inclined in groups either toward the northeast or toward the west and always at angles of fifty to sixty degrees. Traveling from Valenciana to Ovejeras, one can count several thousand of these *Grünstein* beds, alternating with syenite in which quartz is occasionally more abundant than feldspar and amphibole. Within this syenite, one encounters veins of Grünstein, and within the *Grünstein*, crevices filled with syenite. This correspondence between the *deposits* in the veins and the superposed rocks is an interesting phenomenon that might support Mr. Werner's theory of the origin of veins.[1] Near Chichimequillo, columnar porphyry appears to be resting on syenite. It is covered in basalt and basaltic breccia through which trickle springs with a temperature of 96.3 on the Centigrade thermometer.

It remains for me to mention two partial formations that occupy only a very small area: a *compact calcareous* stone (*el caliche*), grayish-black, perhaps belonging to the *transition rocks*,[2] and a calcareous breccia (*frijollilo*). The latter, which I saw in the Animas mine at a depth of over 150 meters, is composed of rounded fragments of compact calcareous stone joined together by calcareous cement. The *Thonschiefer* of Valenciana serves as a

III.191

base for these two partial formations, one of which appears to have originated from the destruction of the other.

According to the observations that I made on site, this is the geological *constitution* of the land at Guanajuato. The vein (*veta madre*) runs through both the clayey schist (*Thonschiefer*) and the porphyry. There are very significant metallic riches in both of these rocks. Its mean direction is h. 8 4/8 on the miner's compass, roughly the same as that of the *veta grande* of Zacatecas and the veins of Tasco and Moran, which are all western veins

1. [Werner,] *Neue Theorie der Entstehung der Gänge*, 1791, p. 60.

2. Between the Secho and Acabuca ravines, the *caliche* beds have both the same direction and the same inclination as the strata of Thonschiefer.

(*Spathgänge*). The inclination of the Guanajuato vein is 45° to 48° toward the southwest. We have already indicated above that it has been worked over a length of more than 12,000 meters; nevertheless, the enormous mass of silver that it has yielded over the past two hundred years, which alone would have sufficed to bring about a change in the price of commodities in Europe, has been extracted exclusively from the section of the vein between the Esperanza and Santa Anita shafts, across a stretch of under 2,600 meters. This section contains the mines of Valenciana, Tepeyac, Cata, San Lorenzo, Animas, Mellado, Fraustros, Rayas, and Santa Anita, all of which have enjoyed widespread celebrity at various points.

III.192

The *veta madre* of Guanajuato bears a great deal of resemblance to the famous *spital* vein of Schemnitz in Hungary. European miners who have had occasion to examine both of these ore *deposits* have debated the question of whether they should be considered true *veins* or *metalliferous layers* (*Erzlager*). If one observed the *veta madre* of Guanajuato only in the Valenciana or Rayas mines, where the *roof* and the *wall* are both of *Thonschiefer*, one would be inclined to accept the latter theory, for, far from *cutting through* or crossing the rock *strata* (*Quergestein*), the *veta*'s direction and inclination are exactly the same as those of its *strata*. But is it possible for a *metalliferous layer* that was formed in the same period as the entire rock mass of the mountain within which it is found to turn from an upper rock into a lower rock, from porphyry to clayey schist? If the *veta madre* really were a *layer*, jagged fragments from its *roof* would not be contained within its mass, as is commonly observed in those sites where the roof is made of *carbon*-bearing *schist* and the wall of *talcose schist*. Within a vein, the *roof* and the *wall* are supposed to predate both the formation of the *crack* and the minerals that subsequently filled it in, whereas a *layer* undoubtedly existed before the rock *strata* that form its *roof*. It is possible, then, to encounter *wall* fragments within a layer, but never pieces detached from the *roof*.

III.193

The *veta madre* of Guanajuato offers the extraordinary example[1] of a crack that formed following both the direction and inclination of the rock strata; toward the southeast, from the Serena ravine, or from the scarcely worked mines of Belgrado and San Bruno, past the Marisanchez mines, it *runs through* porphyritic mountains; to the northwest, between the

1. Mr. Werner, in his Theory of veins (§2), expressly states that "ore deposits *almost always* cut through rock banks" [*Neue Theorie von der Entstehung der Gänge*]. This great mineralogist appears to have meant that it is possible for true veins to exist that are parallel to sheets of clayey or micaceous schist.

Guanajuato shafts and the Cerro de Buenavista and the Cañada de la Virgen, it traverses the *Thonschiefer*, or clayey schist. Its *thickness* varies, like that of all the veins of Europe: when it is not *divided into branches*, it is commonly no more than twelve to fifteen meters wide. Sometimes it *narrows*,[1] even up to one demi-meter. Most often, it is split into three masses or bodies (*cuerpos*) which are separated either by rock beds (*caballos*) or by portions of the gangue nearly devoid of metals. In the Valenciana mine, the *veta madre* was found to be *unbranched* and seven meters wide, from the surface of the III.194 earth to a depth of 170 meters. At that point, it splits into three branches, and its *thickness*, from the *wall* to the *roof* of the *entire mass*, is fifty, sometimes even sixty meters. Of these three *branches* of the vein, there is generally only one that is rich in metals: sometimes, when all three *join together* and *drop*, as at Valenciana near the San Antonio shaft, to a depth of three hundred meters, the vein contains tremendous riches within a *thickness* of over twenty-five meters. In the *pertinencia de Santa Leocadia*, one sees four branches. One *trum*, with an inclination of 65°, separates from the lower branch (*cuerpo bajo*) and cuts through the rock sheets of the wall. Both this phenomenon and the large number of *druses* lined with amethyst crystals that one encounters in the Rayas mines and that go in the most divergent directions, would suffice to prove that the *veta madre* is a *vein* and not a *layer*. Additional proof, no less convincing, could be found in the existence of a vein (*veta del caliche*) mined in the calcareous stone of Animas and parallel to the primary vein of Guanajuato, which has the same silver ore. Does one ever find such *similarity of formation* between two metalliferous *layers* that belong to rocks of very different *ages*?

The small ravines into which the Marfil valley is divided appear to have a formative influence on the wealth of the *veta madre* of Guanajuato. The latter III.195 yielded the most ore wherever the direction of the ravines[2] and the slope of the mountains (*flaqueza del cerro*) were parallel to the direction and inclination of the vein itself. Standing on the hill of Mellado near the shaft that was dug in 1558, one sees that the *veta madre* is generally more abundant in ore toward the northwest, toward the Cata and Valencia mines; and that, to the southeast, toward Rayas and Santa Anita, the output has been at once richer,

1. At the *loading point* of the *Santo Cristo de Burgos* shaft, in the Valenciana mine. (Mr. Alamán thinks that this narrowing does not occur, and that I have been misled by the reports of a few miners.) See the interesting *Report of the United Mexican Mining Association*, 1825, p. 10 [by Lucas Alamán].

2. Those of Acabuca, Rayas, and Secho.

scarcer, and more inconsistent. Additionally, there is in this famous vein a certain *middle region* that can be regarded as a storehouse of great riches; for, both above and below this region, the ore has relatively insignificant *silver content*. At Valenciana, rich ore has been more plentiful between the depths of 100 and 340 meters below the mouth of the gallery. At Rayas, this abundance is evident on the very surface of the soil; according to my measurements,[1] however, the gallery of Valenciana was bored at a level that is 156 meters higher than the mouth of the gallery of efflux of Rayas, which could lead one to believe that the store of Guanajuato's great riches is located, in this part of the vein, between 2,130 and 1,890 meters of absolute elevation above sea level. The deepest *mining operations* of the Rayas mine (*los planes*) have not yet reached the lower limit of this middle region, whereas the *bottom* (*das Tiefste*) of the Valenciana mine, the San Bernardo gallery, has unfortunately already surpassed this perimeter by over seventy meters. The Rayas mine therefore continues to yield extremely rich ore, while at Valenciana, they have sought for a few years now to extract a larger amount of ore in order to compensate for the ore's low intrinsic value. III.196

The mineral substances that constitute the mass of the Guanajuato vein are *common quartz*, amethyst, *carbonate of lime* [calcite], pearly spar, flaky Hornstein, *silver sulfide*, branched *native silver*, prismatic black silver, dark red silver, native gold, *argentiferous galena*, brown blende, spathic iron, and both *copper* and *iron pyrite*. One also finds, although much less often, crystallized feldspar (what Mexican mineralogists call rhomboidal quartz), chalcedony, small masses of fluorite, filamentous quartz (*haarförmiger Quarz*), Fahlerz, and bacillary lead carbonate. The absence of both baryte sulfate and muriatic silver distinguishes the *formation* of the Guanajuato vein from that of Sombrerete, Catorce, Fresnillo, and Zacatecas. In the Guanajuato mines, mineralogists who study regular forms encounter a wide variety of crystals, especially among the silver sulfide and red and black silver ore, among the calcareous spar and the *Braunspath*[2] (browning lime carbonate). III.197

The abundance of water that filters through cracks in the rock and in the gangue varies radically among the different points on the vein. The

1. See my *Recueil d'observations astronomiques*, vol. 324, notes 332–57.

2. On the pearly spar of Guanajuato, see Klaproth's *Beiträge*, vol. IV, p. 198. This variety of *Braunspath* presents microscopic crystals, overlapping and gathered into very thin sticks. The intertwining of these sticks (*parrillas*) is so regular that they consistently form equilateral triangles.

Animas and Valenciana mines are completely dry, although the mining works of the latter occupy a horizontal expanse of one thousand five hundred meters and a perpendicular depth of five hundred meters. Between these two mines, inside which the miner is plagued by both dust and extreme heat,[1] are located the Cata and Tepeyac mines, which have remained flooded due to a lack of mechanical means for drawing out the water. At Rayas, the most extravagant methods are used to dry out the mines: *mule-driven winches* placed inside the passages draw up the water not through pumps but through chains of very shabbily built caissons. It is surprising to see that mines of such tremendous wealth do not have III.198 any *galleries of efflux*,[2] while the nearby ravines of Cata and Marfil and the plains of Temascatio, which are lower than the *bottom* of Valenciana, would seem to invite the miners to undertake works that would serve not only to release the water but also to *haul* ore to the smelting and amalgamation plants.[3]

The *Valenciana* is one of the rare examples of a mine which, for the past forty years, has never provided its owners with an annual profit of less than two to three million francs. It appears that the section of the Guanajuato vein that stretches from Tepeyac to the northwest was only barely mined toward the end of the sixteenth century. After that time, that entire area had remained deserted, and it was only in 1760 that a Spaniard, who had come to the Americas at a very young age, first worked the vein at one of the points that had previously been believed to be devoid of metals (*emborascado*). Mr. Obregón [y Alcocer] (the name of the Spaniard) had no fortune, but since he enjoyed the reputation of III.199 a gentleman, he found friends who, from time to time, advanced small sums to him so that he might continue his work. In 1766, the mining operations had already reached a depth of over eighty meters, yet the costs still eclipsed the value of the metallic output. As passionate about mines

1. Between twenty-two and twenty-seven degrees Centigrade, the outside air temperature being seventeen degrees.

2. In the mining district of Freiberg, which yields only one-seventh of the total amount of silver extracted from the Valenciana mine alone, they have nevertheless managed to bore two galleries of efflux, one of which is 63,213 meters (32,433 toises) long, while the other is 57,310 meters (29,504 toises) long. (See below, vol. III, p. 142 [p. II.79 in this edition].)

3. Mr. Alamán, whose opinions are based on a vast knowledge of the area, thinks that digging a *common gallery* would not only be too costly an undertaking but also impossible due to the lack of unity among the owners of the various mines.

as others are about gambling, Mr. Obregón [y Alcocer] preferred to make all kinds of sacrifices rather than abandon his venture. In 1767, he went into business with a minor trader from Rayas named Otero. Could he have imagined at the time that, within a few years, he and his friend would be the wealthiest private individuals in Mexico and, perhaps, in the entire world? In 1768, they began extracting quite a considerable amount of ore from the Valenciana mine. As the shaft grew deeper and deeper, they approached the region that we have described above as the storehouse of the greatest metallic riches of Guanajuato. In 1771, they extracted from the *pertinencia de Dolores* enormous masses of silver sulfide blended with native silver and red silver. From that time to 1804, when I left New Spain, the Valenciana mine continued to yield an annual profit of over fourteen million livres tournois. There were years so productive that the net profit of the two mine owners rose to a sum of six million francs.

In the midst of his immense wealth, Mr. Obregón [y Alcocer], better known as the count of Valenciana, retained the same simple habits and frank character that had distinguished him in less fortunate times. When he first worked the Guanajuato vein above the San Javier ravine, there were goats grazing on the very hill where, ten years later, he witnessed a city of seven to eight thousand inhabitants grow. Since the death of the old count and of his friend, ▼ Don Pedro Luciano Otero, the ownership of the mine has been shared by several families.[1] At Guanajuato, I met two miner sons of Mr. Otero, each of whom possessed capital in the amount of six and a half million francs in cash, excluding the annual revenue of the mine, which amounted to over four hundred thousand francs.

III.200

One should be all the more astonished by the consistency and uniformity of the output of the Valenciana mine when one considers that the wealthy mines' abundance has significantly decreased, and that the mining costs have increased at an alarming rate since operations reached a perpendicular depth of five hundred meters. Boring and walling the three old *extraction shafts* cost the elderly count of Valenciana nearly six million francs, that is to say:

1. Ownership of Valenciana is divided into twenty-four shares, called *bars*, ten of which belong to the descendents of the count of Valenciana, twelve to Otero's family, and two to the ▼ Santana family.

San Antonio's square shafts, or *tiro viejo*, which have a perpendicular depth of 227 meters and have *four horse-driven winches*	396,000 piasters
The square shafts of Santo Cristo de Burgos, which are 150 meters deep, and have two *horse-driven winches*	95,000
The hexagonal shafts of Nuestra Señora de Guadalupe (*tiro nuevo*), which have a perpendicular depth of 345 meters and have six *horse-driven winches*	700,000
Total costs of the three shafts	1,191,000

III.201 Twelve years ago, they began digging, right in the middle of the rock, a new *extraction shaft* (*tiro general*) that will have the enormous perpendicular depth of 514 meters[1] and will lead to the current bottom of the mine, the *planes de San Bernardo*. This shaft, which will be located near the center of operations, will significantly decrease the number of the nine hundred miners (*tenateros*) employed as beasts of burden for carrying the ore to the III.202 upper *loading points*. The *tiro general* will cost over one million piasters: it is octagonal, with a circumference of twenty-eight and eight-tenths meters. Its walling is stunningly beautiful. During my stay in Guanajuato, people believed that the shaft could reach the vein by 1815, even though, in September of 1803, it was still only 184 meters deep. The boring of this shaft[2] is one of the greatest and boldest undertakings in the history of mining. It is, nevertheless, still questionable whether, to decrease the costs of hauling and extraction, it was indeed useful to turn to a solution that is not only slow but also costly and uncertain.

In a typical year, the operational costs of the Valenciana mine were:

From 1787 to 1791: 410,000 piasters
From 1794 to 1802: 890,000 piasters

1. I am converting the *Mexican varas* using the rule that one *vara* is equivalent to 0.0385 meters and that one toise is equivalent to 2.331 varas. The locals regard the Valenciana mine as the deepest mine dug by humans. In the same period in which I measured the *planes* of San Bernardo, the *Berchert Glück* mine, in Freiberg, Saxony, had reached a perpendicular depth of 447 meters. It is believed that in the sixteenth century, the works of the Saxon miners on the Alter Thurmhof vein extended to a depth of 545 meters.

2. This great work has fortunately been completed (*Notes sur Mexico*, p. 158).

III.203

PERIODS										TOTAL FOR NINE YEARS
	1794	1795	1796	1797	1798	1799	1800	1801	1802	
Income from the sale of ore from Valenciana (in piasters)	1,282,042	1,696,640	1,315,424	2,126,439	1,724,437	1,584,393	1,480,933	1,393,438	1,229,631	13,835,380
Mining costs (in piasters)	799,328	815,817	832,347	878,789	890,735	915,438	977,314	991,981	944,309	8,046,063
Net profit, which was divided among the sharehold-ers (in piasters)	482,713	880,822	483,077	1,249,650	835,702	668,954	503,619	401,456	285,321	5,791,317

Although the costs doubled, the shareholders' profits stayed more or less the same. The following table shows the state[1] of the mine over the last nine years.

This table [the table on the previous page] suggests that in recent years, the shareholders' net profit has been 640,000 piasters, in a typical year.[2] In 1802, circumstances were particularly unfavorable: the majority of the ore were very poor, and their extraction spectacularly costly; in addition, the output fetched very low prices, because the lack of mercury hindered amalgamation, and because all of the mines were saturated with ore. The year 1803 promised more benefits to the owners, and a net profit of over half a million piasters was expected. In Valenciana, I saw 27,000 piasters' worth of silver ore being sold *per week*: the costs amounted to 17,000 piasters. At Rayas, the owners made a larger profit, although the production was smaller: that mine yielded 15,000 piasters' worth of ore per week, while the operational costs amounted to only 4,000 piasters, the effect of the wealth of the ore, of their concentration within the vein, of the relative shallowness of the mine, and of a less costly *draft*.

III.204

To get a sense of the enormous advances that working the Valenciana mine requires, it suffices to recall here that in its current state, it is necessary to account, on a yearly basis, for

	3,400,000 francs {	for daily wages of miners, sorters, masons, and other mine workers.
	1,100,000 francs {	for powder, tallow, wood, leather, steel, and other materials needed at the mine.
Total expenses	4,500,000 francs	

The total consumption of powder alone was four hundred thousand francs per year; that of the steel used in the manufacturing of *stone chisels* and *drill rods* was 150,000 thousand francs. The number of workers who

1. *Estado que manifesta el valor de los frutos que ha producido la mina de Valenciana, costa de sus memorias y líquido producto, a favor de sus dueños; lo presentó Don José Antonio del Maso al Excelentísimo Señor Virey de la Nueva-España, Don José de Iturigarray, el 3 de julio 1802* (unpublished)

2. Over 3,360,000 francs. The profit divided yearly among shareholders in the district of Freiberg (the net profit of the mine owners) amounts to only 250,000 francs.

labor inside the mines is 1,800: if one adds 1,300 persons (men, women, and children) who work at the *horse-driven winches*, in the hauling of ore, and at the *sorting* tables, one finds that 3,100 individuals are employed in the different activities of the mine. The direction of the mine is entrusted to an administrator, who earns a salary of 60,000 francs and through whose hands pass over six million francs each year. Working under this administrator, who is not supervised by anyone, are an *Obersteiger* (*minero*), three *Untersteiger* (*sottomineros*), and nine *master-miners* (*mandones*). These managers visit the underground works daily, carried by men[1] who have a kind of saddle attached to their back, and who are referred to as *little horses* (*caballitos*).

III.205

We shall conclude this summary of the Valenciana mine by presenting, in a comparative table, the state of this Mexican mining operation and that of the famous mine of *Himmelsfürst*[2] located in the Freiberg district. I think I can reasonably claim that this table will attract the attention of those who consider mine administration an important subject of political economy.

Comparative Tableau of the Mines of the Americas and Europe

III.206

TYPICAL YEAR (at the end of the eighteenth century)	AMERICAS Valenciana Mine, the wealthiest MINE IN MEXICO in 1803. (Elevation of the surface ground above sea level, 2,320 meters.)	EUROPE Himmelsfürst Mine, the wealthiest MINE IN SAXONY. (Elevation of the surface ground above sea level, 410 meters.)
Metallic output	360,000 silver marks 10,000 marks of silver	10,000 silver marks
Costs and expenses of the mine (sum total)	5,000,000 francs	240,000 francs
Shareholders' net profit	3,000,000 francs	90,000 francs
Silver contained within one quintal of ore	4 ounces of silver	6 to 7 ounces of silver

(*continued*)

1. On the extraordinary custom of traveling on *man-back*, see my *Vues de las Cordilleras*, plate 5.

2. Everything in the following table that is related to this mine, which I had the opportunity to visit often in 1791, is taken from the work of Mr. D'Aubuisson, [*Des mines de Freiberg en Saxe*], vol. 3, pp. 6–45.

Comparative Tableau of the Mines of the Americas and Europe

Number of workers	3,100 Indians and Mestizos, 1,800 of whom work inside the mine	700 miners, 550 of whom work inside the mine
Miner's daily wages	5 to 6 francs	18 sols
Powder expenditure	400,000 francs (nearly 1,600 quintals)	27,000 francs (nearly 270 quintals)
Amount of ore subjected to smelting and amalgamation	720,000 quintals	14,000 quintals
Veins	One vein, often divided into three *branches* 40 to 50 meters thick (in the *Thonschiefer*)	Five main veins two to three decimeters thick (in the *gneiss*)
Water	no water (in 1803)	Eight cubic feet per minute. Two hydraulic wheels
Depth of the mine	514 meters	330 meters

III.207 In 1803, there were 5,000 miners and workers employed in sorting, smelting, and amalgamation in the entire mining district of Guanajuato; 1,896 *arastres*, or machines for crushing the ore into powder, and 14,618 mules employed in driving the winches and in *milling* the mixture of pulverized ore and mercury in the amalgamation plants. When ore is plentiful, the *arrastres* of the city of Guanajuato grind 11,370 quintals of ore per day. If one recalls that their annual silver output is between five and six hundred thousand marks, this piece of data further demonstrates that the *average [mean] content* of the ore is extremely low.

[To conclude the description of the Real de Minas de Guanajuato, I shall record here a few more general views on the nature and the age of the formations. Having only stayed for a month in this place, doubly interesting for the wealth of its ore deposit and for the geognostic makeup of its mountains, several of my initial insights will need to be rectified by the skilled miners who are today gathered at Valenciana, Rayas, and Villalpando.

The famous vein of Guanajuato, which yielded 556,000 marks of silver in a typical year between 1786 and 1803, runs through a mass of transition schist, or *thonschiefer.* In its lower strata, in the Valenciana mine (at a height of 932 toises above sea level) this rock verges on talcose schist, and I have described it in the first edition of this work as being placed on the border of primary and intermediary terrains [land masses]. A more in-depth examination of the relative positions of the deposits that I noted on site, a comparison of the syenite III.208 beds, which were pierced through when the *tiro general* of Valenciana was being dug, with the beds intercalated within the transition terrains of Saxony, of the Bocchetta [Pass] near Genoa, and of the Cotentin [Peninsula], have since made me absolutely certain that the *thonschiefer* of Guanajuato belong to the oldest intermediary formations. We do not know whether its stratification is both *parallel* and *concordant* with that of the gneiss granite of Zacatecas and Peñon Blanco, by which it is probably supported; for the contact between these formations has not been observed; but, on the great plateau of Mexico, nearly all the porphyritic rocks follow the general direction of the mountain chain (40° north–50° west). Such perfect concordance (*Gleichförmigkeit der Lagerung*) can be observed between the primary gneiss and the transition thonschiefer of Saxony (Friedrichswalde, the Müglitz, Seidewitz, Lockwitz valleys); it demonstrates that the formation of the last layers of primary terrain was immediately followed by the formation of the intermediary terrain. As Mr. de Charpentier has observed, in the Pyrenees the primary terrain has a different (unparallel) deposition, and sometimes even a *transgressive* [*nonconforming*] deposition (*übergreifende Lagerung*) in relation to the intermediary terrain. I shall take this opportunity to recall that the parallelism between the stratification of the two consecutive formations, or the absence thereof, does not alone determine whether the two formations should be joined or not joined together in a single primary, secondary, or tertiary terrain: rather, the question is determined by the combination of all of the geognostic relationships. The transition thonschiefer (Übergangs-Thonschiefer) of Guanajuato is very regularly stratified (direction N 46° W; inclination 45° to the southwest), and the shape of the valleys has no influence over the direction and inclination of the strata. Three varieties can be distinguished, which could be designated as *three formation periods*: a silvery and steatitic [soapstone-like] *thonschiefer* verging on talcose schist (*Talkschiefer*); III.209 a greenish thonschiefer with a silky sheen, resembling chloritic schist; finally, a black thonschiefer with very thin sheets, oversaturated with carbon, which stains the fingers like ampelite [black carbonaceous or bituminous shale] and the marly schist of zechstein but produces no effervescence upon contact with acid. The order in which I have listed these varieties is that in which I observed them, from bottom to top, in the Valenciana mine, up to a perpendicular depth of 263 toises; but in the mines of Mellado, Animas, and Rayas, the over-carburized *thonschiefer* (*hoja de libro*) is covered over by the green and steatitic variety, and it is probable that strata verging on talcose schist, on chlorite, and on ampelite, alternate several times with one another.

This transition *thonschiefer* formation, which I also encountered in the Santa Rosa Mountain near Los Joares, where the Indians collect ice in small man-made basins, is over three thousand feet thick. It contains, in subordinate layers, not only syenite (like the transition *thonschiefer* of the Cotentin Peninsula) but also—what is quite remarkable—both serpentine and an amphibolic schist that is not diorite (*grünstein*). While the main *extraction shaft of Valenciana* was being dug—in the middle of the rock and into the roof of the vein—the following strata were found, from top to bottom, over a depth of ninety-four toises: ancient conglomerate, likely representing red sandstone (*totes liegende*); black transition *thonschiefer*, highly carburized, with very thin sheets; bluish-gray *thonschiefer*, magnesiferrous and talcose; amphibolic schist, greenish-black, lightly blended with quartz and pyrite, devoid of feldspar, not verging on *grünstein* and entirely similar to the amphibolic schist (*Hornblend-schiefer*) that forms layers in the primary gneiss and mica schist; green serpentine of prase verging on olive-green, very fine-grained, and with an uneven fracture, olive-colored on the inside but III.210 bright in the fissures, filled with pyrite, devoid of garnets and metalloid diallage (*Schillerspath*), mixed with talcum and steatite; amphibolic schist; syenite, or a granular blend of large amounts of greenish-black amphibole and yellowish quartz and a small amount of white, flaky feldspar. This syenite splits into very thin layers; the quartz and the feldspar are so unequally distributed that they occasionally form small veins within an amphibolic paste.

Among these eight intercalated layers, the direction and inclination of which are exactly parallel to that of the rock as a whole, the thickest layer is formed by the syenite. It is thirty toises thick; and since I saw—in the deepest works of the mine (planes de San Bernardo) at 170 toises below the syenite layer—the reappearance of a carburized thonschiefer identical to the one through which people have begun to dig the new shaft, there can no longer be any doubt that both the schistous amphibole alternating twice with serpentine and the serpentine, likely alternating with syenite, form beds subordinated to the main thonschiefer mass of Guanajuato. The relationship that we have just indicated between the amphibolic rocks and the serpentine can also be found at other points on the globe, in euphotide formations of different ages: for example, in the Heideberg near Zelle in Franconia, at Kielwig at the northern edge of Norway; at Portsoy in Scotland, and on the island of Cuba between Regla and Guanabacoa.

I found neither remains of organic bodies nor layers of porphyry, grauwacke [greywacke], and Lydian stone in the transition thonschiefer of Guanajuato, which is the richest rock in terms of silver ore that has been found to this point; but this thonschiefer is covered, in a few places, with a very regularly stratified transition porphyry in a concordant deposition (los Alamos de la Sierra); in other places it is covered with *grünstein* and syenite, alternating thousands of times with one another (between the Esperanza and Comangillas); in yet other places, it is covered either with a calcareous conglomerate and a bluish-green calcareous transition rock, slightly clayey and fine-grained (Acabuca ravine), or with red sandstone (Marfil). The relationships between the *thonschiefer* of Guanajuato and the rocks that it supports, some of which (the syenite [masses]) *first manifest themselves* as subordinate banks, suffice to place the thonschiefer among the transition formations; they will especially justify this finding in the eyes of the geognosts who are familiar with the recently published observations on the intermediary terrains of Europe. As for the Lydian stone, there can be no doubt at all that the thonschiefer of Guanajuato contains it at some other, unexplored areas; for I found this substance frequently embedded in large fragments in the ancient conglomerate (red sandstone) that covered the thonschiefer between Valenciana, Marfil, and Cuevas. Ten leagues to the south of Cuevas, between Querétaro and the Cuesta de la Noria, in the middle of the Mexican plateau, one sees, emerging from under the porphyry, a blackish-gray (transition) thonschiefer, relatively non-fissile and verging on both siliceous schist (schistoid jasper, kiesel-schister) and Lydian stone. Perhaps this is pyroxene porphyry that has broken through the *thonschiefer* and has spilled out on top of it. Right near the Noria, fragments of Lydian stone lie scattered in the fields. According to the report of two learned mineralogists, Mr. Sonneschmidt and Mr. Valencia, both the argentiferous-vein-bearing rocks of Zacatecas and a small number of the Catorce veins run through a transition thonschiefer that contains veritable layers of Lydian stone and that appears to rest on syenite masses. Given what I have just reported on the pierced layers in the main shaft of Guanajuato, this superposition would prove that the Mexican *thonschiefer* constitutes, like the thonschiefer of the Caucasus and the Cotentin Peninsula, a single formation with the transition syenite and euphotide, and that it perhaps alternates with the latter. Regarding both the relationships between the schist and the porphyry of Guanajuato and the nature of the rock, which forms the roof of the Veta Grande, one may consult the interesting paper attached to the Report of the United Mexican Association, 1825, a paper in which my illustrious scholarly friend, Mr. Lucas Alamán, recently shared several geognostic insights that diverge from those on which I had thought I could focus.]

III.211

III.212

The famous mines of Zacatecas, which Robertson[1] calls—I do not know why—Sacotecas, are, as we have already observed, older than the mines of Guanajuato: they were first worked in 1748, immediately after operations were launched at the veins of Tasco, Zultepeque, Tlapujagua, and Pachuca, three years after the discovery of the riches of Potosí. They are located on the central plateau of the Cordilleras, which drops steeply toward Nueva Vizcaya and toward the Río del Norte basin. The climate of Zacatecas, like that of Catorce, is noticeably colder than the climate of Guanajuato and Mexico City. Barometric measurements will one day help determine whether this difference in temperature is due to its more northerly location or rather to the height of the mountains.

The nature of Zacatecas and Catorce has been examined by two very learned mineralogists, Mr. Sonneschmidt[2] and Mr. Valencia, the one Saxon III.213 and the other Mexican. On the whole, their observations suggest that, in terms of its geological constitution, the mining district of Zacatecas closely resembles that of Guanajuato. The oldest rocks, which appear *on the surface*, are syenitic: upon them rest [a mass of] *thonschiefer*, whose layers of Lydian stone, *grauwacke* [greywacke] and greenstone (*grünstein*) make it somewhat resemble *transition* clayey schist. It is in this *thonschiefer* that most of the Zacatecas veins are found. The *veta grande*, or the principal vein, has the same direction as the veta madre of Guanajuato: the others are generally oriented from east to west.[3] A mass of porphyry—devoid of metal and forming a number of those boulders, bare and cut steep, that the indigenous peoples call *bufas*—covers the *thonschiefer* at several points, especially on the side of the *City of Jerez*, where a bell-shaped mountain, the basaltic cone of *the Campaña de Jerez*, rises from the midst of these porphyritic formations. Among the secondary rocks of Zacatecas, near the plant of *la Sauceda*, one sees *compact calcareous stone*, in which Mr. Sonneschmidt also discovered Lydian stone, an *ancient sandstone* (*Urfelsconglomerat*) within which are embedded granite fragments[4] and a *clayey and feldspathic* III.214 *agglomerate*, which is easily confused with what German mineralogists

1. [Robertson,] *History of America*, vol. II, p. 389.

2. [Sonneschmidt,] *[Mineralogische] Beschreibung der [vorzüglichsten] Bergswerks-Reviere von Mexico*, pp. 166–237. *Descripción geognostica del Real de Zacatecas, por Don Vicente Valencia* (unpublished).

3. Andrés del Ríos, *Sobre la formación de las vetas* ["Continuación del Discurso sobre las formaciones de las montañas de algunos Reales de Minas,"] *Gazeta de México*, vol. XI, no. 51.

4. In the ravine that leads from Zacatecas to the Guadalupe convent.

refer to as *grauwacke* [*graywacke*]. The presence of Lydian stone in the calcareous stone might lead one to believe that the latter rock belonged to the transition limestone (*Übergangskalkstein*) that seems to appear on the surface in the *Cerro de la Tinaja*, eight leagues to the north of Zacatecas; but I must recall here on both coasts of South America, near the Morro de Nueva Barcelona, I found *Kieselschiefer* forming subordinate layers in a limestone that is undoubtedly secondary.

The wild appearance of the metalliferous mountains of Zacatecas contrasts sharply with the tremendous wealth of the veins contained therein: this wealth has revealed itself—and this is quite a remarkable fact—not in the ravines, nor wherever the veins run through the gentle mountain slopes, but more commonly on the highest summits, at points where the surface of the ground appears to have been tumultuously rent during the prehistoric upheavals of the globe. In a typical year, the Zacatecas mines yield from two thousand five hundred to three thousand silver bars, at 135 marks each.

III.215

The *mass* of the veins of this district contains[1] a wide variety of metals, namely, *quartz*, flaky *Hornstein*, calcareous spar, a bit of baryta sulfate and *Braunspath*; prismatic black silver, referred to locally as *azul acerado*; silver sulfide (*azul plomilloso*), blended with native silver; dusky silver (what the Germans call *Silberschwärze*, and the Mexicans *polvorilla*); pearl-gray, blue, violet, and leek-green muriate of silver (*plata parda azul y* [*and*] *verde*), at relatively shallow depths; a bit of red silver (*petlanque* or *rosicler*) and native gold, especially to the southwest of the city of Zacatecas; argentiferous lead sulfide (*soroche plomoso reluciente y* [*and*] *tescatete*); lead carbonate; black, brown, and yellow zinc sulfide (*estoraque* and *ojo de vivora*); copper and iron pyrite (*bronze nochistle* or *dorado* and *bronze chino*); magnetic iron oxidule; blue and green copper carbonate; and antimony sulfide. The most plentiful metals in the famous vein called the *veta madre* are prismatic black silver (*Sprödglaserz*), silver sulfide or vitreous silver, blended with native silver, and *Silberschwärze*.

The intendancy of Zacatecas also contains the mines of *Fresnillo* and those of *Sombrerete*: the former, barely worked, are located in an isolated group of mountains that rise above the plains of the central plateau. These

1. Sonneschmidt, [*Mineralogische Beschreibung*,] p. 185. The ore that the inhabitants of Zacatecas call *copalillo*, *metal cenzino*, and *metal azul de plata*, appear to this scholar to be blends of galena, silver sulfide, and native silver. I thought it necessary to record here this synonymy of Mexican ore, because it is very important for the traveling mineralogist to know. See Garcés [y Eguía], *Nueva teórica* [*y práctica*] *del beneficio de los metales* [*de oro y plata*], pp. 87, 124, and 138.

plains are covered in porphyritic formations; but the metalliferous group itself is composed of *grauwacke* [greywacke]. According to Mr. Sonn- III.216 eschmidt's observation, the rock there is shot through with an innumerable amount of veins rich in gray and green *muriate of silver.*

The Sombrerete mines, discovered in 1555, became famous for the tremendous wealth of the *Pabellón* vein and the *Veta negra*, which, over the course of a few months, left the Fagoaga (marqués del Apartado) family with a net profit of over twenty million livres tournois. Most of these veins are found in a compact calcareous stone, which, like that of la Sauceda, contains both kieselschiefer and Lydian stone. Especially abundant in this district is *dark red silver*: it has been seen forming the entire mass of the veins, which are over a meter *thick.* Near Sombrerete, the secondary limestone formation mountains soar high above the porphyritic mountains. The Cerro de Papantón appears to have a height of over 3,400 meters over sea level. The countryside surrounding the city of Sombrerete produces large amounts of maize and cereals. The climate is temperate and very wholesome. There is enough wood and running water to meet the needs of mining operations. At three and six leagues from Sombrerete are the old mines of la Noria, Chacuaco, and Chalchihuete. The greatest riches are especially attributed to the Real de la Noria, whose metalliferous deposit (Tajo de Ibarra) has a thickness of over thirty varas.

Today, the ore deposit of *Catorce* occupies the second or third rank III.217 among the mines of New Spain, in order of the amount of silver that they produce: it was not discovered until the year 1778. Both this discovery and that of the Gualgayoc veins, in Peru, commonly called the *Chota* veins, are the most interesting in the last two centuries of the history of mining in Spanish America. The small city of Catorce, the real name of which is la *Purísima Concepción de Alamos de Catorce*, is situated on the calcareous plateau that drops toward the *Nuevo Reino de León* and toward the province of *Nuevo Santander.* From the bosom of these secondary compact limestone mountains rise, as in the Vicentin, masses of basalt and porous amygdaloid that resemble volcanic material and that contain olivine, zeolite, and obsidian. A large number of veins, relatively narrow and quite variable in their width and in their direction, run through the calcareous stone, which itself covers a mass of *transition thonschiefer*; perhaps the latter is superposed on top of the syenitic rock of the *Buffa del Fraile*. The largest number of these veins are *west-facing* (*Spathgange*); their inclination is 25° to 30° toward the northeast. In general, the minerals that form the *gangue* are in a III.218 state of decomposition; they are extracted using pickaxes, rock picks, and stone chisels. Powder consumption is much lower than in Guanajuato and in Zacatecas. These mines also have the great advantage of being almost completely dry, therefore they do not require costly machines for *drawing [draining] out* the water.

In 1773, two impoverished private individuals, Sebastian Coronado and Antonio Llanas, discovered veins at a site today called *Cerro de Catorce Viejo*, on the western slope of the *Picacho de la Variga de Plata*. They worked these veins, which were poor and inconsistent in their output. For three months in 1778, a miner from the *Ojo del Água de Matchuala*, Don Bernabé Antonio de Zepeda, wandered through this group of dry, calcareous mountains. After carefully examining the ravines, he was fortunate enough to find the *crest* or *outcrop* of the *veta grande*, where he dug the *Guadalupe* shaft. From this he extracted a tremendous amount of muriate of silver and *colorados* blended with native silver; he quickly earned over half a million piasters. After that time, the Catorce mines were worked with the greatest intensity: that of the Padre Flores alone yielded, in its first year, one million six hundred thousand piasters' worth. Nevertheless, the vein only revealed great wealth from a perpendicular depth of fifty to one hundred fifty meters. Since 1788, the famous *Purísima* mine, belonging to Colonel Obregón, has almost continuously yielded an annual net profit of two III.219 hundred thousand piasters: its yield in 1798 was 1,200,000 piasters, while the operational mining costs did not exceed 80,000 piasters. The vein of the *Purísima*, which is not the same as that of the *Padre Flores*, at times reaches the extraordinary thickness of forty meters: in 1802, it was worked up to a depth of 480 meters. Since 1798, the wealth of the Catorce ore has sharply declined, with native silver appearing more rarely, and with the *metales colorados* (which are an intimate mixture of muriate of silver, earthy lead carbonate, and red ochre) starting to give way to pyritic and coppery ore. The current output of these mines is around 400,000 marks of silver per year.

The mines of Pachuca, Real del Monte, and Moran enjoy wide celebrity due to their age, their wealth, and their proximity to the capital. Since the beginning of the eighteenth century, only the *Vizcaína vein*, or the Real del Monte vein, has been worked intensively: work was only resumed at the Moran mines a few years ago; and the *ore deposit* of Pachuca, one of the wealthiest in all of the Americas, was entirely abandoned following the horrible fire that took place in the famous *Encino* mine, which alone yielded over thirty thousand marks of silver annually. The fire consumed the framework that propped up the *ridgepole* of the galleries, and most of the miners III.220

suffocated before they were able to reach the *shaft*. In 1787, a similar fire led to the abandonment of the Bolaños mines, which were not drained again until 1792.

The Valley of Mexico is separated from the *Totonilco el Grande* basin by a chain of porphyritic mountains, the highest peak[1] of which is Jacal Peak, with an elevation of 3,124 meters above sea level, according to my barometer-assisted measurements. This porphyry serves as a base for the porous amygdaloid that surrounds the lakes of Texcoco, Sumpango, and San Cristóbal: it appears to be of the same formation as the porphyry that, on the road from Mexico City to Acapulco, directly covers the *granite* between Sopilote and Chilpanzingo, near the village of Acaguisotla and at the Alto de los Cajones. To the northeast of the Real del Monte district, the porphyry is initially concealed beneath the *columnar basalt* of the Regla farm, and further on, in the Totonilco valley, under secondary formation layers. The alpine calcareous stone, of a bluish-gray color (*Alpenkalkstein*), within which is found the famous Danto cavern, also called the *Pierced Mountain* or the *Bridge of the Mother of God*,[2] appears to rest directly upon the porphyry of Moran. Near the Puerto de la Mesa, it contains veins of galena: it is found covered by three other formations of less ancient origin, which, to list them in order of their superposition, are *Jura limestone*, near the Totonilco baths, the *schistous sandstone* of Amojaque, and a *second formation gypsum*, blended with clay. The deposition of these secondary rocks, which I carefully observed, is all the more remarkable in that it is identical to that which has been identified on the Old Continent, according to the excellent observations of Mr. von Buch and Mr. Freiesleben.

III.221

The mountains of the Real del Monte mining district contain layers of porphyry, which, in terms of either their origin or their *relative age*, are quite different from one another. The rock that forms the *roof* and the *wall* of the argentiferous veins is a decomposed porphyry, whose base appears, at some points, to be clayey, and at other points, similar to flaky *Hornstein*. The presence of amphibole is often only noticeable through greenish spots, which are blended with common and vitreous feldspar. At great heights, for example in the lovely oak and pine forest of Oyamel, one finds porphyry with a *pearlstone* [*perlite*] base, within which obsidian is embedded, both in layers and in nodules. What is the relationship between these latter layers,

1. See my barometric land survey, pp. 40–42 [?; p. 339n in this edition], [vol.] II, pp. 290–312 [pp. 456ff in this edition].

2. *Puente de la Madre de Dios.*

which many distinguished mineralogists regard as volcanic material, and these porphyry masses of Pachuca, Real del Monte, and Moran, and in which nature has deposited enormous masses of silver sulfide and argentif- III.222 erous pyrite? This problem, one of the most difficult of all those presented by geology, will only be resolved once a large number of keen and learned travelers have traversed the Mexican Cordilleras and have carefully studied the immense variety of porphyry that are devoid of quartz and that are abundant in both amphibole and vitreous feldspar.

Unlike the mining district of Freiberg, in Saxony, or Derbyshire, in England, or the mountains of Zimapán and Tasco, in New Spain, the Real del Monte mining district does not present a large number of wealthy but relatively narrow veins in a small expanse of terrain. Rather, it bears more resemblance to the Harz and Schemnitz Mountains in Europe, or to those of Guanajuato and Potosí, in the Americas, the riches of which are contained within relatively scarce but large-sized ore deposits. The four veins of the Vizcaína, the Rosario, Cabrera, and the Encino, run through the districts of the Real del Monte, Moran, and Pachuca, over extraordinary distances, without changing *direction*, and without encountering almost any other veins that cross them or *disrupt* them.

The *veta de la Vizcaína*, less thick but perhaps even wealthier than the Guanajuato vein, had been very successfully worked from the sixteenth to the beginning of the eighteenth century. In 1726 and 1727, the two mines of III.223 Vizcaína and Jacal together still yielded 542,700 marks of silver. The large amount of water that seeped through the cracks in the porphyritic rock, combined with the inaccuracy of the draining methods, forced the miners to abandon these works, which had, however, only reached a depth of 120 meters. A very enterprising private individual, ▾ Don José Alejandro Bustamante, was bold enough to begin a gallery of efflux near Moran: he died before finishing this significant work, which is 2,352 meters long from its mouth to the point where the gallery crosses the *Vizcaína* vein. This vein is oriented at six hours; its inclination is 85° to the south; its thickness is between four and six meters. The porphyry of this district is generally oriented at seven to eight hours, with an inclination of 60° to the northeast, as one particularly observes on the road from Pachuca to Real del Monte. The gallery of efflux is initially cut in *the middle of the rock* (*Querschlagsweise*) with an orientation of seven hours toward the west; further on it assumes its *route* along three different veins, eleven to twelve hours, only one of which, the *veta de la Soledad*, has yielded enough silver ore to pay for all of the costs of the undertaking. The gallery was only finished in 1762, III.224

by Bustamante's partner, Don Pedro Terreros. By 1774, the latter, known by his title, Count of Regla, as one of the wealthiest men of his century, had already made a net profit of over twenty-five million livres tournois from the Vizcaína mine. In addition to the two warships that he gave as a present to King Charles III, one of which had 112 cannons, he lent five million francs to the court of Madrid, which have not, to date, been returned to him. He built the main factory of Regla, which cost him over ten million francs; he purchased immense plots of land, and left to his children a fortune that, in Mexico, has only been equaled by that of the Count de la Valenciana.

The gallery of Moran crosses the *Vizcaína* vein at the San Ramón shaft, at a depth of 210 meters below ground level, where the *horse-driven winches* are placed. Since 1774, the owner's profit has decreased from year to year. Rather than extending *investigation galleries* further in order to survey the riches of the vein across a large area, they took the *mining works* even deeper: they worked up to a depth of ninety-seven meters below the gallery of efflux. There, the vein preserved its great riches in silver sulfide, blended with native silver; but the water swelled to such a degree that twenty-eight winches, each of which required over forty horses, were not enough to drain III.225 the mine. In 1783, expenditure rose to 45,000 francs per week. Following the death of the elderly count of Regla, work was suspended until 1791, at which time they were bold enough to restore all of the winches. At that time, the cost of these machines, which drew water, not using pumps but rather bags suspended on ropes, amounted to over 750,000 francs per year. They actually reached the deepest point of the mine, which, according to my measurements,[1] is at an elevation of only 324 meters above the level of Sumpango Lake; but since the ore that was extracted did not compensate for the costs of draining the mine, the latter was once again abandoned in 1801.

It is surprising that no one has thought to replace this wretched means of removing the water through bags with sets of pumps powered by horse-driven winches, hydraulic wheels, or water-column machines. A gallery of efflux beginning in Pachuca, or lower down, near Gasave, in the Valley of Mexico, would have drained the Vizcaína mine, at the San Ramón shaft, up to a depth of 370 meters. The same objective would have been reached,

1. I found the absolute elevation of Sumpango Lake to be 2,284 meters, and that of the San Ramón shaft to be 2,815 meters; yet the deepest point of the Vizcaína mine is 307 meters below the upper opening of this shaft. I have recorded these results here because the locals generally believe that the mining works of Real del Monte have already reached the level of the saline lake of Texcoco.

and at less cost, had Mr. Delhuyar's project been carried out, that is, had the mouth of a new gallery been placed near Omitlan, on the road that leads from Moran to the amalgamation plant at Regla: this gallery would be around 3,800 meters long by the time that it intersected the Vizcaína vein. III.226

The very wise plan that the Count of Regla is following today is to refrain from draining the old works and to identify the *ore deposit* at points where it has never been worked before.[1] Studying the surface of the earth and the undulations [curves] of the terrain at Real del Monte, one observes that, over the past three centuries, the Vizcaína vein has yielded its greatest riches from a single point, a natural dip occurring between the shafts of Dolores, la Joya, San Cayetano, Santa Teresa, and Guadalupe. The shaft from which the most silver ore have been extracted is that of Santa Teresa. To the east and the west of this central point, the vein *narrows* for a distance of over 400 meters: it maintains its original direction, but, devoid of metals, is reduced to an almost imperceptible vein. For a long time, it was believed that the *Vizcaína* vein had disappeared in the rock without a trace, but in 1798 extremely rich metals were discovered at a distance of over 500 meters to the east and to the west of the center of the old works; the San Ramón and San Pedro shafts were therefore dug. It was acknowledged that the vein takes on its previous thickness once again, and that this represented an enormous field for new mining operations. When I visited these mines (in May of 1803), the San Ramón shaft was still only thirty meters deep; there will be a depth of nearly 240 meters up to the floor of the gallery of Moran, which itself is forty-five meters away from the point that corresponds to the intersection of the new shaft and the *top* [*ridgepole*] of the gallery. In its current state, the Count of Regla's mine produces over fifty or sixty thousand marks of silver annually. III.227

At the points where the main operations are located, the Vizcaína vein contains milky quartz, which often verges on flaky *Hornstein*; amethyst; carbonate of lime; small amounts of baryta sulfate, silver sulfide, blended with native silver and sometimes with prismatic black silver (*Sprödglaserz*), dark red silver; galena and iron and copper pyrite. These same silver ore are found close to the surface of the earth, in a state of decomposition, and blended with iron oxide, like the *pacos* of Peru. Near the San Pedro shaft, the pyrite is occasionally richer in silver than the silver sulfide mine.

The Moran mines, once very famous, were abandoned forty years ago, because of the large amount of water that could not be drained. It was in III.228

1. *In unverfahrenem Felde.*

this mining district, which borders that of Real del Monte, near the mouth of the large gallery of efflux of the Vizcaína, that a water-column machine was placed in 1801, the cylinder of which has a height of twenty-six meters and a diameter of sixteen meters. This machine, the first of its kind that has been built in the Americas, is far superior to those that exist in the mines of Hungary. It was built in adherence to the calculations and plans of Mr. del Río, professor of mineralogy in Mexico City, who has visited the most famous mines of Europe and whose knowledge is impressive for both its depth and its breadth. The machine's construction was due to ▾ Mr. Lachaussée, an artist of distinguished talent from Brabant who has also built, for the School of Mines in Mexico City, a truly remarkable set of models intended for the study of both mechanics and hydrodynamics.[1] Unfortunately, this beautiful machine, in which the *valve* regulator[2] is driven by a particular mechanism, was placed in a location where the water necessary for keeping it continually in motion is exceedingly difficult to find. During my stay at Moran, the pumps could work only three hours a day. The machine and the aqueducts cost 80,000 piasters to build. The initial expectations were that III.229 expenses would be half that amount and that there would be a significant amount of water to drive the machine; but since the year in which the water levels were measured was particularly rainy, they were believed to be much higher than they normally are. One can only hope that the new canal, on which they were working in 1803, and which was to be 5,000 meters long, will remedy this lack of water and that the Moran vein (nine and four-sixths hours, inclined 84° to the northeast) will reveal itself, at great depths, to be as wealthy as the mine's *shareholders* assume. When he arrived in New Spain, Mr. del Río had no other objective than to demonstrate to the Mexican miners the effectiveness of this kind of machine and the possibility of building them locally: this objective has been partly met, and it will be met in an even more obvious way as soon as a water-column machine is placed in the Rayas mine, at Guanajuato; in the Count of Regla's mine, at Real del Monte; or in the Bolaños mine, where Mr. Sonneschmidt[3] counted nearly 40,000 horses and mules used to drive the *winches*.

The mines of the *Tasco district*, situated on the western slope of the Cordillera, have lost their former splendor since the end of the last century,

1. See above, vol. I, p. 428 [p. 276 in this edition].

2. ▾ Delius, [*Traité sur la science de l'exploitation*] *des mines . . . de Schemnitz*, translation by Mr. Schreiber, §591.

3. Sonneschmidt, [*Mineralogische Beschreibung*,] p. 241.

for in their current state, the veins of Tehuilotepec, Sochipala, the Cerro del Limón, San Estéban, and Guautla, together yield only around 60,000 marks of silver. It was in 1752 and in the ten subsequent years that the Tasco mines were most intensively and successfully worked. This activity was due to the enterprising spirit of a Frenchman, Joseph de Laborde, who was very poor when he came to Mexico and by 1743 had earned immense riches from the Cañela mine of the *Real de Tlapujahua*. We have written elsewhere about the many reversals of fortune that this extraordinary man experienced. After having built a church in Tasco that cost him 400,000 piasters, he fell into dire poverty because of the rapid decline of the very mines from which he had previously reaped two to three hundred thousand marks of silver annually. The archbishop having permitted him to sell a golden, diamond-encrusted sun with which he had decorated the tabernacle of the church of Tasco, he traveled to Zacatecas with the proceeds of this sale, which amounted to 100,000 piasters. At that time, the mining district of Zacatecas was in such a state of neglect that it provided barely 50,000 marks of silver a year to the mint of Mexico City. Laborde undertook the draining of the famous Quebradilla mine; he lost all the funds that he possessed, without reaching the goal of his undertaking. Finally, with his little remaining capital, he attacked the *veta grande* by digging the *Esperanza* shaft; it was there that he reaped a huge fortune for the second time. The yield of silver from the Zacatecas mines subsequently rose to 500,000 marks per year, and although metals did not remain equally abundant for long, on his death Laborde left a fortune of nearly three million livres tournois. He had forced his daughter to enter into a convent so that he could will his whole estate to his only son who had voluntarily taken the habit. In Mexico, and everywhere else in the Spanish colonies, it is exceedingly rare for children to take up their father's trade. One does not find there, as one does in Sweden, Germany, and Scotland, families in which the miner's profession has become hereditary.

III.230

III.231

The veins of Tasco and the Real de Tehuilotepec run through arid mountains furrowed with very deep ravines. In this mining district, the oldest rock that appears on the surface is primary schist (*Thonschiefer*), which verges on micaceous schist. Its direction is three to four hours, and its inclination is 40° to the northwest, as I observed it in the Cerro de San Ignacio, and to the west of Tehuilotepec, in the Cerro de la Compaña, where Cortés had started an investigation gallery. The micaceous schist probably rests on the granite of Sumpango and on that of the Papagallo Valley; near Achichintla and Acamiscla, it appears to be covered by a porphyritic formation that contains both common and vitreous feldspar and layers of blackish-brown *Pechstein.* In the areas surrounding Tasco, Tehuilotepec, and III.232 Limón, the primary schist serves as a base for the bluish-gray compact limestone, often porous, and belonging to the *alpine formation.* This limestone contains several *subordinate layers*, some of which are of lamellar gypsum, others of schistous clay (*Schieferthon*), loaded with carbon. While journeying from the shores of the Tuspa Lake up to the *Subida de Tasco el Viejo*, we came across fossils of trochites and other univalve shells contained within this calcareous stone. Its stratification is quite pronounced, but its groups of beds follow divergent directions and inclinations. Upon the calcareous stone of Tasco, identical to that which covers the Sopilote plains and the fertile plateau of Chilpanzingo, lies sandstone with calcareous cement.

The mining district of Tasco and of the Real de Tehuilotepec contains a large number of veins which, with the exception of the Cerro de la Compaña, are all oriented from the northwest to the southeast, seven to nine hours. These veins, like those of Catorce, run through the calcareous stone, as well as through the micaceous schist that serves as a base for it. They present the same metals in both types of rock; yet these metals were much more abundant in the calcareous stone. The wealth of the mines has been severely diminished since the owners were forced to mine the veins within the micaceous schist. A highly intelligent and very active miner, Don Vicente de Anza, has taken the Tehuilotepec mines to a depth of 224 meters: III.233 he has dug two magnificent galleries of efflux over 1,200 meters long; but unfortunately he has found that at great depths these veins, which, near the surface of the earth had yielded considerable riches, were as poor in red silver ore as they were abundant in galena, pyrite, and yellow blende.

An extraordinary event that occurred on February 16, 1802, hastened the demise of the miners in the district. The Tehuilotepec mines, like those of Cuautla, have always lacked the water necessary to move the *stamp mills* and the other machines that prepare the ore for the amalgamation process. The most abundant stream used in the plants emerged from a cavern located in the calcareous rock, called the *Cueva de San Felipe*: this stream disappeared on the night of February 16–17, and two days later a new source appeared five leagues away from the cavern, near the village of Plantanillo. A number of studies of great relevance to geology, and to which I shall refer elsewhere, have proven that a series of caverns and natural galleries exists in this area, between the villages of Chamacasapa, Plantanillo, and Tehuilotepec, in the bosom of the calcareous mountains, and that underground rivers, similar to those in the county of Derby, in England, flow through these galleries, which are all interconnected.

The veins of Tehuilotepec generally face west (*Spatgänge*); they have a thickness of two to three meters: separated from the rock by a *verge* of clayey silt, they have several lateral branches which *enrich* the principal vein wherever they lie close to it. Their structure presents the oddity that the metallic ore is rarely disseminated throughout the entire gangue, but is concentrated in a single strip [band], which is sometimes found near the roof, at other times near the wall of the vein. In general, the mineral deposits of Tasco and Tehuilotepec are extremely *inconsistent* in their output: as for the nature of their constituent rock *mass*, I have identified four very different vein formations, namely: III.234

(1) Brown, red, and yellow iron oxide, within which are disseminated both native silver and silver sulfide, in extremely fine particles; cellular brown iron ore, specular iron, small amounts of galena and of magnetic iron, and blue copper carbonate. This formation, analogous to that of the *pacos* of Fuentestiana and Pasco, in Peru, is referred to at Tehuilotepec as *tepostel*: it can be found at shallow depths, near *the surface* (*im ausgehenden*), in the mines of San Miguel, San Estéban, and Campaña, near Tasco, as well as in the Cerro de Garganta, near Mescala. *Tempostel* is generally poorer than the *paco* of Peru: at Tasco, it is even poorer in that the iron oxide is more thoroughly blended with azure copper; it generally contains less than four ounces of silver per quintal.

(2) Calcareous spar, small amounts of galena and transparent lamellar gypsum, within which are encased drops of water with air and filiform native silver. This quite remarkable small formation, which has also been observed in the mountains of Salzburg, is found at a depth of over one hundred meters in the Trinidad vein, which is the continuation of the San Miguel vein, at a point where the *wall* is not made of gypsum but of compact limestone. III.235

(3) Light-red silver, brittle vitreous silver (*Sprödglaserz*), a large amount of yellow blende, brown blende, galena, very little iron pyrite, calcareous spar, and milky quartz. This formation, the wealthiest of all, presents a remarkable phenomenon, namely, that the ores most abundant in silver form spheroid-shaped balls, ten to twelve centimeters in diameter, within which red silver, blended with brittle vitreous silver and native silver, alternates with bands of quartz. These balls, which are frequently encountered only between fifteen and sixty meters of depth, are nested within a gangue of

calcareous spar and browning spar. They have been observed in the three veins of San Ignacio, Dolores, and Perdón, the *masses* of which are filled with *druses*, lined with beautiful crystals of carbonate of lime.

(4) A large amount of argentiferous galena, which is even wealthier in silver in that the *separated pieces* are more fine-grained; a large amount of III.236 yellow blende; few pyrites; quartz and calcareous spar in the mines of Socabón del Re and de la Marquesa.

These veins all run through a plateau that has an elevation of 1,700 to 1,800 meters above the surface of the sea, and that enjoys a temperate climate, ideal for growing cereals from the Old Continent.

An overview of mining in New Spain, especially as it compares to mining in Freiberg, the Harz Mountains, and Schemnitz, leaves one surprised by the relative infancy of an art that has been practiced in the Americas for nearly three centuries, and upon which depends, following a widespread conviction, the prosperity of these overseas establishments. The causes of this phenomenon cannot escape those who, after having visited Spain, France, and the western part of Germany, have seen that in the center of civilized Europe there still exist mountains in which the mining works suffer from all the barbarism of the Middle Ages. The miner's art cannot advance where mining operations are scattered across a large expanse of terrain, where the government gives the owners complete freedom to manage the operations however they please and to rip the ore from the bosom of the Earth with no thought for the future. Ever since the glorious reign of Charles V, Spanish America has been separated from Europe in terms of the communication of discoveries useful to society. The scant knowledge III.237 about the arts of mining and smelting that people possessed in the sixteenth century, in Germany, in Vizcaya, and in the Belgian provinces, had traveled quickly to Mexico and Peru at the time of the initial *colonization* of these countries; but from that time until Charles III's reign, American miners have learned almost nothing from the Europeans, with the exception of *powder blast mining*[1] in rocks that prove resistant to the stone chisel. Both that king and his successor exhibited a highly praiseworthy desire to extend to the colonies all of the advantages that Europe drew from the perfecting of machines, advances in the physico-chemical sciences, and their application to metallurgy. German miners were sent, at the court's expense, to Mexico, Peru, and the kingdom of New Granada; but their wisdom could

1. Powder [blast] mining was not even introduced into the mines of Europe until around 1613. (D'Aubuisson, [*Des mines de Freiberg en Saxe et de leur exploitation*], vol. 1, p. 95.)

be of no use, because the mines of Mexico are regarded as the property of private individuals who manage their operations without allowing the government to exert the least influence therein.

We shall not undertake to indicate in detail here the flaws that we believe to have observed within the mining administration of New Spain; rather, we shall restrict ourselves to a number of general considerations, noting everything that seems to us to be deserving of the European traveler's attention. In most of the Mexican mines, *stone-chisel* work, the procedure that demands the most dexterity on the part of the worker, is very well executed. Ideally the mallet would be a bit less heavy; it is the same instrument that the German miners used in Charles V's time. Small, mobile forges are placed inside the mines to reforge the tips of the stone chisels that have become unusable. I counted sixteen of these forges in the Valenciana mine: in the Guanajuato district, the smallest mines have one or two of them. This arrangement is especially useful in mines that accommodate up to fifteen hundred workers, and in which there is an enormous consumption of steel. I cannot praise the procedure employed in *powder [blast] mining*: the slots that are supposed to hold the cartridges are generally too deep, and the miners do not take sufficient care in removing the part of the rock that should yield to the explosion. These flaws result in a very significant loss of powder. From 1794 and 1802, the Valenciana mine consumed[1] 673,676 piasters' worth of powder per year, and the mines of New Spain currently require 12,000 to 14,000 quintals. It is likely that over a third of this amount is needlessly used. At Chapultepec, near Mexico City, and in the mine of Rayas, near Guanajuato, experiments have been conducted on a draft method proposed by Mr. Bader, a method in which a certain volume of air is left between the powder and the cap. Although these trials have demonstrated the superiority of the new method, the old one has prevailed, because of the scant interest shown by the *master-miners* in curbing abuses and in perfecting the art of mining.

IV.238

III.239

The *lining*, or woodwork surfacing, is carelessly done: it should, however, be of even greater concern to the owners since wood is growing scarcer each year on the Mexican plateau. The masonry-work employed in the

1. In 1799, 63,375 piasters' worth; in 1800, 68,493 piasters' worth; in 1801, 78,243 piasters' worth; in 1802, 79,903 piasters' worth. At Guanajuato, a miner is paid twelve francs for a hole one and a half meters deep; nine francs for a hole 1.9 meters deep, not counting the powder and the tools, which are provided separately. In the Valenciana mine, nearly 600 *two-man holes* are bored every twenty-four hours.

shafts and in the galleries, especially the *lime walling*, merits great praise. The *voussoirs* are crafted with the utmost care, and in this respect the Guanajuato mines rival any of the most perfect features of the mines of Freiberg or Schemnitz. The shafts, and especially the galleries of New Spain present the flaw of having been dug in dimensions (*Ortstosshöhe*) that are entirely III.240 too big and thereby generate exorbitant expenses. At Valenciana one finds galleries extended in order to identify a *barren* vein, and which are eight to nine meters in height. One would be mistaken to assume that this great height facilitates the air circulation: the *airing-out* of a gallery depends exclusively upon the balance and the difference in temperature between two neighboring air columns. It is still believed, and with equally little basis in fact, that to identify the nature of a very thick vein, very wide *investigation galleries* are needed, as though, for ore deposits with a thickness of twelve to fifteen meters, it would not be more worthwhile to extend, from time to time, small *transverse galleries* toward the *wall* and the *roof*, to see if the *mass of the vein* starts to grow rich. It is this absurd convention of digging all the galleries in enormous dimensions that prevents the owners from adding more *investigation works*, indispensable to a mine's preservation and to the long-term duration of operations. At Guanajuato, the width of the diagonal and terraced galleries is ten to twelve meters: the perpendicular shafts are generally six, eight, or ten meters wide. The huge quantity of ore that are extracted from the mines, and the necessity of inserting the cables attached to six to eight *horse-driven winches*, make it essential to dig the shafts in Mexico in larger dimensions than those in Germany; but the attempt that III.241 was made in Bolaños to separate the cables of the winches using a wooden framework, sufficiently proved that the width of the shafts can be decreased without fearing that the ropes would become entangled through their oscillatory movements. It would, in general, be very helpful to use *tuns*, or rectangular parallelepiped crates, instead of leather bags suspended from the cables of the drum. Several pairs of these *tuns*, rolling with their casters along the *channel joists*, could rise and descend within the same shaft.

The greatest flaw that one observes in the mines of New Spain, and which render mining extremely costly, is the lack of connections among the different *works*: they resemble badly constructed buildings in which one must round the entire house to pass from one room to the neighboring room. Because of its wealth, the magnificence of its walling, and the ease with which one enters, via spacious and comfortable staircases, the Valenciana mine elicits legitimate admiration: it nevertheless presents merely a collection of small works, with a shape too irregular to be called *terraced works*; they are veritable sacks with only one opening above, without any lateral connection. I am citing this mine, not as the one that presents the greatest number of flaws in the way its works are distributed, but because it should be seen as the best organized of them all. Since subterranean geom- III.242 etry was entirely neglected in Mexico until the establishment of the School of Mines, no plan exists for the *strucures* already built. With their mazes of *transverse galleries* and internal shafts, two mining operations might be very close to one another without one being able to detect their proximity. The result of this is that, in the current state of the majority of Mexican mines, it is impossible to introduce *wheelbarrow* or *dog haulage* or an economical distribution of the *loading points*. A miner raised in the mines of Freiberg, accustomed to using so many ingenious means of transportation, has difficulty conceiving that, in the Spanish colonies, where the ore are both poor and in great abundance, all of the metal that is extracted from the vein is hauled out *on man-back*. The Indian *tenateros*, who can be considered the beasts of burden of the Mexican mines, must shoulder a weight of 225 to 350 pounds over a period of six hours. In the galleries of Valenciana and Rayas, they are exposed—as we have observed above in our discussion of the miners' health[1]—to a temperature of twenty-two to twenty-five degrees. During this time, they climb and descend several thousand steps through shafts with an inclination of over 30°. These tenateros carry the ore in bags (costales) woven of century plant fiber. So as not to injure their III.243 backs (for the miners are generally naked to the waist), they place a wool covering on top of the bag. Inside the mines, one encounters lines of fifty to sixty of these porters, among whom are men in their sixties and children ten to twelve years old. When climbing the stairs, they thrust their bodies forward and lean on a stick only three decimeters long; they walk in zigzag, to avoid walking the steepest path, and because experience has proven to them (so they say) that their breathing is less restricted if they diagonally cross the air current that enters through the shafts from outside.

One cannot tire of admiring the muscular strength of the Indian and Mestizo *tenateros*, especially when one feels utterly exhausted upon emerging from the deepest part of the Valenciana mine, without having carried even the lightest weight. The *tenateros* cost the owners of the Valenciana mine over 15,000 livres tournois per week: the owners also depend on three

1. See above, vol. I, p. 340 [p. 227 in this edition]. In Paris, the *forts de la halle* generally carry flour sacks that weigh 325 pounds. To be admitted to their guild, one must be able to carry 850 pounds for a period of twenty-five minutes.

men to haul the ore to the loading points, for a worker (*barenador*) who explodes the gangue using powder. These enormous hauling costs would perhaps be reduced by two-thirds if the *mining works* were connected by internal shafts (*Rollschächte*) or by galleries appropriate for *wheelbarrow and dog hauling.* Wisely dug shafts would facilitate ore extraction and air III.244 circulation and would obviate the need for such a large number of *tenateros*, whose energies could be directed to activities that are more socially beneficial and less harmful to the health of individuals. Internal shafts that, by linking one gallery to another, could facilitate ore extraction, could be equipped with man-powered *winches* (*Haspel*) or animal-driven *winches.* For quite some time (and this arrangement should probably be of some interest to European miners) mules have been used inside the mines of Mexico. Every morning at Rayas, these animals descend steps built in a shaft with an inclination of 42°–46°, without guides and in complete darkness. The mules spread out among the various spots where the bucket machines are placed: their step is so sure that, a few years ago, a lame miner was in the habit of entering and leaving the mine on muleback. In the mining district of Peregrino, at the Rosa de Castilla, the mules rest in underground stables, like the horses that I saw in the famous rock salt mine of Wieliczka in Galicia [Ukraine].

The smelting and amalgamation plants of Guanajuato are situated such that two *navigable galleries*, the mouths of which would be near Marfil and Omitlan, respectively, could be used to haul ore, making any *draft* above the level of the galleries unnecessary. Furthermore, the descent from Valenciana to Guanajuato, and from Real del Monte to Regla, is so quick that III.245 one could build iron rails on which carts loaded with ore destined for the amalgamation plant could roll.

We have mentioned above the truly backward convention of *draining* the water from the deepest mines, not by *sets* or *systems of pumps* but by bags attached to ropes that move along the *drum* of a *horse-driven winch.* These same bags are used, as needed, for removing either water or ore; they rub against the walls of the shafts and are extremely costly to maintain. At Real del Monte, for example, one of these skins only lasts seven to eight days: it normally costs six francs to replace, sometimes eight to ten. A water-filled bag, hung from the drum of an eight-horse winch (*malacate doble*) weighs 1,250 pounds: it is made of two skins sewn together. The bags used for the so-called *simple* winches, those that require only four horses, hold only half of this volume, and are made of a single skin. In general, the winches are quite faultily constructed; furthermore people have the bad habit of forcing the horses who drive them to run at an excessive speed. At the San Ramón shaft, at Real del Monte, I found this speed to be ten and a half feet per second;[1] at Guanajuato, in the Valenciana mine, it was thirteen to fourteen feet [per second]: everywhere else, it is above eight feet per second. ▾ Don Salvador Sein, professor of physics in Mexico City, has proven, in a very interesting paper on the gyratory movement of the machines, that despite the Mexican horses' extremely light weight, they do not produce the maximum effect in the winches unless, exerting a force of 175 pounds, they move at a speed of five to six feet per second. III.246

One can only hope that the mines of New Spain will one day have sets of pumps, powered either by *horse-driven winches*, *hydraulic wheels*, or *water-column machines*. Since wood is quite scarce on the ridge of the Cordilleras, and since pit coal has only been discovered in a few northern provinces (New Mexico, for example), *steam engines* can unfortunately only be used in proximity to hot or temperate regions. I think that if the surrounding areas were carefully studied and the water through channels were skillfully redirected into ponds, several sites (if the slope permits) could benefit from ▾ Reichenbach's[2] wonderful water-column engine, which pumps the brine from the Bavarian salt marshes up to a height of 1,500 feet and which is certainly preferable to the water-column machine that uses a pendulum. III.247

It is especially in the draining of water that one senses how indispensable it is to have survey plans made by *subterranean geometers*. Instead of damming and redirecting the water along the shortest path to the machine shaft [shaft where the machines are placed], the water is often made to plunge to the bottom of the mine,[3] only to be removed again at great cost. In addition, in the mining district of Guanajuato, nearly 250 workers died in a matter of minutes on June 14, 1780, because, having neglected to measure the distance between the San Ramón operations and the former *works* of *Santo Cristo de Burgos*, they had carelessly drawn near to the latter mine when they extended an *investigation gallery* toward it. The water that had

1. With the water being drawn from a depth of eighty meters. *The malacate doble* had four *arms*; there is a shaft affixed to the end of each *arm*, to which two horses are attached. The diameter of the horses' circular track was seventeen and a half *varas*. The horses are changed every four hours.

2. Two of these have just been built for the Pollaouen and Huelgoat mines in Brittany, which are mined as successfully as they are intelligently, by ▾ Mr. Junker, and in which there is a 190-foot slope. Each machine will have the power of 300 horses.

3. For example, at Rayas, where they remove water from a depth of 338 *varas*, while they could *simply* collect it from a *sinkhole* to the southeast, at a depth of 180 *varas*.

filled the Santo Cristo works burst forth through the new San Ramón gallery in the Valenciana mine: several workers perished as a result of the sudden compression of the air, which, seeking an escape, hurled pieces of timber from the woodwork lining and chunks of rock across large distances. III.248 This accident would not have happened had there been a plan of the mines to consult when the works were being laid out.

Based on the tableau that we have just sketched of the current state of the mining operations and the wastefulness that characterizes the administration of the mines of New Spain, one should not be surprised to see that mining operations that have long had the highest yields have been abandoned as soon as they have reached a considerable depth, or as soon as the veins have proven less abundant in metals. We have observed above that in the famous Valenciana mine the annual expenses have risen, over a fifteen-year period, from two million francs to four and a half million francs. If there were a large amount of water in this mine, and if draining the latter made *horse-driven winches* necessary, then the profit that would accrue to the owners would likely be zero. Most of the administrative defects that I have just indicated have long been acknowledged by a respectable and enlightened body, the *Tribunal de Minería* of Mexico, by the faculty of the School of Mines, and even by a few indigenous miners who, even without having ever left their homeland, know how flawed the old methods are; but—we shall repeat it here—such changes can only be very slow among a people that have no taste for innovation, and in a country where the government has such little III.249 influence over the mining operations, which generally belong to individuals and not to companies of shareholders. It is, moreover, a belief that the mines of New Spain, because of their wealth, do not demand the same intelligence and economy that is necessary for the maintenance of the mines of Saxony and the Harz Mountains. One must not confuse the abundance of ore with their intrinsic wealth. As most of the ore of Mexico is quite poor, as we have demonstrated above, and as all those who are not dazzled by false calculations know, in order to produce two and a half million marks of silver, it is necessary to extract an enormous mass of metal-bearing gangue. Yet one can easily imagine that in mines where the various *works* are poorly arranged, with no linkages between one another, the costs of extraction must increase at a terrifying rate, the deeper the shafts (*pozos*) become, and the further the galleries (*cañones*) are extended.

The miner's work is entirely free throughout the entire Kingdom of New Spain; no Indian or Mestizo can be forced into mining. It is absolutely untrue, although this claim has been repeated in even the most highly esteemed works, that the court of Madrid sends convicts to the Americas to labor in the gold and silver mines. Russian criminals have filled the mines of Siberia, but in the Spanish colonies this type of punishment has fortunately been unheard of for centuries. Mexican miners are the best-paid of all min- III.250 ers; they earn at least twenty-five to thirty francs for a six-day week, while a day's wage for workers who labor outside, for example in plowing, is seven livres [tournois] sixteen sous on the central plateau, and nine livres [tournois] twelve sous near the coasts.[1] The *tenatero* and *faenero* miners, who are responsible for *hauling* the ore to the *loading points* (*despachos*), often earn over six francs for a six-hour work day.[2] Among Mexican miners, however, trustworthiness is not as common as it is among German or Swedish miners; they know a thousand tricks for stealing particularly wealthy ore. Since they are nearly naked, and since they are searched, in the most indecent manner, as they exit the mine, they hide small pieces of native silver, red silver sulfide, or muriate of silver in their hair, under their armpits, or in their mouth; they even insert clay cylinders containing metal into their anus: these cylinders are called *longanas*, and they have been found up to thirteen centimeters (five inches) long. It is a pathetic spectacle to witness hundreds of workers in the large mines of Mexico, among whom very many are thoroughly honest, all subjected to searches as they exit the shaft or the gallery. A record is kept of the ore found in their hair, in their mouths, or III.251 in other parts of the miners' bodies. In Guanajuato, in the Valencia mine alone, from 1774 to 1787, the total value of these stolen ore, a large part of which were smuggled in *longanas*, amounted to 900,000 francs.

Inside the mines, the ore hauled by the *tenateros* from the mining *work* to the shaft is monitored very carefully. At Valenciana, for example, the amount of metalliferous *gangue* that leaves the mine daily is known to within a few pounds; I say *gangue*, for the *rock* there is never an object of extraction. At the *loading point* of the main shafts, there are two chambers hollowed into the *wall*, within each of which there are two people (*despachadores*), seated at a table, each with a book containing a list of all of the miners employed in hauling. In front of them, near the counter, hang two scales. Each *tenatero* loaded with ore presents himself at the counter; two persons, positioned near the scales, determine the weight of the load

1. See above, vol. II, p. 204 [p. II.116 in this edition]; vol. II, p. 448 [p. II.263 in this edition]; vol. II, p. 17 [p. 546 in this edition].

2. At Freiberg, in Saxony, a miner earns between four pounds and four pounds ten sous for a five-day work week.

by lifting it gently. If the *tenatero*, who, while carrying the ore, has had the opportunity to estimate its weight, believes it to be lighter than does the *despachador*, he says nothing, because this error is profitable for him. If, on the other hand, he believes that the weight of the ore that he is carrying in his III.252 back is greater than is estimated, then he asks that his load be weighed on the scale. The weight that is determined is inscribed in the *despachador*'s book. No matter which part of the mine the *tenatero* comes from, he is paid, per *trip*, one *real de plata* for a load of nine *arrobas*, and one and a half *real* for a load of thirteen and a half *arrobas*. There are *tenateros* who make eight to ten *trips* in a single day, and their pay is based on the *despachador*'s record. This method of accounting is certainly deserving of praise, and one admires not only the promptness but also the order and the calm with which they succeed in determining the weight of so many thousands of quintals of ore yielded in a single day by veins twelve to fifteen meters wide.

These ore, separated from the barren rocks in the mine itself by the master-miners (*quebradores*), undergo three kinds of preparation; that is, on the *sorting tables*, where the women work; under the *stamp mill* and under the *tahonas* or *arrastres*. These tahonas are machines in which the metalliferous gangue is triturated using very hard, gyrating stones, which weigh more than seven to eight quintals. To date, neither *washing in vats* (Sezwäsche), nor washing on *sleeping tables*, nor percussion washing (*Stossheerde*). Preparation under the stamp mills (*mazos*) or in *tahonas*, which I shall also call *mills*, because of their resemblance to certain oil and tobacco mills, varies depending on whether the ore is destined to undergo smelting III.253 or amalgamation. Strictly speaking, the *mills* are not part of the latter technique; however, extremely wealthy metallic grains, called *polvillos*, which have passed through trituration in the *tahona*, are also smelted.

The amount of silver extracted from ore through the use of mercury stands in a ratio of three and a half to one with that produced by smelting. This proportion was obtained from the general table created by the provincial treasuries of the various mining districts of New Spain. In a few of these districts, such as those of Sombrerete and Zimapán, the yields from smelting nevertheless surpass yields from amalgamation.

III.254 I believe that the amounts listed in the preceding [following] table should be increased by one-fifth, to bring them in line with the current state of mining. During peacetime, amalgamation makes gradual gains on smelting, which is generally of poor quality. Since wood is becoming scarcer every year on the ridge of the Cordilleras, the most densely populated part of the country, the decrease in smelting output is welcome news to plants

Silver (plata quintada) *extracted from the mines of New Spain from January 1, 1785, to December 31, 1789:*

PROVINCIAL COFFERS THAT COLLECT THE QUINT	SILVER EXTRACTED BY AMALGAMATION (*MARCOS DE AZOGUE*)	SILVER EXTRACTED BY SMELTING (*MARCOS DE FUEGO*)
Mexico City	950,185	104,835
Zacatecas	1,031,360	173,631
Guanajuato	1,937,895	531,138
San Luis Potosí	1,491,058	24,465
Durango	536,272	386,081
Guadalajara	405,357	103,615
Bolaños	336,355	27,614
Sombrerete	136,395	184,205
Zimapán	1,215	247,002
Pachuca	269,536	185,500
Rosario	477,134	191,368
Total, in marks	7,572,762	2,159,454

that necessitate large fuel expenses. During wartime, mercury shortages put an end to the gains made by amalgamation and force the miners to concentrate on perfecting *smelting* techniques. The chief executive officer of the mines, Mr. Velázquez, still claimed in 1777—therefore prior to the discovery of the wealthy mines of Catorce, where smelting is more or less non-existent—that of all the ore of New Spain, two-fifths were smelted and three-fifths were subjected to amalgamation.

The limits that we have set for ourselves in the drafting of this work do not permit us to enter into more detail about the amalgamation techniques widely used in Mexico. It will suffice to give a general overview of them and to convey the difficulties involved in introducing to the New Continent the method invented in Germany in 1785 by Born, ▼ Ruprecht, and Gellert. Those who wish to gain a more in-depth understanding of the practice of Mexican amalgamation will find the most thorough information in a work published by Mr. Sonneschmidt. That estimable mineralogist, a native of Saxony, resided in New Spain for a period of twelve years. He had the opportunity to subject a wide variety of ore to amalgamation, and his

III.255

firsthand experience enabled him to determine both the advantages and the disadvantages of the various methods that have been employed since the sixteenth century in the mines of the Americas.

The ancients were acquainted with the ability mercury has to combine with gold; they used amalgamation to gild copper and to collect the gold contained within used garments, by reducing the latter to ashes in clay vessels.[1] It also appears certain that before the discovery of the Americas, German miners employed mercury, not only in washes in auriferous lands but also to extract the gold disseminated in veins,[2] either in a native state or blended with iron pyrites and with gray copper ore. But the amalgama- III.256 tion of silver ore, the ingenious technique that is used today in the New World and from which comes the majority of the precious metals that exist in Europe or that have flowed from Europe to Asia, only dates back to 1557; it was invented in Mexico by a miner from Pachuca named Bartolomé de Medina. If one takes into account the documents that exist in the archives of the *Despacho general de Índias*, as well as the studies of Don Juan Díez de la Calle,[3] there can be no doubt as to who is the real author of this invention, which some have attributed[4] to the canon Enrique Garcés, who in 1566 started the mercury mining operations of Huancavelica, and which others have attributed to ▾ Fernández de Velasco, who introduced Mexican amalgamation to Peru in 1571. It is less certain whether Medina, who had been born in Europe, had not already conducted amalgamation experiments before coming to Pachuca. An *Alcalde de Corte* in Mexico City, ▾ Berrio de Montalvo,[5] the author of a paper on the metallurgical treatment of silver ore, insists that "in Spain, Medina had heard that silver could be extracted using III.257 mercury and common salt," but this claim is not supported by any convinc-

1. Pliny, [*Historia naturalis*,] XXXIII, 6. Vitruv[ius], VII, 8. Beckmann, *[Beyträge zur] Gesch[ichte] der Erfindungen*, vol. I., p. 44; vol. III, p. 307; vol. IV, p. 578.

2. For example, in Goldcronach, in Fichtelgebirge, where the position of the former amalgamation mills (*Quickmühlen*) reserved for crushing auriferous ore is still indicated. Precious documents found in the archives of Plassenbourg, which I had the opportunity to study during my long stay in the mountains of Steeben and Wunsiedel, attest to the old age of the amalgamation plants in Goldcronach.

3. [Díez de la Calle,] *Memoria dirigido al Señor Don Felipe IV* (Madrid 1646), p. 49. Garcés [y Eguía], *[Nueva teórica y práctica] del beneficio de los metales*, pp. 76–82. Compare also the *Real cédula espedida en Valladolid, en 4 de Marzo, 1559*.

4. Solorzano [Pereira], *Política indiana*, book VI, chap. VI, n. 17. Garcilaso, part I, p. 225. Acosta, book IV, chap. II. ▾ Lampadius, *Handbuch der Hüttenkunde*, vol. I, p. 401.

5. *Informe al Excelentíssimo Señor Conde de Salvatierra, Virey México, sobre el beneficio descubierto por el capitán ▾ Pedro Mendoza Melendez y Pedro Garcia de Tapia* (Mexico City, 1643), p. 19.

ing evidence. Cold amalgamation was found to be so profitable in Mexico that five years after the discovery of the technique in Medina in 1562, there were already thirty-five plants in Zacatecas[1] in which ores were treated with mercury, despite the fact that Zacatecas is three times farther from Pachuca than are the old mines of Tasco, Zultepeque, and Tlapujagua.

The miners of Mexico do not appear to follow established principles when choosing which ore are subjected to smelting or to amalgamation. One sees the same mineral substances being smelted in one mining district that one would expect to be treated exclusively with mercury in another district. The ore that contains muriate of silver, for example, are either smelted with soda carbonate (*tequesquite*) or are reserved for the techniques of cold and hot amalgamation; often the miner decides upon a particular method merely on the basis of how abundant and easily obtainable mercury is. In general, it is deemed necessary to treat the wealthier *poor ore*, those that contain between ten and twelve marks of silver per quintal, argentiferous lead sulfide, and ore mixed with blende and vitreous [glassy] copper. The following, on the other hand, are amalgamated profitably: *pacos* or *colorados*[2] lacking a metallic shine; vitreous [glassy], red, black, and native horn silver; silver-rich Fahlerz, and all the poor ore that are disseminated in very small particles throughout the gangue. III.258

The ore reserved for amalgamation must be triturated or reduced to a very fine powder, to obtain the greatest possible contact with the mercury. Of all the metallurgical operations, this trituration through *arrastras* or *mills*, which we have mentioned above, is the one that is carried out to the greatest degree of perfection in most of the plants of Mexico. Nowhere in Europe have I seen such fine and evenly grained mineral flour, or *schlich*, as in the great *haciendas de plata* of Guanajuato, which belong to the Count of la Valenciana, Colonel Rul, and Count Pérez Gálvez. If the ore are highly pyritic, they are roasted (*quema[do]*), either outdoors, in piles on wooden beds, as in Sombrerete, or in *schlich*, in reverberatory ovens (*comalillos*). I have encountered the latter in Tehuilotepec. They are twelve meters long; they have no chimneys but are powered by two fires, the flames of which III.259

1. El Conde de Santiago de la Laguna [José de Rivera Bernárdez], *Descripción [breve] de la [muy noble y leal] Ciudad de Zacatecas*, p. 42.

2. ▼ Alvaro Alonzo Barba, *Arte de [los] metales*, 1639, book II, chap. IV. Felipe de la Torre Barrio y Lima, minero de San Juan de Lucanas, *Tratado de azogueria* [*Arte, o cartilla del nuevo beneficio de la plata en todo genero de metales frios, y calientes*] (Lima, 1738). ▼ Juan de Ordoñez, *Cartilla sobre el beneficio azogue* [*Arte o nuevo modo de beneficiar los metales*] (Mexico City, 1758). ▼ Francisco Javier de Sarria, *Ensayo de metalurgia* (Mexico City, 1784).

pass through [traverse] the laboratory. This chemical preparation of the ore is nevertheless quite rare; the sheer volume of the substances that must be treated by amalgamation, as well as the lack of combustibles on the plateau of New Spain, makes roasting both difficult and costly.

Dry pounding is carried out under pestles (*mazos*), eight of which work together and are driven either by hydraulic wheels or by mules. The crushed ore (*granza*) passes through a skin pierced through with holes: it is reduced to a very fine flour via the *arrastras* or *tahonas*, which are called *sencillas* or *de marco*, depending on whether they are equipped with two or four blocks of porphyry or basalt (*piedras voladoras*), which turn in a circle with a circumference of nine to twelve meters. Twelve to fifteen of these *arrastras* or mills are typically arranged in a single row in a single shed: they are set in motion either by water or by mules that are changed every eight hours. Over the course of twenty-four hours, a single one of these machines crushes three to four hundred kilograms of ore. The humid *schlich* (*lama*) that emerges from the *arrastras* is occasionally washed again in pits (*estanques de deslamar*) whose construction, in the mining district of Zacatecas, was recently perfected by Mr. Garcés. When the ore are very wealthy, as in the Rayas mine at Guanajuato, they are only ground to the level of a coarse sand (*xalsonte*) using the *mill* stones, and the wealthiest metallic grains (*polvillos*) are then removed through washing and are reserved for smelting. This operation, which is very economical, is called *apartar polvillos*.

III.260

I was assured that when the silver ore that is very poor in gold is reserved for amalgamation, mercury is poured into the trough, at the bottom of which turn the stones of the *arrastras*. The auriferous amalgam thus forms while the ore is reduced to powder form, and the spinning motion of the *piedras voladoras* encourages the metals to combine. I did not have the opportunity to witness this operation, which is not used at all at Guanajuato. In a few large plants of New Spain, for example at Regla, arrastras are still totally unknown; people are content to rely on pounding; the *schlich* that emerges under the pestles is then passed through sieves (*cedazos* and *tolvas*). This flour preparation method is highly inaccurate: a coarse and unevenly grained powder amalgamates very poorly, and the health of the workers is seriously affected in places where a cloud of metallic dust is constantly produced.

The moistened *schlich* is carried from the mills or *arrastras* into the amalgamation yard (*patio* or *galera*), which is typically paved with slabs. The flour is arranged into heaps (*montones*), each of which contains between fifteen and thirty-five quintals: forty or fifty of these *montones* form a *torte* (*torta*); this is the name given to the mass of humid *schlich* that is left out in the open air, and which is often twenty to thirty meters wide and five to six decimeters thick. For this amalgamation in a paved yard (*en patio*), which is the most widely used technique in the Americas, the following materials are used: muriate of soda (*sal blanca*), iron and copper sulfate (*magistral*), lime, and plant ashes.

III.261

The salt that is used in New Spain is of very inconsistent purity, depending on whether it comes from the salt marshes surrounding the port of Colima on the South Sea shore, or from the famous *laguna del Peñon Blanco*, between San Luis Potosí and Zacatecas. This lake, visited by Mr. Sonneschmidt, is situated at the base of a granite boulder on the slope of the Cordillera: it dries up every year in December. Every year, it yields nearly 250,000 *fanegas* of impure or earthy [muddy] salt (*sal tierra*), all of which are sold to the amalgamation plants. On site, the price of one fanega is half a piaster. The mining districts of the intendancy of Mexico City receive salt from the coasts of Veracruz, as well as from the springs of Chautla: at Tasco, the muriate of soda from Veracruz costs four piasters per quintal.

Magistral is a mixture of pyritic copper (*Kupferkies*) and iron sulfide, roasted for a few hours in a reverberatory oven and slowly cooled: if it is roasted further, what results is an acidic iron and copper sulfide blended with *maximum* oxidized iron. Sometimes,[1] though rarely, the *azogueros* (the name given to the persons responsible for the amalgamation) add muriate of soda to the pyrites while the latter are being roasted, which leads to the formation of both sulfate of soda and muriate (chloride) of copper and iron. I have also seen added to the magistral vitriolic or copperas earths (*tierras de tinta o de alcaparosa*), which are ochreous earths containing *maximum* oxidized iron and iron sulfates. In the mining district of Real de Moran, the preparation of the magistral is carried out using copper pyrites from San Juan Sitacora, which is sold at the rate of ten piasters per *carga*. Lime is obtained by calcining [charring] stone with very pure lime content and then extinguishing it in water; only rarely are alkaline ashes used in place of lime.

III.262

It is through the contact of these different substances—moistened metallic flour, mercury, muriate of soda, iron and copper sulfates, and lime—that the amalgam of silver is formed in the cold amalgamation technique (*de patio y por crudo*). The first step is to mix the salt with the metallic flour and to *stir* (*repasa[r]*) the torta. Depending on the purity of the

1. Garcés [y Eguía, *Nueva teórica y práctica del beneficio de los metales*,] p. 90.

salt employed, an amount of two and a half to twenty pounds is added to each quintal of *schlich*: if the muriate of soda is of below-average purity, 3 III.263 to 4 percent of it is taken. Those metals that are believed to require large amounts of salt, and in which the silver ore occurs in very sizable grains, are referred to as *metales salineros*: the ore, blended with salt (*metal ensalmorado*) is left to rest for several days, so that the salt can dissolve and spread evenly throughout. If the *azoguero* determines that the metals are *hot* (*calientes*), that is, in a state of oxidation, and therefore naturally laden either with iron and copper sulfates that decompose quickly upon contact with air, then lime is added to *cool* the mass. This operation is called *curtir los metales con cal*. If, on the other hand, the *schlich* appear too *cool* (*fríos*)—for example, if they come from ore that display a strong metallic shine, if they contain lead sulfide (*negrillos agalenados*) or pyrites that decompose poorly in humid air—then *magistral* is used instead; this operation is called *curtir con magistral*. Iron and copper sulfate is seen as having the property of *heating* the mass; it is not regarded as having been well prepared if, when moistened and placed in a person's hand, it causes the sensation of heat. In this case, sulfuric acid, which is concentrated in the acidic sulfate, attracts the water and combines with the latter, emitting heat.

We have just described two techniques for the chemical preparation of ore: salting (el ensalmorar) and the method of *tanning* (*curtir*) with lime III.264 or magistral. After a few days of rest, the next step is to *incorporate* (*incorporar*), that is, to mix the mercury with the metallic flour. The amount of mercury used is determined by the amount of silver that is believed to be extractable from the ore; typically, six times as much mercury as the amount of silver contained within the torta is employed in *incorporation* (*en el incorporo*). Three to four pounds of mercury are used for every mark of silver. Along with the mercury, or shortly thereafter, magistral is added to the mass, depending on the nature, or rather, in the barbarous parlance of the *azogueros*, on the *temperature* of the ore, *según los grados de frialdad*. One to seven pounds of magistral are used for every pound of mercury: if the mercury turns the color of lead (*color aplomado*), this is a sign that the torta is working, or that the chemical action has begun. To encourage this action and to increase the contact between the substances, the mass is *beaten* (*se da repaso*) or stirred, either by forcing twenty or so horses and mules to run in a circle for several hours, or by having the *schlich* trod by workers who, for entire days, walk barefoot in this metallic mud. Every day, the *azoguero* examines the condition of the flour; he makes a test (*la tentadura*) of it in a small wooden container (*jicara*); that is, he washes a sample of the schlich with water and determines, based on the appearance of the mercury and the amalgam, whether the mass is *too cool* or *too hot*. When the mercury turns an ashen color (*en lis cenicienta*), and when an extremely fine gray powder breaks off of it and clings to the fingers, it is said that the torta has too much *heat*: it is then cooled through the addition of lime. If, on the other hand, the mercury maintains a metallic shine; if it stays white, covered in a reddish or golden film (*telilla rojiza o de tornasol morado* or *en lis dorada*); if it does not appear to have any effect on the mass, then the amalgam is considered to be too *cold* and is believed to be *heated* (*calentar*) by being mixed with magistral.

III.265

It is thus that, over the course of two, three, or even five months, the levels of *magistral* and of lime in the *torta* are held in balance, for the effects are quite different, depending on the temperature of the atmosphere, the nature of the ore, and the movement to which the *schlich* is subjected. If it is believed that the [chemical] action is too intense and that the mass is working too hard, then it is left to rest further: if it is seen as preferable, rather, to accelerate the amalgamation and to increase the heat, then the *repasos* are repeated more frequently, using either men or mules. If the amalgam forms too quickly and occurs in the form of small globules called *pasillas* or *copos*, then the *torta* is fed (*se ceba la torta*) by adding mercury back with a bit of magistral, sometimes even salt. When the *azoguero*, relying on external indications, determines that the mercury has combined with the silver contained within the ore, and that the *torta has yielded* (*ha rendido*) they pour the metallic mud into vats, some of which are made of wood and others of stone. Small-scale mills equipped with vertically positioned wings rotate in these vats. These machines (*tinas de cal y canto*), which are especially well-built in Guanajuato, resemble the machines set up in Freiburg for filtering out the residues from amalgamation.[1] The earthy and oxidized materials are drained off by the water, while the amalgam and the mercury remain in the bottom of the vat. Since the force of the current also carries off a few globules of mercury, in the large plants one sees poor Indian women busy removing this metal from the washing water. The amalgam gathered at the bottom of the *tinas del lavadero* is separated from the mercury by being pressed through bags; it is then molded into pyramids that are covered with upturned crucible in the shape of a bell. The silver is separated from the mercury through distillation. In the technique that I have just described,

III.266

1. ▼ Fragoso de Siqueira, *Déscription [abrégée] de [. . .] l'amalgamation de Freiberg*, 1800, p. 26.

twelve to fourteen ounces of mercury are typically lost for every mark of silver that is removed, in other words, one and four-twentieths to one and seven-twentieths kilogram of mercury for every one kilogram of silver. In the technique introduced to Saxony by Mr. Gellert and ▾ Mr. Charpentier, the amount of mercury lost is two-tenths of a kilogram for every kilogram of silver, eight times less than in Mexico.[1]

We have described cold amalgamation (*por crudo y de patio*), in which, III.267 rather than being roasted, the ore are exposed to the open air in a courtyard. Medina was only familiar with the use of salt and iron and copper sulfates; but in 1586, fifteen years after his technique was introduced to Peru, a Peruvian miner, ▾ Carlos Corso de Leca,[2] discovered *beneficio de hierro*. He recommended stirring tiny iron plates in with the metallic flour, a mixture that would ensure that nine-tenths less mercury was lost. This technique, as we shall shortly see, is based on the decomposition of the muriate (chloride) of silver by iron, and on the attraction of this metal to sulfur: it is known to the Mexican *azogueros* but is not widely practiced at all. In 1590, Alonzo Barba proposed hot amalgamation, or amalgamation through cooking in copper vats. This technique is called *beneficio de caso y cocimiento*; it is the very one that Mr. de Born proposed in 1786. The loss of III.268 mercury is much less than in *beneficio por patio*, because the copper in the vessels serves to break down the silver chloride: at the same time, the heat encourages the operation, either by energizing the play of affinities, or by setting the liquid, which is beginning to boil, into motion. This amalgamation through cooking is used in several mines in Mexico that are especially rich in horn silver and in *colorados*. Juan de Ordoñez, whose work has been cited above, even recommended amalgamating in stove-warmed hothouses. In 1676, ▾ Juan de Corrosegarra invented a technique that is in very limited use today, called *benificio de la pella de plata*, and in which silver amalgam

1. In a typical year, at the Halsbrück plant, near Freiberg, fifty-eight to sixty thousand quintals of poor ore, which *hold* seven to eight lots of silver per quintal (there are two *lots* in an ounce), are treated through amalgamation. The loss of mercury involved in amalgamation, in the strict sense of the word (*im Anquicken*), and in the washing of residues, is three-quarters of an ounce (or a lot and a quarter) per quintal of ore. In mercury vaporization (*im Ausglühen*), one-quarter lot of mercury is lost for an amount of silver corresponding to one quintal of ore; from which it results, according to Mr. Héron de Villefosse, that for 60,000 quintals of ore, twenty-five and a half quintals of mercury are either lost or *destroyed*. (Lampadius, [*Handbuch der Allgemeinen Hüttenkunde*,] vol. II, p. 178.)

2. ▾ Carvajal y Sande, *Carta de Don Juan Carbajal y Sandi, Presidente de la Real Audencia de la Plata, al Excelentísimo Señor Conde de Chinchon Virey del Perú*, 1736 [*To the Crown, Mar. 18, 1636*].

that has already formed is added to the mercury. It is claimed that this amalgam (*pella*) facilitates silver extraction, and that there is an even smaller loss of mercury since the amalgam spreads with greater difficulty through the mass. A fifth method is the *beneficio de la colpa*, in which, instead of an artificial *magistral*, which contains much more copper sulfate than iron sulfate, relies on the use of *colpa*, which is a natural mixture of acidic iron sulfate and *maximum* iron oxide. This *beneficio de la colpa*, advocated by ▼ Don Lorenzo de la Torre, presents some of the advantages that we have just mentioned in our discussion of iron-based amalgamation.

The technique invented by the miner from Pachuca is one of those chemical operations that have been practiced with a certain degree of success for centuries, without the persons who extract silver from its ore using mercury having the slightest idea about the nature of the substances employed. The *azogueros* speak of a mass of ore as though it were an organic body whose natural heat they merely raise or lower. Similar to doctors who, throughout centuries of barbarism, divided all foods and all remedies into two classes, hot and cold, the *azogueros* see in the ore only the substances that must be heated using sulfates, if they are too cold, or that must be cooled using alkalis, if they are too hot. The practice, already introduced in Pliny's time, of rubbing the metal with salt before applying the gold amalgam, probably gave rise to the use of muriate of soda in the Mexican amalgamation technique: "this salt," say the *azogueros*, "serves to clean (*limpiar*, *castrar*) and to strip off (*desenzurronar*) the silver, which is enveloped in sulfur, arsenic, and antimony, as though in a skin (*tellila* or *capuz*), the presence of which prevents the silver from coming into direct contact with the mercury." The action of the latter metal is made more energetic by the sulfates that heat the mass: it is even probable that ▼ Medina only used the iron and copper sulfates and the muriate of soda simultaneously because he recognized in his first attempts that the use of salt only enhances the technique when applied to ore that contain decomposed pyrites. Having no clear idea about the action of the sulfates upon the muriate of soda, he attempted to *remake the ore*, that is, to add magistral to those that the miner regards as *non-vitriolic*.

III.269

III.270

Subsequent to the beginning of the practice of silver ore amalgamation in Europe and the gathering of scholars at the metallurgical congress of Schemnitz,[1] the muddled theory of Barba and the American *azogueros* has been replaced by more sound ideas, better adapted to the current state of chemistry. I shall attempt here to explain them succinctly, using exclusively

1. To be more precise, Szkleno or Glashütte, near Schemnitz.

(in this part of my work) the chemical nomenclature that the discoveries of▼ Mr. Davy, Mr. Gay-Lussac, and Mr. Thenard have made indispensable. It is assumed that everything that happens in the plant at Freiberg, where, in just a few hours a mass of roasted ore is amalgamated, must occur at a very slow pace in the amalgamation process of Mexico, where the ore are typically not roasted, and where they remain exposed to the open air, to the sun, and to the rain for several months. It is believed that in the moistened mixture of silver ore, mercury, salt, lime, and *magistral*, the latter, which is an acidic iron and copper sulfate, breaks down the sodium chloride upon contact with the oxygen in the air; that soda sulfate and silver III.271 chloride are formed; and that the latter is broken down by the mercury that bonds with the silver. It is accepted that the lime or the potash is added to prevent the overabundant sulfuric acid from acting upon the mercury. According to this explanation, the silver found in the ore in its metallic state, although bonded with sulfur,[1] antimony,[2] iron,[3] copper,[4] zinc,[5] arsenic,[6] and lead, passes to the state of chloride before combining with the mercury.

A Mexican author, Mr. Garcés [y Eguía],[7] whom we have cited on several occasions, believes, on the contrary, that no muriate of silver is formed during the amalgamation procedure; his opinion is that "the muriatic acid only combines with the metals that are already bonded with the silver; the water washes away the soluble muriates of iron and copper," and that "the silver, released from its metallic substances, combines freely with the mercury." This explanation, which appears to be somewhat simplistic, seems contrary to the laws of affinity. If the chlorine released by the action of the III.272 sulfates upon the sodium chloride acts, in circumstances similar to those of *por patio* amalgamation, upon an ordinary silver ore, for example, on prismatic black silver ore containing silver, iron, antimony, sulfur, copper, and arsenic, then silver chloride will necessarily form at the same time as do the chlorides of other metals. Mr. Garcés's theory is equally inapplicable to the amalgamation of silver sulfide ore, which are abundant and widespread throughout most of the veins of Mexico.

1. In Glaserz, red silver, and Weissgültigerz.

2. In red silver, Weissgültigerz, and the Fahlerz of Annaberg.

3. In prismatic black silver (Schwarzgültigerz, Sprödglaserz) and in Fahlerz. Klaproth, *Beiträge*, vol. I, p. 162; [Delius,] *Bergbaukunst*, vol. I, p. 239 [63].

4. In the Fahlerz of Kapalk. Klaproth, [*Beiträge*,] vol. IV, p. 61.

5. In the gray argentiferous copper of Freiberg.

6. In Weissgültigerz.

7. [Garcés y Eguía, *Nueva] teórica [y práctica] del beneficio [de los metales*,] pp. 112–16.

Without entering in this work into an in-depth discussion of the phenomena presented by the contact of so many heterogeneous substances; without resolving the important question of whether cold amalgamation is possible without salt or magistral, I shall limit myself to citing several experiments that Mr. Gay-Lussac and I conducted,[1] which may shed some light on Mexican amalgamation.

It is not true that the sulfur blend completely blocks the silver from bonding with the mercury, and that silver sulfide only yields a cold amalgam if sodium chloride and iron sulfate are added. We observed, on the contrary, that by triturating mercury and artificial silver sulfide, the mercury is swiftly subdued, and one obtains a small amount of silver from the distillation of the amalgam. We blended mercury with vitreous silver ore crushed into powder: after forty-eight hours of contact, a small amount of silver amalgam had formed. Both this experiment and the subsequent ones focused on two or three grams of ore, the air temperature being between ten and twelve degrees Centigrade, and the mixtures having been slightly moistened.

III.273

While copying *de patio* amalgamation, commonly practiced in Mexico, and while cold-blending natural silver sulfide, iron sulfate, sodium and lime chloride,[2] we did not find any trace of silver sulfide, although this mixture remained in contact for one week. On the contrary, we did obtain it when the mass was exposed for a few hours to an artificial temperature of thirty to thirty-four degrees Centigrade. In the hot region of New Spain, the *tortas* exposed to the sun heat up even more. The amalgamation is also observably much slower on the plateaus, where the thermometer drops to the freezing point, than in the deep valleys and the plains near the coasts. It is probable that the silver chloride that swiftly forms at the temperature of thirty-four degrees would take a long time to form at a much lower temperature.

III.274

When one cold-blends sodium chloride, iron sulfate, and mercury, one obtains mercury chloride. This very chloride is also present when one triturates mercury with artificial silver chloride. Mercury chloride was formed from a blend of sodium chloride, acidic iron sulfate, and mercury. It is conceivable that, in the technique for large-scale amalgamation, one part of the

1. At the laboratory of the École Polytechnique in Paris, in the winter of 1810.

2. When we prepared the same blend, but without lime, a cold formation of a small amount of silver chloride occurred after a few days. While preparing the same blend, but adding iron filings, we obtained a large amount of cold silver amalgam. A small amount of this amalgam also formed, quite quickly, when we cold-blended silver sulfide, copper sulfide, lime, and mercury. Sulfur was not released through contact with silver and lime sulfide alone.

mercury is converted into chloride in two distinct ways; that is, through the decomposition of the silver chloride and through the direct action of the magistral and the salt, which is used in excess. Lime, which remedies the latter mode of action, does not remove the sulfur from the silver without heat; for when natural silver sulfide is mixed with lime, no lime sulfide is formed, even though the mixture has been triturated for several days. Lime[1] blocks silver and mercury from combining in the most remarkable way: one

III.275 observes that mercury is subdued with great difficulty when one triturates a blend of lime, silver sulfide, and mercury. Similarly, when creating a paste of silver ore, salt, magistral, and mercury, and when triturating *schlich* until the mercury becomes invisible, one sees the mercury separate from the metallic flour and gather into large masses as soon as one adds lime: globules of mercury, which increase little by little in size, appear everywhere molecules of lime come in contact with the blend. It is because of the peculiar action of lime, well-deserving of chemists' attention, that the *azogueros* say that it *cools* mercury, or that it *prevents the torta from working.*[2]

III.276 Mr. Gay-Lussac and I had no problems whatsoever reproducing on a smaller scale the *beneficio de hierro*, an ingenious technique known in Peru since the end of the sixteenth century and introduced by Mr. Gellert in the plants of Saxony. We have seen that by cold-blending sulfur, natural silver,

1. On the use of lime to break down the last portions of the iron sulfate that forms through the roasting of the ore mixed with *magistral*, and on the utility that could be derived from the ammoniac liquid that dissolves and removes the silver chloride, see a note by ▼ Mr. Rivero [Ustariz, "Sur le traitement des mines d'argent par l'algamation,] *Bulletin de la Société philomathique de Paris*, 1822, p. 86.

2. When chemists still accepted that the action of iron sulfate on common salt caused muriatic (hydrochloric) acid to be released during amalgamation, it must have been assumed that in *schlich* this same acid attacked the silver, despite the fact that silver is found in a metallic state in its ore. No aspects of this assumption were, in fact, contrary to the most widely known facts. When treating vitreous silver with hydrochloric acid, one obtains copious amounts of silver chloride: when this same acid is poured on natural silver sulfide, hydrogen sulfide is released. Mr. Proust observed that the piasters that fell to the bottom of the sea during the famous wreck of the ship *San Pedro Alcantara* were covered, in a short period of time, with a crust of silver chloride half a millimeter thick; I made the same observation during my stay in Peru, during the wreck of the frigate Santa Leocadia, on the South Sea coasts, near Cabo Santa Elena. In the [*Neue*] *Nordische Beyträge*, vol. III, p. 64 [pp. 401–2], Mr. Pallas affirms that in Siberia, on the banks of the Yaik [Ural] River, ancient Tartar coins were found converted into silver chloride through contact with soil laden with sodium chloride. All of the facts would seem to prove that, in many circumstances, hydrochloric acid acts on metallic silver. But such as we must envisage today the decompositions that result from contact with *magistral*, sea salt, and silver ore, it is chloride and not hydrochloric acid that passes from sodium to silver.

salt, *magistral*, lime, and mercury, the amalgam forms in greater amounts when iron shavings are added to the *torta*: in this instance, the iron not only serves to break down the silver chloride, as in the Freiberg amalgamation technique, but also, in particular, to separate the sulfur from the mineralized silver. When the silver sulfide and the iron shavings were allowed to interact for twenty-four hours, the silver was so thoroughly stripped that by adding mercury, we obtained a large amount of silver amalgam in only a few minutes. If one pours hydrochloric acid on the blend, a great deal more hydrogen sulfide is released than when treating silver sulfide alone with the same acid. What type of action is exhibited by the *maximum* oxidized iron contained in the *colorados* or *pacos* that are so easily amalgamated?

The enormous loss of mercury that one observes in American amalgamation comes from several simultaneous causes. If, through the *por patio* technique, all the silver extracted were due to the decomposition of the silver chloride by the mercury, then the amount of mercury that would be lost and the amount of silver from the chloride would be in a ratio of twenty-seven to twenty-five, in those cases where mercury perchloride forms. Another loss, perhaps the most significant of all, is due to the formation of mercury chloride through simultaneous contact with *magistral* and common salt. A third of the mercury is lost because it remains spread throughout an immense mass of moistened *schlich*, and this division of the metal is so great that even the most careful wash is not enough to collect all of the molecules hidden in the residue. Finally, a fourth cause of the loss of mercury can be traced to its contact with salty water, to its exposure to the open air, and to the rays of the sun over the course of three, four, and even five months. These masses of mercury and schlich, which contain a large number of heterogeneous metallic substances moistened by saline solutions, are composed of an endless amount of small *galvanic piles*, the slow but prolonged action of which encourages the oxidation of the mercury and the play of chemical affinities. III.277

Together, these studies suggest that the use of iron would noticeably perfect the technique of amalgamation. If the ore treated contained only vitreous silver, then iron shavings alone would perhaps suffice to strip the silver and separate it from the sulfur, which delays the bonding of the silver with the mercury. But since there are, in all other silver ore, various metals in addition to sulfur that are combined with the silver, the simultaneous use of soda chloride, copper, and iron sulfates becomes necessary for encouraging the release of the chloride, which combines with copper, iron, antimony, lead, and silver. Not only iron chloride but also copper, zinc, and III.278

arsenic chloride remain dissolved: the silver chloride, which is eminently insoluble, breaks down through contact with mercury.

It has long been suggested to pave the ground on which the tortas lie with sheets of iron and copper instead of with slabs. Attempts have been made to stir (*repasar*) the mass by having it be worked by plows in which the plowshare and coulter were made of the two metals we have just named; but the mules suffered too much from this plowing, for the schlich formed a thick and relatively stiff paste. Furthermore, the practice of the schlich being trod by mules, instead of using men, only dates back to the year 1783 in Mexico. *Don Juan Cornejo* brought the idea for this technique to Peru: the government granted him a privilege that he did not enjoy for long, and which brought him only 300,000 livres tournois—a rather unremarkable sum, considering that the costs of amalgamation have dropped by perhaps III.279 one-quarter, since there is no more need to employ the large number of workers who used to walk barefoot on masses of metallic flour.

Amalgamation, as we have described it, serves to remove all the silver from ores treated with mercury, provided that the *azoguero* is experienced and that he is very familiar with the appearance or the external characteristics of mercury, through which it is determined whether the *torta* needs lime or iron sulfate. At Guanajuato, where the plants are the best managed, ore that contain only three-quarters of an ounce of silver per quintal are successfully amalgamated: Mr. Sonneschmidt found only one-sixteenth of an ounce of silver in the amalgamation residue that came from ore of which each quintal[1] contained five to six marks of silver. In the plants of Regla, on the other hand, the *schlich* is often subjected to the wash before the mercury has extracted all of the silver that is found in the torta; and it is believed in Mexico City that the father of the current owner of the famous *Vizcaína* mine threw out an enormous mass of silver into the river along with the residue.

The technique discovered by Medina has the great advantage of simplicity; it does not call for buildings to be constructed, nor does it require fuel, machines, or more than a very small amount of driving force. Using por patio III.280 amalgamation, with mercury and a few mules to drive the *arrastras*, it is possible to extract silver from all the *poor* ore near the well from which they are removed, even in the middle of the desert, provided that the ground is flat enough to set up the *tortas*; but this same technique has the significant disadvantage of being slow and of causing an enormous loss of mercury. Since

1. Sonneschmidt, *Miner[alogische] Beschreibung der [vorzüglichsten]Bergwerks-Reviere*, p. 103.

the latter is extremely divided and since thousands of quintals of mercury are all worked at the same time, it is not possible to collect the oxide and the muriate of mercury that are carried off by the washing water. In the amalgamation method followed in Europe, and that originated from the scholarly research of Mr. Born, Mr. Ruprecht, Mr. Gellert, and Mr. Charpentier, the silver is extracted in a twenty-four-hour period: less time is needed—from sixty to one hundred fifty times less—than in the Spanish colonies, and, as we have demonstrated above, eight times less mercury is consumed.

Is it possible to introduce to Mexico or to Peru the Freiberg technique, which is based upon the roasting of ore and on the gyratory movement of the barrels? In Freiberg, 60,000 quintals of ore are amalgamated per year. It is not easy to estimate with equal precision the enormous mass of ore subjected to the various amalgamation techniques across the whole extent of New Spain every year. In his *Nueva Teoria [. . .] del beneficio de los metales* (p. 121), Mr. Garcés [y Eguía] sets this amount at 7,700,000 quintals, but this author vastly underestimates the wealth of the ore at barely two and two-third ounces per quintal. III.281 The yield of Valenciana mine may provide us with a more satisfying general overview. In an average year, at the beginning of this century (1800–1803), this mine produced nearly 720,000 quintals of ore for smelting and amalgamation, from which 360,000 marks of silver were removed. Yet, recalling that, in New Spain as a whole, smelting ore is in a proportion of one to three and a half to amalgamation ore, and that from 1785 to 1789 the entire *real de minas* of Guanajuato yielded 531,000 *marcos de fuego* and 1,938,000 *marcos de azogue* in *plata quintada*, one finds that in good years, the Valenciana mine has provided its amalgamation plants with 560,000 quintals of ore, containing 284,000 marks of silver. This analogy would suggest that at the time when the total output of silver from smelting and amalgamation from its mines was 2,338,000 marks, the whole of New Spain would have produced at the very least 3,600,000 quintals of ore for amalgamation alone. One barrel can amalgamate 3,000 to 3,600 quintals of ore per year. It is likely that the total mass of ore produced in Mexico in good years—1,860,000 marks of silver removed by mercury—has a *mean wealth* of less than what the Valenciana ore had at the beginning of the century. On the other hand, the perfecting of the concentration techniques[1] (*process of* III.282 *dressing and washing the ore*) may well decrease the overall volume. In this case,

1. On the perfecting of the *process of dressing the ore[s]*, compare the remarks of ▼ Mr. John Taylor, a man who combines a very broad perspective with a deep knowledge of the miner's art, in his *Selections [from the Works of the Baron de Humboldt,] Relating to [the Climate, Inhabitants, Productions, and Mines of] Mexico*, 1824, vol. XXIII, pp. 175–227.

1,200 to 1,500 barrels would suffice for all of Mexico. I shall not pursue these more curious than useful numerical considerations any further but shall recall here that it will be easier to locate a driving force to turn the barrels than it will be to roast the ore in a country that in many areas lacks fuel and where the most productive mines are found on plateaus that are completely lacking in forests.

[In the beautiful plant of Halsbrücke (near Freiberg, in Saxony), which contains twenty *barrels*, the blend of ore most favorable to amalgamation is the one that yields three and a half to four ounces of silver per quintal (of 110 pounds). To the *schlich* is added 10 percent sea salt. Each barrel contains three quintals of water, ten quintals of ore, and fifty to sixty pounds of wrought iron. Five barrels yields one quintal of amalgam, composed of six parts mercury and one part silver. To amalgamate one quintal of ore, one loses one-half to three-quarters of an ounce of mercury. After amalgamation, the water in the barrels contains muriate of soda, muriate of manganese, and soda sulfate. Since the techniques commonly used in Freiberg consistently employ terms comparable to those of the Mexican techniques, it will be useful to list here the numerical results published recently by ▼ Mr. Berthier, professor at the Royal School of Mines, in his excellent paper on the *metallurgical treatment of copper and silver alloys by mercury*. "According to official documents in the Halsbrücke plant, in the third trimester of the year 1822, the following amounts were treated:

III.283

	16,116 quintals	three-eighths of the ores from Freiberg, containing,[1] according to the analysis	6556.15	marks of silver
	327 quintals	seven-eighths of ore from the Upper Erzgebirge containing	490.6	
Total	16,444 quintals	Three-eighths, containing	7047.5	marks of silver
	The raw ore has successively produced:		16444 2/8	quintals
	Roasted ore		15074	quintals

Amalgamated silver	289	quintals
Red amalgam	10624	marks
Impure silver	10415	marks
This impure silver contains, according to the analysis, pure silver	7324.11	
To be extracted for analysis	9.2	
To be extracted as *remedium*	256.15	
Remaining pure silver	7058.10	7058 marks, 10 loths.

[1] 1 quintal = 110 pounds = 53,845 kilograms.

1 pound = 2 marks = 32 lots: 1 lot = 4 gros.

1 écu = 24 gros: 3 écus = approximately 11 francs.

1 measure of wood = 324 cubic Leipzig feet.

1 bushel of coal = 4.7 cubic feet.

1 bagful of charcoal or peat = 12 baskets.

1 basket = 14 cubic Leipzig feet.

Based on the analysis of the ore, there should only be 70,471.5 [marks]. The operation therefore yielded an additional eleven marks and five lots.

The consumption and expenses were as follows:

II.284

OPERATION	CONSUMPTION AND REASON FOR EXPENSES	EXPENSES	
Purchase of		*écus gros*	
ore	Paid for quarrying	57,883.00	
	Royal grant and other contributions	11,919.22	70,054.20
	For analyses	44.23	
	For hauling to the factory	206.23	
Roasting	1,558 quintals of sea salt, at 2 ecus 12 gros	3,895	
	5 measures of wood, at 8 ecus 8 gros	41.16	
	5,236 bushels of coal, at 12 1/8 gros	2,645.6	7,946.19
	For the roasters	1,175.6	
	For unskilled workers	189.15	
Amalgamation	For the sifters	215.4	
	For the molders	533.18	

	6 and three-quarter quintals, 9 pounds of mercury lost in amalgamation at 82 and a half ecus	564	1780.16
	16 quintals of iron plates, at 6 ecus 2 gros	110	
	Labor	357.18	
Distillation	5 quintals 4 pounds of mercury lost in the distillation of the amalgam	54.22	
	5 cartloads 3 baskets of coal, at 5 ecus 12 gros	28.21	159.6
	19 cartloads 3 baskets of coal, at 2 ecus 2 gros	55.8	
	Labor	20.3	
Silver fusion	6 cartloads 2 baskets of coal	33.22	
	6 cartloads 10 baskets of coal for analysis	37.14	77.15
	Labor	6.3	
			écus gros
	Total		80,018.4

These data reveal:

(1) taking into account the value of the ore, at Freiburg, 328 francs of expenses are incurred in the extraction of silver from 1,000 kilograms of ore containing approximately 0.0019 of fine silver; that is, forty francs for all treatment costs, as well as the consumption of 0.454 kilograms of mercury, worth 2.54 francs. (In the operation, 14.78 kilograms are used, but the bulk is collected in distillation);

III.285

(2) to obtain one kilogram of silver, twenty francs in treatment costs are incurred, and 0.364 kilograms of mercury are consumed;
(3) to roast 1,000 kilograms of ore, 32.35 francs are spent, and ninety-five kilograms of sea salt are consumed;
(4) to amalgamate 1,000 kilograms of baked ore, 7.10 francs are spent;
(5) to distill 1,000 kilograms of amalgamated silver, 36.60 francs are spent;

(6) and, finally, that ninety-four francs are spent in smelting 1,000 kilograms of red amalgam

"In these calculations, no mention has been made of either administration costs or maintenance expenses, not to mention the interest on the capital invested in the establishment, because the necessary information was missing; but it appears that these elements would not significantly change the final result presented in the table."

Mr. Berthier has performed a chemical examination on a portion of the output from the Halsbrücke establishment in order to give an exact idea of what takes place in metallurgical work.

"Among the ore, there are some that are highly pyritic and others that are nearly devoid of pyrites. The least pyritic ore are the white or white-grayish ones; they contain a large amount of quartz and baryta sulfate, around 0.15 of iron and manganese carbonate, arsenical pyrites in very small amounts, as well as a bit of copper, and when assayed, they yield 0.0018 of silver; when they are boiled with concentrated sulfuric acid, all of the copper dissolves, but the liquor contains only a trace of silver.

"The blend of the various ore prepared in the plant for eventual roasting is composed (excluding the sea salt, which is added in the proportion of one-eighth to one-tenth) of: III.286

Quartz Barite sulfate, etc.	} 0.278	
Carbonate of lime	0.050	
Carbonate of magnesium	0.030	
Carbonate of manganese	0.042	
Carbonate of iron	0.045	
Carbonate of copper	0.012	
Carbonate of lead	0.040	
Metallic iron Sulfur	0.197 0.194	} Iron persulfate 0.285
Arsenic	0.092	Mispickel 0.198
Silver	0.002	—
	0.982	0.483

"Through roasting, the silver is brought, in its totality, to the state of chloride; for when the roasted ore is digested with ammonia, all of this metal is dissolved, and by saturating the alkali with nitric acid, it is precipitated in a chloride form.

"When analyzed, the mud from the amalgamation barrels showed:

Materials insoluble in acids	0.446
Iron peroxide	0.380
Lime, aluminum, sulfuric acid	0.018
Copper oxide	0.010
Lead oxide	0.028
Water-soluble salts	0.100
	0.982

"When assayed, they exhibited only 0.0002 of silver, which demonstrates that this metal is very thoroughly and precisely removed by mercury.

"The amalgamation water, in which are found all of the soluble salts formed in the operation, contain:

	Sulfate of soda	0.069	0.526
	Muriate of soda	0.019	0.143
Anhydrous salts	Muriate of magnesium	0.009	0.067
	Muriate of manganese	0.036	0.264
	Iron, copper, mercury	0.000	0.000
		0.133	1.000

"It is used, they say, as a fertilizer. It would be easy to extract large amounts of sulfate from it. One might expect to find iron salts in it; but it appears that this metal, constantly agitated by contact with the air, passes completely to a peroxide state, and that this oxide is precipitated by bases that are stronger than it, namely, lime, magnesium, and manganese protoxide.

"In addition to silver, the amalgam contains copper and traces of other metals. The amalgam that flows directly from the tuns is much less coppery than what is collected through the washing of the mud in the vats. In the silver that comes from the distillation of the first amalgam, 0.15 to 0.20 of copper has been found per 0.33 of silver, and in the residue from the distillation of the second, 0.67 of copper has been found per 0.33 of silver. Be that as it may, one notes that a large part of the copper does not pass into the amalgam. It is likely that the only portion of it that combines with the mercury is the one that is still found in a chloride state when the roasting is complete." See Mr. Berthier's paper ["Sur le traitement métallurgique des alliages de cuivre et d'argent"] in *Annales des mines*, vol. XI, pp. 86–91.]

Now that we have discussed the amalgamation commonly practiced in the Americas, it remains for us to tackle a very significant problem: that of the amount of mercury required annually by the mines of New Spain. In general, Mexico and Peru produce larger amounts of silver the more mercury they receive and the lower the prices are. When they do not have this III.288 metal, as often happens during times of war at sea, mining is practiced with less intensity; the ore accumulate in the plants without anyone being able to extract the silver from them. Wealthy mine owners, who store in their warehouses two to three million francs' worth of ore, often lack the money necessary to meet the day-to-day costs of their mines. On the other hand, the more mercury that Spanish America demands, either because of the flourishing state of its mines or because of the amalgamation technique that is practiced there, the more the price of this metal rises in Europe. The few countries that nature has provided with mercury—Spain, the department of Mont-Tonnerre, Carniola, and Transylvania—benefit from this increase; but the silver mining districts, where advances in amalgamation are all the more desirable since they lack the fuel necessary for smelting, experience the negative effects of the large exportation of mercury to the Americas.

New Spain consumes 16,000 quintals of mercury per year. The following table indicates the amount of mercury that is lost in the amalgamation techniques used in different mining districts to remove silver from its ore. A loss (*perdida y consumo*) of 200 marks, or one quintal, of mercury is expected,

---

I.289

For every	125 marks of silver	in the mines of Guanajuato;
	115	in the mines of the intendancy of Guadalajara;
	100	in the mines of Pachuca, Zacatecas, Sombrerete, Guadiana, Durango, Parral, Zichu, Tonala, Comanja, Zerralbo, Temextla, Villalta, Tetela de Tonatla, Alchichica, Tepeaca, Zimapán, Cairo, and Tlapa;

90	in the mines of Chichiapa, Tetela, Tasco, Santa Theresa de Leiba y Baños, Ituquaro, Tehuistla, San Esteban de Albuquerque, and Chiconasi;
85	in the mines of Temascaltepec, Ayuteco, and Cuautla de la Sal;
80	in the mines of Zacualpa, San Luis Potosí, Cuautla, Sultepeque, and Tlalpujahua.

It is on the basis of these data, as well as the amount of silver extracted annually in the various mining districts, that the government determines the distribution (*repartimiento*) of mercury.

In his interesting work *Beschreibung der spanischen Amalgamation oder Verquikkung des in den Erzen verborgenen Silbers, wie sie bey den Bergwerken in Mexico gebräuchlich ist* (Gotha, 1810), Mr. Sonneschmidt confirms that *por crudo y de patio* amalgamation in New Spain does not commonly last less than eight days or more than two months, assuming, nonetheless, that the copper sulfate or *magistral* is of high quality and that an excessively low air temperature does not inhibit the action of the mercury on the silver. One quintal of ore containing three and a half to four ounces of silver costs five to six francs to amalgamate in Mexico, including the mercury lost. Mr. Sonneschmidt estimates the loss of mercury to be ten, twelve, or fourteen ounces per mark of silver; he allows for eight ounces of *consumed* mercury (*azogue consumido*) and three to six ounces of *lost* mercury (*azogue perdido*).

III.290

Having reserved for itself the exclusive right to sell mercury, either Spanish or foreign, since 1784 the court of Madrid has concluded a contract with the emperor of Austria, according to which the latter provides mercury at a price of fifty-two piasters: during peacetime, the court annually ships, via the royal fleet between 9,000 and 24,000 quintals. In 1803, the very practical idea was conceived of supplying Mexico with several years' worth of mercury so that, in the unforeseen case of war, amalgamation would not be hindered by the lack of mercury; but this project (*del repuesto*) shared the fate of so many others that were not carried out. Before the year 1770, when mining was much less intensive than it is today, the only mercury that New Spain received was from Almaden and Huancavelica. German mercury provided by the Austrian government, the bulk of which is from Idria, was only introduced to Mexico following the collapse of the underground works of Huancavelica, and at a time when the Almaden mine,[1] flooded in most of its works, had only a very low output. But in 1800 and 1802, this mine was again flourishing and alone yielded over 20,000 quintals of mercury per year, so that one can imagine the desire to forgo supplying Mexico and Peru with mercury from Germany. There were years in which 10,000 to 12,000 quintals of this mercury were imported through Veracruz. In general, between 1762 and 1781, the amalgamation plants of New Spain consumed[2] the enormous sum of 191,405 quintals of mercury, the value of which, in the Americas, was over sixty million livres tournois.

III.291

As the price of mercury has dropped, mining has increased. In 1590, under the viceroy Don Luis de Velasco II, one quintal of mercury sold for 187 piasters in Mexico City.[3] But over the course of the eighteenth century, the value of this metal had already decreased to such an extent that in 1750, the court distributed it to the miners for eighty-two piasters. From 1767 to 1776, its price was sixty-two piasters per quintal. In 1777, under the administration of Minister Gálvez, a royal decree set the price of Almaden mercury at forty-two piasters and two reales and that of German mercury at sixty-three piasters. At Guanajuato, because of the costly hauling on muleback, the price of both types of mercury rises by two to two and a half piasters per quintal. Because of the difference between the weights used in Germany and in Mexico, the king gains 23 percent on Idria mercury,[4] such that a wise politician should persuade the metropole to sell it at a lower price. Following an old custom, the miners of certain mining districts, for example those of Guanajuato and Zacatecas, are allowed to purchase two-thirds of Spanish mercury and only one-third of German mercury. Other

III.292

1. On these mines and on those of Almadenejos, see the studies of ▼ Mr. Cocquebert de Montbret in [France,] *Journal des mines*, no. 17, p. 396.

2. Through simple oxidation or through chemical combination with muriatic acid.

3. In the *Reglamento por la distribución y venta del azogue* from the year 1590, it is written that "heretofore, one quintal of mercury was sold in Mexico City for 113 *pesos de minas*." The *Real Cédula* of October 17, 1617, set the price at sixty *ducados* per quintal. That price was legally maintained until the end of the seventeenth century and the beginning of the eighteenth century. In 1727, the viceroy, the Marquis de Casa Fuerte [Juan Vázquez de Acuña y Bejarano], tried in vain to persuade the ministry in Madrid to sell one quintal at forty *ducados*, because in Seville it cost only thirteen *pesos*, as opposed to thirty pesos in Mexico City. Finally, the *Real Cédula* of November 24, 1767, lowered the previous price, which was sixty *ducados* or eighty-two *pesos* and six *reales* per quintal, by one-quarter, such that the price for one quintal was set at sixty-two *pesos*. Delhuyar, *[Memoria sobre] el influjo de la minería*, 1825, p. 41.

4. Spain's contract with the Idria mine set the annual purchase at 12,000 quintals.

districts are obliged to take more Idria mercury than Almaden mercury. Since the former is more expensive, people balk at taking it, and the miners pretend that it is impure.

III.293 An impartial distribution of mercury[1] (*el repartimiento del azogue*) is of the greatest interest to the prosperity of the mines of New Spain. As long as this branch of trade is not free, the responsibility for sharing out mercury should be transferred to the *Tribunal de minería*, which has the exclusive right to determine the number of quintals indispensable to the amalgamation plants of the various districts. Unfortunately, the viceroys and the people who surround them fiercely guard the right to administer this branch of the royal revenue themselves. They know very well that distributing mercury, especially that of Almaden, which is one-third less expensive than that of Idria, is tantamount to granting favors; and in the colonies, like everywhere else, it is profitable to favor the wealthiest and most powerful private individuals. The result of this state of affairs is that the poorest miners, those of Tasco, Temascaltepec, or Copala, cannot procure mercury at all, while the large plants of Guanajuato and Real del Monte have it in abundance.

The general superintendence of mines in Spain is responsible for the sale of mercury in the American colonies. Minister Don Antonio Valdés [y Bazán] conceived the bizarre and audacious project of coordinating himself the distribution of mercury to the various mines of Mexico. To arrive III.294 at this goal, in 1789 he ordered the viceroy to have statistical descriptions drawn up of all the mines of New Spain, and to send samples from the veins being mined to Europe. In Mexico City, it was felt to be impossible to carry out the order given by the minister; not a single sample was sent to Madrid, and the distribution of mercury remained, as before, in the hands of the viceroy of New Spain.

The following table[2] demonstrates the influence of the price of mercury on its consumption: both the drop in price and the freedom to trade with all the ports of Spain contributed to advances in mining.

1. On the story of the *repartimiento del azogue* and of the *correspondido* from the government of the viceroy ▼ Don Martín Enriquez de Almansa and the *Reales cédulas* of May 18, 1572, and March 26, 1577, to modern times, see Delhuyar, [*Memoria sobre el influjo de la minería*,] pp. 39–41 and pp. 147–154.

2. *Influjo del precio del azogue sobre su consumo, por el contador del ramo de azogues,* ▼ *Don Antonio del Campo Marín* (unpublished). Compare also Delhuyar, [*Memoria sobre la formación de una ley orgánica*,] pp. 43–45.

III.295

I. *From* 1761 to 1771

PERIOD	PRICE OF A QUINTAL OF MERCURY	MERCURY DISTRIBUTED (QUINTALS)	PRICE OF MERCURY	DUTIES ON GOLD AND SILVER	MARCS-MINTED
1761–1766	82 piasters	35,755	2,957,705 p.	6,685,857 p.	6,435,837
1767–1771	62	42,618	2,803,446	7,528,063	7,242,142
	Difference	6,863	152,259	842,206	806,309

II. *From* 1772 to 1781

PERIOD	PRICE OF A QUINTAL OF MERCURY	MERCURY DISTRIBUTED (QUINTALS)	PRICE OF MERCURY	DUTIES ON GOLD AND SILVER	MARCS-MINTED
1772–1776	62 piasters	53,610	3,390,704 p.	8,965,694 p.	8,961,950
1777–1781	41 Difference	59,221	2,498,051	9,320,159	11,293,374
		5,611	892,653	354,466	2,331,423

As early as the seventeenth century, people in Mexico knew that China[1] possessed mercury mines. It was believed that nearly 15,000 quintals could be extracted from Canton, at thirty-five piasters per quintal. The viceroy Gálvez sent a load of otter furs to be exchanged in the purchase of mercury, but this project, very wise in and of itself, was badly executed. The Chinese mercury, taken from Canton and Manila, was impure: it contained large amounts of lead, and its price rose to eighty piasters per quintal; it was likewise impossible to procure more than a very small amount. Since 1793, this important item has completely disappeared from view. It would nevertheless be very important to reconsider it, especially at a time when Mexicans are experiencing extreme difficulty in procuring mercury from the continent of Europe.

III.296

Together, the studies that I have conducted suggest that Spanish America as a whole—that is, Mexico, Peru, Chile, and the kingdom of Buenos Aires, since the other areas are not familiar with amalgamation techniques—annually consumes over 25,000 quintals of mercury, the total price of which in the colonies amounts to over 6,200,000 pounds. In an interesting table that shows the total amount of each metal mined across the entire globe, Mr. Héron de Villefosse estimates the amount of mercury

1. The oldest *royal cédulas* on the introduction of mercury from China to Acapulco via Manila date from August 15, 1609, and May 16, 1631. This trade was subsequently forbidden until the *bando* of August 21, 1781, which granted duty exemptions to the merchants of the Philippine Islands for the introduction of Chinese mercury into the kingdom of Mexico.

extracted annually from the mines of Europe to be 36,000 quintals. This figure suggests that mercury is the rarest metal after cobalt, and that it is even twice as rare as tin.

What is the amount of ore and silver currently produced by the mines of New Spain? What are the treasures that, since the discovery of the Americas, have flowed into Europe and Asia through the Mexican trade? The detailed information that I drew from the registers of Mexico City, Lima, Santa Fe, and Popayán during my stay in the Spanish colonies has put me in a position to give more exact information about the output of the mines than anything published to this day. One portion of the results that were the III.297 fruit of my research are already recorded in the works[1] of Mr. Bourgoing, Mr. Brongniart, Mr. Laborde, and Mr. Héron de Villefosse, to whom I hastened to communicate my findings immediately after my return to Europe.

As we have seen above, the amount of silver extracted annually from the mines of New Spain depends less on the abundance and on the intrinsic wealth of the ore than on the ease with which the miners are able to procure the mercury necessary for amalgamation. It should not, therefore, be surprising to note that the number of marks of silver that are converted into piasters at the mint in Mexico City is subject to a decidedly irregular variation. When, because of a war at sea or a different type of incident, mercury has been unavailable for a year, and when it arrives in large amounts the subsequent year, then a very significant output of silver follows a very modest period of coin-minting. In Saxony, where the small amount of mercury necessary for the amalgamation techniques is easily obtained, the output of the Freiberg mines is of such admirable regularity that from 1793 to 1799, it was never below 48,300 marks of silver, and never above 50,700. In that III.298 country, the great droughts, which hinder the movement of the hydraulic wheels and halt the draining of the water, exert the same influence on the amount of silver delivered to the mint as does the scarcity of mercury in the Americas.

From 1777 to 1803, the amount of silver extracted from the Mexican mines was almost always in excess of two million marks of silver. From 1796 to 1799, it was 2,700,000 marks, while from 1800 to 1802, it remained below 2,100,000 marks. It would be quite mistaken to conclude from these data that mining has flourished less in Mexico in recent years. In 1801, only

1. Bourgoing, *Tableau de l'Espagne moderne*, fourth edition, vol. II, p. 225. Brogniart, *Traité de minéralogie*, vol. II, p. 351. Laborde, *Itinéraire de l'Espagne*, first edition, vol. IV, pp. 383 and 504. Héron de Villefosse, *Mémoire général sur les mines*, pp. 249–55.

16,568,000 gold and silver piasters were produced, while in 1803, minting rose once again, due to the abundance of mercury, to 23,166,906 piasters.

If one abstracts upon the influence of the accidental causes, one finds that in a typical year, the mines and the washes of the Kingdom of New Spain currently produce 7,000 marks of gold and 2,500,00 marks of silver, the mean combined value of which is 22 million piastras fuertes.

Twenty years ago, this output was only between fifteen and sixteen million piasters; thirty years ago, it was only between eleven and twelve million. At the beginning of the eighteenth century, the amount of gold and silver minted in Mexico City was only between five and six million piasters. The enormous observable increase over the past few years in mining output must be attributed to a large number of causes, all of which have emerged simultaneously, and among which pride of place must be given to the growth of the population on the plateau of Mexico; the progress of enlightenment and of national industry; the freedom of trade granted to the Americas in 1778; the ease of obtaining cheaper iron and steel necessary for the mines; the drop in the price of mercury; the discovery of the Catorce and Valenciana mines; and the establishment of the *Tribunal de minería.* III.299

The two years in which gold and silver output peaked were the years 1796 and 1805: in the former, 25,644,000 piasters were struck at the mint of Mexico City; in the latter, 27,165,888 piasters were struck. To ascertain the effect produced by the freedom of trade, or rather, by the cessation of the galleons' monopoly, one has only to recall that the value of the gold and silver minted in Mexico City between 1776 and 1778 was 191,589,179 piasters, and 252,525,412 piasters from 1779 to 1791; such that the increase [in production] after 1778 amounted to a quarter of the total output.

In the archives of the mint of Mexico City, one finds exact data on the amount of gold and silver[1] minted since the year 1690. Here [on the following three pages] are two tables based on these data: the first shows the value of the gold and the silver, expressed in piastras fuertes; the second presents the *number of marks of silver* delivered to the mint and converted into piasters. III.300

These tables suggest that between 1690 and 1800, the mines of New Spain produced the enormous sum of 149,350,722 marks of silver, at a value of 1,353,452,020 piastras fuertes, or 7,105,623,105 livres tournois, if one values the piaster at 105 sous, in French currency. III.302

1. In 1630, the amount of silver minted was already 601,065 Castilian marks. Delhuyar, [*Memoria sobre el influjo de la minería,*] p. 30.

Gold and Silver Extracted from the Mines of Mexico, and Minted in Mexico City

From 1690 to 1809

YEAR	VALUE IN PIASTERS	YEAR	VALUE IN PIASTERS	YEAR	VALUE IN PIASTERS	YEAR	VALUE IN PIASTERS
1690	5,285,580	1720	7,874,323	1750	13,209,000	1780	17,514,263
1691	6,213,709	1721	9,460,734	1751	12,631,000	1781	20,335,842
1692	5,252,729	1722	8,824,432	1752	13,627,500	1782	17,581,490
1693	2,802,378	1723	8,107,348	1753	11,594,000	1783	23,716,657
1694	5,840,529	1724	7,872,822	1754	11,594,000	1784	21,037,374
1695	4,001,293	1725	7,370,815	1755	12,486,500	1785	18,575,208
1696	3,190,618	1726	8,466,146	1756	12,299,500	1786	17,257,104
1697	4,459,947	1727	8,133,088	1757	12,529,000	1787	16,110,340
1698	3,319,765	1728	9,228,545	1758	12,757,594	1788	20,146,365
1699	3,504,787	1729	8,814,970	1759	13,022,000	1789	21,229,911
1700	3,379,122	1730	9,745,870	1760	11,968,000	1790	18,063,688
1701	4,019,093	1731	8,439,871	1761	11,731,000	1791	21,121,713
1702	5,022,550	1732	8,726,465	1762	10,114,492	1792	24,195,041
1703	6,079,254	1733	10,009,795	1763	11,775,041	1793	24,312,942
1704	5,627,027	1734	8,506,553	1764	9,792,575	1794	22,011,031
1705	4,747,175	1735	7,922,001	1765	11,604,845	1795	24,593,481
1706	6,172,037	1736	11,016,000	1766	11,210,050	1796	25,644,566
1707	5,735,032	1737	8,122,140	1767	10,415,116	1797	25,080,038
1708	5,735,601	1738	9,490,250	1768	12,278,957	1798	24,004,598
1709	5,214,143	1739	8,550,785	1769	11,938,784	1799	22,053,125
1710	6,710,587	1740	9,556,040	1770	13,926,329	1800	18,685,674
1711	5,666,085	1741	8,663,000	1771	13,803,196	1801	16,568,000
1712	6,613,425	1742	16,677,000	1772	16,971,857	1802	18,798,600
1713	6,487,872	1743	9,384,000	1773	18,932,766	1803	23,166,906
1714	6,220,822	1744	10,285,000	1774	12,892,074	1804	24,007,789
1715	6,368,918	1745	10,327,500	1775	14,245,286	1805	27,165,888
1716	6,496,288	1746	11,509,000	1776	16,463,282	1806	24,736,020
1717	6,750,734	1747	12,002,000	1777	21,600,020	1807	22,014,699
1718	7,173,590	1748	11,628,000	1778	16,911,462	1808	21,886,500
1719	7,258,706	1749	11,823,500	1779	19,435,457	1809	26,172,982

Total from 1690 to 1890 in Gold and Silver: 1,499,435,898 piasters

III.301 *Silver Extracted from the Mines of Mexico, from 1690 to 1800*

YEAR	SILVER MARKS	OUNCES	OCHAVAS	YEAR	SILVER MARKS	OUNCES	OCHAVAS	YEARS	SILVER MARKS	OUNCES	OCHAVAS
1690	621,833	4	0	1730	1,146,573	0	0	1770	1,638,391	5	6
1691	731,024	5	2	1731	992,926	0	0	1771	1,506,255	2	2
1692	629,732	6	7	1732	1,026,643	0	0	1772	1,996,689	1	1
1693	329,691	4	6	1733	1,177,623	0	0	1773	2,227,442	6	1
1694	687,121	1	0	1734	1,000,771	0	0	1774	1,516,714	5	5
1695	470,740	3	2	1735	932,001	1	6	1775	1,675,916	0	7
1696	375,366	7	3	1736	1,296,000	0	0	1776	1,936,856	6	2
1697	524,699	5	6	1737	955,545	7	2	1777	2,428,613	4	1
1698	390,560	5	4	1738	1,116,500	0	0	1778	2,334,765	7	2
1699	412,327	7	1	1739	1,005,963	0	0	1779	2,199,548	6	6
1700	397,327	6	2	1740	1,124,240	0	0	1780	1,994,073	4	7
1701	472,834	4	5	1741	1,016,962	0	0	1781	2,311,062	3	0
1702	590,900	0	1	1742	962,000	0	0	1782	2,014,545	1	1
1703	715,206	3	0	1743	1,014,000	0	0	1783	2,709,167	0	3
1704	685,532	5	1	1744	1,210,000	0	0	1784	2,402,965	7	7
1705	558,491	2	2	1745	1,215,000	0	0	1785	2,111,263	7	0
1706	726,122	0	5	1746	1,354,000	0	0	1786	1,978,844	5	6
1707	674,709	2	5	1747	1,412,000	0	0	1787	1,819,141	1	3
1708	675,012	7	6	1748	1,368,000	0	0	1788	2,293,555	5	3

(*continued*)

YEAR	SILVER MARKS	OUNCES	OCHAVAS	YEAR	SILVER MARKS	OUNCES	OCHAVAS	YEARS	SILVER MARKS	OUNCES	OCHAVAS
1709	613,428	4	7	1749	1,391,000	0	0	1789	2,415,821	2	1
1710	789,480	7	3	1750	1,554,000	0	0	1790	2,045,951	6	6
1711	666,598	2	4	1751	1,486,000	0	0	1791	2,363,867	5	3
1712	783,932	3	2	1752	1,603,000	0	0	1792	2,724,105	3	6
1713	763,279	0	5	1753	1,364,000	0	0	1793	2,747,746	4	3
1714	731,861	4	1	1754	1,364,000	0	0	1794	2,488,304	1	0
1715	749,284	4	1	1755	1,469,000	0	0	1795	2,808,380	1	0
1716	767,969	1	6	1756	1,447,000	0	0	1796	2,854,072	6	4
1717	794,204	0	5	1757	1,474,000	0	0	1797	2,818,248	4	4
1718	843,951	6	3	1758	1,500,893	3	4	1798	2,697,038	2	2
1719	853,965	4	0	1759	1,532,000	0	0	1799	2,473,542	2	7
1720	926,390	7	6	1760	1,408,000	0	0	1800	2,098,712	5	1
1721	1,113,027	7	6	1761	1,386,000	0	0				
1722	1,038,109	5	7	1762	1,189,940	2	3				
1723	953,805	5	5	1763	1,385,298	7	4				
1724	926,214	3	3	1764	1,152,063	5	6				
1725	867,037	1	2	1765	1,365,275	7	7				
1726	996,017	1	6	1766	1,318,829	4	1				
1727	956,833	7	7	1767	1,225,307	6	2				
1728	1,035,711	1	7	1768	1,444,583	1	6				
1729	1,037,055	2	4	1769	1,404,564	0	4				

Total from 1690 to 1800, in silver alone: 149,350,722 marks.

The *Mercurio Peruano* (vol. X, p. 133), which records the output of the years 1733–1792, indicates the amounts of gold and silver separately. Here are the results for the average years: III.301

From 1733 to 1742	Silver	8,998,209	piasters
	Gold	434,050	
	TOTAL	9,432,259	
From 1743 to 1752	Silver	11,566,030	
	Gold	455,109	
	TOTAL	12,021,139	
From 1753 to 1762	Silver	11,971,835	
	Gold	462,773	
	TOTAL	12,434,608	
From 1763 to 1772	Silver	12,303,753	
	Gold	770,742	
	TOTAL	13,074,495	
From 1773 to 1782	Silver	17,603,462	
	Gold	733,800	
	TOTAL	18,337,262	
From 1783 to 1792	Silver	19,491,309	
	Gold	644,040	
	TOTAL	20,135,349	
TOTAL for sixty years, in gold and silver		853,361,070	
In an average year during this period		14,222,000	

Since minting in 1805 exceeded even that of 1796, I shall record here the amounts of gold and silver that were minted in Mexico City, month by month, from January 1 to December 1, 1805 [see the table on the following page]. III.303

Minting and seigniorage duties on this sum amounted to 2,073,753 piasters; yet since the price of labor and the production costs were 462,318 piasters, the result was that in 1806, the net profit of the mint of Mexico City, combined with that of the house of separation (*casa del apartado*), was 1,611,434 piasters. III.304

According to a note drafted by Mr. Campo Marin, from January 1, 1772, to December 31, 1803, 648,535,219 piasters in gold and silver were produced in Mexico City; that is: 623,404,405 piasters or 73,104,242 marks of silver, and 25,130,814 piasters or 184,581 marks of gold. In these

III.303

MONTH	GOLD PIASTERS	SILVER		TOTAL	
		Piasters	Reales	Piasters	Reales
January		860,026	5 and three-quarters	860,026	5 and three-quarters
February		1,891,492	4	1,891,492	4
March		2,234,021	4 and a half	2,234,021	4 and a half
April		1,890,883	5 and a quarter	1,890,883	5 and a quarter
May		2,317,683	5 and a half	2,317,683	5 and a half
June		2,045,141	6 and a half	2,045,141	6 and a half
July		2,309,513	6 and three-quarters	2,309,513	6 and three-quarters
August	371,766	2,106,236	One-half	2,478,002	One-half
September	236,304	2,489,358	6 and a quarter	2,725,662	6 and a quarter
October	464,768	2,555,402	1	3,020,170	1
November		2,110,793	5 and a quarter	2,110,793	5 and a quarter
December		286,976	0	2,995,530	0
TOTAL	1,359,814	25,806,074	3 and a quarter	27,165,888	3 and a quarter

In 1806, production was:

In gold	1,352,348 piasters
In silver	23,383,672
	24,736,020

estimates, gold is valued at 136 piasters per mark, and silver *is valued as equivalent to one piaster*, as is the custom at the mint of Mexico City. For the thirty-two years prior to 1803, in a typical year, 20,266,725 and nineteen thirty-second piasters were minted.

For the last 113 years, mining output has constantly been on the rise, with the sole exceptions of the periods 1760–1767 and 1785–1788. This increase is evident if one examines, for every ten-year period, the amount

III.305 Development of Mining in Mexico

I. GOLD AND SILVER					II. SILVER ALONE						
PERIODS				VALUE OF GOLD AND SILVER IN PIASTERS	PERIODS				SILVER		
									Marks	Ounces	Ochavas
From	1690	to	1699	43,871,335	From	1690	to	1699	5,173,099	2	7
	1700		1709	51,731,034		1700		1709	6,109,781	5	2
	1710		1719	65,747,027		1710		1719	7,744,525	2	6
	1720		1729	84,747,027		1720		1729	9,900,203	7	7
	1730		1739	90,529,730		1730		1739	10,650,546	1	0
	1740		1749	111,855,040		1740		1749	12,067,202	0	0
	1750		1759	125,750,094		1750		1759	14,793,893	3	4
	1760		1769	112,828,860		1760		1769	13,279,863	4	1
	1770		1779	165,181,729		1770		1779	19,461,194	0	1
	1780		1789	193,504,554		1780		1789	22,050,440	6	7
	1790		1799	231,080,214		1790		1799	26,021,257	6	3
Total from 1690 to 1799				1,276,232,840	Total from 1690 to 1799				147,252,008	6	6

of precious metals delivered to the mint of Mexico City, as in the following tables, one of which shows the value of the gold and the silver in piasters, while the other shows the amount of silver, expressed in marks. The decrease in royal duties (the conversion of the *quint* into a tithe, and the reduction of the *un y medio* to *un por ciento*); the drop in the price of mercury, the formation of a *body of miners*, communicating its wishes through the organ of the *High Council* or *Tribunal de Mineria*; the advances in public education due to the School of Mines in Mexico City; the declaration of free trade in 1778; the elimination of the *alcabala* duty[1] on mining-related purchases; the ease of the sale (*rescate*) of gold and silver material in the provincial coffers (*tesorias*)[2]; and the fixing of the price of powder (at four *reales de plata* per pound,[3] rather than six) are the primary causes that led to the rapid increases in mining output between 1768 and 1809 [the table above was originally placed here].

III.306 If one focuses on the periods in which the most rapid advances in mining took place, one obtains the following results:

PERIOD	VALUE OF GOLD AND SILVER IN AN AVERAGE YEAR IN PIASTERS	PROGRESSIVE INCREASE	
1690–1720	5,458,830	In 27 years	3,700,000 piasters
1721–1743	9,177,768		
1744–1770	11,854,825	25	2,000,000
1771–1782	17,223,916	19	5,300,000
1783–1790	19,517,081	12	2,300,000
1791–1803	22,325,824	10	2,800,000

This tableau, together with the previous tableaus, demonstrates that the periods in which the mines increased the most in wealth were 1736–1745, 1777–1783, and 1788–1798; but this increase was, in general, so disproportionate to the time elapsed that the total output of the mines was:

1. [Gálvez y Gallardo,] *Reales ordines de 13 de enero 1783*, and *12 de noviembre 1796*.
2. See [Spain,] *[Real] Ordenanza de Intendentes*, 1786, article 152.
3. *Real Orden de 27 de abril 1801*.

4	million piasters in	1695
8		1726
12		1747
16		1776
20		1788
24		1795

This suggests that the output tripled over the course of fifty-two years, and sextupled over the course of one hundred years. From 1809 to 1821, the output of the mines has decreased as follows:

III.307

YEAR	GOLD	SILVER	TOTAL[1]
1810	1,095,504 piasters	17,950,684 piasters	19,046,188 piasters
1811	1,085,363	8,956,433	10,041,796
1812	381,646	4,027,620	4,409,266
1813		6,133,983	6,133,983
1814	618,069	6,902,481	7,520,550
1815	486,464	6,454,799	7,042,620
1816	960,393	8,315,616	9,401,290
1817	856,942	7,994,951	8,849,893
1818	533,921	10,852,367	11,386,288
1819	539,377	9,879,138	12,030,515
1820	509,076	9,879,078	10,406,154
1821	303,504	5,600,022	5,916,226

[1] When the total does not correspond to the values of gold and of silver, the surplus is due to copper minting which, in 1816, for example, was 125,281 piasters (see [Poinsett,] *Notes on Mexico*, 1824, p. 63.)

From 1806 to 1810, the mines of Mexico produced 2,155,927 marks of silver and 9,383 marks of gold per year; but from 1811 to 1815, during the period of civil unrest, the average annual output was only 1,246,586 marks of silver and 3,733 marks of gold.[1] It is believed that from 1816 to 1819, the mean mining output was 1,157,527 marks of silver, and 2,933 marks of gold, but both smuggling and the increase in the number of mints

1. Delhuyar, *[Memoria sobre] el influjo de la minería*, 1825, p. 62.

III.308 in the provinces[1] made the audit very uncertain. Only 5,916,000 gold and silver piasters were produced in the entire year of 1821, while the first eight months of 1825 saw the production of 7,889,044 piasters[2] in the mints of Mexico City, Guanajuato, Zacatecas, Guadalajara, and Durango. Such are the advances that the works of the Mexican mines are making once more!

After gold and silver, it remains for us to discuss the other metals, called common metals, the mining of which is thoroughly neglected, as we indicated at the beginning of this chapter. Copper is found in its native state, as well as in the form of both vitreous and oxidulated copper, in the mines of Ingaran, just south of the Jorullo volcano, at San Juan Guetamo, in the intendancy of Valladolid, and in the province of New Mexico. Mexican *tin* is extracted, through washing, from the alluvial plains of the intendancy of Guanajuato, near Gigante, San Felipe, Robledal, and San Miguel el Grande, as well as in the intendancy of Zacatecas, between the cities of Jerez and III.309 Villa Nueva. One of the most common tin ore in New Spain is concretionary tin oxide, or what British mineralogists call *wood-tin*. It appears that this ore is originally found in veins that run through trappean porphyry; but rather than targeting these veins, the indigenous peoples prefer to extract tin from the alluvial soil that fills the ravines. In 1802, the intendancy of Guadalajara produced nearly 9,200 *arrobas* of copper and 400 of tin.

Iron ore is more abundant than is commonly believed in the intendancies of Valladolid, Zacatecas, and Guadalajara, and especially in the *Provincias internas*. We have explained above[3] the reasons why these ores, the most significant of all, are only worked intensively when maritime wars block steel and iron imports from Europe. We have already mentioned the veins of Tecalitan, near Colima, which were successfully mined ten years ago and were subsequently abandoned again. Fibrous magnetic iron is found, together with magnetic pyrite, in veins that run through *gneiss* in the kingdom of Oaxaca. The western slope of the mountains of Michoacan

1. *Casas provinciales de moneda* in Durango, Zacatecas, Guanajuato, Sombrerete, Chihuahua, and Catorce, several of which were temporarily closed in 1818. The large distances between the principal mines and the capital of Mexico make the mints established in the provinces very useful in terms of domestic circulation.

2. That is, 2,031,023 gold piasters and 5858,021 silver piasters. Over the course of these eight months, the *casas de moneda* of Mexico City and Zacatecas produced, respectively, 4,143,726 and 2,234,179 gold and silver piasters; those of Durango, Guadalajara, and Guanajuato produced 593,666 piasters, 515,799 piasters, and 401,673 piasters. See the note by Don Ildefonso Maniau in *Memoria del ramo de hacienda federal de los estados unidos mexicanos por el año 1726* [1826], no. 95.

3. See above, p. 111 [p. II.62 in this edition].

abounds in compact red iron and brown hematite ore. The former have also been found in the intendancy of San Luis Potosí, near Catorce. I have seen crystallized micaceous iron filling veins near the village of Santa Cruz, east of Celaya, on the fertile plateau that stretches from Queretaro to Guanajuato. The Cerro del Mercado, located near the city of Durango, contains an enormous *mass* of magnetic brown and micaceous iron. I am entering into detail about these localities in order to dispute the theory, proposed by some contemporary physicists, that iron is almost exclusively found in the northernmost region of the temperate zone. We are indebted to Mr. Sonneschmidt[1] for the knowledge of meteoric iron, which is found in several locations in New Spain, for example at Zacatecas, Charcas, Durango, and, if I am not mistaken, on the outskirts of the small city of Toluca.

III.310

Lead, quite scarce in northern Asia, is abundant in the mountains of calcareous formation contained within the northeastern part of New Spain, especially in the district of Zimapan, near the Real del Cardonal and Lomo del Toro; in the kingdom of Nuevo León, near Linares; and in the province of Nuevo Santander, near San Nicolás de Croix. The lead ore are not worked with as much zeal as one might hope, in a country where one-quarter of all silver ore are smelted.

III.311

Among the metals in the most limited use, we shall cite *zinc*, which, in the form of brown and black blende, is found in the veins of Ramos, Sombrerete, Zacatecas, and Tasco; *antimony*, which is common in Catorce and Los Pozuelos, near Cuencame; and *arsenic*, which is found, combined with sulfur as an orpiment, among the ore of Zimapan. As far as I know, *cobalt* has not yet been identified among the ore of New Spain, and *manganese*,[2] which Mr. Ramirez recently discovered on the island of Cuba, appears to me to be generally less abundant in equinoctial America than in temperate climes on the Old Continent.

Mercury, quite far removed from tin in terms of its relative age or the period of its formation, is nearly as scarce as tin in all parts of the globe. For

1. Sonneschmidt, [*Mineralogische Beschreibung*,] pp. 188 and 192. Ten years ago, the mass of Zacatecas still weighed nearly 2,000 pounds. On the meteoric stone that, according to the account of ▼ Cardanus and Mercati, fell to earth between Cicuic and Quivira, see the paper by Mr. Chladni ["Au Catalogue des Météores, à la suite desquels des pierres ou des masses de fer sont tombées"] in the *Journal des mines*, 1809, no.151, p. 79. The geographical position of Cicuic and Quivira, names that recall the fables of El Dorado in South America, is today unknown to us.

2. To the west of the city of Cuenca, in the kingdom of Quito, one finds earthy grey manganese that forms a layer in the sandstone.

centuries, the inhabitants of New Spain have taken the amount of mercury necessary for amalgamation techniques in part from Peru, in part from Europe. The result of this is that they have grown accustomed to regarding III.312 their country as being devoid of this metal. If, however, one focuses on the studies conducted during the reign of King Charles IV, one must admit that few lands present as many signs of cinnabar as does the plateau of the Cordilleras, from 19° to 22° north latitude. In the intendancies of Guanajuato and Mexico City, one finds them everywhere that shafts are dug, between San Juan de la Chica and the city of San Felipe; near the Rincón del Centeno, on the outskirts of Celaya; and from the Durasno and Tierra Nueva to San Luis de la Paz, especially near Chapin, Real de Pozos, San Rafael de los Lobos and La Soledad. Mercury sulfide has also been discovered at Axuchitlan and at Zapote,[1] near Chirangangueo, in the intendancy of Valladolid; at Los Pregones, near Tasco; in the mining district of Doctor, and in the Tenochtitlan valley, south of Gasave, on the road leading from Mexico City to Pachuca. The works that have been established in order to identify these various mineral deposits have been disrupted so often and directed both with such little zeal and, in general, so unintelligently that it would be very imprudent to suggest, as people have dared to do many times, that the mercury ore of New Spain do not deserve to be mined. The interesting theories that we owe to the work of ▾ Mr. Chovell III.313 suggest, on the contrary, that the veins of San Juan de la Chica, like those of the Rincón del Centeno and the Gigante, are quite deserving of the attention of Mexican miners. How could superficial works and mining operations that have barely gotten started be expected to yield a net profit for shareholders in the first few years?

The mercury ore of Mexico are of very different formation: some are found in layers and in secondary terrains, while others are found in veins that run through trappean porphyry. In the Durazno, between Tierra Nueva and San Luis de la Paz, the cinnabar, blended with large numbers of globules of native mercury, forms a horizontal layer (*manto*) that rests on porphyry. This *manto*, which has been pierced by shafts five to six meters deep, is covered in fossilized wood and coal. If one examines the roof of the *manto* from the *surface* down, one first encounters a layer of *Schieferthon*, impregnated with potassium nitrate, and containing petrified plant debris;

1. In the ore of San Ignacio del Zapote, where the cinnabar is consistently blended with *blue copper carbonate*, while at Schemnitz and at Poratich, in Hungary, the *antimonious grey copper* (Graugiltigerz) contains 0.06 of mercury. Klaproth, [*Beiträge*,] [vol.] IV, p. 65.

next, a stratum of schistous coal (*Schieferkohle*), one meter thick; finally, *Schieferthon*, which rests directly on top of the cinnabar ore. Eight years ago, nearly 7,000 quintals of mercury were extracted from this mine in just a few months' time, although this was not enough to cover the operational costs, despite the fact that the ore contains one pound of mercury for each load of three and a half quintals. The carelessness with which the Durazno III.314 mine was worked was all the more harmful in that both the relative lack of solidity of the *roof rock* and its horizontal position resulted in very frequent collapses; today the mine is flooded, and it would not be profitable to resume operations. It has consistently enjoyed great celebrity in the country, not for its wealth, which is inferior to that of the San Juan de Chica veins, but because it lent itself to open mining and because its output was very abundant. There was a futile attempt to uncover a second layer of mercury ore below that of Durazno.

The cinnabar vein of San Juan de la Chica has a width of two or three meters: it runs through the mountain of Los Calzones, and stretches all the way to Chichindara. Its ore are extremely wealthy but relatively scarce; there, I have seen masses of compact and fibrous mercury sulfide, bright red in color, twenty centimeters long and three centimeters thick; the purity of these samples made them resemble the wealthiest output of Almaden and Wolfstein, in Europe. Thus far, the ore of Chica has only been mined to a depth of fifty meters. It is found—and this geological fact is quite remarkable—not in sandstone or in schist, but in a veritable pechstein-porphyry divided into bowls in concentric circles, the interior of which is blanketed in mamelonated hyalite. Both the cinnabar and a small amount III.315 of native mercury can occasionally be seen in the midst of the porphyritic rock, at a considerable distance from the vein. During my stay in Guanajuato, only two mines were worked in all of Mexico: Lomo del Toro, near San Juan de la Chica, and Nuestra Señora de los Dolores, one-quarter league southeast of the Gigante. In the former mine, one load of ore yields two to three pounds of mercury: the operational costs there are very modest. The Gigante mine, from which up to six pounds of mercury are extracted per load of ore, produces seventy to eighty pounds per week. It is worked on behalf of a wealthy private individual, ▼ Don José del Maso [Mazo], who has the merit of having first incited his compatriots, during the last war, to mine mercury and to produce steel. The cinnabar that was extracted from the veins of Fraile Mountain near the *Villa de San Felipe* is found in a porphyry with a base of compact feldspar, which, run through with veins of tin, is undoubtedly more ancient than the *pechstein-porphyry* of Chica.

In their current state, the Americas are dependant on Europe for mercury. It is likely that this dependence will not last long, if the ties that bind the colonies to the metropole remain broken for long, and if human civilization, in its progressive movement from east to west, becomes concentrated in the Americas. The spirit of enterprise and pursuit will grow as the population increases: the more inhabited the country becomes, the more familiar people will become with the natural riches contained within the bosom of the mountains. Even if no single deposit is discovered that equals that of Huancavelica in wealth, several deposits will be mined simultaneously, and their combined output will obviate mercury imports from Spain and from Carniola. These changes will occur with even greater speed as Mexican and Peruvian miners find themselves increasingly hindered by the lack of the metal necessary for amalgamation. But do we consider what silver mining in the Americas would become today if, in the midst of the wars that ravage Europe, the mercury mines of Almaden and Idria were no longer worked?

III.316

I have mentioned the ore deposits of New Spain, which, carefully examined and steadfastly mined, may one day produce quite a considerable amount of mercury. The time is approaching when the colonies, more closely aligned, will also be more attentive to the interests that they share; it is therefore important to give an overview of the indices of mercury observed in South America. Perhaps Mexico and Peru, instead of importing this metal from Europe, may one day provide it to the Old World. I shall limit myself here to the information that I was able to obtain in the field, especially during my stay in Lima. I shall only mention the points where cinnabar has been found, either in veins or in layers. In several places, for example at Portobello and at Santa Fe de Bogotá, large amounts of native mercury, in shallow deposits, have been gathered during the construction of houses. This phenomenon has often attracted the government's attention. People have forgotten that, in a land where, for the past three centuries, goatskins filled with mercury have been hauled on muleback from province to province, some of this metal has, of course, been spilled in the sheds where the beasts of burden are unloaded, as well as in the mercury storehouses established in the cities. In general, mountains contain mercury in its native state only in very small portions, and when several kilograms are found collected in the ground in an inhabited place or on a main road, one must believe that these masses are due to accidental infiltrations.

III.317

In the kingdom of New Granada, mercury sulfide has been detected in three different sites: in the province of Antioquia, in the *Valle de Santa Rosa*, east of the Río Cauca; in *Quindiu* mountain, on the central Cordillera pass, between Ibagué and Cartago, at the far edge of the Vermellon ravine; finally, between the village of *Azogue* and Cuenca, in Quito province. The discovery of cinnabar in Quindiu is due to the patriotic zeal of the famous botanist Mutis, who, in August and September of 1786, commissioned the miners of the Sapo to examine the granitic section of the Cordillera that extends to the south, from *Nevado* of Tolima toward the Río Saldaña. The mercury sulfide ore is not only found in rounded fragments blended with gold particles, in the ancient drift deposit that fills the Vermellon ravine (*quebrada*) at the foot of the *Ibague Viejo* plateau. It is also known from which vein the impact of the water appears to have detached these fragments: this vein runs through the small Santa Ana ravine. Near the village of Azogue, northwest of Cuenca, the mercury is found (as in the department of Mont-Tonnerre) in a quartzose sandstone formation with clayey cement. This sandstone is nearly 1,400 meters thick and contains fossilized wood[1] and asphalt.[2] In the mountains of Guazun and Upar located northeast of Azogue, a vein of cinnabar runs through layers of clay filled with calcareous spar and contained within the sandstone. There one encounters the remnants of an old gallery, 120 meters long and with eleven shafts, very close to one another. The locals believe that this mine [ore] was worked prior to that of Huancavelica and that it was the discovery of the latter that led to its being abandoned. The scholarly studies of ▾ Don Pedro García, and the works that the intendant of Cuenca, Mr. Vallejos, commissioned in 1792, did not demonstrate that the cinnabar vein of Guazun could be successfully mined. At a distance of five leagues to the northwest of the city of Popayán, near Zeguengue, there exists a ravine that is called the mercury ravine (*quebrada del azogue*), although the origin of this name is unknown.

III.318

III.319

In Peru, cinnabar is found near Vuldivui, in the province of Pataz, between the western bank of the Marañón and the Guailillas Missions; at the foot of the great *Nevado* de Pelagato, in the province of Conchucos, to the east of Santa; at the baths of Jesus [Baños de Jesus], in the province of Guamalies, to the southeast of Guacarachuco; near Huancavelica, in the intendancy of the same name; and near Guaraz, in the province of Guailas. According to the ledgers found in the provincial treasury of the city

1. I found two beautiful pieces of it, fourteen decimeters long, at Silcai-Yacu between Delec and Cuenca.

2. At Porche and on the western slope of Coxitambo mountain, I was thoroughly struck by the geological connections between the sandstone formation of Cuenca and Azogue and the mines of Wolfstein and Münsterappel, which I visited in 1790, and which also contain cinnabar, fossilized wood, and petroleum.

of Chachapoyas (between the Río Sonche and the Río Utcubamba), it appears that at the beginning of the conquest, mercury ore were mined in the low-lying mountains that stretch from the Pongo de Manseriche to near Cajamarquillo and the Río Hallaga: but, according to the information that I gathered during my stay in the province of Jaén, it is completely unknown

III.320 today where these mines were located. The cinnabar veins of Guaraz were worked with some success in 1802: up to eighty-four pounds of mercury were removed from the 500-pound mass of ore.

The famous mine of Huancavelica, on the state of which so many false notions have been spread, is located in Santa Barbara Mountain, to the south of the city of Huancavelica, at a horizontal distance of 2,772 *varas* (or 2,319 meters). According to ▾ Le Gentil,[1] the city's elevation above sea level is 3,752 meters (1,925 toises): if one adds to this height the 802 *varas* by which the summit of Santa Barbara Mountain soars above the street level of Huancavelica, one obtains an absolute height of 4,422 meters[2] for this mountain.

III.321 The discovery of the great mercury mine is generally attributed to the Indian Gonzalo Abincopa or Navincopa; it is, however, certain that it dates back to a period much earlier than the year 1567, because the Incas already used cinnabar (*llimpi*) as make-up, and since they extracted it from the Palcas Mountains. The Cerro de Santa Barbara mine was, furthermore, not worked for the benefit of the crown until September 1570, around the same year in which Fernández de Velasco introduced Mexican amalgamation to Peru.

On the outskirts of the city of Huancavelica, mercury occurs in two very different forms: in layers and in veins. In the great mine of Santa Barbara,

1. This height is calculated using Mr. de Laplace's formula, assuming a temperature of ten degrees Centigrade. According to Le Gentil [de la Galaisière], (*Voyage dans les mers] de l'Inde*, vol. I, p. 76), the average barometric elevation at the city of Huancavelica is 18^{po} 1.5^{lines}, which would only give 1,814 toises or 3,535 meters of absolute elevation. The main square in the city of Micuipampa, where I found the barometer of 18^{po} 4.7^{lines} would therefore be at an elevation of eighty-four meters above the street level of Huancavelica (*Recueil d'observations astronomiques*, vol. I, p. 316).

2. This measurement, about which Kirwan has raised a few doubts (*Trans[actions] of the [Royal] Irish Academy*, vol. VIII, p. 32), lends credence to the Ulloa's claim to have witnessed the barometer hover around 17^{po} 2.2^{lines} at the bottom of the *Hoyo Negro* mine, from which it can be concluded that the bottom of the mine was then at a height of 2,159 toises or 4,208 meters above sea level (Ulloa, *Noticias americanas*, p. 279). Hence, a shaft in which the miners worked at a level 500 meters higher than the top of the Peak of Tenerife! In the Cerro de Hualgayoc, I saw galleries whose absolute height surpassed 4,050 meters, and whose temperature (*Mina de Guadalupe* and *Mina del Purgatorio*) were 14.3° and 19.6° Centigrade, when the surrounding air had an average temperature of 7.8°, and when the water in the mines of the area was at 11.2°.

the cinnabar is contained in a layer of quartzose sandstone, which is nearly four hundred meters thick, and which is oriented ten to eleven hours [in arc] on the German compass, with an inclination of 64° toward the west. This sandstone, similar to that of the Paris environs and the Aroma and Cascas mountains, in Peru, resembles pure quartz. Most of the samples that I was able to examine in the Baron of Nordenflycht's geological study [chambers] display almost no clayey cement. The quartzose rock that contains mercury III.322 ore forms a layer in a calcareous breccia from which it is only separated, in its wall and in its roof, by a very thin stratum of schistose clay (*Schieferthon*), which has often been confused with slate or with primary schist. The breccia is covered in a secondary calcareous stone formation, and the fragments of compact limestone embedded in the breccia seem to indicate that the entire mass of Santa Barbara Mountain itself rests on alpine calcareous rock. This latter rock (*Alpenkalkstein*) is indeed visible on the eastern slope of the mountain, near Acobamba and Sillacasa: one still encounters it at very considerable elevations; it is bluish-gray and run through with a large number of small veins of calcareous spar. Ulla observed petrified shells[1] there in 1761 at a height of over 4,300 meters. ▼ Mr. de Nordenflycht also found pectinites and cardiums in a shell bed, between the villages of Acoria and Acobamba, near Huancavelica, at an elevation that surpasses by 800 meters that of the bed of nummulites found by Mr. Ramond at the top of Mont Perdu [Monte Perdido].

The cinnabar does not fill the entire quartzose layer of the great Santa Barbara mine: its forms are peculiar there; sometimes it occurs in small III.323 veins, which *drag* and join together in *clusters* (*Stockwerke*). The result of this is that the metalliferous mass is generally only sixty to seventy meters wide. The native mercury is extremely scarce, but the cinnabar is accompanied by red iron ore, magnetic iron, galena, and pyrite: the slits [crevices] are often lined with lime sulfate, calcareous spar, and fibrous alum (*Federalaun*), with parallel curved fibers: at great depths,[2] the metalliferous layer contains a large amount of orpiment or yellow and red arsenic sulfide. This mixture once cost the lives of several workers who distilled cinnabar ore blended with orpiment, until the government sought to outlaw the mining of the Cochapata *works*, which are the most abundant in arsenic. I suppose

1. We also found these on the ridge of the Andes, near Montan and Micuipampa; *Géographie des plantes*, p. 127. On pelagic shells observed at great heights in Europe and in the Americas, see ▼ Faujas de Saint-Fond, *Essai de Géologie*, vol. II, pp. 61–69.

2. Especially below the depth of 230 varas. Galena is found nearer the surface, and up to forty varas below the San Javier gallery.

that the mofette, called *umpe*, the dreadful effects of which were described by Ulloa, is arsenical hydrogen gas, but it has been much less noticeable than one might believe from the Spanish travelers' accounts.

The great Santa Barbara mine is divided into three levels (*pertinencias*), which bear the names of *Brocal*, *Comedio*, and *Cochapata*. The depth of the mine is 349 varas; its total length, from north to south, is 536 varas. It is esti-

III.324

mated that through distillation, fifty quintals of relatively poor ore yield eight to twelve pounds of mercury. The ore deposit is mined via three galleries, namely, the *Socavón de Ulloa*, the *Socavón de San Francisco Javier*, and the *Socavón de Nuestra Señora de Belem*, begun in 1615 and completed in 1642. The gallery bored by the astronomer Don Antonio Ulloa, who, as governor of Huancavelica, directed the mine works for a few years, is only seventy-five varas long; its mouth is almost at the same level as the main square of the city, and it would need to be extended 2,000 varas farther for it to traverse the *pertinencia de Cochapata*. It is the only gallery that follows the direction of the metalliferous layer; the two others were dug right in the rock. The *Socavón de Belem*, the most useful of these various mining works, is 625 varas long and cuts through the ore deposit at a depth of 172 varas below the summit of Santa Barbara Mountain. The San Javier gallery, completed in 1732, is 112 varas below the Socavón de Belem. All of these galleries, which cost enormous sums of money, because they were built at a width of five varas, are only used for ventilation and for internal hauling, for the mine is completely dry.

From 1570 to 1789, a total amount of 1,040,452 quintals of mercury

III.325

were removed from the great mine of Huancavelica; that is:

From	1570	to	1576	9,137 quintals
From	1577	to	1586	60,000
From	1587	to	1589	31,500
From	1590	to	1598	59,850
From	1599	to	1603	20,000
From	1604	to	1610	19,000
From	1611	to	1615	30,000
From	1616	to	1622	59,463
From	1623	to	1645	96,600
From	1646	to	1648	20,460
From	1649	to	1650	8,342
From	1651	to	1666	109,120

According to this table, the amount of mercury removed from the great mine of Huancavelica in its first ninety-six years totaled 523,472 quintals. In the years listed below, the following amounts were obtained:

From	1667	to	1672	49,026 quintals
From	1673	to	1683	60,000

There is no mention of the mining output from 1684 to 1713 in the treasury archives; but it was this:

From	1713	to	1724	41,283 quintals
From	1725	to	1736	38,882
From	1737	to	1748	65,426

These data suggest that the mine regularly produced between four and six thousand quintals of mercury per year. In the most productive years, from 1586 to 1589, the output rose to 10,500 quintals. III.326

In addition to the cinnabar contained within the layer of quartzose sandstone of the Cerro de Santa Barbara de Huancavelica, cinnabar is also found in this same section of the Cordilleras, especially near Sillacasa, in small veins that run through alpine calcareous stone (*Alpenkalkstein*); but these veins, which are often filled with chalcedony, do not follow regular directions; they cross one another and frequently drag, and form *nests* or metallic *clusters*.[1] For the past ten years, all of the mercury that Huancavelica has provided to the miners of Peru has come exclusively from these latter ore deposits, the metalliferous layer (*Erzflöz*) of the great Santa Barbara mine having been completely abandoned, because of the collapse that occurred in the *pertinencia of the Brocal*. Greed and carelessness were the cause of this unfortunate accident. After 1780, the mine directors had already had difficulty providing the amount of mercury required by the ever-growing demands of Peruvian amalgamation. The deeper the works became, the more impure and mixed with arsenic sulfide the cinnabar was. Since the layer forms a mass of extraordinary size, it was only possible to mine it through the *drifts* and *transverse galleries*.[2] In order to support the roof, *pillars* had been erected at equal distances, as is done in coal and rock salt mines. An intendant of Huancavelica, a lawyer commendable for both III.327

1. *Nidos, bolsas y clavos* (*Zusammenscharende Trümmer*).
2. *In Quer-und Pfeilerbau*; cross-system of mines.

his knowledge and his integrity, had the temerity to order these pillars removed in order to increase the mine's output. Any experienced miner could have easily predicted what the repercussions of this operation would be: the rock, unsupported, succumbed to the pressure; the roof collapsed, and since this collapse was felt throughout most of the upper *pertinencia*, that of the Brocal, it was necessary to abandon work in the two lower *pertinencias*, those of the Comedio and Cochapata. The master-miners accused the intendant of having had these pillars removed in order to ingratiate himself to the court in Madrid, by procuring a large amount of mercury for it in just a few years. The intendant, on the other hand, claimed that he had only acted with the consent of the master-miners, who had thought it possible to replace the pillars by piles of rubble. Instead of making a concrete decision and attacking the metalliferous layer at other points, they lost eight years by periodically sending commissioners [stewards] to the site, preparing a lawsuit, and arguing over empty formalities. When I left Lima, a decision III.328 from the court was still awaited. The great mine remained closed, but the Indians had been permitted[1] since 1795 to mine freely the cinnabar veins that run through the alpine calcareous stone, between Huancavelica and Sillacasa. The annual output of these small operations amounted to 3,200 to 3,500 quintals. Since the law requires all mercury to be delivered to the treasury (*cajas reales*) of Huancavelica, I shall give here the [annual] output from 1790 to 1800, according to the ledgers:

1790	2,021 quintals	37 pounds
1791	1,795	69
1792	2,054	14
1793	2,032	68
1794	4,152	95
1795	4,725	47
1796	4,182	14
1797	3,927	32
1798	3,422	58
1799	3,355	92
1800	3,232	83

1. Prior to 1795, several thousand alpacas and llamas, driven and cared for by trained dogs, carried the mercury ore from the Cerro de Santa Barbara to the furnaces outfitted with aludels that are placed near the city of Huancavelida.

One wonders whether, in the current state of things, it would be prudent to clear away the old works of the great mine of Huancavelica, or whether it is not, in fact, necessary to focus on the exploratory works? According to the papers drawn up by the Baron de Nordenflycht, it is absolutely un- III.329 true that the Santa Barbara mine was exhausted when they were imprudent enough to take away the pillars. In the *pertinencia* of Cochapata, at a depth of 228 *varas*, cinnabar ore have been found that are as wealthy as those of the Brocal; but since the works have been directed for centuries by ignorant men lacking any knowledge of subterranean geometry, the operation has been given a cylindrical shape, the axis of which is inclined from north to south. At the Brocal, near the surface of the earth, the metalliferous layer is almost completely unworked on the south side. Deep down at Cochapata, on the other hand, the galleries have been lengthened only very slightly northward. This peculiar layout of the mining works has led people to believe that the cinnabar runs out toward the bottom of the mine; but if it has been found in less abundance, this is because by continually extending southward, people unwittingly entered into the sterile part of the quartzose sandstone layer.

The relevance of these remarks notwithstanding, it seems somewhat imprudent to recommend that the collapsed mine be cleared out. This operation would incur tremendous costs, and the old works are so poorly laid out that it is impossible to turn them to good account. The metalliferous layer of the Cerro de Santa Barbara extends well beyond Sillacasa, several leagues away, to just above the village of Guachucalpa. If it were worked at the points that have remained intact thus far, one could be almost certain of the success of the operation; for there is no greater proof of the sheer III.330 abundance of mercury in this section of the Cordilleras than the output of the Indians' shallow works. If small veins of cinnabar, exposed in their outcroppings, yield 3,000 quintals in a regular year, it is undeniable that *exploratory works*, sensibly directed, could one day yield more mercury than the plants of Peru require. It is likewise to be hoped that as the inhabitants of the New World learn to take advantage of the natural riches of their land, the perfecting of chemical knowledge will also lead to amalgamation techniques through which less mercury will be lost. It is by decreasing the consumption of this metal and by increasing the output of local operations that the American miners will unwittingly manage to do without mercury from Europe and China.

To complete the portrait of the mineral substances of New Spain, it remains for me to mention pit coal, salt, and soda. *Pit coal*, layers of which

I saw in the Bogotá valley,[1] at a height of 2,500 meters above sea level, appears in general to be quite scarce in the Cordilleras. In the Kingdom of New Spain, it has only been found in New Mexico: it is nevertheless likely III.331 that it is also found in the secondary formations that stretch to the north and to the northwest of the Río Sabina. In general, coal [pit coal] and rock salt are abundant to the west of the Sierra Verde, near Timpanogos Lake, in Upper Louisiana, and in the vast northern regions bounded by the *rocky mountains* (*stony mountains*) of Mackenzie and of Hudson Bay.[2]

In the entire inhabited part of New Spain, there is not a single rock salt mine similar to that of Zipaquirá, in the kingdom of Santa Fe, or Wieliczka, in Poland. Nowhere is muriate of soda collected into beds or into masses of considerable size; it is, rather, disseminated across the clayey terrain that covers the ridge of the Cordilleras. In this respect, the plateaus of Mexico resemble those of Tibet and Tartary. We have observed above,[3] in our discussion III.332 of the Tenochtitlan valley, that the Indians who dwell in the caves of the porphyritic rock called *Peñón de los Baños*,[4] use earth laden with muriate of soda as a wash. It is a widespread theory in the country that this salt is formed, like potassium nitrate, through the influence of the atmospheric air; indeed, it appears that the muriate of soda is only found in the upper ground layer, up to a depth of eight centimeters. The Indians pay the landowners a small sum for the permission to remove this initial muriatiferous layer; they know that after a few months they will find a crust of clay laden with both muriate of soda and lime, potassium and lime nitrate, and carbonate of soda. A distinguished chemist, Mr. [Andrés] del Río, proposed to conduct precise studies on these phenomena, by using the dirt as wash before it has come into contact again with the atmospheric air. The richest salt deposit in Mexico is the lake of *Peñón Blanco*, in the intendancy of San Luis Potosí, the bottom of which displays a layer of clay that contains 12 to 13 percent muriate of soda. It should be noted, however, that were it not for the amalgamation of silver ore, salt consumption would not be very high in

1. Near Tausa, Canoas, and in the Cerro de Suba, on the road from Santa Fe de Bogotá to the rock salt mine of Zipaquirá.

2. There are salt springs on the banks of Dauphin Lake and Slave Lake. It is known that coal deposits exist near the Mackenzie River at 66° latitude; and at the foot of the *Rocky Mountains*, near 52° and 56° latitude. (See the Mackenzie, *Voyages*, vol. III, pp. 332–34, and particularly Major Long's excellent work, which was published [by James Edwin] under the title *[Account of] an Expedition [from Pittsburgh] to the Rocky Mountains*, vol. II, p. 402.

3. [See above] vol. II, p. 91 [p. 351 in this edition].

4. See above, p. 261 [p. II.147 in this edition].

Mexico, because the Indians, who constitute a large part of the population, have not at all abandoned their ancient custom of seasoning dishes with *chile*,[1] or hot pepper [capsicum]. III.333

If one casts a broad glance at the table of the mineral riches of New Spain, far from being struck by the value of the current operations, one is stunned to find that the total output of the mines is not significantly greater. One can easily foresee that this branch of national industry will increase as the country becomes more inhabited, that the poorer landowners shall enjoy the rights of citizenship more freely, and that geological and chemical knowledge shall become even more widespread. Several obstacles have been lifted since 1777, or since the establishment of a *High Council of Mines*, which bears the title of *Real Tribunal general del importante Cuerpo de Mineria de Nueva España*, and which holds its meetings in the viceroy's palace, in Mexico City. Theretofore, the mine owners had not gathered as a corporate body, or at least the court in Madrid did not wish to recognize them,[2] through a constitutional act, as forming an established body.

Mining legislation was once extraordinarily muddled, because at the beginning of the conquest, under the reign of Charles V, a mixture of Spanish, Belgian, and German laws had been transmitted to Mexico, and because these laws, owing to the contrast in local circumstances, were not applicable to these faraway regions. The establishment of the High Council of Mines, the president of which is a famous name in the annals of the chemical sciences, was followed by that of the School of Mines as well as by the drafting of a new law code, published under the title *Ordonanzas de la Minería de Nueva España*. The Council, or *Tribunal general*, is composed of a director, two delegates from the body of miners, a magistrate's assistant, and a judge who is the president of the *Juzgado de alzadas de Mineria*. Under the control of the *General Court* are the thirty-seven *Provincial Mining Councils* or *Diputaciones de Mineria*, the names of which have been listed above. The mine owners (*mineros*) send their *representatives* to the Provincial Councils, and the two general delegates, who reside in Mexico City, are chosen from among the delegates from the administra- III.334

1. Chili or ahi. See above, vol. II, p. 472 [p. 559 in this edition]. If the annual European consumption of muriate of soda is estimated to be six kilograms per head, one would not presume the coppery race's consumption to be more than one half-kilogram.

2. [Lassaga,] *Representación que, a nombre de la Minería de esta Nueva España hacen al Rey Nuestro Señor los Apoderados de ella,* ▼ *D. Juan Lucas de la Sage y D. Joaquín Velásquez de León* (Mexico City, 1774, p. 40). The *cuerpo de minería* was only recognized by the *Real Cédula* of July 1, 1776, which granted it the title of *Importante Cuerpo*.

tive regions. In addition, the body of miners of New Spain has *apoderados* or authorized representatives in Madrid, who negotiate directly with the ministers on the colonies' interests in mining affairs. The students of the III.335 *Colegio de Minería*, whose tuition is covered by the state, are distributed by the *Tribunal* among the administrative centers of the various *Diputaciones*. It is undeniable that there are great advantages to the representative system that has been followed in the new organization of the body of Mexican miners: it maintains the public spirit in a country where the citizens, scattered across an immense area of land, do not have a strong enough sense of their common interests; it makes it easier for the High Council to raise considerable sums of money for any great and practical undertaking. It would be desirable, however, for the director of the Tribunal to have more influence on the progress of mining in the provinces, and for the mine owners, less keen to protect what they call their liberty, to be more enlightened as to their true interests.[1]

The *High Council* has an annual income of over one million livres tournois. Following its establishment, the king granted it two-thirds of the royal prerogative of *seigniorage*, which equals one *real de plata*, or one-eighth of a piastra fuerte for each mark of silver delivered to the mint. This one million in revenue is reserved for the salaries[2] of the Tribunal members, for the maintenance of the School of Mines, and for a relief fund or an advance III.336 (*avios*) fund for the mine owners. These advances, as we have observed above, have been made with rather more largesse than discretion: a miner from Pachuca once obtained 170,000 in a single stroke; the shareholders of the *mina de água* of Temascaltepec received 241,000 piasters, but this aid produced no effect.[3] During Spain's most recent wars with France and England, the *Tribunal* found itself forced to donate two and a half million francs to the court in Madrid and to loan the latter an additional fifteen million francs, six of which have not yet been reimbursed. To meet these extraordinary expenses, it was necessary to ask for loans, and today one-half of the High Council's revenues pays the interest on this capital: the *seignior-*

1. Compare to Delhuyar, *Memoria sobre [la formacion] de una ley orgánica [para gobierno] de la minería de España*, 1825, pp. 15, 46, 47, 49, and 116.

2. These salaries amount to 25,000 piasters. The managing director [director general] only earns 6,000 piasters; the Seminary, or School of Mines, in which creole Spaniards and noble Indians are educated, absorbs only 30,000 piasters per year.

3. See the report drafted for the electors and published with the title *Estado general que manifiesta a los vocales los caudales del Tribunal de Minería, desde 1777 hasta 1788.*

age tax was increased by one-half until the debts assumed by the Tribunal are redeemed. Instead of eight *grains*, the miners are obliged to pay twelve[1] per mark of silver. In this state of affairs, the Tribunal can no longer make advances to miners who, for lack of funds, often find themselves incapable of starting useful undertakings. Large amounts of capital formerly used in mining have been reserved for agricultural development, and the mine owners will likely have need again soon for the kinds of institutions (*bancos de plata, companies refraccionarias*[2] *o de habilitación y avios*) that once advanced considerable sums of money to the miners, at very high interest rates.

III.337

All of the metallic riches of the Spanish colonies are in the hands of private individuals: the government has no other mine in its possession than that of Huancavelica, in Peru, which has long been abandoned. It does not even own the large galleries of efflux, as do several German monarchs. From the king, private individuals receive *concessions* of a certain number of measures along the direction of a vein or a layer; they are only required to pay very modest fees on the silver extracted from their mines, fees that have been estimated at 11 and a half percent for silver and at 3 percent for gold for all of Spanish America.[3]

In New Spain, mine owners pay to the government the *half-quint*[4] or the tithe, the 1-percent duty (*derecho del uno por ciento*), and the *minting* fee, called *derecho de monedage y señoreage*. The latter duty, instituted in 1566 through a law of Phillip II and raised at the end of the seventeenth century,[5] today amounts to three and two-fifths reales per mark of silver, one mark being divided into sixty-eight reales, with a cost of one-half real, the owner of the silver receiving only sixty-four marks. Of these three and two-fifths reales, two and two-fifths are counted as *derecho de monedage*, and one real

III.338

1. *Ocho granos de señoreage, y quatro granos temporalmente impuestos*. In Lima, the Tribunal receives one real per mark.

2. *Real Cédula sobre la compañía refaccionaria propuesta por el Genovés Domingo Reborato, del 12 marzo 1744*. Don José Bustamante, *Informe sobre la habilitación de los mineros*, 1748.

3. Bourgoing, [*Tableau de l'Espagne moderne*,] vol. II, p. 284.

4. The reduction of the *quinto* and *diezmo*, which had such a strong impact on the development of mining, was only granted on a temporary basis at first and for gold alone, in a few districts. On the history of this reduction from the first representations of the *Ayuntamiento de México* in 1530 to the *Real Cédula* of June 19, 1723, see Delhuyar, *[Memoria sobre] el influjo de la minería*, 1825, pp. 33–35. The *1-percent duty* was originally a duty of 1 and a half percent and was based on the *Pragmatica* of June 5, 1552.

5. *Recopilación de leyes de Castilla, de 1598*, book V, tit. XXI, no. 9.

as *derecho de señoreage.* The revenue that the tax office earns is estimated[1] at 200,000 marks of silver, which is equivalent to 1,700,000 piasters:

In *derecho de diezmo*	160,000 piasters
In *derecho de uno por ciento*	16,000
In *derecho de monedaje y señoreaje*	86,750
Total	262,750

That is, around 16 and two-fifths percent. If one deducts from this the profit that the government derives from minting and from the entire *tariff,* one finds that the duties that the mine owners pay amount to only 13 percent. To present a more detailed view of the fees received by the tax office, it is necessary, according to the information that I gathered during my stay in Guanajuato, to make a distinction between pure silver and the III.339 silver that is blended with gold; for if the silver contains less than thirty grains of gold per mark of silver, then the mint does not pay private individuals for the gold. [See table on following page.]

III.340 If the ingot is so rich in gold that half of its weight is composed of this metal, then the assay costs amount to four reales per mark. The examples that we have just given show that the private individual who brings his silver to the provincial coffers of Mexico to exchange it for cash pays the government 12 and a half percent in the first case and 19 and one-eighth percent in the second. It is this tax that incites the mine owners to engage in the fraudulent extraction of precious metals. Despite so many centuries of experience, the court of Madrid has attempted on several occasions to raise[2] the *seniorage* tax, without a thought for how this imprudent measure might discourage private individuals from bringing materials to the mint. As it is for the direct taxes on gold and silver, so it is for the profit that the government seeks in the sale of mercury: mining will only increase as these taxes are decreased, and as mercury, indispensable for amalgamation techniques, is made available at lower prices. It is certainly surprising to learn that one deservedly famous writer,[3] who has the most reasonable ideas about the exchange of metals, has undertaken to defend the seigniorage fee and the tariff.

1. [Juan Lucas de Lassaga and Joaquín Velázquez de León,] *Representación [. . .] de la Minería [. . .] de Nueva España*, 1774, p. 53, §45.

2. [Lassaga and de León,] *Representación de la Minería de Nueva España, sobre la doble exacción del señoreage, de 1766.*

3. Adam Smith, [*Recherches sur la nature et les causes de la richesse des nations,*] vol. III, book IV, chap. VI.

III.339

	An ingot of silver with no traces of gold, removed through amalgamation, weighing 135 marks, at 11 denarii and 22 grains. Value:	1,171 piasters	6 reales		
	1-percent duty plus tithe	127 piasters	6 reales		
	Analyst's fee	4	0		
	Bocado duty collected in the treasury				
Costs		1	0	147	0
	Bocado duty collected at the mint	0	4		
	Seniorage duty	13	6		
	Remainder for the owner			1,024	6

If the silver is the product of smelting and contains below 11 deniers 19 grains, one must add refining costs, which are eight maravedis per mark.

	An ingot of auriferous silver, at 11 denarii 19 and a half grains of silver and 50 grains of gold, weighing 133 marks 2 ochavas.				
	Value in silver			1,133 piasters	3 reales
	in gold			194	0
				1,327	3
	1-percent duty and tithe	123 piasters	6 reales		
	Gold duty, at 3 percent	5	6		
Costs	Analyst's fee	6	0	254	3
	Bocado duty	1	4		
	Apartado	91	7		
	Consumo	12	2		
	Señoreage	13	2		
	Remainder for the owner			1,073	0

III.341 The information that we have provided in this chapter suggests that there is almost no point in raising the question of whether the output of the Mexican silver mining has reached its *maximum point*, or whether it is likely to increase further over the course of centuries. We have seen that three mining districts, those of Guanajuato, Catorce, and Zacatecas, alone produce over one-half of all the silver from New Spain, and that nearly a quarter is extracted from the single vein of Guanajuato. One mine that has only existed for forty years, that of Valenciana, has itself occasionally produced,[1] in a single year, as much silver as the entire kingdom of Peru. It was only thirty years ago that the Real de Catorce veins were first mined, and the discovery of these mines nevertheless resulted in a one-sixth increase in Mexican metallic output. If one considers the vast expanse of land taken up by the Cordilleras and the huge number of ore deposits that have not yet been mined,[2] it is easy to understand that, were New Spain better administered and inhabited by an industrious people, one day it could itself produce the one hundred and sixty-three

1. For example, in 1791.
2. Especially from Bolaños to the Presidio of Fronteras.

million francs' worth of gold and silver produced by the Americas as a whole. In the space of a hundred years, the annual output of Mexican mines has risen from twenty-five million to one hundred and ten million francs. If Peru does not display an equal increase in wealth, this is because the population of that unfortunate country has not grown in centuries because, more poorly governed than Mexico, industry in Peru has encountered more insurmountable obstacles. Furthermore, nature has deposited precious metals at incredible heights, at sites where, because of the high price of supplies, mining becomes extremely costly. In general, there is such abundance of silver in the Andes chain that if one reflects on the number of ore deposits that have remained intact or that have been only superficially mined, one would be tempted to think that the Europeans have barely begun to enjoy the inexhaustible collection of riches contained within the New World. III.342

If one takes a look at the mining district of Guanajuato, which, in a small area of just a few thousand square meters, produces annually between one-seventh and one-eighth of all American silver, one sees that the 550,000 marks that are extracted annually from the famous *veta madre* are the product of only two mines, that of the count of Valenciana and that of the marquis of Rayas, and that over four-fifths of this vein has never been mined. Nevertheless, it is very likely that if the two mines of Fraustros and Mellado were combined and drained, they would form a mine comparable in wealth to that of Valenciana. The theory that New Spain does not produce even one-third of the precious metals that it could in happier political circumstances has long been espoused by all of the learned people who reside in the principal mining districts in the country; it is formally articulated in a paper that the delegates from the body of miners presented to the king in 1774 and that exhibits both wisdom and knowledge of the localities. Europe would be inundated with precious metals if the mineral deposits of Bolaños, Batopilas, Sombrerete, Rosario, Pachuca, Moran, Zultepec, Chihuahua, and many others that were once deservedly famous were all mined simultaneously with all the means that the development of the mining arts has presented. I am not unaware that, by expressing this opinion, I find myself in direct opposition to the authors of several works of political economy, in which they claim that the mines of the Americas are in part exhausted and in part too deep to be worked profitably. It is true that the operational costs of the Valenciana mine have doubled in the space of ten years; but the shareholders' profit has remained the same, and this increase in costs is caused less by the depth of the shafts than by the poor management of the works. People forget that in Peru, the famous mines of Yauricocha or III.343

Pasco, which annually produce over 200,000 marks of silver, are still only III.344 thirty to forty feet deep. It seems to me unnecessary to refute opinions that are contrary to the numerous facts that I have collected in this chapter, and it should not be surprising that the state of mining in the New Continent is judged with such little acumen in Europe, if one considers the relative lack of precision that the most famous political writers have shown in their studies on the state of the mines of their homeland.

But what is the relationship between the output of the Mexican mines and the output of the other Spanish colonies? We shall examine, in order, the riches of Peru, Chile, the kingdom of Buenos Aires, and that of New Granada. It is known that the other main political divisions—namely, the four *capitanías generales* of Guatemala, Havana, Puerto Rico, and Caracas—do not contain mines in operation. I shall not use the vague and incomplete data found in several very recent works; I shall only discuss what I was able to take from the official documents that I was able to obtain.

I. There were coined, at the Lima mint:

From	1754	to	1772	6,102,139	marks of silver and	129,080 marks of gold
	1772		1791	8,478,367		80,846

The total value of the gold and silver[1] amounted to 68,944,622 piasters III.345 in the first of these periods and 85,434,849 piasters in the second, which equals, in an average year,

From	1754	to	1772	in gold and in silver	3,830,000 piasters
	1772		1791		4,496,000

The output of gold has decreased, while that of silver has increased considerably. In 1790, the mines of Peru[2] yielded 534,000 marks of silver and 6,380 marks of gold. From 1797 to 1801, 26,032,653 gold and silver piasters were coined in Lima. Here is the mining output, year by year, from 1780 to 1801.

1. Unanué, *Guía política [eclesiástica y militar del virreynato] del Perú*, 1793, p. 45.

2. *Mercurio Peruano*, vol. 1, p. 59. All the mines of Peru have yielded 2,979,365 marks of silver during the ten-year period from 1776 to 1785. During the following four years:

Coining at the Lima Mint, in Gold and in Silver

From 1780 to	1789, in an average year	3,325,546 piasters
	1790	5,206,906
	1791	5,120,234
	1792	5,605,581
	1793	5,941,706
	1794	6,093,037

The official documents published by the order of the viceroy[1] also show the following: III.346

YEAR	VALUE OF GOLD, IN PIASTERS	VALUE OF SILVER, IN PIASTERS	VALUE OF GOLD AND SILVER, IN PIASTERS
1797	583,724	4,516,206	5,099,930
1798	535,810	4,758,094	5,293,904
1799	496,486	5,512,345	6,008,831
1800	378,596	4,399,409	4,778,005
1801	328,051	4,523,932	4,851,983
TOTAL for five years:	2,322,667	23,709,986	26,032,653

The total output of the five previous years was thirty million, such that one might consider six million as the mean for one year, gold and silver production having been lower in 1800 and in 1801, because of the maritime war, which hindered imports from Europe of mercury, as well as of iron and steel. We shall arrive, however, at a lesser sum, namely, 3,450 marks of gold and 570,000 marks of silver, the combined value of which is 5,300,000 piasters.

The places in Peru most famous for their wealth in precious metals or for the scale of their mining operations, are, following the chain of the Andes from north to south: in the province of *Cajamarca*, the Cerro de III.347

1787	311,634 marks of silver
1788	474,153
1789	389,006
1790	412,117

And from 1788 to 1789, 3,536 marks of gold in an average year.

1. *Razón de lo que se ha acuñado en la Real Casa de Moneda de Lima* (unpublished).

Gualgayoc, near Micuipampa, Fuentestiana, and Pilaneones; in the province of *Chachapoyas*, San Tomas, Las Playas de Balzas, and the Pampas del Sacramento, between the Río Guallaga and the Ucajalé; in the province of *Guamachuco*, the city of Guamachuco (including the Reales of San Francisco, Angasmarca, and the Mina Hedionda), Sogon, Sanagoran, San José, and Santiago de Chucu; in the province of *Pataz*, the city of Pataz, Vuldivuyo, Tayabamba, Soledad, and Chilia; in the province of *Conchucos*, the city of Conchucos, Siguas, Tambillo, Pomapamba, Chacas, Guari, Chavin, Guanta, and Ruriquinchay; in the province of *Huailas*, Requay; in the province of Huamalies, Guallanca; in the province of *Cajatambo*, Chanca and the small town of Cajatambo; in the province of *Tarma*, the Cerro de Yauricocha (two leagues north of Pasco), Chaupimarca, Arenillapata, Santa Catalina, Caya Grande, Yanacanche, Santa Rosa, and the Cerro de Colquisirca; in the province of *Huarochiri*,[1] Conchapata; in the province of *Huancavelica*, San Juan de Lucanas; finally, on the outer edges of the Atacama desert, Huantajaya.

III.348 In this long enumeration, I have followed the old division of Peru into provinces; but now that the border of the kingdom of Buenos Aires passes to the west of Lake Chucuito, between that lake and the city of Cuzco, and that, on one side, the kingdom of Quito and the provinces of Jaén de Bracamoros and Maynas, and, on the other, the governments of La Paz, Oruro, La Plata, and Potosí have been separated from Peru, the latter is divided into seven intendancies, those of *Trujillo*, *Tarma*, *Huancavelica*, *Lima*, *Guamanga*, *Arequipa*, and *Cuzco*, each of which contains several departments or *partidos*.[2] One can only arrive at false conclusions when, as has been done in the most highly regarded works, one compares the output of the mines of the old Peru with that of the current Peru, the latter containing within its borders, since 1778, neither the Cerro del Potosí nor the mines of Oruro and La Paz. Peruvian gold comes in part from the provinces of Pataz[3]

1. The mountains of Huarochiri and Canta contain excellent pit coal; but because of the high cost of hauling, one cannot make use of it in Lima. Cobalt and antimony have also been discovered at Huarochiri.

2. The former provinces of Pataz, Guamachuco, and Chachapoyas are today regarded as *partidos* of the intendancy of Trujillo; those of Cajatambo, Huailas, Conchucos, and Humalies belong to the intendancy of Tarma. The capitals of the seven intendancies are: *Lima*, with 52,600 inhabitants; *Guamanga*, with 26,000; *Arequipa*, with 24,000; *Trujillo*, with 5,800; *Huancavelica*, with 5,200; *Tarma*, with 5,600; and *Cuzco*, with 32,000 (José Hipólito Unanué, *Guía pólitica, ecclesiastica y militar del Virreynato del Perú, para el año 1793*).

3. Among the five mining districts in the *partido* of Pataz, which we have mentioned above, only that of Chilia produces silver.

and Huailas, where it is extracted from the quartz veins that run through primary rock, in part from the washes (*lavaderos*) established on the banks of the Alto Marañón, in the *partido* of Chachapoyas. III.349

Just as nearly all Mexican output comes from the mines of Guanajuato, Catorce, Zacatecas, Real del Monte, and New Viscaya, so too is nearly all Peruvian silver extracted from the great mines of Yauricocha or Lauricocha (commonly referred to as the *Pasco* and the *Cerro de Bombón* mines[1]), those of Gualgayoc or *Chota*, and *Huantajaya*.[2]

The Pasco mines, the most badly worked in the whole of Spanish America, were discovered by the Indian Huari Capca in 1630: they annually produce nearly two million piasters. To form an accurate idea of the enormous mass of silver that nature has deposited in the bosom of these calcareous mountains, at an elevation of over 4,000 meters above sea level, one must remember that the layer of argentiferous iron oxide at Yauricocha has been III.350 worked, with no interruption, since the beginning of the sixteenth century and that in the past twenty years over five million marks of silver have been extracted, without most of the shafts extending below thirty meters and without any of them having reached a depth of 120 meters. The water, very abundant in these mines, is drained not by hydraulic wheels or by horse-driven winches, as in Mexico, but by man-powered pumps. Furthermore, despite the relative shallowness of these wretched excavations, which are referred to as shafts and galleries, these mines are excessively costly to drain: in the Luna mine, a few years ago, these costs amounted to over one thousand piasters per week. The Yauricocha mines would produce the same amount of silver as Guanajuato if hydraulic machines or steam engines[3] were constructed, for which the peat bogs of Lake Gilacocha could be used. The metalliferous layer (*manto de plata*) of Yauricocha appears III.351

1. The high plateau of the Cordilleras, on which is found the small lake *de los Reyes*, to the south of the Cerro de Yauricocha, is called the *Pamba de Bombón*. One must search for the position of Pasco not on the map of La Cruz but on that of the Río Huallaga, drawn up by Father Sobreviela and published in 1791 by the *Sociedad de los Amantes del país de Lima*. It is believed that the city of Pasco is at an elevation of 1,800 toises above sea level. Snow falls there for a large part of the year.

2. Pronounced Guanta-ba-ya.

3. These steam engines were set up several years before the political misfortunes of the Spanish colonies. True pit coal was found in abundance near Pasco, at Rancas, probably at 1,600 toises above sea level, at an elevation above the large-tree vegetation zone. The effect of steam machines on mining output in Yauricocha has been so beneficial that this output has risen, over just a few years, to 480,000 marks of silver. Civil conflicts have interrupted operations: it is hoped that a new mining company in Pasco will merge with the old one.

on the surface for a length of 4,800 meters and for a width of 2,200. The following table, taken from the books of the provincial treasury of Pasco, indicates the number and the weight of the silver ingots that were smelted in Pasco from 1792 to 1801:

Mining at Yauricocha

PERIOD	INGOTS	MARKS OF SILVER
1792	1,052	183,598
1793	1,325	234,943
1794	1,621	291,254
1795	1,550	279,622
1796	1,561	227,514
1797	1,340	242,949
1798	1,478	271,862
1799	1,237	228,356
1800	1,198	281,481
1801	914	237,435
TOTAL for 10 years:	13,276	2,479,014

This table suggests that mining production in Pasco has almost never been below two hundred thousand marks, and that in 1794 and 1801 it nearly rose to the volume of three hundred thousand marks.

The mines of Gualgayoc and Micuipampa, commonly called *Chota*, which I had the opportunity to visit most extensively in 1802, were only discovered in 1771, by a Spaniard, ▾ Don Rodríguez de Ocaño. In the Incas' time, the Peruvians had worked the silver veins in the Cerro de Lin, near Cutervo, at Chupiquiyacu, west of the small city of Micuipampa, where the thermometer drops to the freezing point almost every night, and which is seven hundred meters higher than the city of Quito. Immense riches have been found up to the surface of the Earth, either in Gualgayoc Mountain, which soars like a fortified castle in the middle of the plain, or at Fuentestiana, at Cormolache, and in the Pampa de Navar. In this latter plain, across an area of over one-half square league, everywhere that the sod has been cleared, people have removed silver sulfide and filaments of native silver that cling to the grass roots; silver has often been encountered in masses (*clavos* and *remolinos*), as if portions of this smelted metal had been poured on very soft clay. The output of the Gualgayoc or Chota mines is very uneven, because of the inconsistency of the veins that run through al-

III.352

pine calcareous stone, at Fuentestiana and Cormolache, or hornstone, called *panizo.* This hornstone forms a subordinate layer in the calcareous rock, as was clearly recognized during the digging of the Choropampa shaft, east of the Purgatorio, near the Chiguera ravine. From April 1774 to October 1802, the mines of Gualgayoc, in the partido of Chota, together furnished the provincial coffers of Trujillo with a total of 1,912,327 marks of silver, or, in an average year, 67,193 marks. [See table on the following two pages.] III.353

This table, drawn up at my request in the offices of the intendancy, shows the amount of silver delivered to the *Cayana de Trujillo*, as well as the tithe and 1 and a half percent fee that were paid to the king. Of 11,791 ingots, almost one-eighth, or 1,450, came from the *partidos* of Huamachuco and Conchuco. I was unable to obtain the output of the Cerro de Gualgayoc from the discovery of these deposits in 1771 until 1774. These first years were probably the most productive of all; but since at that time, silver was sent to Lima, the archives of Trujillo could not provide any information about this. It is believed, and rightfully so, that under a more enlightened administration, the Cerro de Gualgayoc would become a second Potosí; indeed, its ore are wealthier than those of Huantajaya, and easier to mine than those of Yauricocha. III.354

The *Huantajaya* mines, surrounded by rock salt layer, are especially famous for the large masses of native silver that they contain in a decomposed gangue: they yield 70,000 to 80,000 marks of silver annually. The native silver there is accompanied by conchoidal muriate of silver, silver sulfide, small-grained galena, quartz, and carbonate of lime. These mines are located in the partido of Arica, five leagues east of the Morro de Tarapaca and the small port of Yquique, in a desert completely devoid of water. The project of bringing in fresh water, for the use of both humans and animals, and seawater, for the amalgamation plants, has been under development for quite some time. In 1758 and 1789, two nuggets of solid silver, one with a weight of eight quintals, the other with a weight of two quintals, were discovered in the mines of Coronel and Loysa. In 1795, the output of the mines of Asiento and Huantajaya was only 72,462 marks of silver.[1] III.355

The small hill on which the Huantajaya mine is located, on the shore of the Pacific Ocean, contrasts sharply with the masses of vitreous silver at the summit of the Cerro de Gualgayoc, at an elevation of 4,080 meters. It proves how vague the systematic ideas are that famous geologists have advanced

1. *Report of the Intendant of Arequipa, Antonio Alvarez y Jiménez* ["Relaciones de la vista del intendente de Arequipa don Antonio Alvárez y Jiménez"] (unpublished [at the time]).

III.353 Output of the Silver Mines of Hualgayoc, Guamachuco, and Conchuco

PERIODS	NUMBER OF INGOTS OF SILVER	THEIR WEIGHT		DUTY IN QUINTALS
		Marks	Ounces	Piasters
1774	182	34,403	4	33,852
1775	300	57,894	5	56,941
1776	432	84,326	1	82,985
1777	302	60,015	3	59,051
1778	327	65,062	3	64,034
1779	324	64,203	7	63,214
1780	306	60,981	0	60,021
1781	308	61,435	4	60,387
1782	429	73,698	6	72,462
1783	329	58,713	6	57,808
1784	335	61,564	0	60,440
1785	397	73,604	2	72,373
1786	398	73,305	6	72,024
1787	450	83,633	0	82,209

1788	404	73,835	5	74,371
1789	469	87,484	0	83,469
1790	645	119,183	5	117,241
1791	575	105,383	2	103,618
1792	731	134,084	4	131,939
1793	406	72,904	6	71,713
1794	480	86,876	1	85,505
1795	434	79,309	4	78,755
1797	378	67,789	3	66,721
1798	501	90,015	4	88,600
1799	607	108,591	6	106,889
1800	392	70,595	6	69,471
1801	255	45,378	3	44,626
1802	267	48,198	6	47,413
TOTAL for 29 [28?] years:	11,791	2,180,457 [2,102,460?]	3 [102?]	2,144,179 [2,068,132?]

about the distribution of metals according to the various climates and latitudes. Ulloa, after having traversed a large section of the Andes, claims that in the Americas silver is particular to the high plateaus of the Cordilleras, called *Punas* or *Paramos*, and that gold, on the other hand, is abundant in the lowest-lying and therefore hottest regions[1]; but that learned traveler appears to have forgotten that in Peru the provinces richest in gold are the *partidos* of Pataz and Huailas, which lie across the ridge of the Cordilleras. The Incas extracted immense amounts of gold from the plains of Curimayo, to the northeast of the city of Cajamarca, at a height of 3,400 meters. It has III.356 also been mined on the right bank of the Río de Micuipampa, between the Cerro de San José and the plain called *Choropampa* or *Plain of Shells* by the indigenous peoples, because of the huge quantity of ostrea, cardium, and other pelagic shell fossils contained within the alpine limestone formation of Gualgayoc. It is there that considerable masses of gold have been found disseminated in branch form and in curved filaments within red and vitreous silver veins, at a height of over 4,000 meters above sea level. As for the drift deposits in which the gold *washes* of Chocó, Sonora, and Brazil are established, is it at all surprising to find them at the foot rather than at the top of mountains? If tin[2] seems to be an exception to this law of nature, it is probably because the granitic layers in which it was originally contained were decomposed in place.[3]

The technique for amalgamating silver ore, followed in Peru since 1571, III.357 is the same as the one used in Mexico. In both countries, *schlich* is treated following the rules prescribed by Medina, Barba, Corso de Leca, and Corrosegarra; but in general, amalgamation is practiced with greater care and intelligence by the Mexican miners at Guanajuato and Zacatecas than by those of Peru. In New Spain, they normally estimate the cost of amalgamating one hundred quintals of ore, with four ounces of silver per quintal, to be eighty-seven piasters and four reales, including a twenty-five-piaster

1. [Antonio de] Ulloa, *Noticias americanas*, 1772, pp. 223 and 236.

2. For example the wash-tin (*Waschzinn*) of the top of Fichtelgebirge.

3. In addition to the three large Peruvian mining operations of Pasco or Yauricocha, Chota or Gualgayoc, and Huantajaya, scattered across the partidos of Huamalies, Cojatambo, and Condesuyo are a large number of mines about which I was unable to obtain positive data. I only know from the tables inserted in the *Guía política del Vireinato del Perú* that in 1795 the output of silver was 1,764 marks in the mines of Urubamba, Calca, Tinta, Paruro, Chumbilicas, Abancay, etc. (intendancy of Cuzco); 34,000 marks in the mines of Caylloma (intendancy of Arequipa); 9,119 marks in the mines of Castrovireyna, Angaraes, Huanta, Tayacaxa, etc. (intendancy of Huancavelica); 25,300 marks in the mines of Huayanca, Chanca, Iulcan, Huayacayan, Tucapa, Capaco, etc. (intendancy of Tarma).

loss of mercury. Since these one hundred quintals yield fifty marks of silver, which, depending on the regular price of silver[1] in mining areas, are worth 362 piasters, the result is that the costs of amalgamation amount to around 24 percent of the value of the silver. In Peru, where mercury from Huancavelica regularly sells for sixty to seventy piasters per quintal,[2] the costs in several mining districts rise to between 30 and 38 percent. In the *Cerro de Gualgayoc*, for example, where the price of labor is three to four reales (forty to fifty sous) per day, one load of *schlich* containing two to three marks of silver costs seven piasters in the amalgamation technique, as follows:

III.358

In roasting,	in wood	8 silver reales
	in daily wages	2
In muriate of soda		6
In lime		4
In daily wages, for crushing the *schlich*		12
In mercury consumption		24
TOTAL		56 reales or 7 piasters

During my stay in the Cordillera of the Andes, there were only two mining districts where Mr. de Born's method, *amalgamation in barrels*, was employed successfully, namely, the Real de Requay, in the province of Huailas, and Tallenga, in the province of Cajatambo.[3] To determine the considerable loss of silver that Peru experiences annually through the ignorance of the *amalgamators*, one has only to observe the simple fact that every day an azoguero extracts fifteen marks per cajón from the same ore which only ten to twelve marks was previously extracted. In the first years following the discovery of the deposits at Yauricocha, only the pacos, or iron oxide blended with native silver and with muriate of silver, were mined. The prismatic black silver ore and the argentiferous gray copper were cast aside among the rubble; similarly, when the small city of Micuipampa was being

III.359

1. Seven piasters and two reales. Garcés [y Eguía, *Nueva teórica y práctica del beneficio de los metales*,] p. 144. At the beginning of the seventeenth century, in Potosí the costs of amalgamating one *cajón* of ore weighing fifty quintals and containing twenty marks of silver were calculated at only thirty piasters, or 20 percent, although one pound of mercury cost one piaster. Barba, [*Arte de los metales*,] p. 118.

2. Campomanes, *[Discurso sobre] la educación popular*, vol. II, p. 132.

3. The mine near Requay, where a German amalgamation plant has been constructed, is called ▼ Ticapamba [Ticapampa] and belongs to Don Juan Ignacio Gamio. The Tallenga plant was set up by ▼ Don Juan Ba[u]tista Ar[r]ieta.

constructed, the walls were built with very wealthy pieces of gangue; only those pieces that were yellowish-brown, or that had an earthy appearance like the *pacos*, were recognized as silver ore. These facts seem less surprising if one recalls that not even forty years have passed since, in one of the most civilized countries of Europe, people used calamine to build roads, without noticing that this substance, soiled with silver, contained zinc.

II. The *Presidencia* or *Capitanía general* of Chile annually produces 1,700,000 piasters' worth of gold and silver.[1] The largest gold mines are those of Petorca, ten leagues south of Chuapa; Yapel or Villa de Cuscus, Llaoin, Tiltil, and Ligua, near Quillota. Gold is also worked in the *partidos* of Copiapó, Coquimbo, and Guasco. The output of the mines of Chile increased considerably toward the end of the eighteenth century. From III.360 1782 to 1786, only 521,644 piasters were minted in Santiago in an average year; in 1789, over 971,000 piasters were minted; in 1790, 721,754 piasters' worth of gold and 146,132 piasters' worth in silver were minted.[2]

III. The large mass of precious metals that the viceroyalty of Buenos Aires produces comes entirely from the westernmost part, the *Provincias de la Sierra*, which, in 1778, were separated from Peru. The annual output, which is nearly all in silver, can be estimated at 4,200,000 piasters. These most productive districts are Potosí, Chaganta, Porco,[3] Oruro, Chucuito, La Paz, Caylloma, and Carangas. In the intendancy of Puno, the Ananca Mountains, near Caravaya, and Azangara, northeast of Lake Titicaca, were famous at the start of the conquest because of the wealth of their gold mines.[4] In 1803, there were plans to resume the old works at Morocollo, in the Pampa Fungosa de la Rinconada, and on the shores of Lake Communi. It was also planned to extend the Veracruz gallery in the famous silver mine of Salcedo, located in the Ycatoca and Cancharani Mountains. The Cerro III.361 de Uspallata,[5] twenty-four leagues northwest of Mendoza, exhibits *pacos*

1. In 1821, output was estimated at only one-half million piasters, including fraudulent extraction, while Father [Alonso de] Molina, whose figures were always exaggerated, estimated it to be four million piasters in 1780. (Caldcleugh, *Travels in South America*, vol. I, p. 353.)

2. ▾ *Bullion Report*, Ap[pendix of Accounts], p. 34.

3. On the silver mines of Porco, worked by the Incas, see Alonzo Barba, *Arte de los metales* (1729 edition), p. 48.

4. *Proclamación del Intendente de Puno, Don José González*. It is claimed that platinum was discovered near Morocollo, but this fact has never been confirmed by trustworthy persons.

5. Since 1824, work has resumed at the famous mines of the Real de Uspallata. The primary groups of this Real are those of San Lorenzo, San Pedro, Ballejos, and San Nicolás.

(iron hydrates) so rich that they yield two or three thousand marks of silver per 5,000-pound *crate* (*cajón*), or forty to sixty marks per quintal.

Even if one counts only the silver on which royal duties were paid, Potosí Mountain[1] alone has produced, from its discovery in 1545 to today, a mass of silver equivalent to 5,750 million livres tournois. Ulloa provided some historical information about this mining operation, which had the greatest impact on the state of trade and on the price of commodities in Europe; III.362 however, he was only able to collect very incomplete materials, basing his calculations on mercury consumption in amalgamation plants. On the basis of official documents, I am now able to publish, year by year, the value of the duties (derechos reales) paid to the provincial treasury of Potosí for the silver delivered to the mint. Since the proportion between these duties and the value of the silver extracted from the mines is known for the various periods, it is possible to deduce the annual output, expressed in piasters, from the following three tables.

As we have observed above,[2] in comparing the current output of the Guanajuato mines of Mexico, these three tables[3] suggest that over the course III.366

1. Potosí, more correctly Potocchi, Potossi, or Potocsi. The ancient name for Huancavelica is Huanca-Villica. Garcilaso [de la Vega El Inca], *Com[entarios] reales*, book VIII, chap. 25. Pedro de Cieza de León, *Crónica del Perú*, chap. 109. The layer of porphyry that crowns Potosí Mountain, the Hatum Potocsi, gives it the shape of a sugarloaf or a basaltic hill. (See above, p. 141 [p. II.79 in this edition].) ▼ Doctor Redhead, who lives in Salta, writes to me that in the city of Potosí, he has never seen the barometer drop below 17^{po} 2^{lines} nor above 18^{po} 0^{lines}, in the old French measurement. Assuming the average barometric height to be 10 lines and the temperature to be 10° R., I find the city of Potosí to have a high elevation of 1,954 toises. Yet if the Cerro del Potosí were elevated above the neighboring plateau, at 1,624 varas or 697 toises (Acosta, [*Historia natural y moral de las Indias*,] book IV, chap. 6; Hernández, [*Quatro libros*,] part I, book XI, chap. 2; ▼ Helms, [*Tagebuch einer Reise durch Peru*], pp. 65–122), its summit would necessarily surpass the perpetual snow limit, since the latitude of the city of Potosí is, according to Mr. Bauzá, 19°38′32″. In an interesting paper that Mr. Redhead published in Buenos Aires in 1819, with the title *Memoria sobre la dilatación del aire atmosférico*, he assigns 2,522 toises to the Cerrro del Potosí, 650 toises to Jujui, and 260 toises to Tucuman; but this scholar himself today regards these estimates as too high.

2. See above, p. 177 [p. II.99 in this edition].

3. The Tesorero of Potosí, ▼ Don Lamberto Sierra, extended this final table to the year 1800. For the royal duties from 1736 to 1800, he reached a total of 18,618,927 piasters. According to this same administrator, the registered output of the Potosí mine from 1556 to 1579 was 49,011,285 piasters; from 1579 to 1736, 611,256,349 piasters; and from 1736 to 1800, 163,682,874 piasters, for a total of 823,950,508 piasters. Mr. Sierra believes (based on somewhat vague insights) that because of the unrecorded output from 1545 to 1556 and because of the fraudulent mining that took place from 1545 to 1800, the sum of 1,647,901,016 piasters can be considered the total result of mining over the 255 years prior to 1800. ([Ignacio Núñez, *An Account, Historical, Political, and] Statistical of the United*

III.362 Royal Duties (Derechos Reales)
Paid on the Silver Extracted from the Cerro del Potosí

Tableau I

First period, from January 1, 1556 to December 3, 1578, during which only the quint was paid.

YEAR	QUINT		YEAR	QUINT		YEAR	QUINT	
	Piasters	Reales		Piasters	Reales		Piasters	Reales
1556	450,734	1	1564	396,158	4	1572	216,117	3
1557	468,534	5	1565	519,944	1	1573	234,972	1
1558	387,032	0	1566	486,014	3	1574	313,778	5
1559	377,031	2	1567	417,107	1	1575	413,487	4
1560	382,428	3	1568	398,381	3	1576	544,614	6
1561	405,655	7	1569	379,906	7	1577	716,087	0
1562	426,782	1	1570	325,467	1	1578	825,505	2
1563	449,965	3	1571	266,200	4			

TOTAL for twenty-three years: 9,801,906 piasters

Tableau II

III.363 ***Second period, from January 1, 1579 to July 19, 1736, during which first the cobos of 1 and a half percent was paid, and then the remaining quint of ninety-eight piasters four reales.***

YEAR	1 AND A HALF PERCENT AND QUINT		YEAR	1 AND A HALF PERCENT AND QUINT		YEAR	1 AND A HALF PERCENT AND QUINT	
	Piasters	Reales		Piasters	Reales		Piasters	Reales
1579	1,091,025	3	1609	1,132,680	4	1639	1,128,738	2
1580	1,189,323	1	1610	1,139,725	4	1640	978,483	2
1581	1,276,872	6	1611	1,299,052	2	1641	940,367	1
1582	1,362,855	7	1612	1,329,701	7	1642	905,797	6
1583	1,221,428	3	1613	1,200,947	6	1643	924,659	0
1584	1,215,558	1	1614	1,269,692	7	1644	871,174	3
1585	1,526,558	1	1615	1,354,412	3	1645	908,414	4
1586	1,456,958	0	1616	1,257,599	0	1646	840,982	0
1587	1,226,328	0	1617	1,071,932	4	1647	891,287	0
1588	1,441,657	0	1618	1,061,264	2	1648	1,123,932	2
1589	1,578,823	7	1619	1,108,744	6	1649	1,067,376	1
1590	1,422,576	1	1620	1,069,599	3	1650	917,845	7
1591	1,562,522	2	1621	1,099,244	1	1651	757,418	6
1592	1,578,449	6	1622	1,093,201	4	1652	796,244	2

(*continued*)

YEAR	1 AND A HALF PERCENT AND QUINT		YEAR	1 AND A HALF PERCENT AND QUINT		YEAR	1 AND A HALF PERCENT AND QUINT	
	Piasters	Reales		Piasters	Reales		Piasters	Reales
1593	1,589,662	1	1623	1,083,641	7	1653	759,904	5
1594	1,403,555	7	1624	1,086,999	0	1654	835,109	4
1595	1,557,221	3	1625	1,024,794	3	1655	754,784	1
1596	1,468,182	5	1626	1,033,868	7	1656	804,071	0
1597	1,355,954	6	1627	1,068,612	3	1657	933,441	4
1598	1,310,911	7	1628	1,172,352	3	1658	877,862	1
1599	1,339,685	2	1629	972,807	0	1659	799,609	1
1600	1,299,028	5	1630	962,250	4	1660	652,728	4
1601	1,477,489	7	1631	1,067,001	6	1661	623,250	7
1602	1,519,152	7	1632	964,370	6	1662	638,167	3
1603	1,478,697	6	1633	1,003,756	0	1663	579,126	7
1604	1,326,231	6	1634	984,414	6	1664	605,450	3
1605	1,532,646	6	1635	946,781	0	1665	655,557	0
1606	1,434,981	5	1636	1,424,758	6	1666	675,729	4
1607	1,414,660	1	1637	1,197,572	4	1667	708,879	2
1608	1,200,488	5	1638	1,174,393	0	1668	691,169	0

(*continued*)

III.364

1669	624,126	4	1692	426,761	7	1715	228,224	One-half
1670	554,614	0	1693	570,870	2	1716	239,287	6 and a half
1671	667,992	3	1694	546,928	3	1717	356,804	1
1672	624,037	6	1695	557,145	1	1718	322,251	1
1673	676,811	0	1696	500,965	3	1719	283,593	3
1674	673,694	7	1697	471,686	4	1720	231,256	7
1675	567,827	5	1698	434,772	1	1721	229,002	0
1676	514,530	4	1699	434,287	0	1722	228,208	5
1677	550,099	3	1700	405,492	5	1723	214,740	3
1678	653,067	1	1701	338,572	4	1724	245,793	4
1679	622,979	5	1702	372,447	1	1725	223,083	3
1680	629,270	0	1703	360,114	6	1726	274,416	1
1681	685,791	0	1704	333,702	0	1727	286,328	3
1682	659,341	0	1705	319,264	7	1728	220,698	1
1683	731,599	6	1706	354,600	1	1729	360,414	7 and a half
1684	719,082	0	1707	364,415	0	1730	303,361	6 and a half
1685	655,256	0	1708	374,183	6	1731	293,497	3
1686	586,835	7	1709	334,080	4	1732	308,137	3 and a half
1687	645,318	1	1710	309,008	1	1733	304,768	3 and a half
1688	646,077	3	1711	246,147	1	1734	273,084	5 and a half
1689	647,189	0	1712	204,931	6	1735	271,621	6
1690	673,097	1	1713	279,913	1	1736	149,567	One-half
1691	593,976	1	1714	265,087	1			
Total for 158 years: 129,417,273 piasters								

Tableau III

III.365 ***Third period, from July 20, 1736, to December 31, 1789, during both 1 and a half percent and a demi-quintal, or 100 piasters, 1 1 piasters, 3 reales.***

YEAR	1 AND A HALF PERCENT AND DEMI-QUINTAL		YEAR	1 AND A HALF PERCENT AND DEMI-QUINTAL		YEAR	1 AND A HALF PERCENT AND DEMI-QUINTAL	
	Piasters	Reales		Piasters	Reales		Piasters	Reales
1736	85,410	2	1754	244,148	2	1772	298,983	1 and a quarter
1737	183,704	3	1755	221,872	4	1773	306,925	3
1738	159,252	7	1756	249,513	7	1774	317,703	4
1739	183,295	6 and a half	1757	244,760	6	1775	332,329	4 and a half
1740	170,229	4	1758	262,835	4 and a half	1776	346,319	5
1741	179,573	6	1759	263,701	6	1777	390,676	5 and a half
1742	161,976	0	1760	272,059	1	1778	351,994	6 and a half
1743	166,131	1 and a half	1761	261,580	7	1779	348,035	4
1744	155,926	3	1762	257,201	7 and a half	1780	400,062	1 and a half
1745	163,140	One-half	1763	279,640	6 and a half	1781	323,109	2

1746	178,080	6	1764	263,092	1 and a half	1782	350,199	2
1747	184,156	5 and a half	1765	281,985	5	1783	400,238	3 and a half
1748	197,022	7 and a half	1766	282,405	One-half	1784	371,362	2
1749	215,283	3	1767	303,650	6	1785	351,777	7 and a half
1750	233,677	5	1768	306,674	7 and a half	1786	332,507	1
1751	238,502	3 and a half	1769	291,075	3	1787	390,836	7 and a half
1752	227,133	5	1770	292,203	3	1788	380,600	1 and three-quarters
1753	244,888	1 and a half	1771	307,765	3 and a half	1789	335,468	6

TOTAL for the 54 years: 14,542,684 piasters

of 233 years, from 1556 to 1789, 788 million piasters' worth of silver—reported to the royal coffers—were extracted from the Potosí mines. If these piasters were all Mexican piasters, at eight reales *de plata mexicana*,[1] the output for these 233 years would amount to 92,736,294 marks. But, as we will shortly see, the mass of silver on which duties were paid was even larger.

The ledgers preserved in the archives of the provincial treasury of Potosí III.367 do not date back further than 1556. It remains for us, therefore, to discuss how much silver was produced by the Potosí mines before that time. This analysis is all the more important in that it is rightfully believed that the years immediately following the discovery of these veins were the ones that yielded the greatest riches.

Ulloa[2] cites a book published in 1634 by ▼ Don Sebastián Sandoval y Guzmán, with the title *Pretensiones del Potosí*, in which the author indicates the quint paid from 1545 to 1633. I sought in vain to procure [a copy of] this work during my stay in Peru. Without knowledge of the partial data that it contains, I cannot analyze the Spanish astronomer's published results. This work is all the more necessary since Ulloa's claims were repeated by Raynal[3] and by all the other writers who discuss the amount of gold and silver taken from the Americas to Europe in the first years of the conquest. According to Sandoval, the quint paid to the royal coffer of Potosí was, in an average year from 1545 to 1564, four million piasters of thirteen and a half *reales de plata*; from 1564 to 1585, it was 1,166,000 piasters; from 1585 to 1624, it was 1,333,000 piasters; from 1624 to 1633, it was 666,000 piasters. From 1564 to 1633, these figures do not correspond III.368 very closely with the annual sums reported in the preceding tables: they are either smaller or greater, but it is particularly the quint of four million belonging to the period before 1564 on which concerns can justifiably be raised.

If this sum were exact, then the silver output from the Potosí mines registered at the royal treasury over the nineteen years from 1545 to 1564

Provinces of Río de la Plata, 1825, p. 285; and ▼ Varaigne [trans.], *Esquisses historiques, politiques [et statistiques] de Buenos-Ayres*, 1826, p. 528.)

1. One must not confuse the three varieties of reales *de plata*; namely: the *real de plata antigua*, with sixty-four maravedis de vellón; the *real de plata nueva* or *provincial*, with sixty-eight maravedis; and the *real de plata mexicana*, with eighty-five maravedis: the latter is the one that is always used in this work. (▼ Damoreau, *Traité [des negociations] de banque*, 1727, p. 115; *Encyclopédie méthodique, Commerce*, vol. III, p. 211 [50].)

2. [Ulloa,] *Noticias americanas, Entretenimiento* XIV, § XVII, p. 256.

3. [Raynal,] *Histoire philosophique* (Geneva edition, 1780), vol. II, p. 229.

would have been 641,250,000 Mexican piasters, converting the piasters of thirteen and a half reales to the piasters of eight reales. On the other hand, the official documents in my possession confirm that the eight-year output from 1556 to 1564 was 28,250,000 in these same Mexican piasters. Sandoval's data would therefore suggest that in its first eleven years, from 1545 to 1556, the Cerro del Potosí would have yielded, in silver on which the quint was paid, 613 million piasters, or, in a regular year, 55,726,000 piasters, which are equivalent to 6,556,000 marks of silver. Although probably quite extraordinary, there is nothing about this outcome that can be considered impossible. It may be surprising to note that a single mountain in Peru was capable of yielding two to three times more silver than all of the mines of Mexico combined, but the various ideas of wealth are merely relative ideas. Mountains might one day be discovered in the middle of Africa that, in terms of their abundance in precious metals, are to the Cordilleras what the latter are to the mountains of Europe. The Valenciana mine III.369 currently yields six to seven times more silver than the whole of Saxony, and the Guanajuato vein, worked over its entire length, would alone be capable of producing over two million marks of silver per year. We have observed above that in five months over 700,000 marks were extracted from the Veta Negra de Sombrerete vein, over a thirty-meter stretch. If one recalls the masses of native silver, red silver, and silver sulfide discovered in recent years at Huantajaya, in Peru, as well as at Batopilas and Real del Monte, in Mexico, one understands what a prodigious amount of silver an ore deposit in the Cordilleras of the Andes can produce when the output is at once abundant and intrinsically wealthy. It is not, therefore, the huge amount of silver that is assumed to have been removed during the first eleven years that leads me to cast some doubt on Sandoval's account, but rather the contradiction between this account and other, well-established historical facts.

Ulloa, Robertson, Raynal, and the authors of the *Encyclopédie méthodique* did not take note of a passage in the Chronicle of Peru [*Crónicas del Perú*], written by Pedro Cieza de León. The author, who writes with that admirable naivety characteristic of all of fifteenth- and sixteenth-century travelers, proposes to give his compatriots an idea of the prodigious wealth of Potosí Mountain. He finds himself all the more in a position to do so since he was present there in 1549—thus four years after III.370 the initial discovery of these famous mines. He reports what he saw himself, while Sandoval writes of a period eighty-nine years earlier. If one can suspect Cieza's numbers of being erroneous, one must nonetheless see them as erring on the side of excessiveness; for a traveler who aims to produce an effect and who hopes to stun his readers is naturally inclined to exaggerate. Let us now examine what the historian of Peru reports[1]: "The wealth of the Cerro del Potosí," he writes, "is so far above anything we have ever encountered that to explain the size of these mines, I shall describe them exactly as I saw them with my own eyes, when I passed through Potosí in 1549, at the time when the licentiate Polo was the city's corregidor. It was in this corregidor's house that the (royal) coffers with three keys were kept. Every Saturday, His Majesty received twenty-five to thirty, sometimes even forty thousand piasters. At that time, people complained that the mines were faring poorly when the quint amounted to only 120,000 *castellanos* per month. This silver only represented the property of Christians, however, for the Indians stole large amounts of it before it could be registered. There was, therefore, no wealthier mountain anywhere in the world, and nowhere has any prince obtained so much III.371 revenue from a single city; for, from 1548 to 1551, the quint returned over three million ducats to the king."

To understand this passage, which contains three distinct estimates, one must recall that *pesos*, or the piasters of that time and at least until 1580,[2] were a purely conceptual currency of 480 maravedis, or around thirteen and a half reales de plata mexicana. One silver mark equaled five and twenty-one twenty-sevenths of these piasters. Five piasters formed a ducat of eleven and a quarter reales. These data suggest that if one includes (as does Cieza) the quint of 30,000 piasters per week and 120,000 *castellanos* per month, the total output of the Potosí mines in 1549 was (in registered silver) either 1,549,000 marks or 1,440,000 marks. According to Cieza, in an average year between 1548 and 1551, this same output amounted to only 7,031,000 Mexican piasters with eight reales de plata, which represent 827,000 marks of silver. This sum contrasts sharply with Sandoval and Ulloa's report, but it corresponds fairly well to the quint of the years with which our first table begins. Some doubt may remain as to whether Cieza is indeed speaking of the entirety of the royal duties levied from 1548 to 1551 or whether he is claiming that, at that time, the quint III.372 equaled three million ducats per year. In the latter case, the annual output would have risen to 21,093,000 Mexican piasters, or 2,481,000 marks of

1. Cieza [de León], *Crónica del Perú*, chap. CVIII (1554 edition), p. 261.

2. Garcilaso [de la Vega El Inca], *Coment[arios] reales*, vol. I, in the second preface, which bears the title "Advertencias acerca la lengua general del Perú"; and vol. II, p. 51.

silver, certainly a very considerable sum, but still quite far from Ulloa and Raynal's estimate. I am inclined to think that the estimate of three million ducats provided by the historian of Peru only corresponds to the total sum of the quints levied over four years, (1) because this estimate is closer to the value of the quint of 1556; (2) because to give the most flattering impression of the mines' wealth, Cieza writes that the quint *occasionally* amounted to 40,000 piasters, which, for the *maximum* annual output at that time, would not be above 2,481,000 marks but barely 2,065,000 marks; (3) because Garcilaso[1] reports that, around that same time, between ten and twelve million piasters entered the Río Guadalquivir from Peru.

If one regards Sandoval's data as exact and combines them both with Cieza's and with the figures contained within the official documents that I have published, one obtains for the output of the Potosí mines the following results, which inspire little confidence:

In an average year, from	1545	to	1548,	23,284,000 marks of silver
	1548		1551,	827,000
	1551		1556,	621,000
	1556		1564,	415,000

Here are the bases for this calculation: Sandoval and Ulloa estimated the output of the Cerro del Potosí from 1545 to 1564 to be 33,750,000 piasters or 3,970,000 marks of silver in an average year. Yet we know from Cieza's chronicle what the output was from 1548 to 1551; the registers of Potosí show the output from 1556 to 1564. If one therefore assumes a decrease in arithmetic proportion over the intermediary period from 1551 to 1556, it is easy to determine what, from the 641,250,000 Mexican piasters or the 75,440,000 marks of silver that Sandoval suggests for the total output of the first nineteen years, belongs to the short interval between 1545 and 1548. III.373

If one assumes that Cieza provided the quint for each of the four years within the period of 1548 to 1551 (which seems just as unlikely), then a similar operation would yield the following annual outputs for the Potosí mines:

1. Garcilaso [de la Vega El Inca, *Comentarios reales*], vol. II, p. 52.

From	1545	to	1548, at	19,146,000 marks of silver
	1548		1551,	2,481,000
	1551		1556,	1,448,000
	1556		1564,	415,000

However one interprets the passage from Cieza's chronicle, it is evident that for either hypothesis the output of the first three years differs so sharply from that of the following years that Sandoval's report should inspire some significant doubt. This is even more the case since, if one examines the III.374 table of quints from 1556 to 1789, one finds a law governing this long series of numbers according to which the latter increase or decrease uniformly. Cieza visited the Potosí mines at the time of their greatest splendor; he expressly states that he is describing the mountain exactly as he found it in 1549, "because this wealth, like anything human, must vary over the course of time, either it must increase or it must decrease." If the output of 1549 had really been eight to ten times smaller than the output of 1546, how could the traveler have kept his silence about this enormous drop in wealth!

Together these discussions lead us to conclude that the total output of registered silver in the eleven years that are missing in the preceding tables, far from being seventy-two million marks (as one might assume from Ulloa and the famous author of the *Recherches philosophiques*), was not above fifteen million marks. Nor shall we lend credence to Solorzano,[1] who vaguely states that from 1545 to 1628, therefore over eighty-three years, the Potosí yielded the sum of 850 million pounds of silver, which is almost twice as much as the mountain produced over two and a half centuries. It is perhaps surprising to see that a writer who had long been III.375 a member of the Audiencia of Lima could be so misinformed; for otherwise how could he assume, over eighty-three years, an annual output of 2,400,000 marks, when the registers preserved in the treasury of Potosí prove to us that during this period the mean size of the output only rarely rose to 800,000 marks!

Furthermore, Acosta,[2] who traveled across both Americas and whose work can only be duly appreciated by those who have visited the same sites themselves, confirms Cieza's account; he recounts that "in the time of the licentiate Polo (thus before 1549), the quint rose *to one and a half*

1. Solorzano Pereira, *De Indiarum jure*, vol. II, book V, chap. I (Lugduni edition).
2. [Acosta,] *Historia natural y moral de las Indias* (Barcelona, 1591), p. 138.

million piasters per year."[1] He adds, "despite the *confusion that reigns in the ledgers of the initial years*, we know both from legend and from the studies conducted by the order of the viceroy Don Francisco de Toledo that the amount of registered silver amounted to seventy-six million piasters, from 1545 to 1574, and to thirty-five million piasters (thirteen reales and one *quartillo*), from 1574 to 1585, which equals one hundred and eleven million piasters in forty years." This sum of 111 million piasters, a purely conceptual currency, presupposes an annual output of only 555 marks, very close to that of the Guanajuato vein. There is no doubt that Acosta is speaking of the entire amount of silver extracted from the mines and declared at the treasury. He clearly states: *se ha metido a quintar, monta lo que se ha quintado* [he has become involved in mining, he records what has been extracted]. Solorzano translates this passage from Acosta's natural history as follows: *ex Potosiensi fodina extracti sunt centum et undecim milliones* [extracted from the Potosí mine are one hundred and eleven million]. III.376

The authors whose works provide exaggerated estimates of the amount of precious metals that flooded Spain from the mid-sixteenth century appear to have confused the value of the mines' output with the quint that was paid on it. Although the official documents that I have reproduced here were unknown to them, they would not have committed this error had they read the works of Acosta, Cieza, and Alonzo Barba[2] more carefully. The latter, who was curate of a parish in the city of Potosí, estimates the amount of silver removed from the Cerro de Potosí from 1545 to 1636 to be only 450 million piasters of eight reales, a sum that presupposes an output of only 4,900,000 piasters or 576,000 marks per year and that contrasts sharply with the 613 million that are gratuitously attributed to the first periods between 1545 and 1556. Nevertheless, Alonzo Barba had no motive for lowering the total output; on the contrary, he seeks to demonstrate that an area of land of sixty square leagues could be covered in the amount of piasters manufactured with the silver of Potosí.

III.377 The following table presents the state of these mines beginning with the period in which the quints were recorded with precision [see following page].

1. Which presupposes an output of 1,490,000 marks. (Herrera [y Tordesillas, *Historia general*,] decada VIII, book II, chap. XIV.)

2. Barba, [*Arte de los metales*,] book II, chap. I.

Yield of the Cerro de Potosí (Hatun-Potocsi)

PERIODS	AVERAGE YEAR		
	OUTPUT IN PIASTERS	MARKS OF SILVER EXTRACTED FROM THE MINES	
		Valuing 1 Piaster at 13 and a Half Reales	Valuing 1 Piaster at 8 Reales
1556 to 1566	2,159,216	428,767	
1585 to 1595	7,540,620	1,497,380	887,073
1624 to 1634	5,232,425		615,580
1670 to 1690	3,234,580		380,538
1720 to 1730	1,299,800		152,918
1740 to 1750	1,850,250		217,676
1779 to 1789	3,676,330		432,510

Since some uncertainty remains about the period in which piasters of thirteen and a half reales (five one-twenty-sevenths of which equal one mark of silver) were no longer used as a unit of currency, I have preferred to give both piaster units up to 1595: one thus obtains the *maximum* wealth that one can assume. One passage in Garcilaso's commentaries, which we have cited above, might lead one to believe, however, that just a few years before 1580, people already used piasters of eight reales de plata. During the entire 233-year period from 1556 to 1789, mining in Potosí never reached such a high degree of splendor as from 1585 to 1606. For several consecutive years, the quint was one and a half million piasters, which assumes an output of 1,490,000 piasters or 882,000 marks, depending on whether one uses a piaster of thirteen and a half or eight reales. This wealth is even more surprising in that, according to Acosta, over one-third of the silver was not registered. Output declined after 1606, especially after 1694. From 1606 to 1688, however, it was never below 350,000 marks. Since the second half of the eighteenth century, the mountain has generally produced three to four hundred thousand marks, and this output is probably still too high for one to suggest, as did one famous author,[1] that the Potosí mines are not worth the trouble it takes to mine them. In their current state, these mines no longer occupy first place among those of the known world, but they can be ranked immediately after the Guanajuato mines. The following amounts of silver were minted in the city of Potosí:

III.378

1. Robertson, *History of America*, book IV, pp. 339 and 399.

1773	marks of gold	231,853 marks of silver
1774		377,958
1775		396,196
1776		480,931
1777		485,328
1778		577,579
1779		544,762
1780	3,532	581,020
1781	1,604	447,994
1782	2,204	410,267
1783	1,841	485,547
1784	1,529	485,344
1785	1,628	428,978
1786	2,451	438,266
1787	1,874	503,544
1788	1,936	420,340
1789		420,340
1790	2,204	468,600

It is believed that 34,400 marks were omitted from the treasury books, such that the total minting from 1773 to 1790 amounted[1] to 8,219,384 marks of silver, or, in an average year during this interval, 1,891 marks of gold and 456,632 marks of silver. The year 1791 alone saw 257,526 piasters in gold and 4,365,175 piasters in silver.

The deeper the Potosí works have gone, the more the *content* of the ore has decreased. In this respect and in several others, the Cerro de Potosí exhibits great similarities with the Gualgayoc mines. At the surface of the earth, in their outcroppings, the veins of Rica, Centeno, and Mendieta, which run through the primary schist, were filled across their entire width with a mixture of silver sulfide, red silver, and native silver. These metallic masses rose in the shape of a crest (crestones), the rock of the *wall* and the *roof* having been destroyed, either through the effect of the water or from some other cause that changed the surface of the globe. The veta del Estaño, on the contrary, only exhibited tin sulfide in its outcroppings, and III.380

1. In 1753 the silver output of the Potosí mine was 2,518,198 piasters.

the muriate of silver ore only became visible at great depths.[1] This mixture of two formations in a single vein also occurs on the Old Continent, for example, in several mines of Freiberg, in Saxony.[2] In 1545 the ore that contained eighty to ninety marks per quintal were quite common: however, one must not suggest, as did Ulloa, that the entire mass of the ore extracted from the mine possessed this degree of wealth. Acosta clearly states that in 1574, the mean content was eight to nine marks and that ore yielding fifty marks per quintal was considered extremely wealthy. Furthermore, ▾ Don Francisco Tejada's report on the mines of Guadalcanal, in Spain, reveals that, in 1607, the mean wealth of the ore from Potosí was not greater than an ounce and a half. Since the beginning of the eighteenth century, one *cajón* contains only three to four marks, at a weight of 5,000 pounds, or forty-eight hundredths to sixty-four hundredths per quintal. The ore from Potosí are therefore extremely poor, and it is only because of their abundance that mining remains in a flourishing state. It is surprising to note that, from 1574 to 1789, the mean wealth of ore has decreased at a rate of 170 to 1; while the amount of silver extracted from the Potosí mines has only decreased by four to one. III.381

From 1545 to 1571, the silver ores of Potosí were only treated through smelting. The *conquistadores*, who possessed only military knowledge, did not know how to conduct metallurgical procedures. They did not succeed in smelting the ore using bellows; they adopted the bizarre method that the indigenous peoples employed in the nearby mines of Porco, which had been worked for the Inca's profit long before the conquest. They set up portable furnaces called *huayres* or guayres in the Quichua [Quechua] language on the mountains surrounding the city of Potosí, everywhere the wind blew impetuously. These furnaces were cylindrical slate tubes, very wide and pierced with a large number of holes. The Indians laid layers of silver ore, galena, and peat coal inside them; the air current that penetrated into the huayre through the holes stirred and stoked the flame. When they noticed that the wind was blowing too hard and that too much fuel was being used, they carried the furnaces to lower-lying sites. The first travelers who visited the Cordilleras all wrote enthusiastically about the impression created by the sight of over 6,000 fires lighting up the mountaintops around the city of Potosí. The Indians extracted the necessary galena for their smelting from a small mountain near the Cerro de *Hatun-Potocsi* called the *Child*

1. Barba, [*Arte de los metales*,] book I, chap. XXXII, p. 56.
2. Werner, [*Neue Theorie von der Entstehung der Gänge*,] p. 248.

or *Huayna-Potocsi*.[1] The Indians resmelted in their huts the argentiferous mattes produced by the mountainside *huayres*, using the old technique of having ten or twelve people blow into the fire at once, through copper tubes one to two meters long and pierced at the lower end with a very small hole. One can easily imagine what an enormous amount of silver must have remained in the scoria without combining with the lead. III.382

Pedro Fernández de Velasco, who, as the Jesuit Acosta clearly states,[2] "had observed in Mexico how silver was extracted from the mine using mercury," proposed to the viceroy of Peru, Francisco de Toledo, that amalgamation be introduced to Potosí. He succeeded in his attempts in 1517; and of the eight to ten thousand quintals of mercury that the Huancavelica mine produced toward the end of the sixteenth century, over six or seven thousand were consumed in the Potosí mines. The ore that had been regarded as too poor to be smelted in the *huayres* were profitably reworked. III.383

The abundance of rock salt mined on the plateau of the Cordilleras near Curahuara, Carangas, and Yocalla greatly facilitate amalgamation at Potosí. According to Alonzo Barba's calculation,[3] an enormous amount of mercury—234,700 quintals—was consumed from 1545 to 1637. From 1759 to 1763, consumption was from 1,600 to 1,700 quintals[4] per year. Toward the end of the sixteenth century, 15,000 Indians were forced to work in the mines and plants of Potosí and over 1,500 quintals of salt were hauled to the city from Yocalla daily; today there are less than 2,000 miners, who are paid a daily wage of fifty sous. Fifteen thousand llamas and the same number of asses are used to haul the ore from *Hatun-Potocsi* Mountain to the amalgamation plants. In 1790 at the Potosí mint, 4,222,000 piasters were struck; that is, 299,246 piasters or 2,204 marks of gold, and 3,923,173 piasters or 462,609 marks of silver.

Reflecting on the history of precious metals and on the interest that it generates among researchers in political economy, one will not be surprised that we have explained in such great detail facts that might shed some light on the amount of silver that was extracted over two and a half centuries from the Potosí mines. It was necessary to compare the accounts of the III.384

1. More precisely, the *Father Mountain* and the *Son Mountain*. The various peaks of the Pichincha volcano bear similar names; and due to the fact that in their works, the French academicians did not distinguish between the old *Rucu-Pichincha* from the Child or *Guagua-Pichincha*, it is today quite difficult to find the location of Bouguer, La Condamine, and Ulloa's *academic station*. (See my *Recueil d'observations astronomiques*, vol. I, p. 308.)

2. Acosta, [*Historia natural y moral de las Indias*,] p. 146.

3. Barba, [*Arte de los metales*,] pp. 12 and 65.

4. Ulloa, *Notícias americanas*.

III.385 *I. Gold Minted at Santa Fe de Bogota*

YEAR	MARKS	OUNCES	OCHAVAS	TOMINES	PIASTERS	REALES	QUARTOS
1789	10,915	2	0	0	1,484,454	0	0
1790	7,343	0	5	0	998,658	5	0
1791	8,318	0	1	4	1,131,251	4	11
1792	8,159	5	3	1	1,109,715	5	24
1793	8,659	3	3	1	1,177,681	5	28
1794	7,307	4	3	4	993,827	6	11
1795	9,310	6	4	4	1,266,272	7	11
TOTAL	60,013	6	5	2	8,161,862	0	0

In an average year: 8,573 (marks of gold) or 1,165,980 piasters.[1]

[1] "La Casa de Moneda de Bogotá dió en el bienio de 1806 y 1807 más de 3,999,000 pesos fuertes." [Francisco José de Caldas,] *Semanario del Nuevo Reyno de Granada*, vol. II, p. 216.

Spanish authors who were among the first to visit the Americas; it was necessary to distinguish between export production and the quint paid to the crown; between piasters, a purely conceptual currency widespread at the beginning of the conquest, and the Peruvian piasters of eight reales. If we had neglected this research, which had never before been undertaken, we would have run the risk of inflating the total amount of silver that has been imported into Europe since 1492 by over fifty-seven million marks, which are equivalent to over two and a half billion francs.

IV. The kingdom of New Granada produces 18,300 silver marks in an average year. The following table indicates how much of it was coined, from January 1, 1789, to December 31, 1795, at the Santa Fe mint, and from 1788 to 1794, at the Popayán mint.

II. Gold Minted at Popayán

YEAR	MARKS	OUNCES	OCHAVAS	VALUE OF THE GOLD	
				Piasters	Reales
1788	7,210	4	3	980,634	3
1789	5,945	2	4	808,562	4
1790	7,123	2	6	968,745	0
1791	6,437	2	0	875,466	0
1792	7,344	5	0	998,869	0
1793	7,026	6	5	955,648	5
1794	6,725	1	0	914,617	0
TOTAL	47,813	0	2	6,502,542	4

In an average year, 6,830 (marks of gold) or 928, 934 piasters.

From 1782 to 1789, the amount of gold minted at Santa Fe in an average year was under 7,000 marks. During this time, the most productive year was 1787, when the output was 981,655 piasters, or 7,218 marks[1]: in 1778, 693,438 piasters' worth of gold was minted. At Popayán from 1778 to 1783, the amount of minted gold did not generally amount to more than 5,800 marks; in 1778, III.386

1. *Relación del gobierno del Excelentís[imo]. Señor Don José de Espeleta, virey del Nuevo Reyno de Granada, para entregar el mando al Señor Don Pedro de Mendinueta, electo virey* [Espeleta, José de, "Relación del Estado del Nuevo Reino de Granada (1796)"]. This unpublished account, which is in my possession, contains the most detailed and the most exact statistical information: it is the work of a very distinguished talent, ▼ Mr. Ignacio Tejada, a native of Santa Fe and secretary of the viceroyalty.

gold production was only 792,838 piasters, but by 1787, it had already risen to 981,655 piasters. The value of the gold ingots exported annually from the port of Cartagena is estimated to be 300,000 to 400,000 piasters. During my stay in Santa Fe de Bogotá, in 1801, the total output of the gold mines of the kingdom of New Granada was estimated to be 2,500,000 piasters; including 2,100,000 piasters of output of the two mints of Santa Fe de Bogotá and Popayán, and 400,000 piasters in exports in ingots and in gold ornaments.

Nearly all the gold that New Granada produces comes from the *washes* established in the alluvial plains. Gold veins have been identified in the mountains of Guamoco and Antioquia[1]; but they have been very neglectfully mined: the greatest riches in stream gold are deposited to the west of the *central Cordillera*,[2] in the provinces of Antioquia and Chocó, in the Río Cauca valley, and on the South Sea coasts, in the *partido* of Barbacoas. Dividing the auriferous terrains into three regions, the output for Chocó is 10,800 marks of gold, or over one-half of the total output of the viceroyalty of Santa Fe; for the province and the southern part of the Cauca valley (between Cali and Popayán), it is 4,600 marks; and for the province of Antioquia and the mountains of Guamoco and Simiti, it is 3,400 marks.[3] This estimate shows that the alluvial plains containing the most gold particles and grains disseminated between *grünstein* fragments stretch from the *western Cordillera* toward the coasts of the Great Ocean.

It is also quite remarkable that platinum has not yet been found in the Cauca valley,[4] nor to the east of the western branch of the Andes, but exclusively on the western bank of the Cauca. These mountains, which are not

III.387

III.388

1. Minerales de Buritica, de Valle de Osos, del Cerro de Morrogacho, de Musingo, de Quiuna, etc.

2. On the division of the Andes into several branches, see my *Vues des Cordillères*, plate V; and my *Relation historique*, vol. III, pp. 202–5.

3. If one includes fraudulent extraction, the output of the gold washes of Antioquia is probably much higher. Even in 1809, Mr. Restrepo estimated it to be 550,000 castellanos (a value of 1,200,000 piasters). *Semanario*, vol. II, p. 67 [?].

4. As this book is being reissued (July 1826), ▼ Mr. Boussingault, whose travels have already enriched the fields of physics, chemistry, and astronomical geography with so many important observations, is delivering a paper to the Academy of Sciences on primary deposits of platinum in Columbia. This scholar has discovered round platinum grains in the auriferous *pacos* from the veins of Santa Rosa de los Osos, ten leagues northeast of Medellín, at a latitude of 6°37′ and at a height of 2,775 meters. These veins run through a terrain of syenite and grünstein (diorite). The deposit of platinum is to the east of the Río Cauca, deep within the Antioquia mountains, and the alluvial plains (auriferous and platiniferous) of Chocó and Barbacoas probably resulted from the simple destruction of a syenite and platinum formation. See Mr. Boussingault's letter to Mr. de Humboldt ["Sur le gisement du platine . . ."] in *Annales de chimie et de physique*, vol. XXXII, p. 204. In the same Río Cauca basin, in the washes of Quilichao, I saw auriferous sands covered with angular fragments of

very tall, separate the famous gold washes of Novita, in Chocó, from those of Quilichao and Jelima, located fifteen leagues north of the city of Popayán: however, not a single grain of platinum has ever been found in the latter washes, which I carefully examined during my journey to Quito. In Chocó, not only gold and platinum but also hyacinth zircon and titanium can occasionally be found. This mixture recalls the sand formations of Expailly, in Velay. A few years ago, near the village of Lloró, a well was hollowed out in an auriferous terrain so that the lower layers could be examined: at a depth of six meters, large trunks of petrified wood were discovered, surrounded by fragments of trappean rocks and gold and platinum particles.[1] III.389

The province of Antioquia, which one can only enter on foot or carried *on man-back*, exhibits gold veins in the micaceous schist at Buritica, San Pedro, and near Armas; but these veins are not worked, for lack of labor. Gold is collected in great abundance in the alluvial plains of Santa Rosa, in the Valle de los Osos, and in that of Trinidad. The number of Negro slaves who specialize in panning for gold (*negros mazamoreros*) was 1,462 in 1770; in 1778, it was 4,896 individuals. The gold of Antioquia, of which the city of Mompox can be considered the main market, has a fineness of nineteen to twenty carats. At Barbacoas, the fineness of the gold is generally twenty-one and a half carats: the washes of the North of Chocó, those of the district of Zitara, produce a finer gold than that of the more southerly district of Novita. The gold from the Indipurdu mines is the only gold with a fineness of twenty-two carats; for the average wealth of the gold from Chocó is between twenty and twenty-one carats. The outputs of the different washes are so consistent in their composition that those who trade in gold particles have only to know the place where the gold was collected to know what its fineness is. The finest gold in New Granada, and perhaps in all the Americas, is that of Giron, which is claimed to have twenty-three carats and three-quarter grain. People assured me that the gold collected at Marmato, west of the Cauca River and south of the ruins of the former Villa de Armas, is a whitish gold that is never above twelve to thirteen carats and is blended with silver, which therefore makes it the true *electrum* of the ancients. Moreover, although the gold in Chocó and Barbacoas is generally accompanied by platinum, no one has ever encountered *aurum platiniferum*, which perhaps only exists in our oryctognostic systems. III.390

diorite. See my *Essai géographique sur le gisement des roches*, p. 141; and *Relation historique*, vol. III, p. 204.

1. Observation by ▼ Don Tomás Valencia, at Popayán.

In Chocó, the richest river in terms of gold is the Río Andageda, which combines with the Quito and Zitara rivers near the village of Quibdó to form the great Río Atrato. All of the terrain between the Andageda, the Río de San Juan (which passes near the village of Noanama), the Río Tamana, and the Río San Augustin is auriferous. The largest piece of gold that has ever been found in Chocó weighed twenty-five pounds. The Negro who discovered it fifteen years ago did not even obtain his freedom. His master offered the *nugget* to the king's cabinet, in the hope that, as a reward, the court would grant him a *title of Castile*, the most ardently desired object for the majority of Spanish creoles: but his only success was in receiving barely enough payment for the value of the metal.

III.391 People claim to have found a fifty-pound piece of gold near *La Paz* in 1830.

Under the government of the archbishop viceroy Góngora, a count was undertaken of all the Negros of Chocó who pan for gold.[1] In 1778, there were only 3,054. In the Cauca valley, there are 8,000. The province of Chocó could alone produce over 20,000 marks of stream gold if, while settling this region, one of the most fertile of the New Continent, the government focused its attention on agricultural development. The richest country in terms of gold also happens to be the one that experiences the most consistent food shortages. Inhabited by wretched slaves from Africa and Indians who suffer under the despotism of the corregidors Zitara, Novita, and Taddo, the province of Chocó has remained in the same state it was in three centuries ago, a thick forest with no trace of farming, no pastures, and no roads. The price of commodities is so exorbitant there that one barrel of flour from the United States commands sixty-four to ninety piasters; a muleteer's food costs one to one and a half piasters per day; during peacetime, the price of one quintal of iron is forty piasters. These high prices should not be attributed to the accumulation of representative signs, which is minimal, but rather to the enormous difficulty of hauling and to the wretched state of affairs in which the entire population consumes without also producing.

III.392 The kingdom of New Granada has extremely wealthy silver veins in the syenite and grünstein mountains of the Vega de Supia,[2] to the north of Quebraloma, between the Cerro Tacón and the Cerro de Marmato. These de-

1. *Relación del Estado del Nuevo Reyno de Granada que hace el Arzobispo-Obispo de Córdoba a su sucesor, el Ex[celentísimo] Fray Don Francisco Gil y Lemos*, 1789. [Caballero y Góngora, "Relación del estado del Nuevo Reino de Granada que hace el Arzobispo Obispo de Córdoba a su sucesor el Exmo. Sr. Dn. Francisco Gil y Lemos. Año de 1789."] (unpublished [at the time]).

2. Mina de los Morenos or Chachafruta. As the crow flies, it is only twenty leagues from Cartago to the Vega de Supia.

posits, which yield both gold and silver, were only discovered ten years ago. A lawsuit between the owners interrupted work at the very moment that the most abundant ore had been found. Work resumed in earnest at the silver mines of Pamplona and Santa Ana, near Mariquita, at the time when the court of Madrid named Don Juan José Delhuyar director of the mines of the viceroyalty of Santa Fe. The argentiferous ore deposit of Santa Ana forms a layer in the gneiss. I visited the mine of La Manta, the output of which contains, on average, six ounces per quintal. Mr. Delhuyar, brother of the director of mines of Mexico, had set up an amalgamation plant with four barrels, similar to the one in Freiberg. The works there were very sensibly managed; but since the amount of silver yielded between 1791 and 1797 was only 8,700 marks, while the total cost[1] was 216,000 piasters, the viceroy ordered the mine to be abandoned. One must hope that in happier times the government will attempt to restart these works, as well as those of Santo Cristo de las Laxas and the Real de Bocaneme, between the Río Guali and the Río Guarinó, which formerly yielded very large quantities of silver. III.393

To summarize the results we have obtained, we find that the total output of the gold and silver mines of the Spanish colonies amounts to 41,400 marks of gold and 3,563,000 marks of silver, Castilian weight. These totals differ very little from those I sent to Mr. Héron de Villefosse and that he included in his interested work on the mineral wealth of the leading powers of Europe. I drew up the following table, using the valuable information that I obtained more recently from Spain and from the Kingdom of New Spain.

NAMES OF THE MAIN POLITICAL DIVISIONS	FINE GOLD, IN CASTILIAN MARKS	FINE SILVER, IN CASTILIAN MARKS	VALUE OF GOLD AND SILVER, IN PIASTERS
Viceroyalty of New Spain	7,000	2,250,000	22,170,740
Viceroyalty of Peru	3,400	513,000	5,317,988
Captaincy General of Chile	10,000	29,700	1,737,380
Viceroyalty of Buenos Aires	2,200	414,000	4,212,404
Viceroyalty of New Granada	18,000	negligible	2,624,760
TOTAL	40,600	3,206,700	36,063,272

1. Costs of underground works, amalgamation costs, and plant construction costs.

III.394 In this table,[1] gold is valued at 145 and eighty-two hundredths piasters, and silver is valued at nine and four-tenths piasters per Castilian mark. It shows the amount of precious metals extracted from the mines and registered in the royal treasuries. It confirms the claim of the count of Campomanes,[2] who in 1775 already estimated Spain's gold and silver imports to be three million piasters; but he indicates only the *minimum* that could reasonably have been produced by the Spanish colonies. Let us examine what additions should be made for the metals that are circulated as contraband. Up to now, people have had very exaggerated notions about the amount of gold and silver on which the quint is not paid: it has been estimated to be either one-half or one-third of the total output, and it has not occurred to anyone that illegal trade is a very different activity, depending on the localities in the different provinces. I shall bring together here the various pieces of information that I was able to gather on site in Mexico, New Granada, and Peru.

New Spain has only two ports through which its products are exported. III.395 The wretched state of the coasts makes smuggling much more difficult than in the provinces of Cumana, Caracas, and Guatemala. The amount of unregistered silver shipped from Veracruz and Acapulco, either to Havana and Jamaica or to the Philippine islands and Canton, is probably under 800,000 piasters; but this fraudulent trade will grow as the population of the United States draws nearer to the banks of the great *Río del Norte*, and as the western coasts, those of Sonora and Guadalajara, are more frequently visited by British or Anglo-American ships. When Mexican trade with China and Japan is freed from the constraints of the loathsome monopoly that hinders it today, an immense amount of silver will flow toward the west, to Asia. Precious metals are goods that are transported to the places where they fetch the highest prices. In Japan,[3] where gold is abundant, the ratio of this metal to silver is eight or nine to one. In China, one ounce of gold can be purchased for twelve to thirteen ounces of silver. In Mexico, the proportion of the two precious metals is around fifteen and three-eighths to one; hence it is much more profitable to send silver to Manila,

1. I think that the overall output of the mines of Spanish America has been overestimated in some otherwise estimable works. In 1793 Malaspina already overestimated it by over forty million piasters (*Bullion Report*, p. 35); [Estela,] *Viagero universal*, 1798 (vol. XX, p. 120), by 38,200,000 piasters; ▼ Mr. Jacob, the learned author of the article "Mexico," in the *Encyclopedia Britannica*, 42,721,000 piasters (Tooke, *[Thoughts and Detail]s on [the] High [and Low] Prices [of the Thirty Years, from 1793 to 1822]*, 2nd edition, p. 380). The output of the mint of Mexico City in an average year from 1800 to 1810 can hardly be estimated at thirty million piasters!

2. [Campomanes, *Discurso sobre la] educación popular*, vol. II, p. 331.

3. ▼ Thunberg, *Voyages [. . .] au Japon* (Langlés's edition), vol. II, p. 263.

Canton, and Nagasaki than it is to send gold. I did not mention gold work (*plata labrada*) exports above because, according to the registers of Veracruz, they do not amount to more than 20,000 or 30,000 marks of silver. III.396

In the kingdom of New Granada, fraudulent exports of gold from Chocó have increased since navigation on the Río Atrato was legalized. Instead of being transported via Cali or Mompox to the mints of Popayán and Santa Fe, gold in powder or ingot form takes the more direct route of Cartagena and Portobelo, whence it flows into the British colonies. The mouths of the Atrato and the Río Sinu, where I was at anchor in April 1801, serve as storage sites for smugglers. Laws allowing for the intermittent introduction of Negroes from Africa and flour from Philadelphia via foreign ships favor this fraudulent trade. According to the information that I was able to obtain from individuals who engage in the gold powder trade (*rescatadores*) in Cartagena de Índias, Mompox, Buga, and Popayán, it appears that the amount of gold produced in Chocó, Barbacoas, Antioquia, and Popayán and that circumvents the quint can be estimated at 2,500 marks.

In Peru, silver on which the quint is not paid is smuggled less from the South Sea coasts, which are frequented by sperm whalers,[1] than from east of the Andes, via the Amazon River. This immense river links two countries in which a great disproportion reigns between the relative value of gold and silver. Brazil is nearly as profitable a market for Peruvian silver as China is for Mexican silver. One-fifth, perhaps even one-quarter of all the silver extracted from the mines of Paso (*Yauricocha*) and Chota (*Gualgayoc*) is smuggled through Lamas and Chachapoyas, down the Amazon River. There are people in Lima who believe that stimulating trade on this river will lead to an increase in silver smuggling. This belief was very harmful to the lovely provinces that stretch along the eastern slope of the Cordilleras and are fertilized by the Guallaga, the Ucayale, the Puruz, and the Beni. People forget that the wild state and solitude of these lands greatly facilitate the smugglers' schemes. We shall estimate the unregistered silver of Peru to be 100,000 marks. III.397

According to Ulloa, in Chile the ratio of the gold on which the quint is paid to the gold that escapes duties is three to two. We shall only include one-quarter of the total output. By estimating the amount of smuggled silver in the kingdom of Buenos Aires at one-sixth or 67,000 marks and by adding nearly 30,000 marks of gold (following Mr. Correia de Serra) for the total output of Brazil, where even now only alluvial deposits are mined, it is possible to present in the following table the total output in gold and silver of the Americas as a whole.

1. See above, p. 93 [p. II.52 in this edition].

III.398 *Annual Yield of the Mines of the New Continent at the Beginning of the Nineteenth Century*

NAMES OF THE MAIN POLITICAL DIVISIONS	GOLD IN		SILVER IN		VALUE OF GOLD AND SILVER MARKS IN PIASTERS
	Castilian marks	Kilograms	Castilian marks	Kilograms	
Viceroyalty of New Spain	7,000	1,609	2,338,220	537,512	23,000,000
Viceroyalty of Peru	3,400	782	611,090	140,478	6,240,000
Captaincy General of Chile	12,212	2,807	29,700	6,827	2,060,000
Viceroyalty of Buenos Aires	2,200	506	481,830	110,764	4,850,000
Viceroyalty of New Granada	20,505	4,714			2,990,000
Brazil[1]	29,900	6,873			4,360,000

[1] This estimate of the amount of gold collected in Brazil was already too high for the period indicated in this table. More exact information, which the author owes to Baron Eschwege, chief executive officer of the mines of Brazil, can be found at the end of book IV.

The current total output of the mines of the New World thus amounts to 17,000 kilograms of gold and 800,000 kilograms of silver, considering that the Castilian mark, on the basis of which the output of the mines in the Spanish colonies is assessed, is at a ratio of 541 to 576 to the French mark,[1] and that the kilogram equals four *marks,* five *gross,* fifteen *grains*, in French weight. All of the tin produced in Europe weighs only three times as much as the mass of silver that is extracted annually from the mines of the Americas. The preceding table also shows that the bulk of the gold that the New Continent sends to the old one is incorrectly attributed to Brazil. The Spanish colonies produce nearly 45,000 marks of gold, while only 30,000 are extracted from the alluvial plains of Brazil. The author of the immortal work on the *wealth of nations*[2] estimates the amount of gold and silver imported annually into Cádiz and Lisbon to be six million pounds sterling, including not only what is registered but also what can be assumed to flow into smuggling. This estimate is two-fifths lower than it should be. III.399

Combining the results we have just obtained for the New World with those generated by the laborious research of Mr. Héron de Villefosse and ▼ Mr. Georgi[3] yields the following results [see table on following page].

In this table, gold is valued at 3,444 francs 44 centimes per kilogram and silver at 222 francs 22 centimes; it indicates the amount of precious metals that enter into circulation annually among the civilized nations of Europe. It is impossible to estimate how much gold and silver is currently being mined across the entire surface of the globe; we have no idea whatsoever what is produced in the interior of Africa, Central Asia, Tonkin, China, and Japan. The trade in gold powder that takes place on the eastern and western coasts of Africa, in addition to the ideas that the ancients have bequeathed to us about the lands with which we are no longer in contact, may lead one to assume that the lands south of the Niger are very rich in precious metals. The same assumption can be made about the high mountain chain that extends to the northwest of Paropamisus toward the border of China. The amount of gold and silver ingots that the Dutch once exported III.401

1. ▼ Bonneville, *Traité des monnaies*, 1806, p. 31.

2. [See above,] vol. II, p. 70. According to ▼ Magens (Post-Scriptum to [Horseley,] *The Universal Merchant*, 1756, p. 15), imports into Spain and Portugal in an average year from 1747 to 1755 equaled 5,746,000 pounds sterling.

3. Georgi, *Geogr[aphisch-]phys[ikalische und naturhistorische] Beschreibung des Russischen Reichs*, 1797, part 6, p. 363. Mr. Georgi's estimate is for the year 1796. The output of the mines of Koliwan doubled, while that of the mines of Nertschinsk decreased by over one-third from 1784 to 1794.

III.400 *Annual Yield of Gold and Silver Mines in Europe, East Asia, and the Americas*

MAJOR NATURAL DIVISIONS	GOLD IN		VALUE OF GOLD, IN FRANCS	SILVER IN		VALUE OF SILVER, IN FRANCS	VALUE OF GOLD AND SILVER, IN FRANCS
	French marks	Kilograms		French marks	Kilograms		
Europe	5,300	1,297	4,467,444	215,200	52,670	11,704,444	16,171,888
Northern Asia	2,200	538	1,853,111	88,700	21,709	4,824,222	6,677,333
Americas	70,647	17,291	59,557,889	3,250,547	795,581	176,795,778	236,353,667
TOTAL	78,147	19,126	65,878,444	3,554,447	869,960	193,324,444	259,202,888

from Japan proves that the mines of Sado, Sourouma, Bingo, and Kinsima are no less wealthy than several mines of the Americas.

Of the 78,000 marks of gold and 3,550,000 marks of silver, in French weight, that have been extracted annually since the end of the eighteenth century from all the mines of the Americas, Europe, and northern Asia, the Americas alone produce 70,000 marks of gold and 3,250,000 marks of silver, thus 90 percent of the total output in gold and 91 percent of the total output in silver. Consequently, the relative abundance of the two metals differs very little between the two continents. The amount of gold extracted from the mines of the Americas is in a ratio of one to forty-six to that of silver in Europe, including Asiatic Russia, this proportion is one to forty.

These results may shed some light on the great problem of political economy that Smith examined in the eleventh chapter of the first book of his work, in which he discusses the causes for the variable proportion between the values of the precious metals.[1] This famous author assumes that for each ounce of gold, just over twenty-six ounces of silver are imported into Europe. If this assumption were correct, the Old Continent should receive from the new one only 1,554,000 marks of silver, instead of the 3,250,000 that it actually receives. Besides, the greater the abundance of gold is in relation to silver, the clearer it should be to us (as it is to Smith) that the proportion between the respective values of the two metals does not depend exclusively on the amounts found on the market. From the discovery of the Americas to today, the value of silver has decreased so much in the western parts of Europe that the value[2] of this metal relative to that of gold, which at the end of the fifteenth century was one to eleven, or one to twelve, is today one to fourteen and a half and even one to fifteen and five-sixths. This change would not have occurred if the increase in the relative total amounts of the two metals had always been as uniform[3] as it is today. Based on the studies that I have just summarized, it is inaccurate to suggest, as has often been done, that the silver deposits of the Americas surpass those of the Old Continent in productivity at a much higher proportion than do their gold deposits. It is true that of the 70,000 marks of gold that the Americas produce in an average year, five-sixths come from the washes established in the

III.402

1. [Smith,] *Wealth of Nations*, vol. II, p. 78.

2. Under Philippe-le-Bel [Philip IV of France], one mark of gold was worth ten marks of silver. In Holland, the rate in 1335 was ten and a half to one. In France it was ten and three-quarters to one in 1388. (▼ Ouder-Meulen, *Recherches sur le commerce*, Amsterdam, 1778, part 2, p. 142.)

3. Nine-tenths.

alluvial plains; but the output of these washes is surprisingly consistent, III.403 and all those who have visited the Spanish or Portuguese colonies know that gold exports from the Americas will necessarily increase with advances in population size and in agriculture.

Until 1545, when mining began in the Cerro de Potosí, Europe appears to have received much more gold from the New Continent than silver. Five-sixths of the spoils that Cortés took from Tenochtitlan, as well as the treasures gathered from Cajamarca and from Cusco, consisted of gold, and the silver mines of Porco, in Peru, and Tasco and Tlapujahua, in Mexico, were only barely worked in Cortés and Pizarro's time. It was in 1545 that Spain began to be flooded with silver from Peru. This accumulation produced an effect that was all the more powerful since Europe was at that time more concentrated, contact was less frequent, and a smaller portion of the metals from the Americas flowed to Asia. From the middle of the sixteenth and the beginning of the seventeenth century, the proportion of gold to silver changed rapidly, especially in the south of Europe. In Holland it was still eleven and three-fifths to one in 1589; but under the reign of Louis XIII, we note that it was already twelve and a half to one in Flanders, thirteen and a half to one in France, and fourteen to one and higher in Spain. Gold extraction increased exponentially in the Americas from the end of the seventeenth century; and although the auriferous terrains have been at least partly known since 1577, III.404 the mining of alluvial deposits nevertheless only began during the reign of ▾ Pedro II [of Portugal]. In the time of Charles V, an amount of gold equal to forty or fifty thousand marks would have sufficed to make a noticeable change in the proportion of gold to silver in Europe. On the contrary, such an influx was barely felt at the beginning of the eighteenth century, when commercial relationships had greatly increased in number. Spread across a vast expanse of land, Brazilian gold was unable to have an impact on the price of silver as it would have done were it accumulated rapidly on a single point of the globe.

We shall now tackle a very important question that has been approached in a variety of ways in works of political economy; namely, how much gold and silver has flowed from the New Continent to the Old Continent from 1492 to the present. Instead of examining the development of mining in the Americas and assessing for each colony the output of the mines in different periods, people have relied upon an estimate of so many million piasters that are assumed to have been randomly introduced into Portugal and Spain over three centuries. It was easy to foresee that calculations based on this principle would lead to results that would vary by billions of livres tournois, so long as annual imports are over- or underestimated by at least ten to twelve million piasters. Even worse is that most famous writers,[1] instead of devoting themselves to new research, simply reproduced the estimates of ▾ Don Gerónimo de Uztáriz, as though citing the personal opinion of a Spanish author were enough to inspire confidence. Before revealing the results that I obtained, let us examine the calculations that have heretofore been presented. III.405

In his excellent *Traité de commerce et de marine*,[2] Uztáriz bases his estimates on those of ▾ Don Sancho de Moncada and Don Pedro Fernández de Navarrete. The former, a professor at the University of Alcalá, somewhat vaguely claims that "according to a statement prepared for the king, from 1492 to 1595 two billion piasters in gold and silver extracted from the mines of the Americas entered into Spain," that "at least the same amount has entered unregistered," and that "even with so much gold and silver, it would be difficult to find two hundred million piasters, one hundred liquid and one hundred in furniture." To these two billion, Uztáriz adds what came into Spain from 1595 to 1724, the value of which he estimates to be 1,536,000,000 piasters; such that, according to that author, the total output in gold and silver of Spanish America from 1492 to 1724 was 5,536,000,000 piasters.

It is easy to prove that this calculation is not well-founded: four billion piasters distributed across 103 years, from 1492 to 1595, suggests a mining output of thirty-eight million piasters in an average year. Yet we know from the history of mining in the Americas that the amount of gold and silver introduced into Spain from 1492 to 1535 was very small; at most, it can be estimated at 130 to 140 million. If, however, for this first period, one allows for twelve million piasters per year, as Uztáriz does for the period from 1595 to 1724, one would find that the annual output from 1535 to 1595 should have been at least fifty-eight million. These estimates are all four to five times too high, as one is easily persuaded by consulting the ledgers of Potosí and by recalling that until the beginning of the eighteenth century, the annual production of the mines of New Spain never exceeded three million piasters. What is more, in describing the wealth of the New Continent's mines, Garcilaso and Herrera clearly state that toward the end of the sixteenth century, ten to twelve million piasters per year entered into Spain via the mouth of the Guadalquivir. By no means can estimates given in round III.406

1. ▾ Forbonnais, Raynal, Gerboux, and the judicious author [Cornelius van der Ouder-Meulen] of *Recherches sur le commerce* (Amsterdam, 1778).

2. Paris edition, 1753, p. 11. ▾ Toze, *Kleine Schriften*, 1791, p. 90 [101 and 107].

numbers in the billions be considered the result of rigorous analysis; they represent no more than approximate calculations, and each author has thus felt it necessary to propose a different amount.

III.407 On ▾ Dávila's authority, Solórzano[1] claims that, from their discovery in 1492 to 1628, Spain received from the Americas fifteen billion registered piasters, a sum that differs by half from the one adopted by Uztáriz. On the other hand, we find in Navarrete's political treatise[2] that from 1519 to 1617, 1,536,000,000 piasters were brought from the Indies, according to registers. This estimate assumes a total sum of piasters for a ninety-eight-year period that is below what Solórzano and Dávila suggest for a 136-year period, a contradiction that is all the more striking in that one of these periods is subsumed by the other.

In the first editions of his famous work on the institutions of the Indies,[3] Raynal estimated the amount of gold and silver imported into Europe from the Americas since the discovery of the New World to be nine billion piasters; in 1780, he reduced this sum to five billion. He supposes that, on an average of eleven years from 1754 to 1764, annual imports into Spain of registered gold and silver amounted to only 13,984,185 piasters; whereas we know from the ledgers preserved at the mint of Mexico City that in this same period III.408 New Spain alone produced, per year, nearly twelve million piasters. I do not know how such a sagacious author, who has generally drawn from excellent sources, was so misled regarding the trade in precious metals. Extensive work seems to have gone into the tables that Raynal presents: he gives separate estimates for the amounts of gold and silver from each part of the colonies, yet despite this appearance of precision, several of his calculations are quite flawed. He claims that "each year beginning in 1780, Spain took from the American continent 89,095,052 pounds of gold and silver, or 16,970,484 piasters," because, in an average year, taken from the period from 1748 to 1753, the following amounts were imported [see table on the following page].

It is surprising to note that Raynal confuses the mining output of 1750 with III.409 that of 1780; over this thirty-year period, Mexican silver exports increased by a quarter, and the mines of South America, far from becoming exhausted, have in fact become more productive. In 1780, the mint of Mexico City alone coined a total of 17,514,263 piasters, while the Abbot Raynal estimates the total output of the mines of Spanish America to be only eighteen million piasters. He

1. [Solórzano Pereira,] *De Indiarum jure*, vol. I, p. 846. *Hist[oria] magna Matritensis*, p. 472.

2. [Fernández Navarrette,] *De la conservación de las monarquías*, discurso 21.

3. Compare the changes made in [*Histoire philosophique*,] book VIII, § 42; book IX, § 54.

	LIV. TOURN.	PIASTERS
From New Spain	44,196,047	8,418,294
From Cartagena or New Granada	14,087,304	2,683,296
From Lima or Peru	25,267,849	4,812,924
From Buenos Aires or the kingdom of La Plata	5,304,705	1,010,420
From Caracas	239,144	45,551
TOTAL, in an average year	89,095,049	16,970,485

should have known, from the account of a man deeply knowledgeable about Spanish trade,[1] that already in 1775 this total output amount had risen to thirty million piasters, or 157,500,000 livres tournois per year.

As for the amount of precious metals that Spain has received from her colonies since the discovery of the Americas, Raynal sets this at 25,570,279,924 livres [tournois] or 4,870,529,509 piasters. This calculation, which would inspire more confidence were the sum expressed in round numbers, is quite exact; it demonstrates that, even when relying on the most inaccurate data, one can occasionally have the good fortune of compensating for this flaw and obtaining results that approximate the truth.

In his classic work on the causes of the wealth of nations,[2] Adam Smith estimates silver imports from the New Continent into Cádiz and Lisbon to be six million pounds sterling, or twenty-six and a half million piasters per year; this estimate was two-fifths too low, even for his time, 1775. The British author followed Magens's calculations, according to which, in 1748 and 1753 Spain and Portugal had received, on average, 5,746,000 pounds sterling or 25,337,000 piasters of precious metals per year. Following Magens, if one includes four million piasters for gold imports from Brazil, one obtains twenty-one million piasters for the Spanish colonies alone and, therefore, three million piasters more for 1750 than Raynal cites for 1780. Smith's scholarly commentator, ▼ Mr. Garnier,[3] who conducted his research with the greatest precision, estimates that in 1802 the output of the gold and silver mines of Spanish America was 159 million livres tournois or 30,285,000 piasters, a sum that is much closer to the truth than all those found in other works of political economy.

III.410

1. Campomanes, *Discurso sobre la educación popular de los artesanos*, vol. II, p. 331 [337].

2. [Smith, *Wealth of Nations*,] book I, chap. I (Paris edition), vol. 2, p. 70.

3. [Smith, *Recherches sur la nature et les causes de la richesse des nations*, trans.by Germain Garnier,] vol. V, p. 137.

In his *History of America*, Robertson values the amount of precious metals imported into Spain from 1492 to 1775 at the staggering sum of two billion pounds sterling, or 8,880,000 piasters. What is more, this justly famous author regards his calculation as being based on quite moderate sums, despite the fact that he estimates the mines' annual output over 283 consecutive years to be four million pounds sterling, and the total contra-

III.411 band during this period to be 968 million.[1] If one compares these figures to those contained within Uztáriz's work, one sees that the Spanish author uses sums that are lower by a half.

In [Oudermeulen's] *Recherches sur le commerce*, published in Amsterdam in 1778,[2] the total amount of gold and silver from Spanish America from 1674 to 1723 is estimated at 672 million piasters. If one extends this estimate over the 283 years from 1492 to 1775 and adds a third for contraband, one obtains a total of 5,072,000,000 piasters for the metals imported into Spain. The same author values the gold from Brazil since its discovery at 1,350,000,000 piasters, a sum that seems twice as high as it should be, as we shall demonstrate in the next part of this discussion.

In his studies on cash levels in France, Necker[3] estimates the amount of gold and silver shipped to Cádiz and Lisbon from 1763 to 1777 to be 1,600,000,000 livres tournois, or 304,800,000 piasters. According to this estimate, total imports of precious metals from both Americas would have amounted to only twenty-one and a half million piasters per year, while solid information reveals that Spanish imports alone totaled over

III.412 thirty million piasters.[4] On the other hand, in his *Discussions sur les effets de la démonétisation de l'or*,[5] Mr. Gerboux values European gold and silver:

From	1724	to	1766	at	4,000	million livres tournois
	1766		1800		4,000	
	1789		1803		1,500	

1. [Robertson,] *History of America*, vol. IV, p. 62.

2. Book I, chap. 10 (vol. I, part 2, p. 124).

3. Necker, *Sur [la législation et] le commerce des grains*, book II, chap. 5. *De l'administration des finances [de la France]*, vol. III, chap. 8, p. 71.

4. *Encyclopédie méthodique, Économie politique*, vol. II, p. 324.

5. Gerboux, [*Discussion sur les effets de la démonétisation de l'or*,] pp. 36, 66, 69, 70.

which suggests that annual imports between 1724 and 1803 were twenty-one million piasters.

If one brings together the results of these various calculations, which are based on mere conjecture, one finds that the amount of precious metals that have flowed from Spanish America to Europe—and have been registered—is, according to

AUTHOR'S NAME	PERIOD	PIASTERS
Uztáriz	1492–1724	3,536 million
Solorzano	1492–1628	1,500
Moncada	1492–1595	2,000
Navarrete	1519–1617	1536
Raynal	1492–1780	5,154
Robertson	1492–1775	8,800
Necker	1763–1777	304
Gerboux	1724–1800	1,600
▾ Author of *Recherches sur le commerce*	1492–1775	5,072

In order to avoid, where possible, the innumerable possibilities for error, I shall take a different approach than the one followed by the writers I have just cited. I shall first present the amount of gold and silver that we know, from the ledgers of the mint and the royal treasury, to have been extracted year after year from the mines of both Mexico and Potosí. Using the historical knowledge that I have acquired about the state of the American mining operations, I shall add what was produced in each metalliferous region of Peru, Buenos Aires, and New Granada in different periods, and I shall distinguish what was registered from what was exported through fraudulent trade. Rather than valuing the total output of smuggling at one-third or one-quarter of the combined registered metals, as has heretofore been done, I shall make partial estimates based on the location of each colony and its relationship to the neighboring countries. When attempting to determine the size of a distance that cannot be measured precisely, one is sure to commit less fatal errors if one divides the total area into several parts and if one compares each one of these to objects of a known size.

III.413

III.414

I. Registered Amount of Gold and Silver from the Mines of the Americas, from 1492 to 1803

A. SPANISH COLONIES	PIASTERS
According to the aforementioned ledgers, from 1690 to 1803, the Kingdom of New Spain supplied the Mexico City mint with	1,353,452,000
The mines of Tasco, Sultepeque, Pachuca, and Tlalpujahua, are almost the only ones that were worked immediately after the destruction of the city of Tenochtitlan in 1521, and from that memorable period until 1548. Since the amount of gold and silver minted at the beginning of the eighteenth century did not exceed five million piasters per year, I am calculating Mexico's total output, from Hernán Cortés's conquest to1548, at	40,500,000
In 1548, mining started at Zacatecas; in 1558, at those of Guanajuato; and, almost simultaneously, so did amalgamation, invented by Medina. From 1548 to 1600, one can include at least two million, and from 1600 to1690, three million per year	374,000,000
From their discovery in 1545 to 1803, the mines of Potosí furnished 1,095 and a half piasters' worth [of gold and silver], or 128,883,000 marks; that is, from 1545 to 1556, nearly	127,500,000
From 1559 to 1789, according to the ledgers of the aforementioned treasury,	788,258,500
Adding because of the value of the *peso de minas,* from 1556 to 1600,	134,000,000
The output of Potosí, from 1789 to 1803,	46,000,000
III.415 The mines of Pasco or of Yauricocha, discovered in 1630, yielded nearly 300 million piasters, or 35,300,000 marks, through 1803; that is, from 1630 to 1792, 200,000 silver marks per year	274,400,000
From 1792 to 1801, according to the ledgers,	21,501,600
Output of the Cerro de Yauricocha, from 1801 to 1803,	3,400,000
The mines of Hualgayoc, discovered in 1771, yielded nearly 170,000 silver marks per year through 1802	4,300,000
From 1774 to 1802, for the mines of Hualgayoc, Guamachuco, and Conchuco	185,339,900

III.416

	PIASTERS
Adding, for 1803,	504,000
I am estimating the output of the mines of Huantajaya, Poco, and other, smaller Peruvian mines, from the sixteenth century through 1803, at150,000 or 200,000 marks of silver per year	350,000,000
Chocó was settled in 1539; the province of Antioquia, settled by anthropophagic peoples, was conquered in 1541. Work at the alluvium mines of the Sonora and Chile started quite late. If one includes 12,000 marks of silver for the total yield of the Spanish colonies, not including the Kingdom of New Spain, one can add	332,000,000
Registered gold and silver from the Spanish colonies from 1492 to 1803	4,035,156,000
B. PORTUGUESE COLONIES	
For the first sixty years, Raynal assumes twice the output of today. On the basis of the fleet registers, he concedes that from the discovery of mines in Brazil to 1755, the total value of the gold that arrived in Europe was	480,000,000
From 1756 until 1803, excluding the annual output of 32,000 marks	204,544,000
Registered gold in the Portuguese colonies from the discovery of Brazil until 1803	684,544,000

This sum, at which I believe I must stop, differs by over sixteen billion francs from the one presented by Robertson. It is not surprising that it is similar to the estimates made by several other writers, for the figures generated by political economy are not unlike the positions determined by astronomers. Once the longitude of a place has been determined, one is sure to find, among the large number of maps on which all points are randomly placed, at least one that indicates its true position. III.417

My research suggests that of the 5,706,700,000 piasters or 29,960,175,000 livres tournois produced in gold and silver from 1492 to 1803, or over the course of 311 years, we owe:

III.416

II. Unregistered Gold and Silver from the Mines of the New Continent, from 1492 to 1803

A. SPANISH COLONIES	PIASTERS
I am including for New Spain, where clandestine mining was quite prevalent in the middle of the eighteenth century, one-seventh	260,000,000
For Potosí, one-quarter of the total output, because of the huge scope of smuggling when mining started	274,000,000
Pasco, Hualgayoc, and the rest of Peru, where silver flows via the Amazon River toward Brazil	200,000,000
For the gold of Chile, New Granada, and the Kingdom of Buenos Aires	82,000,000
B. PORTUGUESE COLONIES	
For Brazilian gold	171,000,000
Unregistered gold and silver from 1492 to 1803	987,000,000

Recapitulation

VALUE OF GOLD AND SILVER FROM THE MINES OF THE AMERICAS, FROM 1492 TO 1803		PIASTERS
Registered (no. I)	from the Spanish colonies	4,035,156,000
	from the Portuguese colonies	684,544,000
Unregistered (no. II)	from the Spanish colonies	816,000,000
	from the Portuguese colonies	171,000,000
TOTAL		5,706,700,000

POLITICAL DIVISION	PIASTERS	LIVRES TOURNOIS
To the Spanish colonies	4,851,200,000	25,468,800,000
To the Kingdom of New Spain	2,028,000,000	10,647,000,000
To the kingdoms of Peru and Buenos Aires	2,410,200,000	12,653,550,000
To the kingdom of New Granada	275,000,000	1,443,760,000
To Chile	138,000,000	724,500,000
To the Portuguese colonies	855,500,000	4,491,375,000
TOTAL	5,706,700,000	29,960,175,000

Since the Cerro de Potosí, by virtue of its location, is part of the Cordilleras of Peru, I have brought together in this table the mines situated on the ridge of the Andes chain from 6° to 21° southern latitude over a length of 500 leagues. The metalliferous region of Mexico, contained between 16° and 31° northern latitude, today produces twice as much silver as the two viceroyalties of Peru and Buenos Aires; this region is only 450 leagues long. The following table indicates the proportion of gold to silver extracted from the mines of the New Continent, from their discovery to 1803. III.418

POLITICAL DIVISION	MARKS, CASTILIAN POUNDS	PIASTERS
Gold	9,915,000	1,348,500,000
From the Portuguese colonies	6,290,000	855,500,000
From the Spanish colonies	3,625,000	493,000,000
Silver	512,700,000	4,358,200,000
Total		5,706,700,000

According to this rough estimate, the mass of silver that the Cordilleras of the Americas have produced over the past three centuries has a total weight of 117,864,210 kilograms; it would form a solid sphere with a diameter of 27,8 meters or eighty-five and thirteen-twentieths Paris feet. Recalling that the iron extracted from the mines of France alone amounts to 225 million kilograms[1] per year, one sees that, in regard to the relative abundance or the distribution of substances across the Earth's external crust, the ratio of silver to iron is more or less the same as that of magnesium to silica or that of baryta to aluminum. III.419

One must not, however, confuse the total amount of precious metals extracted from the mines of the New Continent with what has actually flowed into Europe since 1492. In order to determine this latter sum, it is indispensable to have values for (1) the gold and silver found at the time of the conquest among the indigenous peoples of the Americas and which eventually became the spoils of the conquerors; (2) what has remained in circulation on the New Continent; and (3) what passed directly to the coasts of Africa and Asia without ever reaching Europe.

The conquerors found gold not only in the regions that still produce it today, such as Mexico, Peru, and New Granada, but also in the countries

1. Héron de Villefosse, *Mémoire general sur les mines*, p. 240.

whose rivers we currently consider very poor in auriferous sands. The indigenous peoples of Florida, Saint-Domingue, and the island of Cuba, as well as those of Darien and the coasts of Paria, had golden bracelets, rings, and necklaces; but it is likely that the bulk of this metal did not come from III.420 the countries in which these peoples were encountered at the end of the fifteenth century. In South America, as in Africa, contacts based on trade existed even among the hordes furthest removed from civilization. Coral and pearls from pelagic shells have often been seen among men who dwelled far from the coasts. During our journey on the Orinoco, we confirmed that the famous Mahagua stone, the jade of the Amazon, comes from Brazil to the banks of the Caroni, inhabited by Carib Indians, through an established trade between different tribes of savages. In addition, it should be noted that the peoples whom the Spanish encountered in Darien or on the island of Cuba had not always dwelled in these same lands: in the Americas, the direction of the great migrations was from northwest to the southeast, and wars often forced entire tribes to leave the mountains and settle in the plains. One can thus imagine that gold from the Sonora or from the Río Cauca valley could be found among the savages of Darien or at the mouth of the Magdalena River. Moreover, the smaller the population, the more deceptive the appearance of wealth. The accumulation of gold is especially striking in countries where all the metal that the people possess is converted into decorative objects. One must not therefore judge the so-called wealth of the mines of Cibao, the coast of Cumana, and the Isthmus of Panama on the basis of the first voyagers' accounts; one must recall that rivers become less aurifer- III.421 ous over the centuries as their slopes diminish. A horde of savages who settle in a valley untouched by humankind finds grains of gold accumulated over thousands of years, while in our day, even the best-maintained washes yield barely a few scattered particles. These considerations, to which I must limit myself here, will serve to shed light on the oft-debated question of why the very regions that, immediately after the discovery of the Americas, and especially from 1492 to 1515, were considered eminently rich in precious metals, in our day produce almost none, despite the fact that arduous and quite well-organized studies were conducted in many of them.

To focus our ideas on the spoils of gold and silver that the first conquerors brought to Europe before the Spanish began working the mines of Tasco, in Mexico, and those of Porco, in Peru, let us take a look at the facts reported by the historians of the conquest. I have examined these facts carefully, and I have attempted to assemble all the passages in which the riches that fell into the hands of Europeans are valued in *pesos ensayados* or in *cas-*

tellanos de oro; for it is only these data—and not the vague and oft-repeated statements about "enormous amounts of gold" or "immense treasures"—that are capable of yielding satisfying results.

In 1502, ▼ Ovando sent to Spain a fleet of eighteen ships, commanded by Bovadilla and Roldán, and loaded with a large amount of gold. The majority of these ships were destroyed in the famous storm that, during Christopher Columbus's fourth voyage, nearly cost him his life off the island of Saint-Domingue. The historians of the time regard this fleet as one of the richest, and yet they all agree that its load of gold did not surpass 200,000 pesos,[1] which, if converted into *pesos de minas* of fourteen reales, makes the modest sum of 1,750,000 livres tournois, or 2,560 marks of gold. The presents that Cortés received during his sojourn in Chalco amounted to only 3,000 *pesos de oro*,[2] or a weight of thirty-eight marks of gold. When Montezuma gathered his vassals to take the oath of fidelity to Emperor Charles V, whom they were led to believe to be a direct descendant of Quetzalcoatl,[3] the Buddha of the Aztecs, Cortés demanded a tribute of gold: "I pretended," he wrote to the emperor, "that Your Highness had a great need for certain works that he needed to carry out." The quint on this tribute, paid to the army coffers, rose to 32,400 pesos;[4] which leads one to conclude that the amount of gold collected through the general's stratagem was equal to 2,080 marks. According to Cortés's own claim, during the taking of Tenochtitlan, the spoils that fell into the hands of the Spaniards did not exceed a weight of 130,000 castellanos or 2,600 marks of gold;[5] on Bernal Diaz's authority, it amounted to 380,000 pesos, which equals 4,890 marks.

III.422

III.423

The two periods in the conquest of Peru when the Spaniards collected the greatest riches were Atahualpa's trial and the plundering of Cuzco. According to Garcilaso, the Inca's ransom, divided in 1531 among sixty

1. Herrera [y Tordesillas, *Historia general*,] decade I, book I, chap. I (vol. I, p. 126).

2. *Cartas de Hernán Cortés* [Lorenzana, *Historia de Nueva-España*], letter I, §18, p. 72.

3. See my *Vues des Cordillères et monumens des peuples indigènes de l'Amérique*, plate VII.

4. *Cartas de Hernán Cortés* [Lorenzana, *Historia de Nueva-España*], letter I, §29, p. 98.

5. [Lorenzana, *Historia de Nueva-España*], Letter III, §51, p. 301. The expression "*se fundió mas de* 138,000 *castellanos*" [138,000 castellanos were smelted] is unclear. It is unknown whether Cortés is referring to castellanos as a weight or as an imaginary currency. Like the Abbot Clavijero, I have followed the former theory (*Storia antica del Messico*, vol. I, p. 232); the latter theory would make the spoils equal to only 1,660 marks of gold for Herrera [y Tordesillas] plainly states that "*castellano y peso es uno*," and, in his view, one *peso de minas* is equivalent to fourteen reales; one *peso ensayado* is equivalent to thirteen reales (de plata) and one quartillo. [*Historia general*,] decade VIII, book II, chap. X, vol. V, p. 41.

cavalrymen and one hundred infantrymen, amounted to 3,930,000 golden ducats and 672,670 silver ducats. If one converts these sums into marks, one obtains 41,987 marks of gold, and 115,508 marks of silver, for a combined value of 3,838,058 piasters of eight reales of *plata mexicana*, or 20,149,804 livres tournois.[1] These treasures, which had been collected in a house whose ruins were still visible during my stay in Cajamarca in 1802, had III.424 served as decorations in the Sun temples of Pachacamac, Huailas, Cuzco, Huamachuco, and Sicllapampa. Gómara[2] values Atahualpa's ransom at only 52,000 marks of silver and 1,326,500 *pesos de oro*, or 17,000 marks of gold. Whenever numbers are involved, it is very rare for the sixteenth-century authors to be in agreement. According to Herrera,[3] the spoils of Cuzco were worth over two million pesos, or over 25,700 marks of gold.

These data make it likely that the conquests of Mexico and Peru did not place over 80,000 marks of gold into Spanish hands. The bulk of the treasures were buried by the indigenous peoples or cast into the lakes.[4] What was found of it, little by little, during excavations in the *huacas*, paid for the king's quint and was mixed up with the gold extracted from the mines. To these 80,000 marks of gold, we shall add what was taken, in small portions, from the Antilles islands, the Paria and Santa Marta coasts, as well as those of Darien and Florida, and counting two thousand marks per year, we shall have another sum of 106,000 marks of gold.

III.425 The amount of cash that is in circulation today in the New World is much smaller than is commonly assumed. To determine this with some degree of precision, one must recall that cash in France[5] is valued at two and a half billion livres tournois, 450 million in Spain,[6] and 920 million in Great Britain[7] and that, rather than being linked to the size of the popu-

1. Garcilaso [de la Vega El Inca, *Historia general del Perú*,] p. 11, book I, chaps 28 and 38 (vol. II, p. 27 and 51). ▼ Father Blas Valera counts 4,800,000 ducados.

2. [López de Gómara,] *Historia de las Indias*, 1553, p. 67.

3. [Herrera y Tordesillas, *Historia general*,] decade V, book VI, chap. III.

4. In Lake Tezcoco, in Mexico; in Lake Guatavita, to the northwest of Santa Fe de Bogotá; Lake Titicaca and the Orcos Valley Lake. It is the latter lake that is assumed to contain the famous golden chain that the Inca Huayna-Capac commisioned at the birth of his son Huescar and that so captured the imagination of the first colonists of Peru.

5. In 1784, Necker valued it at 2,200,000,000 livres tournois; in 1791, Mr. Arnould valued it at two billion; in 1801, ▼ Mr. Des Rotours valued it at 2,290,000,000; in 1805, Mr. Peuchet and Mr. Gerboux valued it at 2,550,000,000.

6. In 1724, Uztáriz valued it at one hundred million piasters; in 1782, ▼ Mr. Músquiz, minister of finance, cited in Mr. Bourgoing's work, claimed that it was eighty million piasters.

7. Adam Smith assigns it a maximum value of only thirty million pounds sterling; [*Wealth of Nations*], vol. III, p. 31.

lation, the total amount of gold and silver that remain in circulation in a country depends on trade activity, the well-being and the civilization on its inhabitants, and the number of goods that must be represented by monetary signs. Assuming that the value of the existing precious metals, in the form of either cash or finely wrought gold and silver, is

	LIVRES TOURNOIS
In the United States, including British Canada	180 million
In the continental Spanish colonies[1]	480
In Brazil	120
In the Antilles	25

One obtains a sum total of 153,333,000 piasters or 805 million livres tournois.

[1] In these estimates, we have followed the principles set by Adam Smith and Necker, using as a base the number of inhabitants, the total amount of taxes paid to the government, the wealth of the clergy, and trade-related activity. What makes these calculations even more imprecise is that a large number of Negroes and indigenous peoples are mixed in with the Whites.

A very small amount of the gold and silver removed from the mines of the Americas passes directly to Africa and Asia without first reaching Europe. We shall value the amount of precious metal that have flowed from Acapulco to the Philippine islands since the end of the sixteenth century at 600,000 piasters per year.[1] The expeditions from Lima to Manila have been quite rare, even in recent years. The ships sent from the Antilles, and formerly from the ports of the United States, to the western coasts of Africa for the Negro Trade, exporting firearms, eau de vie [brandy], hardware items, and European linens, also export silver in specie form; but these exports are compensated for by the purchase of gold powder on the Guinea coasts; and by the lucrative trade that the Anglo-Americans conduct with several parts of Europe. III.426

If we now deduct from the 5,706,000,000 piasters removed from the mines of the New Continent from its discovery by Christopher Columbus to the present day,

1. I am not unaware of the fact that ▼ Lord Anson found the sum of 1,357,454 piasters in the galleon from Acapulco that fell into his hands (Anson, *Voyage [around the World]*, p. 384); but it would be impossible to value total annual imports at over 600,000 piasters if one considers that the galleon stopped departing every year after the end of the sixteenth century.

153	million piasters that exist, either in specie or in gold and silver works, in the civilized parts of the Americas, and
133	million piasters that have been transported from the western coasts of America to Asia,
286	million piasters

III.427 we find that over the past three centuries Europe has received 5,420,000,000 piasters from the New World. On the other hand, if the 186,000 marks of gold that have passed in the form of spoils into the hands of the conquerors are valued at twenty-five million piasters, then it results from these combined calculations that the amount of gold and silver imported from the Americas to Europe between 1492 and 1803 is *five billion four hundred forty-five million piasters*, or *twenty-eight billion five hundred eighty-six million livres tournois*.

This calculation, like all those presented by Forbonnais, Uztáriz, Necker, and Raynal, is based partly on facts and partly on mere conjecture. It is easy to understand that the exactness of the results depends on whether a large number of facts have been employed and whether the conjectures are based on an intimate knowledge of both the history and the current state of the mining operations of the New World. Those of my readers who are accustomed to this type of research may judge whether the numbers at which I have arrived present a greater degree of probability than those that have heretofore been adopted in the most highly esteemed and widely available works.

III.428 If one distributes the 5,445,000,000 piasters across the 311 years that elapsed from the discovery of the New World to 1803, one finds imports of seventeen and a half million piasters in an average year. The historical research that I have heretofore been able to conduct leads me to believe that the treasures of the Americas flowed to Europe in the following progression:

PERIOD	GOLD AND SILVER IMPORTS TO EUROPE, FROM THE AMERICAS, IN AN AVERAGE YEAR	NOTES CONCERNING THE HISTORY OF MINES
1492–1500	Piasters 250,000	Discovery of the Antillean islands; gold washes at Cibao; Alonzo Niño's expedition to the coast of Paria; Cabral's voyage. Ships did arrive every year in Spain, and those of Ovando were regarded as immensely rich, although they carried only 2,560 silver marks.

1500–1545	3,000,000	Exploitation of the Mexican mines of Tasco, Zultepec, and Pachuca; Peruvian mines at Porco, Carabaya, and Chaquiyapu (or La Paz); spoils made at Tenochtitlan, Cajamarca, and Cuzco; conquest of Chocó and Antioquia.
1545–1600	11,000,000	Mines at Zacatecas and at Guanajuato in New Spain; Cerro of Potosí, in the Cordilleras of Peru; smooth occupation of Chile and the *provincías internas* in Mexico.
1600–1700	16,000,000	The mines of Potosí begin to wane, especially from the middle of the seventeenth century; but the mines of Yauricocha are discovered. The output of New Spain increases from two million to five million piasters per year; gold washes at Barbacoas and Chocó.
1700–1750	22,500,000	Work at the alluvium mines in Brazil; Mexican mines at Vizcaína, Jacal, Tlalpujahua, Sombrerete, and Batopilas; Spain's gold and of silver imports are eighteen million piasters in an average year.
1750–1803	35,300,000	Final period of the splendor of Tasco; mining at Valenciana; discovery of the mines at Catorce and Cerro in Hualgayoc; gold and silver imports are forty-three and a half million piasters toward the beginning of the nineteenth century.

We have observed above that the ratio of gold to silver, which was ten to one before the discovery of the Americas, has gradually risen to around sixteen to one. It would be important to know the amount of gold and silver that have flowed from one continent to the other at different periods; but we lack exact data on this subject, and the little we know about it is limited to the following facts. III.429 III.430

Until the year 1525, Europe more or less only imported gold from the New World; from that time to the discovery of the mines of Brazil, near the end of the seventeenth century, imported silver prevailed in weight over imported gold at a rate of sixty or sixty-five to one. It was in the first half of the eighteenth century that the trade in precious metals experienced an extraordinary revolution: the output of the silver mines was relatively consistent, but Brazil, Chocó, Antioquia, Popayán, and Chile produced such large amounts of gold that Europe took from the Americas perhaps no more than thirty marks of silver for each mark of gold. In the second half of the last century, silver once again flooded the market. In an average year, the mines of Mexico provided Spain with two and a half million marks of silver, instead of the six hundred thousand that they had produced between 1700 and 1710. Since gold production did not continue to increase in the same proportion, what happened was that between 1750 and 1800, the amount of imported gold in Europe was at ratio of one to forty to that of imported silver.[1] The mines of New Spain III.431 have, so to speak, counterbalanced the effects that the abundance of Brazilian gold might have produced. One should not, in general, be surprised that the proportion between the value of gold and that of silver has not always varied to a significant degree whenever one of the two has been predominant in the mass of metals imported into Europe from the Americas. The full impact of the accumulation of silver appears to have been felt prior to 1650, when the ratio of gold to silver in Spain and Italy was already at one to fifteen. Since that time, both the population and the trade relations of Europe have increased so sharply that variations in the value of precious metals have depended on a large number of factors at once, but especially on exports of silver to the East Indies and China, as well as its use in the manufacture of plate.

If, as Mr. Héron de Villefosse suggests, Europe today produces 215,000 marks of silver per 5,300 marks of gold, or forty marks of silver per one mark of gold, it appears, on the contrary, that in the fifteenth and sixteenth century this proportion was more favorable to silver. The output of the mines and the gold washes decreased in Germany and Hungary at the same time as silver mines were being more profitably worked. The mines of Freiberg alone, which, in the sixteenth century, yielded only 16,000 marks per year, today produce over 50,000. I would be tempted to believe that, even without the discovery of the Americas, the value of gold would have increased in Europe.

III.432 To conclude this chapter, let us examine what has become of the treasures removed from the New Continent. Where today are the twenty-eight billion livres tournois that Europe has received over the past three centuries from Spanish and Portuguese America? Forbonnais estimated that, of the twenty-seven and a half billion livres tournois that, in his account, had flowed from

1. Magens found the proportion of gold to silver to be one to twenty-two and two-fifths between 1753 and 1764, and one to twenty-six and four-thirteenths between 1753 and 1764. In 1803, Mr. Gerboux estimated it to be one to twenty-nine and one-sixth.

one continent to the other from 1492 to 1724, fully half had been absorbed by trade with India and the Levant; that one-quarter had been used in plate manufacturing or dispersed through being melted down and broken up into jewelry; and that the rest had been converted into cash. He estimated that in 1766, 7,500,000,000 livres tournois of precious metals were in circulation in Europe, without including in this sum the output of the mines of Spanish America subsequent to 1724, nor the cash that might have existed in Europe before the discovery of the New World. In an interesting paper on monetary legislation, Mr. Gerboux has attempted both to verify and to extend Fourbonnais's calculations. He believes that the current level of cash in Europe is 10,600,000,000 livres tournois, or 219 million piasters, and that before 1492, it was only 600 million livres tournois, or 114 million piasters.

It is surprising to note that such an enlightened financier as Mr. Necker suggested in 1775 that the cash in France formed nearly one-half of the minted silver in the whole of Europe, and that Europe in its entirety possessed only 4,500,000,000 livres tournois in cash. ▼ Mr. Démeunier (in the *Encyclopédie méthodique*), Mr. Gerboux, and Mr. Peuchet proved how imprecise this claim is.[1] Mr. Necker himself greatly modified it in his work on financial Administration. III.433

On the other hand, Mr. Gerboux's estimate, which places the current level of cash at ten billion six hundred million livres tournois, appears quite high when one focuses on the population of this part of the world. It is generally believed that the amount of precious metals that existed in old France is known with some degree of certainty, and that, for the year 1803, it can be valued—taking into account the losses caused by the monetary law of November 30, 1785, and by the collapse of the colonial trade—at 1,850,000,000 livres tournois. Estimating the population at that time to be 26,363,000, one obtains sixty-nine livres tournois per inhabitant. But, according to the recent studies by Mr. Hassel, Europe as a whole contains 182,600,000 inhabitants, of which Russia, Sweden, Norway, Denmark, and the Slavic and Sarmatian lands embrace nearly sixty-two million. If one allows for fifty-five livres tournois per individual in Great Britain, as well as in the west and south of Europe, and for thirty livres tournois in the other countries less advanced in civilization,[2] one finds that the total amount of cash in Europe cannot be III.434

1. Démeunier, *Économie politique*, vol. II, p. 325. Gerboux, [*Discussion sur les effets de la démonétisation de l'or*,] pp. 75 and 92. Peuchet, *Statistique de la France*, p. 474. Necker, *De l'Administration des finances*, vol. III, p. 75.

2. In 1805, the amount of actual cash in the Austrian monarchy was valued at 250 to 300 million florins, assuming a population of 25,548,000 inhabitants (Hassel, *Statis[tischer]*

above 8,603,000,000 livres tournois (1,637,000,000 piasters), a sum nearly equivalent to half the debt of Great Britain.[1] The result of this is that although the population of France is in a ratio of one to five to that of Europe, the amount of precious metals that it contains is in a ratio of one to three and a half to the amount spread throughout Europe as a whole.

We have seen above that the mines of Russian Asia and of Europe today produce twenty-one million livres tournois, or four million piasters per year. Information provided by Dutch authors reveals that four to five thousand marks of gold powder arrive annually in Europe from the coasts of Guinea. We value the output of the mines of Europe and imports from northern Asia and Africa since the discovery of the Americas at only six million livres III.435 [tournois] per year; and the result of this, assuming the current level of cash to be 8,603,000,000, and that of 1492 (according to Mr. Gerboux) to be 600 million, is that since the end of the fifteenth century, 22,450,000,000 livres [tournois] have been taken to the East Indies, converted into plate, and dispersed by being melted down again. Distributing this sum across a period of 213 years, one finds, in an average year, a loss in gold and silver of seventy-two million livres [tournois] (13,700,000 piasters). It has been proven above that imports from the Americas in this same period amounted to ninety-two livres [tournois] (seventeen and a half million piasters) per year.

Since it was only relatively recently that serious statistical research began, it is impossible to know in detail the value of gold and silver exports to Asia in the sixteenth and seventeenth centuries. We shall thus restrict ourselves to shedding some light on the current state of things and to observing the periodic ebb and flow that has marked the spread of precious metals from one continent to the other. If one recalls that since the end of the eighteenth century, Europe has annually received 80,000 marks of gold and nearly four million marks of silver, in Castile pounds, one is surprised not to experience more acute effects from the accumulation of metals in the Old World.

Umriss [der sämmtlichen Europäischen und der vornehmsten aussereuropäischen Staaten], p. 29). How, then, could the Abbot Raynal suggest that level of cash in Portugal was only eighteen million livres tournois and that of Brazil was twenty million? (*Hist[oire] philos[ophique]*, vol. II, pp. 434 and 450) Today, Brazil has four million inhabitants, among whom there are 1,500,000 blacks [Nègres]. How can one assume, in a country where even the Indians enjoy greater affluence than in the Spanish colonies and where there are very populous cities, only ten livres tournois for each free person, while for the northern part of Europe, between thirty and forty are attributed?

1. Playfair, *Statistical Breviary* (1801, p. 37). In 1802, the debt was 562 million pounds sterling; in 1820, it was 640 million.

European gold and silver flow into Asia through three main channels: (1) through trade with the Levant, Egypt, and the Red Sea; (2) through maritime trade with the East Indies and China; (3) through the Russians' trade with China and Tartary. The trade with the Levant and the northern coasts of Africa requires a considerable quantity of ducats, piasters, and German ecus, the exporting of which decreases the level of cash in Europe. It is not, however, believed possible to assign a value of over four million piasters per year to this loss, because the balance of trade with the Levant today favors England,[1] to the sum of two and a half million piasters. According to the tables published by Mr. Arnould,[2] it was unfavorable to France by three to four million. Spain, the nations of the North, and especially Germany, are obliged to settle accounts in cash in the ports of the Ottoman Empire and the Barbary Coasts. In the Austrian monarchy alone, silver exports to Turkey and the Levant are valued at one and a half million piasters. III.436

The East Indies and China are the lands that absorb the bulk of the gold and silver extracted from the mines of the Americas. I cannot accept, as does Mr. Gerboux, that prior to 1760, this absorption amounted to eight million piasters per year, and that, from that period to 1803, it gradually decreased to five million.[3] Although people commonly form somewhat exaggerated ideas about the loss that Europe experiences in its trade balance with Asia, it is nonetheless certain that specie exports are far greater than the sum indicated by the estimable author whom we have just cited. III.437

European luxury today demands eleven times more tea than in 1721; but trade with the countries located beyond the Ganges has undergone significant changes since the English [British] established a great empire in India. The factories of Great Britain currently provide the Asia trade with over 11,460,000 piasters' worth of goods per year.[4] According to valuable information contained in Lord Macartney's travel narrative,[5] in 1795 the British imported to

1. According to Mr. Playfair's tables, in 1800 Great Britain had a trade surplus of 600,000 pounds sterling with the Levant; it had a trade deficit of 60,000 pounds sterling with Turkey (*Commercial [and Political] Atlas*, 1801, plate XIII).

2. [Arnould,] *De la balance du commerce*, vol. III, note ii.

3. Gerboux, [*Discussion sur les effets de la démonétisation de l'or*,] pp. 36 and 70. Consult also Mr. Garnier's study on trade with India, in his commentary on Smith, vol. V, pp. 361–75, and Toze, [*Kleine Schriften*,] pp. 124–50.

4. Playfair, [*The Commercial and Political Atlas and Statistical Breviary*,] chart III.

5. Macartney, *Voyage* (French edition), vol. V, pp. 47 and 58. According to the table presented on p. 73, imports of silver by the British East India Company between 1775 and 1795 would have amounted to only 3,676,000 pounds sterling. (I am valuing the pound sterling at four and four hundred and nine-thousandths piasters, or 463 sous tournois.)

Canton 4,410,000 piasters' worth of products from their manufacturing industry and goods from India. In exchange, they received 6,614,000 piasters'

III.438 worth of Chinese goods and products. Assuming that China's trade balance was less favorable for the other nations of Europe than for the English [British], it would appear that precious metal imports in China, via Canton, Macao, and Amoy, can be valued at four to five million piasters in a regular year.[1]

Let us now examine more closely the state of trade in Canton. In 1795, Lord Macartney still estimated the total weight of the tea purchased by all the nations of Europe at only thirty-four million pounds, of which the British alone exported twenty million. But according to the interesting information that ▾ Mr. de Sainte-Croix[2] has shared, the following amounts were imported from Canton:

YEAR	BY ALL OF THE NATIONS OF EUROPE, AND BY THE ANGLO-AMERICAS	BY THE BRITISH ALONE
1804	411,149 pikles	279,063 pikles
1805	353,480	245,021
1806	357,506	258,185
Average year	374,045 pikles	260,756 pikles
Or (a pikle being equal to 120 French pounds)	44,885,000 livres	31,290,900 livres

III.439 Tea exports thus increased by one-quarter between 1795 and 1806. No one would accept, however, that the loss in specie experienced by Europe increased at the same rate; for imports of linens and woolens from England and China alone rose from 600,000 piasters to three million piasters from 1787 to 1796.

According to Mr. de Guignes, who had the rare fortune of penetrating into the heart of China, the amount of cash introduced into Canton by the English [British] did not exceed three million piasters in 1787. If Great Britain

1. Raynal, [*Histoire philosophique*,] vol. I, p. 674.

2. [Renouard de Sainte-Croix,] *Voyage commercial et politique aux Indes orientales*, 1810, vol. III, pp. 153, 161, and 170. The price of a pic or pikle of Bou tea in Canton is between twelve and fifteen taels (at seven francs forty-one centimes per tael): other kinds of teas are much more expensive, cangfous costs tweny-five to twenty-seven taels; saoutchou, forty to fifty; haysuen, fifty to sixty. (De Guignes, *Voyage à Pékin*, vol. III, p. 248. Zach, *Allgemeine Geographische Ephemeriden*, 1798, pp. 179–91.)

did not possess a large part of the East Indes, its loss in specie would be more than twice this amount, for nearly four million piasters are paid annually to the Chinese through the trade between the Indies, that is, through the cotton of Surat and Bombay, through tin (calin) from Malacca, and through opium from Bengal. The Dutch settled their balance with 1,300,000 piasters; the Swedes and the Danes together did so with one million.[1] From 1784 to 1788, in general France lost 6,968,000 livres tournois or 1,327,000 piasters in its trade with the East Indies.[2] These partial data are in close agreement with the general result that we obtained above for silver exports to China. III.440

It is more difficult to assign a value to the loss that Europe experiences in its relations with Asia as a whole through trade around the Cape of Good Hope. According to Mr. Playfair's studies,[3] in 1800 the part of this loss that results from trade with the British was 2,200,000 pounds sterling or 9,701,000 piasters. It is true that the same author estimates the value of exports from across Hindustan at thirty million piasters: but this vast country does not only benefit from trade with Europe, but also from trade with other parts of western Asia, specifically the neighboring islands. Although we recognize the extreme uncertainty of these trade balance calculations, these open accounts between one nation and another, we are forced to resort to them in order to obtain results that come close to the truth. The information that we have just presented suggests that gold and silver exports from Europe along the trade route around the Cape of Good Hope amount to over seventeen million piasters. In this calculation, we have taken into account the current state of trade in Madagascar, Moka, and Bassora [Basra?], as well as the auriferous copper from Japan provided by Dutch traders in Nagasaki,[4] and the treasures that the [British East] Indies Company employees bring back to England from Bengal. These treasures have been valued by ▾ Mr. Dundas at over four million piasters per year. III.441

If some part of China had the misfortune of being subjugated by some warlike people, who were also masters of Mexico, Peru, and the Philippine islands, this conquest would result in a lesser amount of precious metals flowing into the Americas and Europe than is generally believed. In the accounts of Macartney, Barrow, Mr. de Guignes, and other learned travelers, we see that gold and silver are no more common in China than in

1. De Guignes, [*Voyage à Pékin*,] vol. III, pp. 206, 207, 210, 215.

2. Arnould, *De la balance du commerce*, vol. III, no 13.

3. *Trade to and from the East Indies* ([in Playfair, *The Commercial and Political] Atlas*, plate [chart] III, p. 13).

4. Thunberg, *Voyage au Japon*, vol. II, p. 8.

most European countries. It is true that the annual state revenue is valued at 1,584,000,000[1] francs (301,714,000 piasters); but the bulk of this sum is paid in agricultural products and Chinese industrial products, and, according to Mr. Barrow,[2] the amount of silver returned to Peking in the form of specie is only thirty-six million ounces, which are valued at 52,914,000 III.442 piasters. The Chinese believe that large sums are sent annually to Mukden [now Shenyang], capital of the land of the Manchu Tartars: but this theory is not based on actual facts. A few mandarins possess tremendous riches. The first minister of the Emperor Tchienlong [Qianlong] was stripped of ten million taels, or 74,500,000 livres tournois in cash, which he had accumulated by harassing his neighbors[3]; nevertheless, the emperor often lacks silver. Whatever silver Europe loses in its balance of trade with China is distributed across a large population. A large amount of gold and silver is converted into wire and blades.[4] Cash is very slow to accumulate, and this effect has only been felt in the last twenty years, through the increase in the price of commodities.[5]

It remains for us to consider a third export route for precious metals from Europe to Asia, namely, through the Russian trade. The tables published by ▾ Count Romanzoff show us that between 1802 and 1805, the government of Irkutsk imported 2,035,900 rubles' worth of tea and 2,434,400 rubles' worth of cotton fabric from China. In general, Russia's balance of III.443 trade with China, Bukhara, the country of Khiva, and the Kirghiskaisaks, was in deficit for the Russian empire during the same period, at 4,216,000 rubles per year.[6] These data reveal that, if smuggled goods are valued at one-sixth of specie exports through the route of the Caspian Sea, the Caucasus, Orenburg, Tobolsk, Tomsky, Irkutsk, and Kyakhta, then these exports cannot amount to more than four million piasters.

By drawing on the sources that should be regarded as the most reliable, we have just discovered[7] that of the

1. According to Lord Macartney; 710 million francs according to Mr. de Guignes ([*Voyage à Pékin*,] vol. III, p. 102).

2. Barrow, *Travels in China* (French edition), vol. II, p. 198.

3. Barrow, [*Travels in China* (French edition),] vol. II, p. 173 [174].

4. Macartney, [*Voyage dans l'intérieur de la Chine*,] vol. IV, p. 286.

5. [Macartney, *Voyage dans l'intérieur de la Chine*,] vol. III, p. 105; vol. IV, p. 231.

6. [Rumiantsev,] *Tableau du commerce de l'empire de Russie*, translated by Mr. Pfeiffer, 1808, numbers nine and ten. ▾ Olivarius, *Le Nord littéraire*, 1799, no. 7, p. 202.

7. See the sketch of a diagram that presents the ebb and flow of metals from one continent to the other, on plate XIX of the *Mexican Atlas*.

43,500,000	piasters that Europe today receives annually from the Americas, roughly the following amounts flow to:
	4,000,000 to Asia, because of the Levant trade;
25,500,000	17,500,000 to Asia, by the route around the Cape of Good Hope;
	4,000,000 to Asia, by way of Kyatkha [Mongolia] and Tobolsk.
18,000,000	gold and silver from America remain in Europe.[1]

[1] These results are not applicable to the period in which this new edition of the *Political Essay* is appearing. The East Indies no longer take in precious metals from the Americas. See the considerations presented at the end of book IV.

From these eighteen million piasters or 94,500,000 livres tournois, one must subtract what is lost by being melted down and broken up into small jewels and trinkets, as well as what is used in plate, gold braid, and gilt. It has been noted at the Paris mint that the years between 1709 and 1759 saw an increase in plate at a proportion of one to seven. Prior to 1789, Mr. Necker believed it possible to value at three to four million piasters what was consumed annually in gold ornaments and silver objects, gold braid, and embroidered fabrics manufactured in France.[1] It is evident that a portion of these metals comes from the melting down of old plate and gold braid; the consumption by goldsmiths of silver in ingots is nevertheless also substantial;[2] and if one adds what is lost in transport and through the continual rubbing that occurs from daily use, it will be possible to estimate (as do Forbonnais and other writers on political economy) the amount of precious metals that are lost in Europe, or that are converted into plate and bars, at one-third of the total mass that is not absorbed by the Asian trade; in other words, six to seven million piasters per year. On the other hand, the mines of Europe and Siberia produce nearly four million piasters annually. According to these calculations, which are rough by nature, the increase in cash, gold, and silver in Europe appears to be only fifteen million piasters or 78,700,000 livres tournois. Those who have lived for a long time in the north and the east of Europe and who have closely followed

III.444

1. Necker, [*De l'administration des finances de la France*,] vol. III, p. 74. Peuchet, [*Statistique élémentaire de la France*,] p. 429.

2. Smith, [*Wealth of Nations*,] vol. II, pp. 69 and 73.

III.445 the inroads that civilization has made among the lowest classes of people in Poland, Norway, and Russia, will have no doubt that this accumulation of cash is a reality. Its effects can barely be felt, because capital in Europe as a whole increases only by 1 percent a year.

The overview that we have presented in this chapter on the current state of the mines of the New World and of Mexico in particular should inspire some concern that the total number of representative signs might rapidly increase as the mountain peoples of both the Americas emerge from the deep lethargy in which they have long slumbered. We would digress from the main purpose of this work if we discussed whether the societies' interests would actually be negatively impacted by this accumulation of cash. It suffices to observe here that the danger is not as great as it appears at first glance, because the amount of commodities that enter into trade and that must therefore be represented increases along with the number of representative signs. It is true that the price of wheat has tripled since the treasures of the New Continent started flowing into the old one. This rise, which was not felt through the middle of the sixteenth century, happened suddenly, between 1570 and 1595, when silver from Potosí, Porco, Tasco, Zacatecas, and Pachuca began to spread throughout the whole of Europe; but also, between that memorable period in III.446 the history of trade and 1636, the discovery of the mines of the Americas had its full impact on the reduction of the value of silver. The price of wheat has not actually risen to the current day, and if a few authors have made claims to the contrary, this is because they have confused the nominal value of coins with the true proportion that exists between silver and commodities.

Whatever theory one adopts about the future effects of the accumulation of representative signs, if one considers the peoples of New Spain in terms of their trade relations with Europe, one cannot deny that in the current state of things, the abundance of precious metals has a powerful influence on national prosperity. It is this abundance that puts the Americas in the position of paying in silver for foreign manufactured goods and of partaking in the delights of the most civilized nations of the Old Continent. Despite this real advantage, let us hope that the Mexicans, enlightened as to their own interests, remember that the only capital whose value grows over time are agricultural products, and that nominal wealth becomes illusory whenever a people does not possess these raw materials, which either serve for the subsistence of man or exercise his industry.

[So as not to interrupt the thread of the general discussions in chapter XI, and so as to not to add items belonging to entirely different periods into a discussion of such important matters for the study of political economy, I shall record III.447

here the information that I have obtained about the amounts of gold that Brazil and the Ural Mountains today place into trade; the relationship between the gold and silver materials converted into currency or into gold work and silverware; the changes that British trade with the East Indies has undergone; and the accumulation of precious metals in Europe. This information will also, I dare say, shed some light on the question of whether the growth of mining activity that has become apparent over the last few years in the Americas will soon have a decisive impact on the price of commodities in our midst.

On the Output of the Mines of Brazil, Compared with That of the Gold Mines of the Urals

Gold production has experienced great changes over the past half-century: almost unbeknownst to Europe, it has decreased in Brazil at such a fast rate that it is surprising that this decrease has not had a more noticeable impact on trade and on the relative value of precious metals. The recent discoveries of auriferous terrain in the Urals, on the border of Europe and Asia, hold promise that the losses inflicted to the west will be compensated; but these geographic changes in the sources of gold are nearly as important as the variations in abundance that the latter have experienced. Great Britain, the nation that uses the most gold in the civilized world, is in a very different mercantile position vis-à-vis Brazil and Russia. Although precious metals likely find their way to wherever need calls them, their local distribution is not the same in a country that already numbers fifty-four million inhabitants, and on an American coast that, through both legal and illegal exchanges, sends all its gold directly to the western parts of Europe. I shall not undertake a discussion of these political effects here; it suffices for me to examine the *numerical elements*, the knowledge of which is indispensable to those who wish to obtain correct results. III.448

We have seen above that at the beginning of the eighteenth century, shortly before the civil conflicts, Spanish America produced—according to the *quint* registers, therefore excluding the amount absorbed by smuggling—45,300 Castile marks or 10,400 kilograms in an average year, of which:

New Granada	4,700 kilograms
Chile	2,800
Mexico	1,600
Peru	800
Buenos Aires	500
	10,400 kilograms

This is six times larger than the output of Europe and Asiatic Russia, which, prior to 1817 (the year in which ▾ Mr. de Schlenew began mining the auriferous sands of the Urals), could be valued at only 1,600 to 1,700 kilograms.

In Brazil, the period of the gold washes' greatest wealth was from 1752 to 1761; at that time, the *quinto do ouro* in the Capitania of Minas Gerais in an average year totaled 104 arrobas. It was at this same time that the four great casas de fundições were founded in Vilarica, Sabará, São João del Rei, and Vila do Príncipe, two of which have been closed due to the deplorable state of Brazilian gold mining today. I shall combine here the data presented in a famous paper in the annals of the Parliament of Great Britain, the *Bullion Report*,[1] with those provided to me on the current state of affairs by the ▾ Baron Eschwege, former director of the gold and diamond mines of Minas Gerais. Here are the totals for the *quinto do ouro* in this same Capitania during this period:

III.449

1751	53 *arrobas*	34 *marcos*
1752	107	50
1753	118	29
1754	117	57
1755	114	57
1756	110	53
1757	88	7
1758	116	59
1759	97	69
1760	111	36
1761	102	33
1762	83	7
1763	99	55

1. *Bullion Report for the House of Commons*, Ap[pendix of Accounts], p. 30. Eschwege, *Journ[al] von Brasilien*, vol. I, p. 218. Guilherme barão de Eschwege, *Notícias [e reflexões] estad[ísticas] a respeito da província de Minas Gerais, estadísticas* 1825, pp. 10–13. I recall that one arroba (at sixty-four marcos) contains 294,912 grains (Portuguese weight) or, according to ▾ Mr. Franzini, 14.656 kilograms; for one arroba = thirty-two arrateis; one arratel = sixteen oneas; one onea = eight oitavas; one oitavas = three escrúpulos or seventy-two grãos. In the *Bullion Report*, one hundred arrobas of quinted gold are valued at 182,186 pounds sterling. The Brazilian government purchases one arroba of gold at a price of 12,000 cruzadas; but the intrinsic value of an arroba is 15,000 cruzadas.

1764	93	49
1765	85	49
1767	84	15
1771	80	54
1775	74	50
1777	70	2
1800	30	3
1810	26	0
1811	24	47
1813	20	39
1817	18	0
1818	9	2
1819	7	24
1820	2	30

These precise data suggest that the quinto was III.450

1752–1762 on average	104 *arrobas*
1763–1773	90
1774–1784	69
1785–1794	45
1810–1817	22
1818–1820	6

These figures are all related to the Capitania de Minas Gerais, where in 1821 the government purchased gold as it was removed from the washes, and despite this precaution, a total mass of only twenty-three arrobas could be obtained. After the most productive period, between 1752 and 1761, when the amount of registered *and quinted* gold fluctuated between 440 and 590 arrobas (between 6,424 and 8,614 kilograms), the washes have declined to such an extent that from 1785 to 1794, the registered output was no more than 225 arrobas (3,285 kilograms). As Mr. Eschwege writes to me, "Today gold mining is in such an appalling state that it has entirely ceased in the province of Saint Paul [São Paulo]. From 1788 to 1793, in an average year the province of

Goiás yielded only eight arrobas of *quint*. In the province of Mato Grosso, the washes are almost completely neglected, such that in 1824, the entire output of Brazilian gold,[1] including contraband, likely did not exceed 600,000 cruzadas (584 kilograms). In the time of the greatest prosperity, in the middle of the last century, this total output (including what escaped payment of the quint) likely amounted to twelve and a half million cruzadas (around 12,000 kilograms), or eight and a quarter million of our German thalers. It may well be that this estimate was too high, even for the period between 1752 and 1762."

III.451

These data suffice to rectify my mistaken impression, in the first edition of my work, regarding Brazilian gold output. I had shared this misapprehension with the majority of writers on political economy (Malte-Brun, *Géographie*, vol. V, p. 675; ▾ Lowe, *Present State of England*, 1822, p. 267; Caldcleugh, *Trav[els] in South America*, vol. I, p. 58; Héron de Villefosse, [*De la richesse minérale*,] p. 240) in suggesting, on the basis of an otherwise very informative paper by Mr. Correia de Serra, that in 1810 the quint was not twenty-six arrobas (or 379 kilograms) but 51,200 Portuguese ounces, or 1,465 kilograms. This quint presupposed an output of 7,300 kilograms. In the important work by Mr. Tooke (*On High and Low Price*, vol. II, p. 2), this output is still valued by Mr. Jacobs at 1,736,000 piasters in an average year (1810–1821), while the official documents in my possession suggest that the average quint of these ten years was only fifteen arrobas; in other words, there was *quinted output* of 1,095 kilograms (755,000 piasters). ▾ Mr. John Allen had already reminded the Committee of the Bullion Report (on the occasion of a few critical comments on Mr. Brogniart's overview of gold and silver production in the two continents) that the Brazilian washes had undergone a dire change of circumstances since 1794. In his *Histoire des plantes les plus remarquables du Brésil* (1824, pp. 9 and 23), Mr. Auguste de Saint-Hilaire describes this very abandonment of the gold mines: "The former miners are becoming farmers," he writes, "and while it is undeniable that there is more smuggling than there was before, it is also certain that gold extraction has decreased infinitely more than smuggling has increased." Here is the value of one mark of gold and silver in Portugal and Brazil since the thirteenth and fourteenth centuries:

1. ▾ Mr. Balbi says that prior to 1815 (at the beginning of the nineteenth century), the quint yielded no more than 270,000 cruzadas in Minas Gerais and 90,000 cruzadas in the rest of Brazil, for a total of 360,000 cruzadas. The number of individuals employed in the washes, which, in times of prosperity, was 80,000, had decreased to 16,000 (*Essai statistique sur le [royaume de] Portugal*, vol. I, p. 305). These 360,000 cruzadas are equivalent to a quint of twenty-four arrobas, both of these values being slightly lower than those we have obtained.

GOLD		SILVER	
Under king D[om] Sancho,		Under king D[om] Pedro I,	
1211	6,480rees	1367	945rees
D[om] Pedro I,		D[om] Fernando,	
1367	7,380	1383	900
D[om] João III,		D[om] João I,	
1557	30,000	1483	2,600
D[om] João IV,		D[om] Manoel,	
1656	55,000	1521	2,280
D[om] Pedro II,		D[om] João IV,	
	85,000	1656	3,600
D[om] João V,		D[om] Pedro II,	
	96,000	1706	5,600
		D[om] João V,	
		1750	5,600

III.452

This latter value has been maintained to the present day (Eschwege, *Journ[al] von Brasilien*, vol. II, p. 158). Research conducted at the mint of Lisbon in 1777 revealed that the amount of minted gold in circulation in Portugal and Brazil versus that of silver stood in a ratio of eight to one.

At the same time as the gold washes of Brazil experienced such a steep decline, new riches in gold were discovered, virtually on the border of Europe and Asia. From the middle of the eighteenth century to the beginning of the nineteenth, the gold mines of Berosov, near Yekaterinburg, were the only famous mines in the Urals. But in 1814, the councilor of state, Mr. Schlenew, revealed the existence of this immense deposit of auriferous and platiniferous alluvia, which, on the right bank of the Melkovka, yielded six pood [pud] twenty-six pounds of pure gold[1] and twenty pounds 82 solotnik of silver (from the washing of 1,602,184 pud of sands) from 1814 to 1816. This discovery was followed by others on the left bank of the Melkovka and near Cheremshanka. Today, the metalliferous deposit stretching from Verkhoturye to the Ural River—that is, from the foundry at Bogoslov to the Polkov mine on the Uy, over a length of 1,000 Wersts

III.453

1. One pud contains forty pounds; one pound contains ninety-six solotnik. One pud is equal to 16.38 kilograms.

(105 to an equatorial degree)—has already been discovered. It is especially between Nizhny Tagilskoy and Kuschtymskoi that one hundred pud of sands yield five solotnik of gold. In 1823, across the entire Urals chain, 20,686,000 pood [pud] of auriferous sands were submitted to the washes, which produced 112 pud twenty-three pounds or 1,845 kilograms of gold. The output for 1824 was valued at 286 pud or 4,700 kilograms of gold.[1] This is equal to the amount produced in the whole of New Granada before the war for independence, when the washes of Chocó were worked most intensely; it is twice as much as the gold from Chile. In the Urals, 11,500 workers, for the most part children, were put to work.[2] In a single day, gold *nuggets* of five, nine, and sixteen pounds were discovered at the washes of Slatust. As early as the year 1819, in the mines of Neiwin, gold panners from the Urals had begun separating out *grains of white gold* whose nature was unknown. These were sent to Yekaterinburg in 1822, but the assays were inconclusive, and the grains were not identified as platinum. This important discovery was only made in 1823, by ▼ Mr. Lubarsky, professor at the corps of mining III.454 students of Saint Petersburg. This scholar observed the presence not only of platinum but also of osmium and iridium. No palladium was isolated.[3]

The information that I have just provided about the mines of Brazil suggests that at the beginning of the nineteenth century, the Americas as a whole[4] placed into trade only 11,000 kilograms of quinted gold.

Exactly how much gold is mined today in Europe (Hungary, Transylvania, Gastein in Salzburg, Goslar in the Harz Mountains, Ädelfors, and

1. The most interesting observations about these auriferous and platiniferous sands come from ▼ Mr. Erdmann's scholarly studies. See the informative work that he has published under the title *Beiträge zur Kenntnis des Inneren von Russland*, 1826, vol. II, pp. 127, 136, 265, and 267. Compare also ▼ Gilbert, ["Eine literarische Notiz zu Aufsatz III und VII,"] *Annalen der Physik*, 1823, vol. XLV, p. 227. *Leipziger Zeit[ung]*, 1824, number 284. Schtscheglew, *Neueste Entdeckungen in der Physik*, 1825.

2. According to Hermann, prior to Mr. de Schlenew's discovery, the gold mines across the Ural Mountains produced twenty pud or 328 kilograms of gold annually. Since the output of gold from the mines of Kolyvan and Nertchyusk is valued at twenty-two pud thirty-four pounds, until 1810 the average output of European and Asiatic Russia was estimated at 40 pud or 656 kilograms of gold. See Hassel, *[Vollständige und neueste] Erdb[eschreibung] des Russischen Reichs in Asien*, 1821, pp. 290 and 296.

3. Compare also ▼ Laugier ["Examen du platine trouvé en Russie"] in *Annales de chimie et de physique*, vol. XXIX, p. 289.

4. Even in 1825, Mr. Caldcleugh values the gold of Brazil at 900,000 pounds sterling. According to ▼ Mr. Schmidtmeyer (*Travels [into Chile]*, p. 78), at the beginning of this century Brazilian gold output was 980,870 pounds sterling. This latter estimate, according to information provided by the director of the mines of Brazil, Mr. Eschwege, is around thirteen times too high.

Falun in Sweden) and in Asiatic Russia? Mr. Héron de Villefosse found the following amounts of gold for 1807–1809:

2,600	mark for Hungary
2,500	for Transylvania
118	for Salzburg
10	for the Harz Mountains
5	for Sweden
5,233 marks	

But in his *Statistique [du Royaume] de la Hongrie*, printed in Buda, Mr. Schwartner values the average output (1780–1788) at only 1,250 mark of gold and 58,500 mark of silver for Lower Hungary and at 350 mark of gold and 13,500 mark of silver for Upper Hungary, which would suggest a total of only 1,600 marks of gold and 72,000 marks of silver for the whole of Hungary. Mr. Beudant's discussions[1] led us to believe that since 1788, mining outputs have often exceeded the totals suggested by Mr. Schwartner, but that in Hungary and Transylvania they are far from reaching the sums obtained by Mr. Héron de Villefosse in his excellent and useful work on mineral riches. By including 4,200 marks of gold for Hungary, Transylvania, and Salzburg; by valuing production in the Russian Empire,[2] before the discovery of the auriferous terrains at the foot of the Urals, at forty-one pud or 672 kilograms; and by adding only 280 pud or 4,592 kilograms for the output of the washes in the Urals, one finds the total gold mining of Europe and Asiatic Russia to be 6,272 kilograms, which is three to four times greater than in 1810, when the first edition of my work was published. Such has been, even on the Old Continent, the progress of industry applied to the search for precious metals! Today, the whole of the Russian Empire annually produces, without a doubt, over 5,300 kilograms of gold (at a value of 3,360,200 piasters), that is, within one-quarter of the amount produced by the washes of Brazil in the middle of the last century, when they were worked most intensively.

III.455

Let us now examine the necessary modifications to the discussions in chapter XI (pp. 400–423 [pp. II.236ff] in this edition) as a result of (1) the

1. [Beudant,] *Voyage minéralogique [et géologique] en Hongrie*, 1822, vol. I, p. 410; vol. III, p. 122. Malte-Brun, *Précis de la géographie universelle*, vol. VI, p. 308.

2. From the seventeenth century until 1810, that empire placed into trade a total of 1,727 pud or 28,322 kilograms of gold.

new knowledge that we have obtained regarding the decrease in gold output in Brazil and (2) the discovery of the auriferous terrains of the Urals. First, we find that, from 1800 to 1810, the *quinted gold* of Brazil in an average year was five by twenty-eight arrobas = 140 arrobas or 2,044 kilograms, and that if one values illegal mining in this period at four-fifths (as does Mr. de Eschwege), one obtains a figure of nearly 3,700 kilograms for the true production of Brazil.

III.456 These data suggest that *gold output at the beginning of the nineteenth century* was

10,400 kilograms of gold	in Spanish America
3,700	in Brazil
1,700	in Europe and Asiatic Russia
15,800 kilograms	

In the table charting gold production at the beginning of the nineteenth century in the Americas, p. 633 [not in this edition], as well as in the table for Europe, northern Asia, and the Americas, p. 634 [not in this edition] (quarto edition), the total error was slightly over *one-fifth*; but since gold is only a very small part of the forty-three and a half million piasters assumed to be the total gold and silver output in the Portuguese and Spanish colonies at the beginning of the nineteenth century, this value of forty-three and a half million will be reduced by less than *one-twentieth. The table published* on p. 634 (quarto edition) *should present a sum total of* 247,800,000 *francs, instead of* 259,202,888, *or nearly* one twenty-third *fewer.* Since, during my travels on the Río Negro and the banks of the Amazon, I only visited the borders of the Portuguese colonies and did not venture into Brazil itself, I was unable to bring back any precise data on Brazilian gold. When the first edition of my *Political Essay* was published, I was obliged to adhere to the results of studies by Mr. Héron de Villefosse, Correia de Serra, and all of the authors who had theretofore written. One can imagine that errors in the estimate of gold and silver outputs in both the Americas and the Old Continent which amount to only one-twentieth or one twenty-third are more than masked by fluctuations from one year to the next in the wealth of the mines and in fraudulent mining activity. What is more, once a liberated Spanish America, enjoying the delights of domestic peace, returns, through increased mining output, to the same point where it was before the war of independence, the discovery

III.457 of the auriferous terrains of the Urals will bring the gold and silver output of Europe, northern Asia, and the Americas back to within one-fortieth of the sum indicated in the table on p. 634 [not in this edition]. In terms of gold, we will have (1) for Spanish America, 10,400 kilograms once again; (2) for Brazil, not 3,700 kilograms, a sum that was more typical of the years 1800–1810, but rather, subsequent to 1822, according to Mr. de Eschwege, a maximum of 600,000 cruzadas or 584 kilograms; (3) for Russia, Hungary, and Transylvania, 6,272 kilograms, if we include 286 pud for the Urals (as in 1824); for a combined total of 17,256 kilograms, or between one-eighth and one-ninth less than what we had assumed in the first edition. The total value of the gold and the silver of Europe, the Americas, and northern Asia placed annually into trade by European peoples will be, for the period when the mining operations of Spanish America once again equal their 1805 wealth, 253 million francs. The table (p. 634 [not in this edition]) shows 259,202,888 francs.

Since the first edition of the *Political Essay*, a thoroughly praiseworthy author, Mr. John Crawfurd,[1] has shed a great deal of light on gold production in the Indian Archipelago. He values it at over 4,700 kilograms per year (at a value of 2,980,000 piasters), that is:

Montradak (Borneo)	88,362 British ounces
Sumatra	35,530
The rest of the Indian Archipelago	30,973
	154,865 ounces

If we wished to join Mr. Crawfurd in valuing the gold powder from the central African trade at 14,000 kilograms (which nevertheless seems to me to be an extremely high estimate), we would obtain the following results for gold from the known world:

Spanish America and *Brazil*, assuming that gold and auriferous silver mining in Spanish America has already reached its pre-war normal state of

II.458

Independence	11,000 kilograms
Europe and Asiatic Russia	6,272
Indian Archipelago	4,700
Africa?.	14,000
	35,972 kilograms

1. [Crawfurd,] *Hist[ory] of the Indian Archipelago*, 1820, vol. III, pp. 470 and 487.

In 1805, the amount of silver that was extracted annually from the mines[1] was 870,000 kilograms, which would place the ratio of gold output to silver output, once the mines of the Americas have returned to their normal state, at 1:242; but this is quite far off from the actual ratio between the amounts of gold and silver *that flow into Europe*, either from the Americas or from Asia. This ratio, which I formerly assumed[2] to be one to forty, appears to me today—if one considers the estimate of African gold to be quite excessive and if one recalls that only a very small portion of the gold from the Indian Archipelago reaches the markets of Europe—to be one to forty-seven, perhaps even one to fifty. Gold is therefore much less expensive in trade than it should be according to its actual scarcity, probably because, with the exception of England and Portugal, it is less in demand than silver, not in the form of gold work but as a means of exchange. To this factor one must add others that resulted in a very old accumulation of gold and that cannot be discussed here.

I shall presently address the question of how much registered gold and silver was removed from the mines of Spanish America and Brazil from III.459 1492 to 1803. Without detracting from Raynal's estimate, which covers the interval between the discovery of the mines of Brazil and 1755, we find, using the account of the state of the quint that Mr. de Eschwege published, that from 1756 to 1777 Brazil produced around 138,000 kilograms of *registered gold*, and around 90,000 kilograms from 1777 to 1803; the total for these fifty-seven years is 228,000 kilograms or 144 and a half million piasters. This suggests that according to the *quint registers*, the gold from the Portuguese colonies in the Americas totaled not 684 and a half million piasters but rather one-tenth less; that is, 624 million piasters. I think, nevertheless, that this decrease should not at all change the definitive result of 5,706,700,000 piasters,[3] since, according to the unanimous accounts of all those persons who have resided in Brazil and who are familiar with the his-

1. See above, p. 400 [p. II.236 in this edition].

2. See above, p. 401 [p. II.237 in this edition]. The transformations in the ratio of gold to silver were very judiciously discussed by ▼ Mr. Mushet and Mr. Charles Grant in the *Bullion Report*, pp. 209, 233, 865, and Appendix [of Accounts], p. 21. The same work sets (p. 38) the gold output of the Americas as a whole in the mid-eighteenth century at 67,095 Castilian marks, and the silver output at 2,000,000 Castilian marks; which would place the ratio of their abundance not at one to sixty-two, but one to thirty. See below, p. 463 [p. II.276 in this edition].

3. See above, p. 416 [p. II.244 in this edition].

tory of trade in that country, my estimate of fraudulent mining (one-quarter of quinted Brazilian gold) is far too low, even for past centuries when communication with the coasts and therefore with foreign nations was very difficult.

I shall observe at the end of this note that the three tables on pp. 393, 398, and 400 [pp. II: 231, 234, and 236 in this edition] show gold and fine silver; but that the two tables on pp. 300 and 301 [II.170–71 in this edition], drawn up at the mint of Mexico City, present Castilian mark of *piaster silver* or silver of a very similar title; for, according to these tables, silver minting in 1796, 1797, and 1799 was (in Castilian marks) 2,854,072; 2,818,248; and 2,473,542, respectively, while the lists printed in Mexico City give the following production amounts for these same three years: 24,346,772; 24,041,180; and 21,096,031 piasters. In the calculations shown on pp. 178, 180, 367, 383, 417, and 423 [pp. II: 99, 100, 216, 225, 244, and 249 in this edition], following local custom, I converted the piasters into Castilian marks by dividing by eight and a half, with the result that I also obtained only marks of silver for piasters, or 0.903. *The combined mass of fine silver removed from the mines of the Americas over the past three centuries would form a sphere with a diameter of* III.460 twenty and forty-seven hundredths *meters*. The Castilian mark is equal to 0.229881 kilograms. From each Castilian mark are coined eight and a half piasters, and since this mark corresponds to 229.881 grams in the new French weight measurement, the weight of one piaster is 27.045 grams. Since the title of one piaster should be 10 den[arii] 20 gr[ains], or 0.903, it is worth—if considered perfect in terms of both weight and title—five francs forty-three centimes. One Castilian mark of fine gold is worth 145 and eighty-five hundredths piasters; one mark of fine silver is worth nine and four-tenths piasters. We have valued one kilogram of fine gold above at 3,444 francs 44.444 centimes, and one kilogram of fine silver at 222 francs 22.222 centimes. Given that, in the mines and mints of the Americas, people do not always count in mark of gold and silver of the same title, one is at something of loss whenever one works with papers in which the title is not specified. The margin of error cannot, however, exceed one-tenth, an amount that seems less noticeable when one takes several years' worth of averages and when one reflects on the mass of precious metals on which the *quint* is not paid.

On the Relative Amounts of Precious Metals Minted and Converted into Gold Work and Silverware Objects

"I have touched above (p. 444 [p. II.261 in this edition]) on the important question of how much gold and silver is extracted from the mines of both continents and converted annually by the Europeans into gold work objects. Since old plate is melted down repeatedly, and since the bulk of new plate is merely the product of a change in form, one can form only a vague sense of the amount of precious metals that is added each year to the total amount of worked gold and silver over the past few centuries. Mr. Necker thinks that around 1770, this increase was ten million per year in France alone.[1] Mr. Peuchet suggests that at the time of the revolution, the annual III.461 manufacturing of gold work objects, gold braid, and jewelry was twenty million. Here are the most recent data.

"In 1809, production in France was:

	Gold Plate	Silver Plate
In the departments	1,608 kilograms	21,326 kilograms
In Paris	1,026	40,541
	2,634	61,867

"In 1810, 1,213 kilograms of gold and 47,403 kilograms of silver were manufactured in Paris alone. These figures represent only the materials on which duties were received by the government: one can expect that despite the activity of and surveillance over employees, there is always either one-third or one-quarter that is not registered at the mint. These considerations suggest that, despite the obstacles placed on exports by the maritime war, annual French gold work manufacturing amounts to

In gold	3,300 kilograms, or	11,365,000 francs
In silver	80,000	17,760,000
	Total value	29,125,000

"It would be interesting to have similar data for Britain, Germany, Russia, and Italy. In the absence of these data, we assume that gold work production is in a ratio of one to four to that of Europe as a whole, and we find that the total value of European manufacturing must equal 120 million francs per year.

1. See also Gerboux, *Sur la démonétisation de l'or*, p. 70.

"I shall not discuss what portion of these precious metals does not come from the melting down of old plate; but I think that one can deduce from the data we have just presented that the amount of gold and silver removed from the mines of Europe and Siberia (chapter XI, p. 400 [p. II.236 in this edition]) is quite far from replacing the precious metals Europe used in the production of plate, gold braid, and gilt; that are wasted by extreme splitting; or that are accidentally lost."

The preceding reflections are found in the Supplement to the first edition of this *Political Essay*. I have since become concerned with three other problems for which it would be more prudent, given the lack of sufficiently robust materials, not to seek a solution. Here is an overview of these problems: III.462

A. By what amount is the annual output of the mines greater—due to both the actual loss of metal and the burgeoning civilization of peoples—than what is siphoned annually into gold work and silverware production in Europe? Mr. Lowe[1] thinks that two-thirds of the mines' output is absorbed by gold work and silverware production. ▼ Mr. Sismondi[2] claims that "the vast majority of the mines' output is funneled into minting,[3] that is, into the increase in cash levels." Nevertheless, this philosophical writer expressed some doubt himself when he adds, "I do not know whether the work of the mines counterbalances the amount of precious metals that are lost through daily use or are siphoned off into plate."

B. What is the value of the gold and silver used annually in European gold work and silverware production? What is the proportion between what is melted down and what is added annually?

C. What is the stock in trade of goldsmithing in Europe and what is the ratio of this stock to existing plate?

By no means do I believe that these problems are capable of being resolved at this point; but in order to inspire new studies, I shall gather a few facts and likely suppositions here.

When speaking of the output of mines, one must always begin with the period that immediately preceded the political unrest in Spanish America. This output has since decreased: first by two-thirds, then by half, but a few years of peace has sufficed for it to return to 1809 levels. Both the results that I have published in chapter XI and the rectifications that I have just added to gold from Brazil and the washes of the Urals seem to prove III.463

1. [Lowe,] *The Present State of England*, 1822, vol. II, p. 263; and Appendix, p. 89.
2. [Sismondi, *Nouveaux principes d]'économie politique*, vol. II, p. 55.
3. [Sismondi, *Nouveaux principes d'économie politique*, vol. II,] p. 60.

that Europe will receive from the American, European, and Siberian mines (once the silver mines of Spanish America are again at the level of prosperity where I saw them):

870,000 kilograms of silver (value: 193 million francs).
17,300 kilograms of gold[1] (value: 59 and a half million francs).

According to this estimate, which can be regarded as sufficiently precise, the relative abundance of these metals is in a ratio of one to fifty and not in a ratio of one to sixty-two or one to sixty-eight, as the *Committee* of the *Bullion Report* assumes.[2] The balance of trade between the two metals depends not only on their relative production rates but also on demand and on the unequal rates of consumption between them. There is a greater *demand* for silver than for gold as a sign of exchange and as a material in plate manufacturing.

It is believed that the level of cash in ancien régime France (2,400,000,000 francs) is in a ratio of around one to four (more precisely 1:3.6) to that of Europe as a whole (8,600,000,000 francs). In a report to the Emperor Napoleon,[3] ▼ the Duc de Gaëte estimated the level of cash in France in 1813 at 3,479,156,000 francs. At that time (under the Empire), France had a population of forty-four III.464 million inhabitants. The interesting *Recherches statistiques sur la Ville de Paris*, published in 1823 by the Count de Chabrol [de Volvic],[4] reveal

(1) That the raw materials employed in a regular year in gold work and silverware objects and in jewelry manufactured in Paris are:

in gold	5,332,388 francs' worth (nearly 1,550 kilograms)
in silver	9,220,500 francs' worth (nearly 41,500 kilograms).

1. That is, 10,400 kilograms from Spanish America; 600 kilograms from Brazil; 6,300 kilograms from Europe and Asiatic Russia. See above, pp. 455 and 457 [pp. II: 269 and 271 in this edition].

2. [*Bullion Report*, Appendix of] Accounts, p. 43. In China, the balance of trade between gold and silver was one to ten until 1730; at present, it is one to sixteen ([*Bullion Report*,] Minutes, p. 234).

3. See the discussion above (pp. 425 and 443 [pp. II: 250 and 261 in this edition]) on Mr. Necker's results. These attempts to assign a value to cash levels in Europe are very old. See the extract from the interesting work by ▼ Gregory King (1688) in Tooke, *[Thoughts and Details] on [the] High and Low Price[s]* (2nd edition), Appendix to part I, p. 2. King assumes a cash level of 225 million pounds sterling in Europe at the end of the seventeenth century, of which seventy-seven million pounds sterling are in plate.

4. [Chabrol de Volvic, *Recherches statistiques sur la Ville de Paris*], Table no. 85 [125, p. 33].

Another table gives a twenty-three-year average of 4,459 marks of gold and 114,116 marks of silver, or one-third less of each metal, which, according to a note added to the table, appears to derive from the fact that the smallest numbers are taken from the registers of the *Bureau de Garantie*, without any consideration for fraud.

(2) That manufacturing in Paris is twice that of the departments. In this supposition, gold work and silver plate manufacturing in France has been valued, by combining the approximate data that we have just gathered, at a minimum of 2,300 kilograms of gold and 62,300 kilograms of silver.[1]

One would be tempted to include for Europe as a whole,

Gold	9,200 kilograms	31,684,800 francs
Silver	250,000	55,500,000
		87,184,800

(3) Of the gold work and silver plate manufactured in Paris, two-thirds is used in the departments. Exports to Germany, Switzerland, and the rest of Europe represent only one-tenth of the total production in Paris.

(4) The value of the *inventory* of all manufactured and unsold items in the whole of France in 1819 was sixty-four million francs, that is: in Paris, twenty million francs; in the departments, forty-four million francs; or, for Paris and the departments combined, 7,000 kilograms of gold; 218,000 kilograms of silver. This figure likely represents the stock in trade, the total amount that remains in store because more is manufactured each year than can be sold. This suggests that the stock in trade of Europe as a whole is perhaps 28,000 kilograms of gold and 872,000 kilograms of silver: in other words, if all the shops were stripped at once, *three years'* worth of gold and *one year's* worth of silver from the Americas would be necessary to replenish their stocks. It should be noted that estimates of annual manufacturing in Paris and the stock in trade suggest that the weight of gold currently in use is in a ratio of one twenty-seventh to one-thirtieth to that of silver. Since the mines of the Americas, Siberia, and Europe together produce forty-seven to fifty times more silver than gold, and since gold work and silverware manufacturing consumes up to one-thirtieth of gold, it tends, through its greater demand for gold, to inflate the price of this metal. The relative use of the

III.465

1. In the first edition of this work, I had arrived at a total of 3,300 kilograms of gold and 80,000 kilograms of silver.

two metals in plate cannot, therefore, serve to explain why the balance of trade between gold and silver is only one to sixteen.

Should we assume, as do several scholars better versed than I am in areas of political economy, that precious metals are much less prevalent in Europe in the form of cash than in the form of plate? Here are the data that I have heretofore been able to gather: the level of cash in Europe is at least 8,600,000,000 francs, while the total value of the stock in trade cannot be greater than four by sixty-four million or 256 million francs. For the claim we have just cited to be correct, it would thus be necessary for the amount of plate that is in the hands of private individuals across Europe to be thirty-four times greater than the stock in trade. Since, according to the inventory, this stock amounted to sixty-four million francs in France, the total value of plate in France would be over two billion francs, which translates into sixty-three francs per person! I must admit that there

III.466

is something strange about this result when one reflects on the number of individuals who certainly have in their possession a few francs but no plate. Is this compensated for by the number of rich families and by the multitude of watches and place settings, 400,000 of which are manufactured in Paris alone every year? The total value of plate in France (two million francs) would still be one-fifth below that of the cash that is assumed to be in circulation in the kingdom; but as we shall soon see, other considerations lead one to believe that plate manufacturing annually absorbs less than one-fifth of the output of the mines.

If current cash levels are greater than the total value of plate, then Mr. Lowe's claim, namely, that plate annually absorbs more of the mines' output than does minting, is unacceptable. On the basis of several conjectures about families' wealth and their annual gold work and silverware needs, the estimable author of *The Present State of England* finds (1) that 144 million francs' worth of the mines' output is funneled annually into the manufacturing and repair of plate in Europe; (2) that these 144 million francs represent three-quarters of total annual plate manufacturing.

Let us compare this claim with what we have identified for Europe as a whole, on the basis of a comparison with manufacturing in France. We have suggested that Europe likely uses 31,700,000 francs' worth of gold plate and 55,500,000 francs' worth of silver plate.

According to the Count de Chabrol [de Volvic]'s tables, the ratio of labor costs to material costs in gold plate is seven to five; in silver plate, it is five to nine. One therefore obtains a total value of over 144 million francs for annual plate manufacturing in Europe.

Using a completely different approach, Mr. Lowe obtains a total of eight million pounds sterling for Europe and the United States, and he thinks that by subtracting the value of old plate from the total value of manufacturing, one is left with six million pounds sterling in gold and silver that gold- and silversmithing inject annually into Europe and the United States from the mass of precious metals that are removed from the mines. These 144 million francs III.467 would represent not two-thirds but rather fifty-seven hundredths of the total value of the annual output of the mines of both continents. But if one distinguishes, as we have done above, between the actual mass of gold and silver used in gold and silver ornaments and the commercial value of these objects, then the result, even in Mr. Lowe's supposition, is that gold- and silversmithing annually consumes an amount of precious metals equal to one-third of the mines' output. Necker and Gerboux believed that gold- and silversmithing in France in 1789 was (in terms of the value of the precious metals) twenty million francs, of which ten million francs came from newly added gold and silver. We have seen above that even today France consumes approximately 2,300 kilograms of gold and 62,300 kilograms of silver in plate manufacturing, with an intrinsic value of 21,750,000 francs. If Necker's estimates are correct, the tables published by the Count de Chabrol would therefore prove that the state of gold- and silversmithing has not changed significantly. For Paris, these tables show the following: value of the metals: 14,552,000 francs; value of the manufactured items (including labor costs): 27,400,000 francs.

If one followed Necker's assumption that half of the precious metals used in gold- and silversmithing are *newly added*, one would obtain a total for Europe as a whole of forty-four million francs deducted from the output of the mines, which would appear to prove that barely one-fifth or one-sixth of the mines' output is absorbed annually into the production and repair of plate.

The table of the amount of gold and silver minted in Mexico City from 1690 to 1800 that I have drawn above (p. 300 [p. II.170 in this edition]) is an official document that has not appeared in any work prior to mine. I find it all the more important in that this table, together with that of Potosí (1559–1789) encompasses half of the precious metals that the Spanish colonies have shipped to Europe from the discovery of the Americas to the present day. In the *Balanza del comercio de Vera-Cruz*, printed in 1811, it is stated (p. 23) that III.468 from 1590 to 1809, the *Real Casa de moneda* minted a total of 1,523,005,095 piasters in gold and silver. This is 23,569,197 piasters more than in the unpublished table (1690–1800) that was provided to me from the archives and to which I have added the years 1800–1809. Numerical errors can easily occur when such large sums are added. I have already remarked elsewhere that the combined output for 110 years was 1,294,918,514 piasters, and not 1,298,217,472 piasters, as the official document indicates. Minting for the year 1804, which, according to the *Notes on Mexico*, p. 65, had been 27,080,000 piasters, was actually 3,633 marks or 494,210 piasters in gold, and 2,756,657 marks or 23,512,079 piasters in silver; total, 24,007,789 piasters.

The following remark is added to the table containing the marks of silver minted from 1690 to 1800 at the mint of Mexico City: "The 94,645,996 marks of silver minted from 1690 to 1777 included 1,771 marks coming from Peru: 6,964 marks of *moneda de martillo* and 851,285 marks of old coins removed from circulation in 1772 and 1776 and converted into piastras fuertes, such that 1690 to 1778, there were, strictly speaking, only 93,785,975 marks of silver coming from the mines of Mexico. This reduction was not incorporated into the sum of 147,252,008 marks of silver recorded above (p. 305); for 1690–1799, it shows a total volume of 146,391,988 marks."

On the Activity of the Mints of France Compared to That of the Mint of Mexico City

If the sixteen mints of France together produce[1] less than the mint of Mexico City, one must not seek the reason for this in the lack of raw materials. In III.469 Paris, each press is capable of striking 2,500 forty-, twenty-, two-, and one-franc coins per hour, or 3,000 half-franc coins and 2,000 five-franc coins.

In April 1796, the mint of Mexico produced a total of 2,922,185 piasters, a sum that rose to 3,065,000 piasters in December 1792.

This amount was partly produced in gold and partly in silver. Valuing the piaster at five francs forty-three centimes, the 3,065,000 [piasters] would equal, in French currency,	16,642,950 francs
Over thirteen days in January 1811, gold and silver manufacturing in Paris rose to 7,999,454 francs, which would amount, for twenty-six days, to	15,992,908 francs
Twelve mints in France coin—if the materials are deposited with precision—1,000,000 francs in silver, which, in twenty-six days, amounts to	26,000,000 francs

One notes that gold is not included in this latter estimate. If it actually occurred, gold minting would yield a much greater sum than the twenty-six million silver francs.

1. See below, vol. IV, chap. XII, p. 24 [p. II.302 in this edition].

In his work on financial administration, Mr. Necker records the amount of gold and silver minted from 1726 to 1780. We shall give here a precise report of the overall manufacturing of all the mints of France from 1726 to 1809.

Gold minting from 1726 to 1785 was 986,643,888 livres tournois. Over two-thirds of this gold was melted down in the subsequent nine years, for gold manufacturing between 1785 to 1794 totaled 751,281,504 francs.

Silver minting from 1726 to 1794 was 2,072,022,441 livres tournois.

The total value of the various gold, silver, billon, copper coins, and bells manufactured in all the French mints from 1726 to 1794 was 3,849,026,184 livres tournois.

From 1795 to 1802 a total of 106,237,255 francs' worth of five-franc pieces of the *Hercule et la Liberté* variety were coined. III.470

Gold minting from 1802 to 1809 equaled 173,219,700 francs, and silver minting equaled 259,454,874 francs, or, in an average year over the last eight years, over fifty-four million francs. These data suggest that in the space of eighty-three years, from 1726 to 1809, the total value of gold, silver, and copper manufacturing in France was 4,410,396,000 francs.

According to the official information published by ▾ Mr. Tarbé [des Sablons], former division head in the ministry of manufacturing and trade, the following amounts of new coins were minted in France through December 31, 1825 [see table on the following page].

This sum is more or less equal to the total amount of cash in circulation in 1789. Indeed, manufacturing of gold specie (of the title set by the edict of January 1726) through 1785 totaled 967,407,923 francs. Since these were ordered to be melted down by the declaration of October 30, 1785, the value of the new louis that were minted was 738,157,152 francs.

Silver specie, at their 1726 value	1,956,402,112
Combined	2,694,559,264
Since 1803, there were used for the production of new coins,	
in 1785 louis	134,162,144 francs
In écus of 3 and 6 francs	753,443,289
In coins of 24, 12, and 6 sous	4,988,218
Combined	892,593,651

Gold, before the restoration	528,024,440 francs	933,897,200
under Louis XVIII	389,333,060	
under Charles X	16,539,700	
Silver, under the Republic	106,237,255	1,638,812,173
under the Empire	887,830,055	
under Louis XVIII	614,836,109	
under Charles X	29,908,754	
Coins of 30 and 15 sous		27,278,019
Billon, 10-centime coins before the restoration		3,286,932
	All together	2,603,274,334

To these final sums, calculated not on the basis of the nominal value of the melted-down coins but rather on the exchange rate with no deductions, must be added both the total amount of minting fees and the weight that the coins lost through extended circulation. The remainder of the metals used in minting comes from foreign specie and from ingots and material brought to the mint. III.471

Based on the difference between the amount of melted-down coins and the amount of minted coins, the following amounts can be assumed to remain in circulation:

In the old louis of 1726	231,250,771 francs
In the louis of 1785	603,994,608
In old silver coins	1,197,970,905
Combined	2,033,215,984

excluding the discrepancy that occurred when melted-down coins were purchased at prices lower than their nominal value.[1]

A large part of this surplus was presumably melted down either for the minting of foreign currencies or for gold- and silversmithing both in France and in other foreign countries where emigration contributed to the scattering of these coins.

On the Changes in the Accumulation of Precious Metals in Europe. III.472

The development of the manufacturing industry in Europe and the ever-growing need for Western products in several parts of Asia have brought about important changes in trade with India and China. The loss[2] in pre-

1. The gold coins struck in London from 1797 to 1810 had a total value of 8,960,000 pounds sterling (*Bullion Report*, [Appendix of] Acc[ounts], p. 27). From 1819 to 1825, gold and silver minting in Britain was 17,000,000 pounds sterling. In general, total minting from 1760 to 1819 in London is valued at 744,501,600 pounds sterling in gold and 5,092,300 pounds sterling in silver. (See [Powell, *Statist[ical] Illustr[ations]*, p. 47, and [Great Britain,] *Reports by the Lords Committees Appointed [a Secret Committee] to Inquire into the State of the Bank of England*, 1819, p. 370.) "The circulating medium of England as far as it consists of notes of the bank of England or of the country bank-notes from 1810–1818 was from forty-two to forty-eight millions of pounds sterling."

2. Mr. Félix de Sainte-Croix, a traveler who resided for a long time in India, China, and the Philippine islands, and whose active curiosity has seized on everything related to both the manufacturing industry and European trade, thinks that the sums poured into Indies by

cious metals that the European peoples have experienced through the various routes of the Levant, Asiatic Russia, Hindustan [Indostan], China, and III.473 the Indian archipelago, has considerably decreased since the publication of the first edition of this work; and it is a widely shared theory today that Great Britain floods Europe with gold and silver from the Indian peninsula. I shall record here some numerical data that will serve to rectify—or rather

5,200,000	piasters, from the	British trade		
	2,000,000		Anglo-American trade	
	600,000		Spanish trade	
	400,000		Danish trade	
	8,200,000			
Europeans brought to China,				
	In	1804		6,117,600 piasters
		1805		5,293,000
		1806		3,384,998
Europe lost, according to Mr. de Sainte-Croix,				
	By way of Canton and Macao			2,500,000 piasters
	By way of Emoui [Amoy]			800,000
	By way of Cochinchina			500,000
				3,800,000

the various trading nations and converted into *rupees* still amounted, in an average year from 1804 to 1806, to eight to nine million piasters, which includes

If one adds to this sum the eight to nine million piasters converted into *rupees* in India, as well as European and Anglo-American trade with Japan, the great Archipelago of Asia, Persia, Muscat, Moka, Mozambique, and Madagascar, one finds a cash loss for this period of sixteen to seventeen million piasters. On the cash loss brought about by the India and China trade during the years 1788–1809, see the tables of the *Bouillon Report*, 1810, *Accounts*, p. 15.

to make applicable to the present day—the discussions presented in chapter XI, pp. 435–445 [pp. II.256ff in this edition].

Here, first, is the state of the Asian trade[1] in 1824 and 1825:

YEARS ENDING ON JANUARY 5	FOREIGN AND COLONIAL OFFICIAL VALUE	MERCHANDISE BRITISH AND IRISH OFFICIAL VALUE	OFFICIAL VALUE	TOTAL EXPORTS DECLARED VALUE
In the East Indies and in China, not including the Cape of Good Hope				
	liv. st.	liv. st.	liv. st.	liv. st.
1824	604,047	3,751,391	3,753,469	4,357,516
1825	710,575	3,684,305	3,490,325	4,200,900
At the Cape of Good Hope				
1824	59,661	375,663	334,967	394,628
1825	30,966	245,455	245,054	276,020
In the British East Indies				
1824	285,247	4,600,665	3,678,120	3,963,367
1825	324,374	4,843,560	3,827,489	4,151,863
Reexports from the British West Indies abroad (declared value)				
1824	1,519,350			
1825	1,014,152			liv. st.
Average of the value of the exports to the East Indies and China, including the Cape of Good Hope[1]				4,614,532
Average of the exports of the British West Indies, minus that of reexports				2,790,864
Balance in favor of the East Indies trade				1,823,668

[1] *Asiati[ck] Journ. [sic: Researches]*, 1826, p. 483. Compare Staunton, *Miscellaneous Notices relating to China*, 1822, pp. 126, 191, 354, 384. [Young, *An Inquiry into the Expediency of Applying the Principles of*] *Colonial Policy [to the government] of India*, p. 94.

1. *Asiat[ic] Journ[al]*, 1826, p. 483. Compare with Staunton, *Miscellaneous Notices relating to China*, 1822, pp. 126, 191, 354, and 384. [Young, *An Inquiry into the Expediency*

Tea consumption is certainly increasing on the continent of Europe, but it is increasing less rapidly than one might believe. In an average year from 1807 to 1812, Great Britain provided the other nations with over 3,790,000 pounds, and in an average year from 1822 to 1826, it reexported a total of 4,118,000 pounds. Thus, consumption outside of Great Britain increased by only one-sixth in the intervening twenty years [see table below].

III.475 It is especially since 1809 that exports in cash have more or less completely ceased in the [British East] India Company trade with the ports of Calcutta, Bombay, and Madras. In recent years, the value of the gold and silver that have flowed annually from India into Britain through the trade balance has been two million pounds sterling. In addition, one and a half million pounds sterling has been brought back by persons who have resided for a long time in India[1] or who have been sent to Britain for the children's education. As for trade with the Philippine islands and between the United States and Canton, these are two means by which Asia gains[2] in cash.

The great activity that has recently been deployed in gold and silver mining in the Americas, the introduction of more powerful machines, and the political changes on the New Continent have given rise to the idea that a large accumulation of gold and silver in Europe could have a sharp impact on the price of commodities. I think that this effect will be quite slow and extremely subtle. When, in the sixteenth century, the deposits at Potosí and their greater wealth were discovered, a small amount of silver, shipped in its entirety to one part of Europe that was small in size and more or less the only civilized region, triggered a rapid revolution in the price of things. Today, civilization has spread across an area six times greater in size. The lower classes of society everywhere enjoy a relatively high level of comfort. The total amount of precious metals required by minting and by gold- and silversmithing has undergone staggering growth. In Europe's population of 208,000,000 III.476 inhabitants, each individual possesses a certain amount of silver in cash or in silverware. The tremendous activity of trade uses a mass of precious metals that cannot be compared to what was demanded by sixteenth-century trade. There is no doubt that the output of the mines

of Applying the Principles of] Colonial Policy [to the government] of India; p. 194.

1. [Young, *An Inquiry into the Expediency of Applying the Principles of] Colonial Policy [to the Government] of India*; 1822, p. 58.

2. *Bullion Report, Minutes*, p. 253.

Amount of Tea Imported and Consumed in Great Britain from 1805 to 1826, and Averages of These Amounts for Ten-Year Periods[1]

YEARS ENDING ON JANUARY 5	IMPORT IN POUNDS	TEA DELIVERED FOR INTERNAL CONSUMPTION, MINUS SUBSIDIZED EXPORTS IN POUNDS
1807	22,155,557	20,979,128
1808	12,599,236	20,859,929
1809	35,747,224	20,859,929
1810	21,717,310	19,869,134
1811	19,791,356	19,083,244
1812	21,231,849	20,702,809
1813	23,318,153	20,018,251
1814	30,383,504	20,443,236
1815	26,110,550	19,224,154
1816	25,602,214	22,378,345
1817	46,234,380	21,843,903
1818	31,467,073	20,619,455
1819	20,065,728	21,859,482
1820	23,750,413	22,881,957
1821	30,147,994	22,366,547
1822	30,731,105	22,494,828
1823	27,362,766	23,559,495
1824	29,046,887	23,810,967
1825	31,682,007	23,908,629
1826	29,345,778	24,150,372
Average annual consumption	from 1807 to 1816	20,280,754
	from 1816 to 1826	22,750,063

[1] *L.C.*, 1826, page 775. On the history of the tea trade, see Crawfurd, *Hist[ory] of the Indian Archipelago*, vol. III, pp. 522–32. *Oriental Herald*, 1824, vol. I, p. 106; [Powell,] *Stat[istical] Illust[rations]*, pp. 61–64. The indications of the quantities sold by the [British East] India Company do not at all comport in the way one might hope with the works that I have just cited.

of Mexico could double or triple in the space of one century. If all the silver removed from the mines over a twenty-year period flowed into Europe at once, if the ports of Spain, Britain, and France, received all at once twelve to fifteen million kilograms of silver, the effect on prices would be immense; but a progressive accumulation, a slow growth from the 537,000 kilograms that Mexico produced at the beginning of the century to the 800,000 to one million to one and a half million would barely be felt. The past can be cited in support of this theory; in an average year before the war of independence, the mines of Mexico provided Spain with two and a half million marks of silver, while from 1700 to 1710, they provided barely 700,000 marks. The relative abundance of gold and silver has experienced the most extraordinary fluctuations in Brazil and the Urals; the absorption of precious metals in Asia has significantly slowed; and nevertheless all of these phenomena, which can be considered great and memorable events in the *metals trade*, have only had a weak effect on the price of things on a continent where the level of cash exceeds 8,600,000,000 francs and which has heretofore annually consumed over forty-five million francs to produce and repair gold and silver ornaments and other objects. The impact of paper money, the production of which appears to be inexhaustible, also tends to mask the changes that might be triggered by the accumulation of precious metals.

END OF THE THIRD VOLUME

BOOK V

The State of Manufacturing and Trade in New Spain

CHAPTER XII

The Manufacturing Industry—Cotton Cloth—Woolens—Cigars—Soda and Soap—Gunpowder—Coins—The Exchange of Products—Trade in the Interior—Roads—External Trade through Veracruz and Acapulco—Obstacles to This Trade—Yellow Fever

When one considers how little progress factories have made in Spain, despite the numerous encouragements they have received since the ▼ Marquis de la Ensenada took office, it comes as no surprise that everything related to factories or the manufacturing industry has made even less progress in Mexico. The nervous and suspicious policies of the European peoples, the legislation and the colonial system of the moderns, which hardly resemble those of the Phoenicians and the Greeks, have created insurmountable obstacles to businesses that could otherwise ensure great prosperity to these distant possessions and an independent existence from the homeland. Principles that call for vines and olive trees to be uprooted are not conducive to manufacturing. For centuries, a colony has only been considered useful to the metropole insofar as it has provided an abundance of raw materials and consumed large amounts of provisions and merchandise delivered to it from the fatherland. IV.1 IV.2

It has been easy for various trade nations to adapt their colonial system to small islands or factories established on the seacoast of a continent. The inhabitants of Barbados, Saint Thomas, or Jamaica are too few in number to offer much manpower for the production of cotton cloth. Furthermore, the position of these islands has always facilitated the exchange of their agricultural products for goods from the European manufacturing industry.

This was not the case for the continental possessions of Spain in the two Americas. Above 28° northern latitude, Mexico is 350 leagues wide. The plateau of New Granada is connected with the port of Cartagena by a great river that is difficult to navigate going upstream. Industry is stimulated IV.3

when towns of fifty to sixty thousand inhabitants on the ridge of mountains are located at long distances from the coastline; when a population of several million can only receive merchandise from Europe when it is transported on muleback for five to six months through forests and deserts. The new colonies were not established among completely barbarous people. Before the Spanish arrived, the indigenous peoples living in the Cordilleras of Mexico, Peru, and Quito were already clothed. Those who knew how to weave cotton cloth or spin the wool of llamas and vicuñas easily learned to make fabric. Consequently, this industry was established in Cuzco in Peru and in Tezcoco in Mexico a few years after the conquest of those countries, as soon as European sheep had been introduced in America.

In taking the title of Kings of the Indies, the Kings of Spain considered these distant colonies more as an integral part of their monarchy and as provinces dependent upon the crown of Castile than as colonies in the sense that the trading peoples of Europe have attributed to this word since the sixteenth century. It was understood early on that these vast areas, whose coastal regions were generally less inhabited than the interior, could not be governed in the same way as islands scattered throughout the Antillean Sea. IV.4 These circumstances forced the crown of Madrid to adopt a less restrictive policy and tolerate what it saw was impossible to prevent by force. Consequently, there arose a more equitable legislation than the one that governed most of the other colonies on the New Continent. In those colonies, for instance, the refinement of raw sugar is not allowed: the proprietor of a plantation must buy back the product of his own land from the refiner in the homeland. No law prohibits the establishment of sugar refineries in the possessions of Spanish America. If the government does not encourage factories, and if it uses the same indirect methods to prevent the establishment of silk, paper, and crystal factories, no decree by the *audiencia*, or royal *cédula* mandates that these factories cannot exist overseas. In these colonies, like everywhere else, the spirit of the law must not be confused with the politics of those who exercise them.

Only half a century ago, two citizens motivated by the purest patriotic ardor, the ▼ Count de Gijón and Marquis de Maenza, envisioned the project of bringing a colony of European workers and artisans to Quito: pretending to applaud their efforts, the Spanish minister did not believe it necessary to refuse them permission to set up workshops. But he was able to obstruct the two entrepreneurs' plan to such an extent that when they ultimately realized IV.5 that secret orders had been given to the viceroy and the *audiencia* to make their business fail, they willingly abandoned the idea. I tend to believe that a similar event could not have taken place at the time that I resided in those lands, because for the past twenty years, the Spanish colonies have undeniably been governed according to more equitable principles. Virtuous men have occasionally raised their voices to alert the government as to its true interests: they have made it known that it would be more useful to the homeland to stimulate manufacturing industries in the colonies rather than allow the treasures of Peru and Mexico to trickle away in payment for foreign merchandise. This advice would have been heeded had not the minister too often sacrificed the interests of the people of a great continent for those of a few coastal towns belonging to Spain. For it was not the workers on the Peninsula—industrious, not boisterous men—who hindered the progress of manufacturing in the colonists, but rather monopolistic merchants whose political influence is enhanced by great wealth and supported by intimate knowledge of the intrigues and immediate needs of the court.

Despite all of the obstacles, these factories have not failed to make some progress over the past three centuries, during which time the Biscayans, Catalans, Asturians, and Valencians have established themselves in the New World and brought industries there from their provinces. Factories specializing in unrefined goods have been able to function at very low cost wherever raw materials are abundant and transportation increases the cost of merchandise from Europe and East Asia. In times of war, the absence of connections with the homeland and prohibitive trade regulations for neutral parties favored the establishment of factories for decorative fabrics, fine cloth, and all kinds of refined luxury goods. IV.6

The value of what is produced by the manufacturing industry in New Spain is reckoned at seven or eight million piasters annually. Until 1765, cotton and wool were exported from the intendancy of Guadalajara to support factory production in Puebla, Querétaro, and San Miguel el Grande; since then, factories have been set up in Guadalajara, Lagos, and nearby towns. In 1802, the entire intendancy, which has more than 630,000 inhabitants and the coastline of which is bathed by the South Sea, provided[1] 1,601,200 piasters' worth of cotton and wool fabric, 418,900 piasters' worth of cured hides, and 268,400 piasters' worth of soap.

In discussing above the different varieties of *gossypium* grown in the warm and temperate regions, we have demonstrated how important local cotton manufacturing could be for Mexico. In peacetime, manufacturing

1. *Estado de la Intendencia de Guadalajara, comunicado en 1802, por el Señor Intendente al Consulado de Vera-Cruz* (unpublished official document).

in the intendancy of Puebla annually yields 1,500,000 piasters' worth of IV.7 products for domestic trade This production does not come from the combined factories, however, but from the large number of weavers (*telares de algodón*) scattered throughout the towns of Puebla de los Angeles, Cholula, Huexocingo, and Tlascala. In Querétaro, a good-sized town located on the road between Mexico City and Guanajuato, over 200,000 livres of cotton are used annually to make *mantas* and *rebozos*; the production of *mantas* or cotton fabrics amounts to 20,000 pieces costing thirty-two varas each. In 1802, there were more than 1,200 weavers[1] of cotton cloth and striped cotton fabric in Puebla. In that town as well as in Mexico City, the printing of decorative fabrics (both those imported from Manila and the ones made in New Spain) has made some advances in the last few years. In the port of Tehuantepec in the province of Oaxaca, the indigenous peoples dye cotton cloth purple by rubbing it against the shell of a *murex* [tropical sea snail] attached to granite rocks. Following an ancient custom, the cloth is washed in seawater, which in that area is rich in muriate of soda, to intensify the color.

The oldest cloth factories in Mexico are the ones in Tezcoco: they IV.8 were mainly established by viceroy ▼ Luis de Velasco II, the son of the famous commander-in-chief of Castile, the second viceroy of New Spain, in 1592. This branch of national industry gradually passed entirely into the hands of Indians and Mestizos [métis] in Querétaro and Puebla. I visited the factories in Querétaro in August in the year 1803: there are large factories there called *obrajes* and small ones called *trapiches*. I reckoned that there were twenty *obrajes* then and over 300 *trapiches*, which together used over 63,900 arrobas of Mexican sheep wool. According to exact statements drawn up in 1793, there were at that time in the Querétaro *obrajes* alone 215 trades and 1,500 workers who had produced 6,042 pieces or 226,522 varas of cloth (*paños*); 287 pieces or 39,718 varas of regular woolens (*xerguitillas*); 207 pieces or 15,369 varas of baize (*bayetas*); and 161 pieces or 17,960 varas of serge (*xergas*). Seven arrobas of wool are usually reckoned for each bolt of cloth and *bayeta*; six arrobas per bolt of *xerguitilla,* and five arrobas per bolt of *xerga.* The value of the fabrics and woolens from the *obrajes* and *trapiches* in Querétaro is now over 600,000 piasters or 3,000,000 francs annually.

When visiting these workshops, the traveler is dismayed not only by IV.9 the extreme imperfection of the technical processes in finishing the dyeing

1. *Informe del Intendente Don Manuel de Flon, Conde de la Cadena* (unpublished).

but especially by the insalubrity of the place and the poor treatment of the workers. Free men, Indians, and people of color are mixed there with convicts whom justice consigns to the factories as day-workers. Both groups are thin and haggard, half-naked, and dressed in rags. Each workshop is like a dark prison: its double doors are always closed and the workers are not allowed to leave the premises. Married men can visit their families only on Sundays. All are mercilessly whipped for the most minor infringement on the established order of the factory.

It is difficult to imagine how the *obreja* owners can treat free men this way, or how the Indian worker can endure the same treatment as the convict; these so-called rights are in fact acquired underhandedly. Factory owners in Querétaro use the same ruses employed in many cloth factories in Quito and on farms where, in the absence of slaves, manpower is extremely rare. They choose from among the local people those who are the most abject but show some aptitude for work. The latter are then advanced a small sum of money: the Indian, who likes to get drunk, spends it in a few days. Having become indebted to the master, he is locked in the workshop on the pretext of working off his debt. His day's labor is reckoned at one and a half reales or twenty sous tournois. Instead of receiving the money, he is readily provided with food, brandy, and old clothes, for which the factory owner makes a 50 to 60 percent profit. In this way, the most assiduous worker is always in debt, and the same rights that one believes one acquires over a bought slave are applied to him. I have known many people in Querétaro who agree with me over these egregious abuses. Let us hope that a government that will protect its people will also turn its attention to these vexations, which are so opposed to humanity, the laws of the land, and the development of Mexican industry.

IV.10

With the exception of some cotton fabric blended with silk, there is almost no silk industry now in Mexico. At the time of Acosta's voyage near the end of the sixteenth century, silkworms brought from Europe were raised near Panuco and in the Misteca. Even then, excellent taffetas[1] were made with Mexican silk. We have already seen above that it is not the *bombyx mori* but an indigenous caterpillar that provides the raw material for the silk handkerchiefs made by the Indians of Mixteca and in the village of Tlistla near Chilpanzingo.

1. Acosta, [*Historia natural y moral de las Indias*,] book IV, chap. 32, p. 179. See also above, chap. III, p. 66 [p. II.38 in this edition].

IV.11 New Spain has no linen or hemp factories; there are also no paper factories. Tobacco production is controlled by the state; the cost of producing cigars and tobacco powder is over 6,200,000 livres tournois in an average year. The factories in Mexico City and Querétaro are the largest. Here is the state of all production for the years 1801 and 1802:

TOBACCO PRODUCED IN NEW SPAIN	IN 1801 PIASTERS	IN 1802 PIASTERS
Tobacco value, according to sales prices	7,825,913	7,686,834
Production costs	1,299,411	1,285,199
Labor	798,452	794,586
Tobacco price at which it is bought from the workers in Mexico	629,319	594,229
Net revenue (*líquido*) of the crown for tobacco sales	3,993,834	4,092,629

When I traveled through Querétaro, I visited the main cigar factory (*fábrica de puros y cigarros*), which employs more than 3,000 workers, among whom are 1,900 women. The workrooms are clean but poorly aired, and as a result, extremely hot. In this factory 130 reams (*resmas*) of paper and 2,770 livres of tobacco leaves are used daily. In the course of the month of July 1803, 185,288 piasters' worth of cigars were produced; that is, 2,654,820 small boxes (*cajillas*) of cigars, the sale price of which is 165,926 IV.12 piasters, and 289,799 boxes of *puros* or cigars that are not wrapped in paper. Production costs in the month of July alone were 31,789 piasters. It appears that the royal factory in Querétaro annually produces over 2,200,000 piasters' worth of *puros* and *cigarros*.

Solid soap production is a major trade factor in Puebla, Mexico City, and Guadalajara: the first of these cities produces almost 200,000 arrobas annually. In the intendancy of Guadalajara, its value is reckoned to be 1,300,000 livres tournois. The abundance of soda, which is found almost everywhere on the interior plateau of Mexico at an elevation of 2,000 to 2,500 meters, greatly enhances this production. *Tequesquite*, which we have had several occasions to mention,[1] covers the surface of the earth (especially in October) in the Valley of Mexico, on the shores of lakes Texcoco,

1. See above, vol. II, p. 183 [p. 400 in this edition]; and Del Río, *Elementos de Oryctognosia*, p. 154.

Zumpanango, and San Cristóbal; on the plains around the city of Puebla; on the plains between Zelaya and Guadalajara; in the valley of San Francisco near San Luis Potosí, between Durango and Chihuahua; and in the nine lakes scattered throughout the intendancy of Zacatecas. We do not know if the *tequesquite* originates from the decomposition of volcanic rock in which it is contained, or to the slow action of limestone on the muriate of soda. In Mexico City one can buy 1,500 arrobas of *tierra tesquesquitosa*, in other words, a clayey soil highly impregnated with carbonate and a small amount of muriate of soda for sixty-two piasters. Purified in soap factories, these 1,500 arrobas will yield 500 arrobas of pure carbonate of soda. As a result, in the present state of production, it costs fifty sous tournois to make one quintal. Mr. Garcés, who successfully uses carbonate of soda in smelting muriates of silver, has proven in a specific report that by perfecting technical procedures, one could produce carbonate of soda for less than thirty sous tournois per quintal in the soda factories of Mexico City called *tequesquiteras*. Since the cost of Spanish carbonate of soda in France is generally from twenty to twenty-five livres per quintal in peacetime, it is conceivable that despite the difficulties of shipping, Europe may one day obtain soda from Mexico, just as it has long already imported potash from the United States of North America.

IV.13

The town of Puebla was formerly known for its excellent faience (*loza*) and hat factories. We have seen above that up to the beginning of the eighteenth century, these two industrial sectors enlivened trade between Acapulco and Peru. Today there is almost no communication between Puebla and Lima, and the faience factories have so decreased in number because of the low cost of European crockery and porcelain brought in through the port of Veracruz, that of the forty-six factories that existed in 1793, there were no more than sixteen producing faience and two producing glass in 1802.

IV.14

In New Spain as in most European countries, gunpowder manufacturing is controlled by the state. To conceive of the enormous amount of gunpowder that is made and sold as contraband, one has only to recall that despite the flourishing state of the mines, the King has never sold more than 3,000 to 4,000 quintals of gunpowder per year[1] to miners, whereas a single mine, the Valenciana, requires 1,500 to 1,600 [quintals]. My research suggests that the ratio of the amount of gunpowder made at the crown's expense

1. For only 255,455 livres in 1801; for 339,921 livres in 1802. See above, pp. 204 and 238 [pp. II: 116 and 134 in this edition].

to what is sold illegally seems to be one to four. Since potassium nitrate and sulfur are prevalent almost everywhere in the interior of New Spain, and the contraband producer can sell his powder to a miner for eighteen sous tournois per livre, the government should either reduce the price of the product from the factory or turn over the gunpowder trade completely to the free market. How else could one prevent fraud in such a vast country, in mines that are far from towns and scattered across the ridge of the Cordilleras, in the middle of the wildest and most solitary places?

IV.15

The royal powder factory, the only one in Mexico, is near Santa Fé in the Valley of Mexico, three leagues from the capital, surrounded by hills of clayey breccia studded with fragments of trappean porphyry. The buildings are very beautiful, built in 1780 according to the plans of Mr. Costansó, chief of the corps of engineers, in a narrow valley that provides, in abundance, the water needed to move the hydraulic wheels; the Santa Fé aqueduct passes through the valley. All the mechanized parts—especially the wheels, the axels of which are seated on friction pulleys, as well as bronze cams that move the set of *pestles*—are arranged with keen intelligence. It would be desirable that the riddles [sieves] that make the *pellets* also be turned by either water or horse power: eighty Mestizo [métis] boys, who are paid twenty-six sous per day, are currently assigned to this task. The buildings of the old gunpowder factory, established near the castle of Chapultepec, are only used for refining potassium nitrate. Sulfur—which is abundant in the Orizaba and Puebla volcanoes in the province of San Luis near Colima, and especially in the intendancy of Guadalajara, where the rivers carry along considerable masses of it mixed with pumice fragments—comes in purified form from the town of San Luis Potosí. In 1801, over 786,000 livres were produced at the royal gunpowder factory in Santa Fé; in 1802, more than 750,000 livres, a portion of which was exported to Havana. It is a pity that this beautiful building, where more than half a million pounds of gunpowder is usually stored, does not have an electrical conductor. There are only two conductors in this vast country, both of which were built in Puebla by an enlightened administrator, the Count de la Cadena, despite the imprecations of the Indians and a few ignorant friars.

IV.16

In discussing the gunpowder factory in Santa Fé, I must not fail to mention a historical fact that is repeated in many works, although there is no very solid basis for it. It is reported that the brave Diego Ordaz managed to enter the crater of the Popocatépetl volcano to gather sulfur, and in so doing enabled the Spanish to make the gunpowder they needed for the siege of Mexico City. The very letters that the commanding general sent to the Emperor Charles V prove that this claim is false. When in the month of October of the year 1519, the army of Spanish and Tlaxcalteca marched from Cholula to Tenochtitlan; they crossed the Ahualco Cordillera, which connects the Sierra Nevada or Iztaccihuatl to the summit of the Popocatépetl volcano. The Spanish followed more or less the same route as the letter mail travels from Mexico City to Puebla via Mecameca, which appears on the map of the valley of Tenochtitlan. The army suffered at the same time from the cold, impetuous winds that constantly sweep across this plateau. In describing the march to the emperor, Cortés[1] wrote: "Having seen smoke coming from a very high mountain, and wishing to provide your majesty with a detailed report of all the marvels that this country contains, I chose ten of my bravest comrades in arms and ordered them to climb up to this peak and there discover the secret of the smoke (*el secreto de aquel humo*), so that they could tell me how it was possible and from whence it came." IV.17

Bernal Díaz confirms that Diego Ordaz was a member of this expedition and that the captain was able to reach the edge of the crater. He may have boasted about his exploit afterwards, since other historians relate that the emperor allowed him to include a volcano in his coat of arms. López de Gómara,[2] who followed the accounts of conquistadores and religious missionaries in composing his work, does not refer to Ordaz as the leader of the expedition; but he does confirm vaguely that two Spaniards visually measured the size of the crater. Cortés, however, explicitly writes that his "men climbed very high," that "they saw much smoking rising," but that "none of them was able to reach the summit of the volcano, because of the enormous amount of snow covering it, the severe cold, and the whirlwinds of ashes that covered the travelers." Above all else, a terrifying noise that they heard as they approached the summit caused them to turn back. Cortés's account tells us that the goal of Ordaz's expedition was in no way to extract sulfur from the volcano and that neither he nor his comrades saw the crater in 1519. "They brought back only snow and pieces of ice, whose appearance astounded us, because this country lies below 20° latitude, on the same parallel as the island of Hispaniola (San Domingo) and should thus, according to navigators' opinions, be very warm." IV.18

It is evident from Cortés's third and fourth letters to the emperor that, after taking Mexico City, the general sent other missions to discover the

1. Lorenzana, [*Historia de Nueva-España*,] p. 70; Clavijero, [*Storia antica del Messico*,] vol. III, p. 68.

2. Gómara, *Conquista de México* (Medina del Campo, 1553), folio 38.

summit of the volcano that seemed to draw his attention even more since the locals insisted that *no mortal was allowed to approach this abode of evil spirits*. After two unsuccessful attempts, the Spanish finally succeeded in viewing the crater of Popocatépetl in 1522: it appeared to them to be three-quarters of a league in circumference, and they found a little sulfur on the edge of the precipice that the vapors had deposited. In discussing the tin from Tlaxco that was used to melt down the first cannons, Cortés[1] reported that "there is no lack of sulfur for making gunpowder, because a Spaniard had extracted some from a mountain, from which smoke constantly issues, by descending tied to a rope to a depth of seventy to eight arm lengths." He adds that this means of obtaining sulfur is very dangerous, and that for this reason, it would be wiser to have it brought from Seville.

IV.19

A document preserved among the Montaño family, which Cardinal Lorenzana confirms having seen with his own eyes, proves that the Spaniard to whom Cortés referred was Francisco Montaño. Did this intrepid man actually enter the crater of Popocatépetl itself, or did he take the sulfur, as some in Mexico City imagine it, from a lateral crevice of the volcano? We shall have the opportunity to discuss this matter in another work, in which we shall give a geological description of New Spain. Mr. Alzate[2] claims, with no basis, that Diego Ordaz took sulfur from the crater of the old Tuctli volcano, east of Lake Chalco, near the Indian village of Tuliahualco. It is true that smugglers look for sulfur there to use in making gunpowder, but Cortés clearly referred to Popocatépetl by the term "the constantly smoking mountain." Be that as it may, it is certain that it was after the reconstruction of the town of Tenochtitlan and not during the siege, as Solís[3] claims, that soldiers from Cortés's army climbed to the summit of Popocatépetl,[4] where no one has ventured since. If La Condamine[5] had known the absolute elevation of this volcano, which I found to be 5,400 meters, he would not have thought himself the first person to reach an elevation of 4,800 meters above sea level on the ridge of the Cordilleras of the Americas. Furthermore, ▼ Ordaz's and Montaño's expeditions recall the bold act of a Dominican priest, Blas de Iñena, who had himself lowered in a wicker basket by a chain, armed with

IV.20

1. *De allí (de la Sierra que da humo) entrando un Español setenta y ochenta brazos atado a la boca abajo se ha recado (el azufra) que hasta ahora nos hemos sostenido.* (Lorenzana, [*Historia de Nueva-España*,] p. 380.)

2. *Gazeta de Literatura de México*, 1789, p. 52.

3. Solís, *Conquista de México*, p. 142.

4. Lorenzana, [*Historia de Nueva-España*,] p. 318.

5. Bouguer, *Mesure de la Terre*, p. 167. La Condamine, *Voyage*, p. 54.

a ladle and an iron bucket, to a depth of 140 arm lengths inside the crater of the Granada volcano, called the Cerro de Masaya, located near Lake Nicaragua, to collect lava which he believed to be gold: he lost his iron bucket, which was melted by the excessive heat, and was barely able to save himself. But in 1551, the court of Madrid granted formal permission[1] to Juan Alvarez, deacon of the chapter house of the town of Léon, to open the volcano and gather "the gold inside it." Admittedly, no naturalist traveler in our time has undertaken such dangerous activity in the name of science as what was attempted at the beginning of the sixteenth century to gather sulfur and gold from the mouth of burning volcanoes. IV.21

We shall conclude the chapter on manufacturing in New Spain by discussing the silversmith's art and the minting of coins which, when considered only in relation to industry and the perfection of manual arts, are noteworthy topics. There are few countries with a greater number of pieces of jewelry, vases, and religious ornaments: even the smallest towns have jewelers, whose workshops employ workers from all castes, whites, Mestizos, and Indians. The Academy of Fine Arts and the Schools of Drawing in Mexico City and Jalapa have contributed greatly to broadening the taste for beautiful antique styles. Silver services valued at prices from 150,000 to 200,000 francs, whose elegance and fine workmanship rival the most beautiful pieces in this style from the most civilized parts of Europe, have been made in Mexico City. The amount of precious metals that have been converted into dishware in an average year from 1798 to 1802 is 385 marks of gold and 26,803 marks of silver.[2] The following pieces of jewelry, taxed at one-fifth of their sale price [quint], were declared at the mint building: IV.22

YEARS	GOLD MARKS	SILVER MARKS
1798	402	19,823
1799	484	26,762
1800	412	30,887
1801	379	30,860
1802	249	25,692
Total	1,926	134,024

1. Gómara, *Historia de las Indias*, fol[io] 112.

2. Castilian weight. It is useful to mention that each time that the opposite is not specifically indicated, the word *mark* used in this work refers to the *Castilian mark*.

Architecturally, the mint of Mexico City, the largest and the richest in the entire world, is a very simple building attached to the viceroys' palace. Directed by an enlightened administrator and friend of the arts, the ▼ Marquis de San Román,[1] there is almost nothing extraordinary about this institution in terms of the perfection of its machinery or chemical processes. But it is very worthy of the traveler's attention because of the order, activity, and economy that dominates all aspects of minting there. This interest is enhanced by other considerations that are apparent even to those who have IV.23 no concern for the speculations of political administration. Indeed, it is impossible to walk through this relatively unspacious building without recalling that more than ten billion livres tournois have been issued from it in less than 300 years and without reflecting on the powerful influence that these treasures have had on the destiny of the peoples of Europe.

The mint of Mexico City was established fourteen years after the destruction of ancient Tenochtitlan, under the first viceroy of New Spain, Antonio de Mendoza, by a royal *cédula* dated May 11, 1535. At first minting was done commercially, at the expense of a few private individuals to whom the government had entrusted the task. Their lease was not renewed in 1733. Since that time, royal officials manage all operations on the king's account. The number of workers employed at the mint is between 350 and 400. There are so many machines that in the space of a year and without working overtime, over thirty million piasters can be struck; that is, approximately three times as many as are usually struck at the sixteen mints in France. In Mexico City, in April 1796 alone, the sum of 2,922,185 piasters were minted; in the month of December 1792, more than 3,065,000 piasters. In 1810 in Paris, the busiest month for production was March, when the mint struck 1,271,000 piasters' worth of five franc pieces. From the year IV.24 1726 to 1780, the production of gold and silver pieces increased:

IN FRANCE'S SIXTEEN MINTS[1]	IN MEXICO CITY'S MINTS
2,446,000,700 livres	3,364,138,060 livres

[1] Necker, *De l'administration des finances*, vol. III, p. 59.

To give an idea of the activity at the mint in Mexico City, we reproduce here one of the tables that the government publishes annually to notify the public of the state of the mines, which is seen as the regulator of public prosperity. I shall choose the year 1796, when 25,644,000 piasters were

1. *Juez Superintendente de la Real Casa de Moneda.*

struck, although this number was 24,593,000 in 1795, and 25,080,000 piasters in 1797.

MONTH OF THE YEAR 1796	GOLD	SILVER		GOLD AND SILVER	
	Piasters	Piasters	Reals	Piasters	Reals
January					
February		2,078,958	7	2,078,958	7
March	246,578	2,071,001	One-half	2,317,579	One-half
April		2,922,185	1	2,922,185	1
May	252,240	2,538,847	4 and a half	2,791,087	4 and a half
June		1,907,980	3	1,907,980	3
July	117,008	2,028,329	6	2,145,335	6
August		1,551,143	2	1,551,143	2
September	161,312	2,257,900	3 and a half	2,419,212	3 and a half
October		2,455,057	3	2,455,057	3
November	110,112	2,685,903	1 and a quarter	2,796,015	1 and a quarter
December	410,1544	1,849,467	Three-quarters	2,260,011	Three-quarters
Total	1,297,794	24,346,772	One-half	25,644,566	One-half

IV.25

The workshops at the mint in Mexico City contain ten rolling mills operated by sixty mules, fifty-two cutters, nine fitting counters, twenty indentation machines, twenty pendulums, and five mills for amalgamating the *scrapings* and *filings* called *mermas*. Since a pendulum can strike over 15,000 piasters in ten hours, one should not be surprised that with such a large number of machines, they are able to produce daily 14,000 or 15,000 marks of silver; regular work, however, is not more than 11,000 or 12,000 marks. These facts, which are based on official documents, suggest that the combined silver produced by all European mines would not suffice to run the mint of Mexico City for more than fifteen days.

IV.26

The cost of minting coins, including employees' pensions and the loss from the *scrapings,* is one silver real or thirteen sous tournois per mark. This loss of the *mermas*, which was formerly reckoned at one-third of 1 percent has now been reduced by half, since instead of three marks, no more than one mark three ounces are lost for every 1,000 marks converted into coins. The king's revenue from this output is estimated in the following manner: if minting does not exceed fifteen million piasters per year, then the profit is only 6 percent of the amount of gold and silver struck. On the contrary, it is reckoned at 6 and a half percent when production is eighteen million piasters, and at 7 percent when the mines produce even more, as they have in the last twenty years. Indeed, we shall see below that the mint of Mexico City, together with the *separation house* (*casa del apartado*) operate at an annual profit of almost eight million francs.

The separation house where the gold and silver that come from ingots of auriferous silver are separated formerly belonged to the Marquis de Fagoaga's family. This important establishment was restored to the crown only in IV.27 1799. The building is very small and old: it has recently been partially rebuilt, which cost the government more than if it had been replaced by a new building that was not located in the middle of the city and in which acidic vapors could be better dispersed. Several persons who are interested in keeping the separation workshops where they currently stand suggest that the nitrous acid vapors that permeate through one of the most densely populated neighborhoods of the city are useful in decomposing the miasma that rise from the surrounding lakes and marshes. These ideas have found favor since acidic fumigations have been performed in hospitals in Havana and Veracruz.

The *casa del apartado* contains three types of workshops that are intended for 1) glass blowing; 2) making nitrous acid; and 3) the separation of gold and silver. The procedures followed in these different workshops are as imprecise as the construction of the glass ovens and *galleys* that are used to make retorts and distill brandies. The *frit* for the glass (*pasteladura*) is composed of 0.46 parts quartz taken from mining seams in Tlalpujahua, and 0.54 parts soda, which the Indians of Xaltocan and Peñol extract from the incineration of Sesuvium portulacastrum and several species of Chenopodium, Atriplex, and Gratiola, which will be described in Mr. Sesse's and IV.28 Mr. Cervantes's *Flora mexicana*, and from European Salsola soda, which is grown in the Valley of Mexico, either to be eaten as a vegetable or to be reduced to ashes. This soda from Xaltocan is mixed with large amounts of potassium sulfate and lime, so that the carbonate of soda that forms a powdery crust on clay-like soil would be much more suitable for making glass. The frit is not melted in clay pots, as it is in Europe, but in crucibles of highly refractory porphyritic rock taken from a quarry near Pachuca. Over 15,000 francs' worth of wood are consumed annually in the glass-making ovens: a single retort costs the factory nearly fourteen sous, and over 50,000 of them are broken every year.

The nitrous acid used for the separation process is made by using vitriolic earth (*colpa*) containing a mixture of aluminum, iron sulfate, and red iron oxide to decompose crude saltpeter. This *colpa* comes from around Tula, where a mine is worked at the expense of the *color and dyeing farm.*[1] The royal gunpowder factory provides the *separation house* with first-boil saltpeter. Each retort is loaded with eight livres of *colpa* and again as many of unrefined potassium nitrate: the distillation process lasts from thirty-six to forty hours. The ovens are round and have no grills. The nitrous acid, which is the by-product of the decomposition of saltpeter saturated with muriate, perforce contains a large amount of muriatic acid, which is removed by adding silver nitrate. One can estimate the enormous amount of silver muriate that is produced in this establishment when one recalls that a sufficient amount of nitrous acid is purified there to separate 7,000 gold marks annually. Silver muriate is melted with lead shot and decomposed by heat. Instead of first-boil saltpeter, it would probably be more profitable to use refined saltpeter to distill spirits. Until now the slow and difficult method of purifying the acid using silver nitrate has been employed because the *royal establishment of the apartado* has been forced to buy its saltpeter from the *royal factory of powders and saltpeters*, which will only deliver refined saltpeter at a price of 126 francs per quintal. IV.29

The separation of gold and silver reduced to little grains or pellets to multiply points of contact is made in glass retorts arranged in long rows of round *galleys* [metal trays] five to six feet long. These galleys are not heated by the same fire, but two or three flasks form, so to speak, a separate oven. The gold that remains at the bottom of the flasks is melted into ingots weighing fifty marks each, while the silver nitrate is decomposed by the heat during distillation in the retorts. This distillation, through which the nitrous acid is recovered, is also performed in a galley and takes eighty-four to ninety hours. The retorts must be broken to recover the reduced and crystallized silver; they could probably be saved by precipitating the silver through the copper, but another step would then be necessary to decompose the copper nitrate that would replace the silver nitrate. In Mexico City, the costs of separation are reckoned at two to three reales *de plata* (twenty-six to thirty-nine sous tournois) per mark of gold. IV.30

It is surprising not to find students from the Royal School of Mines employed either at the mint or the separation house. These two great establishments, however, must anticipate useful reforms by benefiting from advances in mechanics and chemistry. Furthermore, the mint is in a part of the city where it would be easy to use running water to turn the rolling mills by

1. *Estancia real de tintes y colores.*

using hydraulic wheels. This machinery is far from the level of perfection it has recently attained in England and France. Improvements will be all the more advantageous since production consumes an enormous amount of gold and silver: the piasters struck in Mexico City may be considered the raw material that sustains activity in most of the mints in Europe.

Not only have the gold works and silverware that we have mentioned above been perfected in Mexico, but noticeable progress has also been made IV.31 there in other branches of industry that depend upon luxury and wealth. Candelabras and other valuable decorations were recently executed in gilded bronze for the cathedral in Puebla, whose bishop has an income of over 550,000 livres. Although the most elegant carriages that circulate in the streets of Mexico City and Santa Fé de Bogotá at 2,300 to 2,700 meters above sea level have come from London, some rather fine ones have also been made in New Spain. Cabinetmakers create pieces of furniture remarkable for their lines, color, and polished wood from the equinoctial region near the coasts, especially from the forests of Orizaba, San Blas, and Colima. It is interesting to read in the Mexico City gazette[1] that harpsichords and pianos are made as far away as the *provincias internas*, for example in Durango, 200 leagues north of the capital. The local people show unflagging patience in making small, carved bibelots of wood, bone, and wax. In a country whose vegetation provides the most valuable materials[2] and where artisans may select the nuances of color and shape they seek from among the roots, medullated lengths of branches, and fruit pits, the Indians' handiwork IV.32 could become an important commodity for export to Europe. The considerable sums of money that this branch of industry brings to the inhabitants of Nuremberg and to the mountain people of Berchtesgaden and the Tyrol is well-known; but they can use only pine, cherry, and walnut wood to make boxes, spoons, and children's toys. Americans in the United States send boatloads of furniture to the island of Cuba and to other Antilles islands, the wood for which comes largely from the Spanish colonies. This branch of industry will pass into the Mexicans' hands when, animated by a noble desire to emulate others, they begin to take advantage of the resources of their own land.

We have heretofore discussed agriculture, mining, and manufacturing as the three main sources of trade in New Spain. We must now present a

1. *Gazeta de Mexico*, vol. V, p. 369.

2. Swietenia, Cedrela, and Cæsalpinia wood; trunks of Desmanthus and Mimosa, whose center is a reddish color tending toward black.

portrait of the exchanges that are made either in the interior of the country or with the metropole and other parts of the New Continent. We shall therefore discuss, in order, domestic trade, which transfers the overstock of one province of Mexico to another; external trade with the Americas, Europe, and Asia; and the influence of these three branches of trade on public prosperity and the increase of national wealth. We shall not renew rightful complaints about the obstructions to trade and the prohibitive system that are the basis of the Europeans' colonial legislation: it would be difficult to add to what was been said on this subject when significant problems of political economy occupied everyone's mind. Rather than attack the principles whose falseness and unfairness have both been recognized, we shall limit ourselves to gathering facts and demonstrating how trade relations between Mexico and Europe will become important once they are freed from the obstructions of an odious monopoly that is disadvantageous to the metropole itself. IV.33

Domestic trade includes both the shipping and production of goods in the hinterland, as well as smuggling along the coastline of the Antillean Sea and the Pacific Ocean. Such trade is not animated by interior navigation on rivers or artificial canals: like Persia, most of New Spain lacks navigable rivers. The Río del Norte, which has virtually the same width as the Mississippi, irrigates land that might be highly suitable for cultivation but which offers in its present state only a vast desert. This great river is home to no more domestic trade activity than the Missouri, the Casiquiare, and the Ucayali, which flow across the savannas and uninhabited forests of South America. Between 16° and 23° latitude in Mexico, in that part of the country where the population is most concentrated, only the Río de Santiago could be made navigable at little cost. This river[1] is as long as the Elbe and the Rhone: it feeds the plateaus of Lerma, Salamanca, and Celaya, and it might serve to transport flour from the intendancies of Mexico City and Guanajuato to the western seacoast. We have proven above[2] that if it is necessary to abandon the project of creating an internal waterway between the capital and the port of Tampico, it would be very easy to dig canals in the Valley of Mexico, from the northernmost point, the village of Huehuetoca, as far as its southern extremity, the small town of Chalco. IV.34

Since connections with Europe are only made via the two ports of Veracruz and Acapulco, anything that is imported or exported must neces-

1. The Río Santiago, the former Río Tololotlan, is more than 170 leagues long.

2. See chap. III, vol. 1, p. 277 [p. 187 in this edition]; chap. VIII, vol. II, pp. 136–45 [pp. 375ff in this edition].

sarily go through the capital, which has, as a result, become a focal point of domestic trade. Located on the ridge of the Cordilleras and dominating, as it were, the two seas, Mexico City is sixty-nine leagues as the crow flies from Veracruz, sixty-six leagues from Acapulco, seventy-nine leagues from Oaxaca, and 440 leagues from Santa Fe in New Mexico. Because of the position of the capital, the most frequented and most important routes for trade are 1) the route from Mexico City to Veracruz via la Puebla and Jalapa; 2) the route from Mexico City to Acapulco via Chilpanzingo; 3) the IV.35 route from Mexico City to Guatemala via Oaxaca; 4) the route from Mexico City to Durango and Santa Fé in New Mexico, commonly called *el camino de tierra dentro*. One may consider the roads that lead from Mexico City either to San Luis Potosí and Monterrey, or to Valladolid and Guadalajara as ramifications of the main road through the *provincias internas*. When one looks at the physical constitution of the country, one sees that no matter what advances in civilization may occur, these routes will never be replaced by natural or artificial waterways like the ones in Russia, from Saint Petersburg to the heart of Siberia.

Roads in Mexico either run on the central plateau itself, from Oaxaca to Santa Fé, or else they lead from this plateau to the seacoasts. The former preserve connections between the towns located on the mountain ridge, in the coldest and most densely populated part of the kingdom; the latter are intended for external trade, that is, the connections that exist between the interior and the ports of Veracruz and Acapulco. Furthermore, they facilitate the exchange of products between the plateau and the scorching plains on the seaboard. The roads on the plateau, running from south-southeast to north-northwest, which one might call *longitudinal*, given the overall configuration of the country, are very easy to maintain. We shall not repeat IV.36 here what we have already related[1] in preceding chapters about the vastness and continuity of the high Anahuac plain, where one finds neither crevices or ravines, or about the gradual decline of the plateau from 2,300 meters to 800 meters absolute elevation. Carriages may travel from Mexico City to Santa Fé over an area of land that would exceed the length of the Alps chain, if the latter stretched uninterrupted from Geneva to the shores of the Black Sea. In fact, one can travel in four-wheeled coaches on the central plateau in all directions, from the capital to Guanajuato, Durango, Chihuahua,

1. *Introduction*, vol. I, p. 155 [p. 106 in this edition]; chap. III, vol. I, p. 250 [p. 172 in this edition]; chap. VIII, vol. II, pp. 183, 233, 248, and 255 [pp. 400, 426, 434, and 438 in this edition].

Valladolid, Guadalajara, and Perote. But because of the poor condition of the roads at present, circulation is not set up for transporting merchandise: beasts of burden are preferred, and thousands of horses and mules ply the roads of Mexico[1] in long queues (*requas*). Several mestizos and Indians are hired to guide these caravans: preferring the roaming life to any sedentary occupation, they spend the night in the open air or under hangars (*tambos*, or *casas de comunidad*) that are built in the middle of villages to accommodate travelers. The mules graze freely on the savannas, but when severe droughts destroy cereal grains, the mules are given corn, either the plant [*zacate*] or its grain.

The roads leading from the interior plateau toward the seacoast, which I call transversal roads, are the most strenuous and deserve the special attention of the government. The roads from Mexico City to Veracruz and Acapulco, from Zacatecas to Nuevo Santander, from Guadalajara to San Blas, from Valladolid to the port of Colima, and from Durango to Mazatlán, passing through the western branch of the Sierra Madre, belong to this class of roads. The roads that connect the capital with the ports of Veracruz and Acapulco are naturally the most traveled. The total value of precious metals, agricultural products, and the European and Asian merchandise that flows back and forth over these two highways is 320,000,000 francs annually. These treasures travel across a route that resembles the road leading from Airolo to the Saint Gotthard hospice. From the village of Vigas as far as Encero, the Veracruz road is hardly more than a narrow, twisting pathway, and there is scarcely a more torturous one in all of the Americas, with the exception of the road by which merchandise from Europe travels from Honda to Santa Fé de Bogotá, and from Guayaquil to Quito. IV.37

Products from the Philippines and Peru arrive via the road from Mexico City to Acapulco: it is built on a slope of the Cordilleras that is less steep than the road leading from the capital to the port of Veracruz. A mere glance at the *dips* in the Mexican Atlas will suffice to demonstrate how justified this claim is. On the route from Europe, as we have seen above,[2] one remains on the central plateau at an elevation of 2,300 meters above sea level from the Valley of Mexico all the way past Perote. One then descends extremely rapidly as far as the gully of *Plan del Río*, west of Rinconada. Conversely, over the Acapulco road, which we call the Asia route, the descent IV.38

1. See chap. VII, vol. I, p. 459 [p. 293 in this edition]; chap. I, vol. III, p. 59 [p. II.34 in this edition].

2. Chap. III, vol. I, p. 263 [p. 179 in this edition].

begins as early as eight leagues from Mexico City on the southern slope of the basaltic Guarda Mountain. With the exception of the section that goes through the forest of Guchilaque, it would be easy to make this road passable for carting; it becomes narrow and treacherous as one approaches the capital, especially from Cuernavaca to Guchilaque, and from there to the summit of the tall mountain called *la Cruz del Marqués*. The main difficulties that obstruct connections between the capital and the port of Acapulco arise from the sudden high waters of the two rivers, the Papagallo and the Río de Mescala. These torrents, which during droughts are less than sixty meters wide, swell to 250 to 300 meters wide during the rainy season. During this time, *loads* are often held up for seven to eight days on the bank of the Papagallo, which the mule drivers do not attempt to ford. I have seen the remains of several pillars built with enormous cut stones that the current had swept away before the arches could be completed. In 1803, there was a new attempt to build a large stone bridge over the Río Papagallo; the government had allotted almost half a million francs to this project, which is infinitely important for Mexico's trade with the Philippine Islands. The Río de Mescala, which farther west is called the Río de Zacatula, is almost as dangerous as the Papagallo: I have crossed it on a raft built after the ancient Mexican custom of tying reeds over dried squash hulls, where two Indians guide the raft by holding it with one hand and swimming with the other.

IV.39

The construction and improvement of a new road leading from Mexico City to the port of Veracruz have lately become a cause of concern for the government. A felicitous rivalry has arisen between the new trade council established in Veracruz as the *real tribunal del consulado* and the former *consulado* in the capital; the latter is slowly beginning to emerge from the lethargy of which it has long been accused. After constructing, at their own expense, a fine highway on the heights of Tiangillo and Las Cruces, which separate the Toluca basin and the basin of Mexico City, the merchants of Mexico City preferred to have the Veracruz road pass through Orizaba. Conversely, the merchants of Veracruz, who have country houses in Jalapa and maintain major trade relations with that town, insisted that the new road, suitable for transportation by cart (*camino carretero*), be routed through Perote and Jalapa. After discussions that lasted for several years,[1] the *consulado* of Veracruz took advantage of the arrival of the viceroy José de Iturrigaray, who recognized the usefulness of the Jalapa road and put Mr. García Conde, an intelligent and active engineer, in charge of the project.

IV.40

1. See chap. VII, vol. II, p. 216 [p. 417 in this edition].

The old road from Mexico City to Jalapa and Veracruz crossed the high Apa plains without bordering on the large town of La Puebla de los Angeles: the Abbot Chappe describes this road in [the narrative of] his journey through California, and that learned man determined several points using barometric measurements.[1] At that time, merchandise and local products were sent from Mexico City to Perote and Jalapa via the canal that separates Lake Texcoco and Lake San Cristóbal; via Totolcingo and Teotihuacan; via the former Otumba battlefield, the inns at Apa, Piedras Negras, San Diego, Hongito, Vireyes, and Tepeyacualco. Over this road, it was forty-three leagues from Mexico City to Perote, and seventy-four leagues from Mexico City to Veracruz. At that time and until 1795, it took two days to travel from the capital to Puebla by taking a long detour toward the northeast via Otumba and Irolo and then heading southeast via Pozuelos, Tumbacaretas, and San Martín. Finally, during the Marquis de Branciforte's administration, a new and very short road was opened via the Venta de Chalco and the small chain of porphyritic mountains of Córdoba, Tesmelucos, and Ocotlán. The advantages of these more direct connections between the capital, the town of Puebla, and the fortress of Perote will be easily recognized if one consults the third and the ninth map in my Atlas of New Spain. IV.41

The new road from Mexico City to la Puebla still presents the small problem of crossing the mountains that separate the Tenochtitlan from the Cholula basin. Conversely, the plateau that extends from the foot of the volcanoes of Mexico City to the Orizaba and Cofre Mountains is a single, arid plain covered with sand, bits of pearly stone, and crusts of salt. The road that runs from Puebla to Veracruz via Jalapa goes through Cocosingo, Acaxete, and Perote. It is as if one were traveling over land leveled by a long period underwater. When the sun warms these plains, they present, at the same elevation as the Saint Bernard pass, the same phenomena of extraordinary suspension and refraction of light that are usually observed only near the seacoast of the Ocean.

The superb road that the *consulado* of Veracruz had built from Perote to Veracruz may rival the roads of the Simplon and Mont Cenis: it is wide and solid, with a very gentle rise. The path of the old route, which was narrow and paved with basaltic porphyry, and that seems to have been built around the middle of the eighteenth century, was not followed: steep climbs were carefully avoided, and the complaint that the engineer extended the road too far will disappear once carriage trade has replaced the transportation of IV.42

1. *Voyage de Chappe,* published by Mr. de Cassini, p. 216.

merchandise on muleback. It will probably cost over fifteen million francs to build this road, but one hopes that such fine and useful work will not be interrupted: it is a subject of the highest importance for those who are the farthest removed from the capital and the port of Veracruz, for as soon as the road is completed, the cost of iron, mercury, brandy, paper, and all other merchandise from Europe will markedly decrease. Mexican flour, which until now was more expensive in Havana than flour from Philadelphia, will be preferred to the latter; sugar and leather exports from the country will increase considerably, and the hauling of products by cart will require fewer mules and horses than are currently necessary. These changes will have a double effect on food commodities: the famines that until now have almost periodically ravaged Mexico will become rarer, not only because the consumption of corn will decrease, but especially because the farmer, moti-

IV.43

vated by the hope of selling his flour in Veracruz, will allot more land for planting corn.

During my stay in Jalapa, in February 1804, the new road built under the direction of Mr. García Conde had been started in the sites presenting the greatest challenges; in other words, at the gully called the *Plan del Río*, and at the *Cuesta del Soldado.* Porphyry columns will be placed along the road to indicate, in addition to distances, elevations above sea level. These inscriptions, which are nowhere to be found in Europe, will be of particular interest to travelers crossing the eastern slope of the Cordillera and will serve to reassure them that they are approaching the happy, elevated region where they need no longer fear the scourges of the *black vomit* and yellow fever.

The old Jalapa road runs from Rinconada to the east over the former Veracruz road, commonly called *la Antigua.* After crossing, below this village, the eponymous river which is almost 200 meters wide, one follows the beach via Punta Gorda and Vergara, or if the tide is high, one takes the Manga de Clavo road, which meets the coast only at the port of Veracruz itself. It would be advantageous to build a bridge over the Río de la Antigua near la Ventilla, where the riverbed is only 107 meters wide; the Jalapa road would then be shortened by more than six leagues, and it would run

IV.44

directly, without meeting the old Veracruz road, from Plan del Río via the Ventilla bridge, Paso de Ovejas, Ciénaga de Olocuatla, and Loma de San Juan toward Veracruz. This change is all the more desirable since it is the section from Encero to the coast that is the most dangerous to the health of the inhabitants in the interior of Mexico when they descend from the Perote plateau and the heights of Jalapa. The stifling heat that dominates this arid, lifeless plain has a strong effect on those whose nervous system is not used to such extreme irritation. This heat, together with the fatigue of the journey, predisposes the organs to accept the noxious miasma of yellow fever more easily; shortening the section of the road that crosses the arid plains of the seaboards would reduce the ravages of this plague.

The road from Mexico City to Veracruz that runs via Orizaba has the least traffic: it goes through Nopaluca, San Andrés, Orizaba, Córdoba, and Cotastla. The group of porphyritic mountains that connect the summits of the Peak of Orizaba and the Coffre de Perote prevent the engineer from tracing a road in a straight line from the capital to the port of Veracruz. On the Jalapa road, travelers circle around the high Coffre Mountain on its northern side; on the Orizaba and Córdoba road, one circles the Peak of Orizaba on its southern slope. One of these roads diverts to the north, the other to the south; the longest detour is made via Orizaba. The latter road would be shortened considerably if instead of going to Veracruz via Cotastla and Venta de Xamapa, one crossed over the mountainous country known as the *Sierra de Atoyaque*. According to an estimate made by the *regidores* of Villa de Córdoba, the cost of building this road would be 1,416,800 piasters.

IV.45

The main items of domestic trade in New Spain are (1) the products and merchandise imported or exported through the two ports of Veracruz and Acapulco, which we shall discuss later; (2) the exchanges transacted between the various provinces, especially between Mexico proper and the *provincias internas*; (3) some products from Peru, Quito, and Guatemala that cross the country before being exported via Veracruz to Europe. Without the high consumption of provisions by the mines, domestic trade would be relatively nonexistent between provinces that enjoy more or less the same climate and thus have the same products. The elevation of the land gives the southern regions of Mexico the average temperature that is necessary for growing European plants. So as we have seen above, the same latitude produces bananas and apples, sugar cane and corn, manioc and potatoes. The nutritious grains that thrive in the wintry weather of Norway and Siberia cover the fields of Mexico in the Torrid Zone: consequently, the provinces located below 17° and 20° latitude rarely require flour from Nueva Vizcaya. Fortunately, corn farming stimulates domestic trade much more than farming European cereals. Since it rarely happens that the corn harvest is equally good over a great expanse of land, a part of Mexico lacks corn while it is plentiful in another part, and the cost of a *fanega* varies in the two neighboring intendancies, often by from nine to twenty livres tournois:[1] in

IV.46

1. See chap. IX, vol. II, p. 413 [p. 528 in this edition].

fact, the corn trade is an item of great importance for the provinces of Guadalajara, Valladolid, Guanajuato, Mexico City, San Luis Potosí, Veracruz, Puebla, and Oaxaca.

Thousands of mules arrive in Mexico City every week from Chihuahua and Durango, bringing leather and tallow in addition to silver ingots, a little wine from Paso del Norte, and flour: in exchange, they bring back woolen goods from the factories in Puebla and Querétaro, merchandise from Europe and the Philippine Islands, iron, steel, and mercury. In our discussion of the connections between the coasts of the South Sea and the Atlantic Ocean,[1] we have observed how useful it would be to bring camels to Mexico. The plateaus where the important roads run are not so high that the cold IV.47 there might be harmful to these animals; they would suffer less than horses and mules do from the aridity of the land and the lack of water and pastures to which the beasts of burden are exposed north of Guanajuato, especially in the desert that separates Nueva Vizcaya from New Mexico. ▾Juan de Reinaga, a Biscayan, had introduced camels[2] in Peru around the end of the sixteenth century; they were still commonly used in Spain for some time even after the destruction of the Moorish empire, although it appears that they did not propagate in Peru. Furthermore, in those barbarous times, the government did not support the introduction of these useful animals, yielding to the conquerors' (*encomenderos*) insistence that breeding beasts of burden there would prevent them from hiring out the indigenous peoples to travelers and merchants for the hauling of provisions and merchandise in the interior of the country.

During wartime, when navigation around Cape Horn is dangerous, a large part of the 80,000 loads[3] *(cargas)* of cacao exported annually from the port of Guayaquil crosses through the Isthmus of Panama and Mexico. The cost of transportation from Acapulco to Veracruz is ordinarily two piasters per *carga*, and this route is preferable whenever cacao from Guayaquil costs IV.48 more than twenty piasters per fanega in Havana. The purchase price on the hillsides of Quito is usually from four to five piasters; the sale price in Cádiz varies between twenty-five and thirty-five piasters, and despite the protracted length of the voyage around Cape Horn, carriage from Guayaquil to Spain is not more than seven or eight piasters per fanega.

1. Chap. II, vol. I, p. 230 [p. 160 in this edition].

2. Garcilaso [de la Vega El Inca, *Historia general del Perú*,] vol. II, p. 328.

3. One *carga* is eighty-one livres; one *fanega* is equivalent to 110 livres in the Castilian weight.

Copper from Guasco, known as Coquimbo copper, often takes the same route as cacao from Guayaquil: in Chile, this copper costs only six or seven piaster per quintal; in Cádiz its regular price is twenty piasters. But since this price rises to thirty-five or forty in wartime, Lima merchants who trade in products from Chile find it advantageous to send copper to Spain via Guayaquil, Acapulco, Veracruz, and Havana. These awkward connections will cease as soon as an active and trade-protective government commissions a good road from Panama to Portobelo and as soon as the isthmus can provide the number of pack animals necessary for transporting products from Quito, Peru, and Chile.

The same reasons that force the inhabitants of Guayaquil to send their cacao in wartime through the kingdom of Mexico also compel merchants in Guayaquil to send the indigos from their country, whose rich color surpasses all other known indigos, over the Tehuantepec road and the Río Huasacualco to Veracruz. Here we must discuss, more amply than we did IV.49 above,[1] the project of a canal in the intendancy of Oaxaca connecting the two oceans, a project that is worthy of the government's undivided attention.

During his first stay in Tenochtitlan, Cortés had already recognized the great importance of the Huasacualco[2] River, as is evident in his third letter to Emperor Charles, dated from *Villa Segura de la Frontera* on October 30, 1520. Keen to discover either a more secure port than the one at Veracruz or a passage from one ocean to the other, which he calls *the secret of a strait*, the Spanish general asked Montezuma for "information on the state and configuration of the eastern coastline of the Anahuac empire. The monarch replied that he himself was unfamiliar with this coastline, but that he would have the entire seaboard with its bays and rivers painted and would provide the necessary guides to accompany the Spanish who intended to explore these regions. The following day, Cortés was presented with a drawing of the entire coastline drawn on a cloth. On this map the pilots recognized the mouth of a great river that they assumed was identical to the opening that they had seen on the coastline (when they arrived at Veracruz), near the Sanmyn Mountains[3] in the province of Maxamalco." Guided by this infor- IV.50

1. Chap. II, vol. I, pp. 203 and 210 [pp. 146 and 150 in this edition]; chap. VIII, vol. II, p. 192 [p. 405 in this edition].

2. In Mexico, the names Huasacualco, Guasacualco, and Goaxacoalosa are written indiscriminately. Corrupting all Mexican names, Cortés calls this river Quacalco.

3. These mountains may be the range of San Martín and the Tuxtla volcano. See chap. VIII, vol. II, p. 205 [p. 412 in this edition], and *Cartas de Hernán Cortés* [*Historia de Nueva-España*], pp. 92 and 351. I have already shown elsewhere [in *Views of the Cordilleras*] that

mation, Cortés sent a small detachment of ten men under Diego Ordaz's orders to identify this river in 1520. At its mouth, the pilots found a depth of only two and half fathoms, but after sailing twelve leagues upstream against the current, they saw that the river was five to six fathoms deep everywhere. The banks of the Huasacualco were then much more populated than they are now.

After the conquest of Mexico City, ▾ Gonzalo de Sandoval conquered the province of Tehuantepec in 1521. Although the pilot ▾ Andrés Niño[1] had observed that there was no strait from the coastline of Nicaragua to the isthmus of Tehuantepec, this isthmus was considered to be no less important, because the proximity of the two oceans and the Huasacualco River

IV.51 offered the first Spanish conquerors the ease of having the material necessary for shipbuilding delivered from Veracruz to the Pacific Ocean coast. Hernando de Grijalva's expedition, which set sail for California in 1534, left from Tehuantepec, just as the ships on which Cortés embarked from Chametla[2] had been built at the mouth of the Río Chimalapa from material transported over the Río Huasacualco. One of these ships was lost crossing the San Francisco sandbar after leaving the *Laguna de Santa Teresa*.

The port of Tehuantepec, which hardly deserves to be called a harbor, has been little frequented since the end of the sixteenth century. Trade from the South Sea has been concentrated in Acapulco and the vessels used for connections with the Philippine Islands have all been built either in Manila or the port of San Blas. The sea also recedes daily from the coastline of Tehuantepec, and anchorage becomes worse and worse from year to year; the sand that the Chimalapa River brings down increases the height and extent of the sandbar. Four leagues now separate Villa de Tehuantepec from the sea, going via la Hacienda de la Zoleta: the best anchorage is at Morro del Carbón, in the salt marshes and the Laguna de Santa Teresa.

IV.52 A happy circumstance toward the end of the last century led the two viceroys Bucareli and Revillagigedo to turn the attention of the government once again to the Isthmus of Tehuantepec and the Río de Huascualco. In

---

in a collection of hieroglyphic manuscripts preserved in the viceroys' palace in Mexico City, there are maps of the valley and lakes of Tenochtitlan painted on cotton canvas by the Aztecs. It has been confirmed to me that the inhabitants of the village of Tetlama near Cuernavaca as well as the people of Tlaxcala have topographical maps that were made before the conquest. Gómara refers to a road map from Xicalanco to Nicaragua drawn up by the inhabitants of Tabasco and presented to Cortés. *Conquista de Mexico*, fol[io] 200.

1. Gómara, *Historia*, fol[io] 113; and *Conquista*, fol[io] 87.

2. See chap. VIII, vol. II, p. 258 [p. 439 in this edition].

1771, a few cannons cast in Manila were found in Veracruz among the artillery of the castle of San Juan de Ulúa. Since it was known that before the year 1767, the Spanish did not sail around either the Cape of Good Hope or Cape Horn en route to the Philippine Islands, and that since Magellan and ▼ de Loayza's first expeditions had departed from Spain, all trade with Asia was conducted by galleon from Acapulco, it was impossible to imagine how the cannons had crossed the continent of Mexico in transit from Manila to the castle of Ulúa. The extremely difficult road from Acapulco to Mexico City and from there to Jalapa and Veracruz made it highly unlikely that they had taken that route. After some research, it was learned from the Tehuantepec Chronicle[1] written by ▼ Father Burgoa, as well as from the tradition preserved among the inhabitants of the isthmus of Huascualco, that the cannons that were cast on the island of Luzon and unloaded at the San Francisco sandbar had sailed up the bay of Santa Teresa and the Río Chimalapa; that they had been transported to the Río del Malpaso via the Chivela farm and the Tarifa forest; and that after having embarked once again, they were brought down the Río Huasacualco to its mouth on the Gulf of Mexico. IV.53

It has since been rightly observed that the same road that was frequented at the beginning of the conquest could become very useful again in opening a direct connection between the two seas. Viceroy Antonio Bucareli gave orders to two skilled engineers, ▼ Mr. Agustín Cramer and Mr. Miguel del Corral, to examine in the utmost detail the terrain between the Huasacualco sandbar and the Tehuantepec harbor, instructing them to verify as well whether, as was vaguely assumed, one of the three small rivers—the Ostuta, Chicapa, or Chimalapa—might connect with the two seas through its tributaries. I drew up my map of the Isthmus of Tehuantepec based on the travel journals of these two engineers, the first of whom was the king's lieutenant at Ulúa castle. They had discovered that no river emptied its waters both into the Great Ocean and the Atlantic Ocean; that the Río Huasacualco did not originate (as had been confirmed to the viceroy) very near the town of Tehuantepec, but that by traveling upstream beyond the cataract on that same river, even as far as the former *desembarcadero* of Malpaso, one was still more than twenty-six leagues away from the South Sea coast. They observed that a chain of low-lying mountains divides the water between the Antillean Sea and the Gulf of Tehuantepec. This small Cordillera extends from east to west from the Cerros de los Mixes, formerly inhabited by a savage IV.54

1. Burgoa, *Palestra historial ó Cronica de la Villa de Tehuántepec*. Mexico City, 1674.

and warlike tribe,[1] toward the high plateau of Portillo de Petapa. The engineer Cramer confirms, however, that south of the village of Santa María de Chimalapa, the mountains form a group rather than an uninterrupted chain and that "there is a lateral valley there where a canal connecting the two seas might be dug." This canal, which would connect the waters of the Río de Chimalapa and the Río del Paso (or Malpaso), would be only six leagues long; boats could sail up the Río Chimalapa, which provides easy navigation, from Tehuantepec as far as the village of San Miguel. From there, they could continue on to the Río del Paso through the canal proposed in Count de Revillagigedo's day. That river empties into the Río de Huascualco near the *Bodegas de la fábrica*; it is extremely difficult to navigate because of the seven rapids (*raudales*) that are found between its sources and the mouth of the Río de Saravia.

It would be very important to have this terrain reexamined by well-trained engineers to decide whether (as Mr. Cramer believed) the *canal between the two seas* might be built without sluices or *inclined planes*, and whether the riverbeds of the Paso and the Chimalapa could be deepened by using gunpowder to explode rocks. Rich in livestock, the isthmus could IV.55 provide valuable products for trade in Veracruz. The beautiful Tehuantepec plains could be irrigated through drainage channels from the Río de Chimalapa: in their present state, these plains already produce a small amount of indigo and cochineal of superior quality.

Before the felling of cedar and mahogany wood (Cedrela odorata and Swietenia mahagoni) began on the islands of Cuba and Pinos, builders' yards in Havana took lumber from the dense forest that covers the northern slope of the Cerros de Petapa and Tarifa. At that time the isthmus of Tehuantepec was heavily trafficked. The ruins of several houses that are still seen on both sides of the Huascualco River date from that time. Cedar and mahogany were sent out from the Bodegas de Malpaso.

To avoid the seven *rapids* on the Río del Paso, a new port (*desembarcadero*) was established at the mouth of the Río Saravia in 1798. Salted meat (*tasajo*) from Tehuantepec, indigo from Guatemala, and cochineal from Oaxaca were carried over this route to Veracruz and Havana. A road from Tehuantepec to the new port of La Cruz was opened via Chihuitán, Llano Grande, Santa Maria Petapa, and Guchicovi; the distance over this road is thirty-four leagues. Products intended for Havana do not travel downriver to the mouth of the Río Huasacualco or the small eponymous fort for fear of

1. *Cartas de Cortés* [Lorenzana, *Historia de Nueva-España*,] p. 372.

exposing the canoes to the north winds during the rather long passage from the Huascualco sandbar to the port of Veracruz. Merchandise is unloaded at Paso de la Fábrica; from there it is carried on muleback through the village of Avayucán to the banks of the San Juan River, where it embarks again on very large dugout canoes and is transported via the Tlacotalpan sandbar to the port of Veracruz. IV.56

In the past few years, Cedrela trunks needlessly felled by some commissioners of the Royal Navy have encumbered the Tarifa and Petapa roads. These trunks, the best in the forest, are rotting away with no thought of transporting them to Havana. The inhabitants of the Spanish colonies are inured to these measures, which bring no results: they attribute them to the lack of seriousness with which the ministry accepts and abandons projects. Shortly before my stay on the banks of the Orinoco, the *comisionados del Rey* went upriver the mouth of the Río Caroni to take stock of all the trees that might be useful for shipbuilding. They measured their diameter and height and marked so many trunks of Cedrela, Laurus, and Caesalpinia that all the shipbuilder's yards in Europe would not have been able to use them in ten years. No trees were cut and this long and difficult labor had no other outcome but to cause expense to the government.

IV.57 If new studies were to prove that the construction of a canal on the isthmus of Tehuantepec would not be advantageous, then the government should at least encourage the inhabitants of this province to improve the road through Portillo de Petapa to the new port of la Cruz. A portion of the products from the kingdom of Guatemala and the intendancies of Oaxaca and Tehuantepec could always be brought over this route to Veracruz. Upon my departure from New Spain in 1804, it cost thirty piasters per load to haul goods on muleback from Tehuantepec via Oaxaca to Veracruz. The mule drives took three months to cover a road that is not even seventy-five leagues long as the crow flies. By transporting products via the isthmus route and the Huascualco River, it would cost only sixteen piasters to transport one load, and since it takes only ten days to go from Paso de la Fábrica to Veracruz, one would gain nearly seventy days over the entire journey. In 1803, the consulado of Veracruz, who has shown the most commendable enthusiasm for opening this new road to domestic trade, abolished the duty of 5 percent to which the merchandise shipped on the Río Huasacualco was subjected. This duty was known by the ridiculous name of the *duty of a hot country* (*derecho de tierra caliente*). I believe that it would be important to publish in the greatest detail anything that is related to the proposed connections between the two seas. The topography of the isthmus of Tehuantepec

IV.58 is completely unknown in Europe; the information I have just given should dispel any doubt that this point on the globe is equally deserving of the government's attention as the Río Chamaluzon, Lake Nicaragua, the isthmus of Panama, Cúpica Bay, and the Raspadura ravine in Chocó.

Depending on the position of the coastline, *external trade* in New Spain is naturally comprised of trade from either the South Sea or the Atlantic Ocean. The ports on the eastern coastline are Campeche, Huasacualco, Veracruz, Tampico, and Nuevo Santander, if one can use the word "port" to describe harbors surrounded by shallow water or mouths of rivers blocked by sandbars that provide slight shelter against the furor of the north winds. We have discussed above[1] in chapter three the physical causes that give the Mexican coastline facing Europe its special character. We have also already discussed the futile attempts made since 1524 to discover a safer port than Veracruz. The vast seaboard that extends from Nuevo Santander to the north and northwest is still very unfamiliar, and one might repeat today what Cortés wrote to Emperor Charles V three years after the siege of Tenochtitlan, "that the secret of the coast extending from the Río de Panuco to Florida remains to be discovered."[2]

IV.59 For centuries, almost all maritime trade in New Spain has been concentrated in Veracruz. A glance at the eleventh plate of our Mexican Atlas shows that the pilots in Cortés's squadron were correct in comparing the port of Veracruz to a sack with a hole in it. Between *Punta Gorda* and the small *Cabo Mocambo*, the Island of Sacrifices, near which vessels are quarantined, and the shallows of *Arecife del Medio, Isla Verde, Anegada de Dentro, Blanquilla, Galleguilla,* and *Gallega* form with the mainland a sort of cove that opens to the northwest. When the north winds (*los nortes*) blow at full force, it happens that vessels moored at the foot of the castle of San Juan de Ulúa lose anchor and drift eastward: pushed from the channel that separates the Island of Sacrifices from Isla Verde, they are blown by winds to the port of Campeche in twenty-four hours. During a storm eighteen years ago, a ship of the *la Castilla* line, moored by nine cables to the bastion of castle Ulúa tore out the bronze rings attached to the bastion wall and ran aground on the coast in the port itself, near the shallows of *los Hornos* west of Punta Mocambo. By an extraordinary circumstance, the great quarter circle that the unfortunate Chappe used to make his observations, which the Academy of Sciences in Paris had asked him to return to verify the divi-

1. See chap. III, vol. I, pp. 286–291 [pp. 191ff in this edition].

2. *Cartas de Cortés* [Lorenzana, *Historia de Nueva-España*], pp. 340 and 382.

sions, was lost on this vessel. Good mooring in the port of Veracruz is found between Ulúa castle, the town, and the la Lavandera shallows. The water is up to six fathoms deep near the castle, but the channel through which one enters the port is scarcely four fathoms deep and 380 meters wide. IV.60

According to customs declarations, and taking the average of several years of peacetime, the main items[1] of *exportation from Veracruz* are:

Gold and silver either as ingots or converted into coin and plate, *seventeen million piasters.*

Cochineal (*grana*, *granilla*, and *polvos de grana*), approximately four thousand *zurrones*, or four hundred thousand kilograms, at a value of *two million four hundred thousand piasters.*

Sugar, five and a half million kilograms, *one million three hundred thousand piasters.*

Flour, at a value of *three hundred thousand piasters.*

Mexican indigo, eighty thousand kilograms, at a value of *two hundred eighty thousand piasters.*

Salted meats, dried legumes, and other comestibles, *one hundred thousand piasters.*

Tanned hides, *eighty thousand piasters.*

Sarsaparilla, at a value of *ninety thousand piasters.*

Vanilla, *sixty thousand piasters.* IV.61

Jalap, one hundred twenty thousand kilograms, *sixty thousand piasters.*

Soap, *fifty thousand piasters.*

Campeche wood, *forty thousand piasters.*

Tabasco peppers, *thirty thousand piasters.*

In wartime, indigo from Guatemala and cacao from Guayaquil are very important commodities for the Veracruz trade. We are not listing them in this table, however, because we have decided to restrict the latter to products that are indigenous to New Spain.

Imports to Veracruz include the following articles:

Fabrics (*ropas*), linen and cotton cloth, woolens and silks, at a value of *nine million two hundred thousand piasters.*

1. Compare to vol. II, pp. 201, 437, 475 [pp. 410, 540, and 560 in this edition]; vol. III, pp. 16, 46, 48, 52, 59, 72, 394 [pp. II: 11, 27–28, 30, 34, 41, and 232 in this edition]; vol. IV, p. 11 [p. II.296 in this edition].

Paper, three hundred thousand reams, *one million piasters.*
Brandy, thirty thousand casks [barriques], *one million piasters.*
Cacao, twenty-four fanegas, *one million piasters.*
Mercury, eight hundred thousand kilograms, *six hundred fifty thousand piasters.*
Iron, two and a half million kilograms, *six hundred thousand piasters.*
Steel, six hundred thousand kilograms, *two hundred thousand piasters.*
IV.62 Wine, forty thousand barrels, *seven hundred thousand piasters.*
Wax, two hundred fifty thousand kilograms, *three hundred thousand piasters.*

We roughly estimate in an average year,

Export via Veracruz	22 million piasters
Import via Veracruz	15 million piasters
Trade activity	37 million piasters

We shall first present here the state of trade in Veracruz, published by the *consulado* at the end of the years 1802 and 1803.

Tableau I
Veracruz Trade Balances in 1802

A. Spain's Imports in Mexico, in Agricultural Products and Products of Domestic Industry

AGRICULTURAL AND INDUSTRIAL PRODUCTS	AMOUNTS		VALUE IN STRONG PIASTERS
Spirits	29,695	*barriques*	1,283,914
White wine	40,335	*barriques*	683,079
Red wine	21,657	*barriques*	331,882
The same bottled	13,159	*bottles*	8,542
Vinegar	3,374	*barriques*	48,149
Raisins	2,501	*quintals*	27,417
Almonds	2,590	*quintals*	81,545
Olives	9,519	*jars*	22,205
Oil	32,099	*arrobas*	96,297
Saffron	5,187	*livres*	99,765
Aromatic plants	185	*quintals*	2,009

IV.63

AGRICULTURAL AND INDUSTRIAL PRODUCTS	AMOUNTS		VALUE IN STRONG PIASTERS
Capers	202	*barils*	2,714
Nuts	227	*quintals*	3,240
Figs	320	*quintals*	2,491
Oregano	2,450	*livres*	306
Cumin	242	*arrobas*	1,992
Grapes	1,170	*cruches*	3,510
Sardines	93	*barils*	1,347
Anchovies	10	*arrobas*	50
White paper	274,211	*rames*	885,884
Ruled paper	7,906	*rames*	4,577
Thread	376	*quintals*	11,451
Corks	699	*milliers*	5,177
Canteens (*frasqueras*)	492		20,583
Ham	142	*arrobas*	1,380
Fine liqueurs	852	*arrobas*	11,766
Soap	119	*quintals*	1,785
Earthenware	3,041	*dozens*	4,651
Beer	71,876	*bottles*	45,779
Cider	1,920	*bottles*	968
Sausages	3,368	*livres*	1,684
Vermicelli	233	*quintals*	4,823
Grindstones	513		1,282
White iron	289	*cases*	10,115
Iron bars	42,440	*quintals*	382,480
Manufactured iron	42,440	*quintals*	78,882
Steel	7,020	*quintals*	132,392
Cordages	459	*quintals*	6,442
Canvas, woolens, *tercios*,	5,651		2,210,552
cottons, silks, *cajones*	3,293		3,889,891
Gauze, in *baules*	899		606,130
Cajones toscones	3,415		520,182
	Total value, in piasters		11,539,210

[IV.64] *B. Spanish Imports in Mexico, in Agricultural Products and Products from Foreign Industry*

AGRICULTURAL AND INDUSTRIAL PRODUCTS		AMOUNTS		VALUE IN STRONG PIASTERS
Butter		15,884	*livres*	4,678
Cheeses		259	*quintals*	10,344
Wine		16,920	*bottles*	12,690
White paper		87,665	*ramas*	328,714
Steel		7,050	*quintals*	126,605
Earthenware		9,234	*dozens*	23,085
White iron		996	*cases*	32,400
Tin trunks		12		390
Coarse linen		50	*pieces*	2,000
Candles		337	*livres*	270
Cod		340	*quintals*	8,500
Cloves		14,737	*livres*	47,204
Pepper		37,465	*livres*	22,657
Cinnamon		199,965	*livres*	661,569
Canvas, woolens, cottons, silks, and gauze in	*tercios*	18,529		6,572,108
	cajones	501		3,889,891
	baules	24		8,533
	cajones toscones	5,200		595,458
	Total value, in piasters			8,851,640

C. America's Imports (Spanish Colonies) from Mexico

AGRICULTURAL AND INDUSTRIAL PRODUCTS	AMOUNTS		VALUE IN STRONG PIASTERS
Wax	20,571	*arrobas*	322,359
Coffee	344	*quintals*	6,060
Cacao from Caracas	1,984	*fanegas*	106,234
Cacao from Maracaibo	18,709	*fanegas*	687,928
Id. from Tabasco	6,952	*Id.*	315,902
Starch	1,746	*arrobas*	2,550
Wood from Campeche	28,019	*quintals*	38,958
Indigo	4,910	*pounds*	4,910
Salted fish	6,586	*arrobas*	15,185
Tortoise shells	570	*pounds*	4,910
Salt	18,699	*fanegas*	33,316
Bags (*costales*)	130,800		42,388
Straw hats	5,084	*dozens*	7,948
Strings (*henquen*)	1,964	*arrobas*	6,065
Ropes	259	*pieces*	2,842
Harpoons (*tiburoneras*)	1,057	*arrobas*	2,379
Blankets	716		2,229
Hammocks	325		846
Cinchona	1,030	*pounds*	5,150
Shoes	62 and a half	*dozens*	302
Miscellaneous articles			1,224
Total value, in piasters			1,607,729

IV.65

D. Mexican Exports to Spain

AGRICULTURAL AND INDUSTRIAL PRODUCTS		AMOUNTS		VALUE IN STRONG PIASTERS
	Grana fina	43,277	*arrobas*	3,303,470
Cochineal	*Granilla*	2,355	*Id.*	50,472
	Polvos de grana	1,322	*Id.*	14,615
Indigo		1,480,570	*pounds*	3,229,796
Vanilla		1,793	*milliers*	65,076
IV.66 Sugar		431,867	*arrobas*	1,454,240
Roucou		195	*arrobas*	1,419
Cotton		8,228	*Id.*	28,644
Pepper from Tabasco		2,920	*quintals*	15,622
Wood from Campeche		17,389	*Id.*	23,644
Cacao from Soconuzco		1,724	*pounds*	1,078
Coffee		272	*quintals*	4,360
Greenbrier		461	*Id.*	2,988
Jalap		2,921	*Id.*	68,760
Balm		48	*arrobas*	1,200
Cinchona		700	*pounds*	612
Pelts				14,626
Tortoise shells		439	*Id.*	2,290
Miscellaneous articles				3,516
Copperplate		670	*quintals*	15,745
Gold coins and gold articles				62,663
Silver articles				52,622
Silver coins				25,449,289
Total value, in piasters				33,866,219

E. Mexican Exports to Other Parts of Spanish America

AGRICULTURAL AND INDUSTRIAL PRODUCTS	AMOUNTS		VALUE IN STRONG PIASTERS
Flour	22,858	*tercios*	404,051
Sugar	7,265	*arrobas*	22,195
Cacao from Guayaquil	631	*fanegas*	15,821
Wax	368	*arrobas*	6,426
Wood from Campeche	6,219	*quintals*	7,773
Furs	2,300		2,403
Tallow	1,675	*arrobas*	6,711
Food			100,461
Woolens			9,062
Tar	403	*barils*	1,012
Bags	7,690		2,419
Standard earthenware	239	*cases*	2,019
Gold leaf			7,041
Soap	1,946	*Id.*	55,832
Pita [agave]	1,235	*arrobas*	9,504
Leather			82,353
Miscellaneous articles			66,912
Copper plate	895	*quintals*	20,542
Copper ornaments	13,947	*pounds*	5,844
Lead	330	*quintals*	2,779
Silver works			15,417
Silver coins			3,730,171
Gold coins			4,400
Total value, in piasters			4,581,148

IV.67

IV.68 RESULTS:
Trade Balance of Veracruz in 1802

		PIASTERS		PIASTERS
Spain's imports	in domestic products	11,539,219		20,390,859
	in foreign products	8,831,640		
Spain's exports				33,866,219
Export surplus				13,475,360
The metropole's trade with Veracruz				54,257,078
				piasters
America's imports				1,607,729
America's exports				4,581,148
Export surplus				2,973,419
America's trade with Veracruz				6,188,877
				piasters
Total imports				21,998,588
Total exports				38,447,367
Total trade of Veracruz				60,445,955

In 1802, Veracruz's trade was carried out on 558 ships, of which:

From	Spain	148	To	Spain	112
	America	143		America	155
Arrived in the port of Veracruz		291	Departed for Veracruz		267

IV.69 OBSERVATIONS

I. "The *Consulado de Veracruz* publishes annually these states of trade to inform merchants about consumption in New Spain and guide their speculations: he regrets that he cannot show in greater detail the value of canvas, woolens, cottons, and silks in crates (*cajones* and *baules)* that are not opened at the customs office. It is true that the *cajones arpillados* generally contain silks; the *cajones toscos* contain hardware, spices, crystal, glassware, crockery, hats, shoes, or boots; the *tercios arpillados* contain bolts of linen and cotton, sheets, and baize; and finally the *baules* contain silk and cotton stockings, blond-lace and other laces, handkerchiefs, dresses, and other luxury goods."

2. "Merchandise and products imported for governmental use (*para la Real Hacienda*), which would have increased the total amount of imports to *twenty-one and a half million piasters*, are not included in this inventory, since the government has received 150,000 reams of paper for its cigar factories, 34,000 quintals of mercury, and other articles whose value is two million piasters. Exports of gold and silver coin minted on behalf of the crown totaled nineteen and a half million piasters, on which twelve and a half were sent to Spain, and seven [million] to other Spanish colonies in the Americas."

3. "Locally manufactured products have been in great demand: it was not possible to fill all orders. This fact should encourage manufacturers to boost productivity in their workshops."

4. "Imports of European brandies would have been much greater without the increasing consumption of sugarcane-based spirits made in Mexico. Sherry and Rioja are the most sought after."

5. "There are still grounds for complaints about damage caused by the poor packing of merchandise sent to the Americas: the example given by Cádiz is not emulated by other ports on the Peninsula." IV.70

6. "Most of the indigo exported through Veracruz is from the kingdom of Guatemala: this valuable product arrives in wartime over the Oaxaca road; one hopes that it will continue to be exported through Veracruz in peacetime, if the government frees up trade on the Río Huascualco."

7. "Despite the large number of ships that arrived this year in Veracruz, *there has not been a single shipwreck*, nor any other horrible event on the sea, out of two hundred sixty voyages from Europe to the Americas. The black vomit, a merciless disease that has been rife from the month of April to October, has cut down fifteen hundred individuals, Europeans as well as inhabitants of the cold regions of Mexico. This disease has posed tremendous obstacles to domestic trade, and mule drivers are afraid to approach the port of Veracruz."

8. "We must not consider all ships indicated in the table under the heading of ships coming from the Americas as vessels used in trading with the American colonies: Spanish vessels often take on silver in Mexico and then sail to Havana and Caracas where they load sugar and cacao."

9. "In the course of the year 1802, one hundred ninety-six lawsuits were adjudicated by the tribunal of the consulado; only a single lawsuit remains to be decided."

***Veracruz*, February 19, 1803.**

IV.71 Tableau II

Trade Balance of Veracruz in 1803

A. Imports of National Agricultural Products from Spain to Mexico

NAMES OF GOODS AND COMMODITIES	QUANTITIES		VALUE IN STRONG PIASTERS
White wine	7,597	*hogsheads*	142,367
Red wine	17,520	*Id.*	267,870
Bottled wine	23,455		8,974
Vinegar	705	*Id.*	8,583
Brandy	31,721	*Id.*	1,105,859
Olive oil	12,479 and a half	*arrobas*	37,722
Saffron	17,174 and a half	*pounds*	344,087
Almonds	1,298	*quintals*	34,825
Filberts	255 and a half	*Id.*	4,201
Olives	21,611	*jars*	30,609
Capers	193	*barrels*	5,609
Aromatic herbs	68	*quintals*	659
Linseed oil	125	*Id.*	250
Dried grapes	1,107	*Id.*	12,749
Figs	631	*Id.*	1,604
Prunes	36 and a quarter	*Id.*	797
Candied fruits	259	*arrobas*	380
Hams	147	*Id.*	1,341
Sausages	175	*doz.*	350
Spices			1,287
Fruits preserved in brandy	600	*canteens*	300
	Total value in piasters		2,010,423

NAMES OF GOODS AND COMMODITIES		QUANTITIES		VALUE IN STRONG PIASTERS
White paper		137,958	*reams*	502,812
Ruled paper		6,644	*Id.*	3,171
Thread		111 and three-quarter	*quintals*	3,029
Corks		1,192	*thousand*	5,912
Ordinary delft		11,482	*doz.*	11,126
Wax candles		233	*arrobas*	4,916
Canteens		77		2,626
Fine liqueurs		373	*arrobas*	4,409
Beer		14,134	*bottles*	12,035
Vermicelli		746	*quintals*	12,532
Salt fish				5,006
Paving stones (slabs)		6,307		4,857
Chairs		400		1,100
Steel		4,052 and a half	*quintals*	75,769
Bar iron		45,640	*Id.*	564,816
Wrought iron		3,064	*Id.*	53,995
Nails		142 and a half	*Id.*	1,183
Canvas, woolens, cottons, and linen, muslins, silks and stockings in	*Tercios arpillados*	4,405		2,513,868
	Cajones arpillados	2,570		3,685,524
	Cajones tocos	1,513		352,116
	baules	937		783,578
Total value in piasters				8,604,380

C. Imports of Agricultural Products and Products of Foreign Industries from Spain to Mexico

NAMES OF GOODS AND COMMODITIES		QUANTITIES		VALUE IN STRONG PIASTERS
Butter		3,660	*pounds*	2,747
Cheese		52 and a half	*quintals*	1,840
Sausages		884	*pounds*	1,295
IV.73 Cod		200	*quintals*	5,000
Beer		1,455	*bottles*	850
Canvas		48	*pieces*	1,536
Canteens (*frasqueras*)		273		13,250
Delftware				66,256
Iron		100	*quintals*	700
Cinnamon		20,512	*pounds*	68,713
Cloves		6,176	*Id.*	18,419
Pimento		380	*Id.*	380
White Paper		18,182	*reams*	64,163
▼ Grand aigle paper		24	*Id.*	528
Steel		5,966 and a half	*quintals*	108,561
White iron		553	*cases*	14,742
Genoa sharpening stones		1,500		1,125
Canvas, woolens, silks, linens, muslins, and stockings in	*Tercios arpillados*	13,348		5,884,467
	Cajones arpillados	470		570,461
	Cajones tocos	5,260		971,908
	baules	101		81,545
Total value in piasters				7,878,486

D. Imports from America (Spanish Colonies) to Mexico

NAMES OF GOODS AND COMMODITIES	QUANTITIES		VALUE IN STRONG PIASTERS
Cocoa from Maracaibo	7,965	*fanegas*	235,040
Id. from Tabasco	13,551 and a half	*Id.*	470,229
Coffee	474	*quintals*	10,720
Havana wax	24,470	*arrobas*	455,760
Campeche wax	582 and a half	*Id.*	6,281
Campeche wood	38,444 quintals	*quintals*	57,045
Starch	1,711	*arrobas*	4,079
Rice	619 and a half	*Id.*	466
Pitch	338	*hogsheads*	2,028
Tar	548	*Id.*	2,760
Sacks (*sacas*)	21,697		5,421
Sacks (*costales*)	132,811		35,450
Straw hats	3,082	*doz.*	2,413
Packthread	3,329 and a half	*arrobas*	7,685
Matches	442 and a quarter	*Id.*	2,187
Blankets and hammocks	883		1,490
Salt	31,783	*fanegas*	47,037
Salt fish	4,000	*arrobas*	14,050
Cables			4,250
Tortoise shell	826	*pounds*	3,150
Various articles			5,887
Total value in piasters			1,373,428

E. Exports from Mexico to Spain

IV.75

NAMES OF GOODS AND COMMODITIES		QUANTITIES		VALUE IN STRONG PIASTERS
Cochineal	grana	27,251	*arrobas*	2,191,399
	granilla	1,573	*Id.*	40,226
	Polvo de grana	786	*Id.*	7,048
Indigo		149,069	*pounds*	263,729
Vanilla		968 and a half	*milliers*	31,625
Sugar		483,944	*arrobas*	1,495,056
Cocoa from Guayaquil		3,995 and a half	*fanegas*	98,794
from Caracas		480 and a half	*Id.*	17,298
from Maracaibo		1,739 and a half	*Id.*	53,936
from Soconuzco		3,959	*pounds*	2,599
Campeche wood		26,635 and a half	*quintals*	49,019
Hides and furs				22,549
Pimento de Tabasco		5,755 and a half	*Id.*	36,981
Cotton seed		17,327	*Id.*	35,910
Roucou [Annatto]		374	*arrobas*	3,838
Wood for furniture				14,345
Sarsaparilla		4,912 and a half	*quintals*	86,980
Jalap		2,281 and a half	*Id.*	61,971
Balms				5,000
Silver				7,356,560
Gold				142,229
Total value in piasters				12,017,072

F. Exports from Mexico to Other Parts of Spanish America

NAMES OF GOODS AND COMMODITIES	QUANTITIES		VALUE IN STRONG PIASTERS
Flour	19,496	*tercios*	275,905
Sugar	6,348	*Id.*	19,826
Cocoa from Guayaquil	459 and a half	*fanegas*	12,429
Campeche wood	6,871	*quintals*	11,792
Hides	3,000		3,161
Cochineal	152	*arrobas*	12,160
Skins			71,905
Cotton	5,974	*Id.*	11,397
Soap	1,766	*crates*	44,350
Gold leaf			1,650
Serge	14,732	*varas*	4,705
Aniseed	1,022 and a half	*arrobas*	1,802
Delfware	692	*crates*	2,220
Baize	1,300	*varas*	1,673
Misc. articles			40,496
Comestibles			83,267
Wrought copper	14,444	*pounds*	8,849
Tin	58 and a half	*quintals*	1,483
Lead	100	*Id.*	900
Silver			1,834,146
Gold			21,730
Total value in piasters			2,465,846

IV.76

IV.77 Results

VERACRUZ TRADE BALANCE IN 1803

		PIASTERS	PIASTERS
Imports from Spain	of domestic products	10,614,803	
	of foreign products	7,878,486	18,493,289
Exports to Spain			12,017,072
Difference in favor of imports			6,476,217
Total trade of the mother country with Veracruz			30,510,361
Imports from the Americas			1,373,428
Exports from the Americas			2,465,846
Difference in favor of exports			1,092,418
Total trade of the Americas with Veracruz			3,839,274
Total imports			19,866,717
Total exports			14,482,917
Total amount of Veracruz trade			34,349,634

In 1803 the Veracruz trade was carried on by 419 vessels of which:

103 came from Spain and 82 were bound for Spain: total 214

111 came from the Americas; 123 were bound for the Americas: total 205

OBSERVATIONS

1. "Since the balance table drawn up by the Consulado de Veracruz received the approval of the court as well as that of all state entities, everything related to trade in New Spain has continued to be publicized widely. Not included among the import and export items are 50,000 quintals of mercury, 280,000 IV.78 reams of paper destined for the tobacco factory, 4,000 quintals of iron sent off in warships, 12,300 quintals of copper plate, and five million piasters sent to Spain, as well as 1,200,000 piasters shipped to the Antilles for the maintenance of fortresses, because all these items were exported and imported on behalf of the government."

2. "There were three shipwrecks this year on the Island of Cancún and the shoals of Alacran: the insurance company founded on July 17 covered 746,000

piasters' worth of losses in six months. Political circumstances in Europe and the fear of a naval war have blocked trade in Veracruz, leading to a significant reduction in activity from the preceding year."

Veracruz, January 28, 1804.

These tables of the Veracruz trade, published by the *Consulado*, suggest that the combination of the merchandise imported on behalf of the government and the items on which merchants have speculated would yield:

VERACRUZ TRADE	IN 1802 VALUE		IN 1803 VALUE	
	In piasters	In liv. tournois	In piasters	In liv. tournois
Exports	57,947,000	304,221,750	20,922,000	109,840,500
Gold and silver	41,800,000	256,200,000	15,554,000	31,658,500
Agricultural products	9,147,000	48,021,750	5,368,000	28,182,000
Imports	24,100,000	126,525,000	22,975,000	120,618,750
Total trade	82,047,000	430,746,750	43,897,000	230,459,250

There was extraordinary trade activity in one of these years because, after a long naval war, Europe began to enjoy the benefits of peace; the trade table for the other year was less impressive because, beginning in June, fear of an imminent war halted exports of precious metals and agricultural products from New Spain. IV.79

The printing of the first edition of this work had been entirely completed when I received via Spain the trade inventory, printed in Veracruz in the years 1804, 1805, and 1806. Mexico continued to enjoy peace until 1805; since then, the naval war and other political circumstances have thoroughly hindered trade relations. Although what has resulted from this position is such an extraordinary state of affairs that the *trade balance* cannot inform us of any increase or reduction in national wealth, it seemed of interest to me to record here the most recent statistical figures I was able to obtain for this part of the Spanish colonies in the Americas.

Veracruz Trade in 1804

		PIASTERS	PIASTERS
Imports from Spain	in domestic products	10,412,324	14,906,060
	in foreign products	4,493,736	
Imports from the Americas			1,619,682
Exports from Veracruz	to Spain	18,033,371	21,457,882
	to the Americas	3,424, 511	
Total trade movement			37,983,624

Among the national products imported to Veracruz from Spain, there were 48,735 casks of brandy, valued at 1,235,130 piasters; 43,162 casks (value: 837,776 piasters) of red and white wine; 20,946 arrobas (value: 78, 456 piasters) IV.80 of oil; 19,721 livres (value: 287,057 piasters) of saffron; 79,200 bottles (value: 78,456 piasters) of beer; 236,381 reams (value: 486,583 piasters) of paper; 73,827 quintals (value: 812,707 piasters) of iron; 3,108 quintals (value: 53,052 piasters) of steel; and more than six million piasters' worth of silks, woolens, canvas, muslins, and hats packed in crates that the merchants are not expected to open at the customs house.

Among the foreign products imported from Spain, there were four million piasters' worth of silks, fabrics, linens and other textiles; 47,236 livres (value: 163,171 piasters) of cinnamon; 28,167 pounds (value: 85,952 piasters) of cloves [clove spikes], and 2,997 quintals (value: 51,477 piasters) of steel.

Among the American products imported from other Spanish colonies into Veracruz, there were: 27,814 arrobas (value: 576,836 piasters) of wax from Havana; 1,928 arrobas (value: 461,845 piasters) of cacao from Tabasco; 8,141 fanegas (value: 2,055 piasters) of cacao from Caracas; 49,535 quintals (value: 100,219 piasters) of timber from Campeche, and 18,496 fanegas (value: 37,845 piasters) of salt.

Among the local products exported from Mexico to the metropole, there were 381,509 arrobas (value: 1,097,505 piasters) of sugar; 11,737 arrobas (value: 1,220,193 piasters) of fine cochineal, the product of a very modest harvest; 867 arrobas (value: 24,414 piasters) of *granilla*; 464 arrobas (value: 5,816 piasters) of powdered cochineal; 189,397 livres (value: 367,302 piasters) of indigo; 37,797 quintals (value: 77,485 piasters) of timber from Campeche; 1,818 quintals (value:

62,411 piasters) of jalap; 7,169 quintals (value: 96,734 piasters) of sarsaparilla; 1,014 thousands (value: 111,195 piasters) of vanilla; and 3,786 fanegas (value: 124,819 piasters) of cacao from Tabasco. Additionally, 18,801 fanegas (value: 460,585 piasters) of cacao from Guayaquil were exported. Exports of minted coins were valued at 16,847,843 piasters. Havana received 26,371 *trosos* (value: 417,709 piasters) of Mexican flour. IV.81

In 1804, 107 ships from Spain and 123 ships from Spanish colonies in the Americas reached Veracruz. Not included in this inventory are 13,500,000 piasters exported on behalf of the King of Spain and 20,000 quintals of mercury imported on behalf of the government.

Trade in Veracruz in 1805: Imports from Spain of local products: 1,514,473 piasters (including 60,617 reams or 582,769 piasters' worth of paper alone); in foreign products and merchandise, 574,963 piasters. Imports from the Americas, 1,262,907 piasters (including 19,964 arrobas or 547,304 piasters' worth of Havana wax alone). Exports to Spain, 10,200 piasters, to the Americas, 330,546 piasters. Exports via neutral ships, 562,048 piasters. Total trade activity, 4,355,137 piasters. The number of ships that entered Veracruz: from Spain, twenty-seven; from the Americas, seventy-seven.

Trade in Veracruz in 1806: Imports from Spain of Spanish products: 1,815,579 piasters; of foreign products, 327,295 piasters. Imports from the Americas: 1,499,244 piasters. Imports via neutral ships, 3,485,655 piasters. Exports to Spain, 803,037 piasters; to the Americas, 574,191 piasters; via neutral ports, 4,101,534 piasters. Thus, total imports, 7,137,773 piasters. Total exports, 5,478,762 piasters. Overall trade activity, 12,616,535 piasters. In 1806, eight ships from Spain entered Veracruz, ninety ships from other Spanish colonies, and from neutral ports, thirty-seven.

Together, this trade inventory and the preceding ones suggest that in an average year during the three years of peace (1802, 1803, and 1804), total imports from Veracruz (not counting fraudulent trade) amounted to 20,700,000 piasters; exports, not including gold and silver (minted or worked), 6,500,000 piasters. IV.82

	MILLIONS		MILLIONS
1802 imports	21 and a half	Exports	9
1803	23		5 and one-fifth
1804	17 and a half		5

These figures show that in the present state of its civilization and manufacturing, the vast country of New Spain needs between *one hundred* to *one hundred ten million* francs' worth of products and foreign merchandise. By granting full freedom of trade with China and India to Acapulco and San Blas, Mexico could acquire cotton fabric, silks, paper, spices, and perhaps even mercury directly from Asia; these circumstances would reduce imports from Europe by more than *twenty million* francs. The more extensive relations between the Americas and Asia become, the less gold and silver the New Continent will pour annually into trade with Europe. The effects of this trade revolution will be felt sooner among us than those produced by the establishment of new factories and the late awakening of local industry.

For centuries, trade between Mexico and the metropole had never been as hindered as it was in 1805. The value of exports from Veracruz to Spain that year was only 10,200 piasters, whereas in an average year it is twenty-two million piasters. Consequently, since the year 1805, the cost of paper, iron, and steel has almost tripled:

IV.83

PRICE OF	1802	1803	1804	1805	1806
	Piasters	Piasters	Piasters	Piasters	Piasters
White paper/ream	3 and three-tenths	3 and six-tenths	3 and five-tenths	9 and seven-tenths	8 and two-tenths
Iron/quintal	9	11	10	19	24
Steel/quintal	18 and eight-tenths	18	17	40	30

According to customs' ledgers, in 1806, the period when neutral ships were allowed entry into the port of Veracruz, Mexico received 1,079, 714 piasters' worth of linen cloth (*bretañas, bramantas, caserillos, listados, ruanes, platillas, creas*, and *estopillas*); 1,554,647 piasters' worth of cottons and muslins (*acolchados, cambray, musolinas, mahones, zarazas*, and *pañuelos de Bayaja y Madras*); and 164,989 piasters' worth of woolens.

Despite the rise in cost of iron and steel, mining has continued at the same pace as before the onset of the most recent war. The mint of Mexico City struck 24,007,789 gold and silver piasters in 1804; 27,165,888 piasters in 1805; and 24,736,020 piasters in 1806. Of the 24,007,789 piasters struck in 1804, there were 23,513,079 piasters or 2,756,657 marks of silver and 494,710 piasters or 3,633 marks of gold.

We shall include here the overall trade balance of Mexico in 1824, according to the inventory that was published in the year 1826 by order of the government.

From the Port of Alvarado IV.84

		PIASTERS	PIASTERS	PIASTERS
Imports at domestic ports	domestic products	206,096	284,087	
	foreign products	80,991		
Imports at American ports	domestic products	878,737	4,360,568	11,058,291
	foreign products	3,481,831		
Imports at foreign ports	foreign products	6,413,636		
Exports from domestic ports	domestic products	176,311	202,042	
	foreign products	25,731		
Exports from American ports	domestic products	3,022,422		4,098,650
Exports from foreign ports	domestic products	874,186		
				15,158,941
From the Port of Veracruz				
European imports	foreign products	1,023,739		1,617,646
European exports	domestic products	593,907		
Grand total				16,776,587

The value of goods imported on behalf of the Mexican Republic and of loans contracted with Great Britain has not been included in this balance.

In 1824, Mexican trade deployed 388 ships, of which

Coming	from domestic ports	39	Going	to domestic ports	80
	from the Americas	76		to the Americas	100
	from Europe	61		to Europe	31
		176			212

The following tableau presents the trade balances of Mexico through the port of Veracruz in the period between 1796 and 1820. Mr. von Humboldt owes his access to this document to Mr. Thomas Murphy's friendship. The gold and silver shipped on behalf of the kings of Spain is not included.

IV.85 *Total Trade Balance of Veracruz, from 1796 to 1820*

	WITH SPAIN									
	IMPORTS			EXPORTS						
YEARS	Domestic Products	Foreign Products	Total in Piasters	Gold and Silver	Cochenille in Arrobas	Value in Piasters	Sugar in Arrobas	Value in Piasters	Other Products	Total in Piasters
1796	3,647,068	2,902,757	6,549,825	5,453,843	6,112	439.609	346,361	1,347,231	63,659	7,301,342
1797	381,336	139,136	320,472	9,604	838	54,471	60,835	159,834	14,741	238,649
1798	1,407,253	392,482	1,799,735	1,104,177	12,220	804,903	79,563	212,691	108,588	2,230,359
1799	3,834,398	1,676,036	5,510,434	2,744,647	40,602	2,703,471	150,881	479,062	384,290	6,311,470
1800	1,903,577	1,224,417	3,187,994	4,197,946	5,150	379,256	87,570	287,277	331,587	5,196,066
1801	1,647,473	371,229	2,018,702	274,882	3,848	298,258	9,148	25,157	229,045	827,342
1802	11,539,219	8,851,640	20,390,859	25,564,574	43,277	3,303,470	431,867	1,454,240	3,543,935	33,866,219
1803	10,614,803	7,787,486	18,493,289	7,498,759	27,251	2,191,399	483,944	1,495,056	831,858	12,017,072
1804	10,412,324	4,493,736	14,906,060	14,275,420	11,737	1,220,193	381,509	1,097,505	1,440,253	18,033,371
1805	1,514,47 3	574,963	2,089,436	10,200						10,200
1806	1,825,579	327,295	2,152,834		4,154	425,400	25,857	64,642	312,995	803,037
1807	3,662,053	694,032	4,356,085		2,823	282,300	5,288	13,220	317,230	612,733
1808	2,367,538	655,646	3,023,184	4,420,488	7,373	737,400	19,917	39,834	719,529	5,917,231
1809	10,252,698	6,914,607	17,167,305	16,338,812	21,560	2,587,200	241,246	482,492	2,416,722	21,825,226

1810	10,806.384	6,336,846	17,143,230	9.446,943	20,415	2,449,800	119,726	269,383	629,887	12,796,013
1811	5,200,413	4,970,419	10,170,832	6,227,250	11,215	1,211,220	95,016	237,450	401,322	8,077,332
1812	2,616,7 18	1,366,673	3,983,391	3,772,230	7,664	766,400	12,236	30,575	122,451	4,649,659
1813	3,241,439	2,353,665	5,595,104	9,237,545	6,381	724,080	7,657	19,142	56,739	10,037,369
1814	2,660,123	5,882,180	7,942,303	7,240,921	7,993	959,160			122,305	8,322,386
1815	3,080,375	5,758,261	8,838,636	4,326,207	21,006	2,520,720			325,190	7,172,117
1816	2,748,294	4,793,276	7,541,57 0	3,556,247	11,343	1,476,420			192,434	5,225,101
1817	2,398,825	2,109,956	4,508,781	5,113,194	14,640	1,903,200			98,871	7,115,265
1818	1,794,658	751,179	2,545,837	2,271,949	4,961	545,710			18,504	2,836,163
1819	3,693,923	2,464,256	6,158,179	4,552,765	21,704	2,430,848			81,214	7,064,827
1820	5,068,856	4,462,140	9,530,996	7,305,678	15,956	1,675,380	7,100		366,064	9,371,972
	107,779,800	78,345,313	186,125,113	144,849,172	330,415	32,090,221	2,565,720	7,739,731	13,129,396	197,853,520

	WITH THE OTHER COUNTRIES									
	IMPORTS			EXPORTS						
YEARS	Domestic Products	Foreign Products	Total in Piasters	Gold and Silver	Cochenille Arrobas	Value in Piasters	Sugar Arrobas	Value in Piasters	Other Products	Total in Piasters
1796										
1797										
1798										
1799										
1800										
1801										
1802										
1803										
1804										
1805	7,968	554,080	562,048	67,399	848	72,080	37,332	93,347		232,826
1806	380,386	3,105,269	3.485,655	3,151,905	6,339	633,900	75,862	189,655	126,024	4,101,534
1807	496,663	9,627,232	10,133,895	19,287,710	11,334	1,113,400	69,236	173,090	831,897	21,406,097
1808	583,686	4,437,634	5,021,311	5,385,889	2,242	224,100	39,280	78,560	189,884	5,887,433
1809										
1810										
1811										

1812										
1813										
1814										
1815										
1816										
1817		1,731,567	1,731,567	26,350					5,358	31,708
1818		674,207	674,207	141,724	1,662	183,260			37,261	362,245
1819										
1820		363,951	363,951	267,534					3,080	270,614
	1,468, 697	20,503,940	21,972,637	28,328,511	22,428	2,226,740	221,710	534,652	1,202,554	32,292,457

	WITH THE AMERICAS						
	IMPORTS			EXPORTS			
YEARS	Domestic Products	Foreign Products	Total in Piasters	Gold and Silver	Misc. Products	Total	Grand Total
1796	1,419,216		1,419,216	1,269,144	734,901	2,004,045	17,277,428
1797	1,713,372		1,713,372	23,928	1,160,499	1,184,427	3,656,920
1798	1,447,108		1,447,108	500,945	640,024	1,140,969	6,618,171
1799	1,211,428		1,211,428	1,614,944	789,542	2,404,486	15,437,818
1800	1,526,206		1,521,206	297,022	565,382	862,404	10,767,670
1801	1,468,246		1,468,246	589,489	553,742	1,143,231	5,457,521
1802	1,607,729		1,607,729	3,749,988	831,160	4,581,148	60,445,955
1803	1,373,428		1,373,428	1,855,876	609,970	2,465,846	34,349,635
1804	1,619,682		1,619,682	2,645,182	770,329	3,424,511	37,983,624
1805	1,169,907	93,000	1,262,907		97,720	97,720	4,225,137
1806	1,472,989	26,255	1,499,244		574,191	574,191	12,616,535
1807	1,690,838	555,694	2,246,532		488,503	488,503	39,243,845
1808	1,873,013	495,720	2,368,733	2,076,687	717,224	2,793,911	25,011,806
1809	1,643,018	1,620,183	3,263,301	5,454,688	997,619	6,452,907	48,708,039
1810	2,043,870	1,243,632	3,287,502	2,164,929	955,810	3,120,739	36,347,484

1811	533,322	643,024	1,176,346	981,387	807,970	1,789,357	21,213,867
1812	1,206,797	50,578	1,257,375	288,807	187,947	476,754	10,359,178
1813	1,925,859	411,259	2,337,118	1,699,688	363,041	2,062,729	20,032,320
1814	955,455	771,869	1,727,321	1,923,066	152,871	2,075,937	20,067,950
1815	1,158,393	989,238	2,147,631	1,852,325	166,441	2,018,766	20,177,150
1816	1,417,918	1,046,536	2,464,454	1,357,730	92,475	2,450,205	17,681,330
1817	1,025,493	1,419,758	2,445,251	1,222,001	150,863	1,372,864	17,205,436
1818	1,032,278	1,513,092	2,545,370	1,397,956	134,171	1,523,127	10,495,949
1819	884,150	3,056,867	3,941,017	1,423,062	196,893	1,619,955	18,783,978
1820	1,2044,093	2,412,677	3,656,770	1,156,679	94,435	1,251,114	24,445,417
	34,658,808	16,349,382	51,008,190	36,354,523	12,833,723	49,388,246	538,640,163

Recapitulation			
Imports from	Spain	186,125, 311	
	foreign ports	21,972,635	259,105,940 piasters
	the Americas	51,008,199	
Exports to	Spain	197,853, 520	
	foreign ports	32,492,457	279,534,223 piasters
	the Americas	49,388,246	
Total trade			548,640,163 piasters

IV.85 Among its members, the *Consulado* of Veracruz includes men who are distinguished for their enlightenment as well as for their patriotic fervor: it is both a trial court (*tribunal*) that adjudicates contentious trade disputes and an administrative council responsible for maintaining the port and roads, hospitals, the city police, and anything else related to the development of trade. The council is composed of a *prior*, two *consuls*, an associate judge, a trustee, and nine councilmen: lawsuits are adjudicated *gratis* based on oral testimony and without lawyers' intervention. The Consulado of Veracruz deserves credit for the Perote road project (which cost over 480,000 francs per league in 1803), the improvement of hospitals, and the construction of a beautiful revolving lighthouse, built according to the plans by Mr. *Mendoza y Ríos* in London. This lighthouse consists of a very tall tower, placed at the far end of the castle of San Juan de Ulúa, which (including its lantern) cost nearly half a million francs; the air-current lamps, outfitted with reflectors, are attached to a triangle that turns by means of a clock mechanism, making the light disappear each time one of the machinery's acute angles faces the entrance of the port. At the time of my departure from Veracruz, the *Consulado* had two new equally useful projects: IV.86 providing the town with potable water and the construction of a jetty-like pier that would resist the force of the waves. We had the opportunity to examine the first of these projects in our discussion of the Río de Jamapa dam.[1]

Across Spanish America, there is a marked antipathy between the inhabitants of the plains or warm regions and those who dwell on the plateau of the Cordilleras. This antipathy strikes the European traveler who ascends the Magdalena River, traveling from Cartagena of the Indies to Santa Fe de Bogotá or who traverses the chain of the Andes en route from Guayaquil to Quito, from Piura and Trujillo to Cajamarca, or from Veracruz to the capi-

1. Chap. VIII, vol. II, p. 213 [p. 415 in this edition].

tal of Mexico. The coast dwellers accuse the mountain people of being cold and lifeless, while the plateau dwellers reproach those on the coast for being trivial and inconsistent in whatever they undertake. One might almost say that people of different origins had settled in the same province, since this small area of land combines not only the climate and products but also all of the national prejudices of both northern and southern Europe. These prejudices stoke the rivalries one observes between the merchants of Mexico City and those in Veracruz. Since they are closer to the seat of government, the former know how to take advantage of their central position. Upon his arrival in New Spain, a viceroy finds himself between the various factions of IV.87 the legal profession, the clergy, mine owners, and merchants in Mexico City and Veracruz; each faction tries to cast suspicion on its opponents, accusing them of being restless and progressive and of harboring a secret desire for independence and political freedom. The metropole has unfortunately always believed that its security lies in the internal dissensions in the colonies. Far from placating individual animosities, Spain was pleased to see this rivalry arise between the indigenous peoples and the Spanish, and between the whites living on the coast and those settled on the interior plateau.

While the port of Veracruz receives 400 or 500 ships annually, despite the poor mooring that it offers between its shoals, the port of Acapulco,[1] which is one of the most beautiful in the known world, barely receives ten. Trade activity in Acapulco is restricted to the galleon from Manila, known incorrectly as the *ship (nao) from China*, smuggling between the coastline of Guatemala, Zacatula, and San Blas, and four to five ships that are sent annually to Guayaquil and Lima. Trade in the western part of Mexico is impeded by its extreme distance from the coastline of China, the monopoly held by the Royal Company of the Philippines, and the tremendous difficulty of sailing toward the coastline of Peru against the current and winds.

The port of Acapulco forms an immense basin cut into granite rocks, IV.88 over 6,000 meters wide, with a south-southwest opening. I have seen few sites in either hemisphere that present a wilder appearance that is, I would almost say, at once doleful and romantic. The shape of the rock masses reminds one of the jagged crest of Montserrat in Catalonia: it is composed of large-grained granite, like that found in Fichtelberg and Carlsbad in Germany. This granite is stratified but the layers slant irregularly, sometimes southward and sometimes to the southeast. Furthermore, this rocky coastline is so steep that a ship of the line squadron might graze by it with-

1. See above, chap. III, vol. I, p. 286 [p. 191 in this edition].

out running any risk, because the water is ten to twelve fathoms deep at all points.

The small island of La Roqueta or Grifo is positioned in such a way that one may enter the port of Acapulco through two canals, the narrower of which, called *Boca Chica*, forms a channel that runs from west to east and is only 240 meters wide between Punta Pilar and Punta Grifo. The second channel or *Boca Grande* lies between the island of La Roqueta and Punta de la Bruja and is one and a half miles wide: at its center there is an average depth of twenty-four to thirty-three fathoms of water. The port itself is commonly distinguished from the large cove called *Bahia,* where the sea is strongly felt from the southwest because of the width of the Boca Grande. This port in-

IV.89

cludes the westernmost part of the *Bahia*, between *Playa Grande* and the *Ensenada de Santa Lucia*. There, ships find excellent anchorage near land, in six to ten fathoms of water. We moored there on the frigate *Orue* in March of 1803, thirty-three days after our departure from Guayaquil.

On examining the narrow isthmus that separates the port of Acapulco from the *Bay of la Langosta* and the *Abra de San Nicolás*, one might say that nature attempted to make this place into a third channel like the Boca Grand and the Boca Chica. This isthmus, which is 400 meters wide at its broadest point, is very interesting from a geological point of view. We clambered over strangely shaped exposed rocks, which were barely sixty meters high and appeared to have been rent asunder by the prolonged action of the earthquakes that are frequent on this coastline. In Acapulco one observes that the tremors spread in three different directions, coming sometimes from the west, through the isthmus that we have just mentioned; sometimes from the northwest, as though coming from the Colima volcano; and sometimes from the south. For some years now, the final tremors are the strongest, preceded by a hollow sound that is as frightening as it is extremely long. These earthquakes, which are felt toward the south, are attributed to underwater volcanoes, since what is experienced here is what I observed several times at night in Callao de Lima: in the midst of mild and calm weather, the sea becomes

IV.90

suddenly and terrifyingly agitated, without the slightest wind.

Despite the width of the bay of Acapulco, its only shallows are not even forty meters wide; they are named after Saint Anne because they were discovered in 1781 when the ship the *Santa Ana*, which participated in the Lima trade, was unexpectedly lost. *Las Bajas*, the rocks that we scraped as we entered through the Boca Grande, the *Farallón del Obispo*, and the small island of San Lorenzo near the Punta de Hicacos present no danger because they are visible shelves: these masses of rocks, which one may approach without fear of running aground, may be considered the remnants of the former coastline. The small port of *Marqués* is southeast of Punta de la Bruja: it forms a cove about one mile wide with an entrance between eighteen and twenty fathoms deep; its interior is eight to ten fathoms deep. There is no traffic in the cove because of its proximity to Acapulco; it is a wild and lonely place, where a populous city would grow were it located on the eastern coastline of New Spain.

Landing in the ports of Realejo, Sonsonate, Acapulco, and San Blas is very dangerous in winter, or in other words, during the rainy season that lasts from May to December on the entire western seaboard of the Americas,[1] from the island of Chiloé to California. The beginning and the end of winter are IV.91 the most fearful times. Great storms are felt[2] in the months of June and September, when one finds as angry and squalling a sea on the coastline of Acapulco and San Blas as one finds in winter near the island of Chiloé and on the coast of Galicia and Asturias. The Great Ocean deserves the name Pacific Ocean only from the parallel of Coquimbo to the parallel of Cabo Corrientes, between 30° southern latitude and 5° northern latitude. Serene weather generally prevails in this region; mild winds from the south-southwest and the southeast blow all year long there, with no noticeable seasonal influence. South-southwesterly winds[3] and even winds from south-southeast, which are generally called *bendavales*, prevail in the eastern part of the Great Ocean in winter, or in other words from May to October, between 5° northern latitude and the Bering Strait. The *brisas*, or north and northwest winds, blow IV.92 in summer, that is, from November until the end of April. The *bendavales* are harsh and stormy, accompanied by low-lying thick clouds that drop rain showers lasting twenty to twenty-five days, especially in August, September, and October. Such abundant rain destroys the harvests, while the southwest wind uproots the largest trees. Near Acapulco I saw a silk-cotton tree (Bombax ceiba) with a trunk over seven meters in circumference that had been struck down by the *bendavales*. The *brisas* are weak and often interrupted by dead calm; they blow when the sky is clear and serene, as do all the winds that take the name of the hemisphere where they prevail.

1. Except for Guayaquil, where the rains last from December to April and May. It rains torrents in Guayaquil, while drought prevails not only in Panama but even north of Cabo San Francisco at Atacames. I shall have the opportunity to address elsewhere these seasonal contrasts in the Cordilleras and on the coastline, often at different points on the same coast. Suffice it to say here that in general it has been incorrectly suggested that droughts and rain alternate everywhere in the tropics according to laws that have been observed in the Antilles.

2. See above, chap. III, vol. I, p. 291 [p. 193 in this edition].

3. *Vientos de tercer quadrante.*

Near Acapulco—and this knowledge is very important for the pilots who navigate in the area—the northern *monsoons* always slant northwest; the northeast wind[1] that blows out at sea and in more northerly latitudes is very rare there, and the true west wind is feared for its terrible violence. It is likely that the width of the continent and the rising air current that forms over such intensely hot terrain cause these eastward atmospheric movements and that IV.93 this effect becomes unnoticeable the farther one moves from the continent. The regularity of the monsoons and the changes in wind direction, which depend on the influence of the seasons, are only noticeable at a distance of 4° or 5° longitude from the coastline. Farther west, the Great Ocean presents the same phenomena as the Atlantic Ocean, since one finds the trade wind (which one might call the *earth rotation wind*) all year long there between the edges of the tropics; depending on the name of the hemisphere in which it blows, this wind slants either to the north or to the south. Ships coming from Chile or Lima occasionally move into longitudes that are too far to the west, for fear of landing east of Acapulco; there they await in vain for the northwest wind, which only blows near the coastline. The northeast wind forces them to move up to the twentieth parallel to approach the continent, which stretches to the southeast and northwest: only there, forty leagues from land, do they find the northwest wind that leads them to the port. These same west winds force the Acapulco galleon, as it returns to Manila, to take a southerly route as far as 12° or 14° latitude. On these parallels, at 103° longitude, and thus over 200 leagues west of the coastline of Guatemala, the galleon encounters the trade winds (east and east-northeast) that accompany them to the Marianas.

There is little trade activity between Acapulco and the ports of Guayaquil IV.94 and Lima. The main commodities are copper, oil, some Chilean wine, a very small amount of sugar and quinine from Peru, and lastly, cacao from Guayaquil, which is intended either for domestic consumption in New Spain or as stores for Havana and the Philippine Islands, or for shipment to Europe in wartime. There is almost no cargo loaded onto the ships returning to Guayaquil and Lima, merely some woolens from factories in Querétaro, small amounts of cochineal, and some smuggled merchandise from the East Indies. The length and extreme difficulty of the journey from Acapulco to Lima pose the greatest obstacles to exchanges between the inhabitants of Mexico and Peru. One

1. The land wind (*terral*) that blows during the night until eight or nine o'clock in the morning in Sonsonate, Realejo, and Acapulco is an east and northeast wind; thanks to this mild wind, it is possible to sail back up in the summer if one has had the misfortune of landing east of Acapulco.

can easily sail in six or eight days from Callao de Lima to Guayaquil; it takes three, four, or five weeks to go from Guayaquil to Acapulco. But traveling from the northern to the southern hemisphere, or from the coast of Mexico to those of Quito and Peru, one must simultaneously fight the currents and the winds. There are only 210 nautical leagues from Guayaquil to Callao, but it often takes twice as long to make this crossing from north to south as it does to go from Acapulco to Manila via a route over 2,800 nautical leagues long. It often happens that it takes as many weeks to go from Guayaquil to Callao as it takes days to return from Callao to Guayaquil.

There are three things to fear in the passage from the Peruvian coasts to the coasts of New Spain: the dead calm, which reigns especially around the equator; the furious winds known as Papagallos, which we mentioned at the end of chapter three; and the danger of landing east of Acapulco. The dead calm is all the more dreadful because as long as it lasts, the currents have full sway. What is more, the Spanish ships used in the South Sea trade are so poorly built that even in mild winds they are tossed about by the currents. The areas where these currents are felt most strongly are the Galapagos Islands, which Mr. Collnett was the first to survey with some degree of accuracy. There have been instances where ships built in Guayaquil responded so poorly to their helm that they sailed among these islands for two months without being able to leave, while running the risk at every moment of being carried by the currents[1] to the shore, which is ringed by reefs, in the midst of a dead calm. Peruvian navigators try to cross the line 7° or 8° east of the Galapagos Islands group. The English and the Anglo-Americans[2] who are lured here by whale hunting dread this archipelago much less than the Spanish. They put in there either to collect turtles, which provide a tasty and nutritious sustenance for the sailors, or to put sick sailors ashore. Since the fishing ships (*whalers*) are of fine construction, they drift much less in mild and gentle winds.

IV.95

IV.96

After escaping from the calm that prevails below the equator between Cabo San Francisco and the Galapagos archipelago, Peruvian ships encounter another region, between 13°30′ and 15° northern latitude and 103° and 106° western longitude, that is dangerous because of the frequent calm spells there in February and March. The year before we visited this area, a dead calm lasting twenty-eight days, together with the lack of water that followed it, had forced the crew of a ship newly built in Guayaquil to abandon its valuable cargo of cacao and to escape in a rowboat to seek land, eighty

1. Vancouver, [*Voyage de découvertes*,] chap. III, p. 402.

2. See chap. I, vol. III, p. 95 [p. II.53 in this edition].

leagues away. Such incidents are not rare in the South Sea, where navigators have the deplorable habit of setting sail with only a few casks of water, to save space for their merchandise. The calm spells that prevail on the parallel of 14° north, which are comparable only to those in the Gulf of Guinea, are even more fearsome in that they are encountered at the end of the crossing.

In sailing from Callao and Guayaquil to Acapulco, one attempts to land IV.97 west of the port because of the winds and currents, whose direction is very regular near the coast. One usually tries to set a course for the Zihuatanejo headlands, located over forty leagues west-northwest of Acapulco, slightly west of the Morro de Petatlán. Since these cliffs are very white, they are visible four leagues away at sea. After sighting them, one skirts along the coast, setting the helm southeast, toward the point [or morro] of Satlán and the beautiful beaches of Sitiala and Cuyuca, which are covered with palm trees. The port of Acapulco is identifiable only by the knolls (*tetas*) of Cuyoca and the great Cerro de la Brea or Siclata. This mountain[1] is thirty-eight miles from the port but is visible from the sea; it is located west of the Alto del Peregrino and navigators use it, like the Peak of Orizaba, the Campaña de Trujillo, and the Silla de Payta, as a signal. From the coastline of California and Sinaloa as far as Acapulco, and usually even as far as Tehuantepec, the current runs from northwest to southeast from December to April, in the season that is usually called *summer*; in winter, from May to December, the current runs northwest, most often west-northwest. Because of the motion of the ocean waters, which is felt only forty leagues from the coast, the crossing IV.98 from Acapulco to San Blas in summer takes twenty to thirty days, whereas the return in winter takes only five to six days.

On the western coastline of the New Continent between 16° and 27° northern latitude, a navigator who has no means of determining his longitude can be fairly certain that if the latitude reading puts him farther north than the *loch*, then his ship has drifted westward with the currents. On the other hand, his longitude will be farther east than his *estimate*, if the observed latitude is less than the *estimated* latitude. South of the parallel 16° north and in the entire southern hemisphere, however, these rules become very dubious; I have become convinced of their uncertainty, having carefully compared, on a daily basis, the *estimated point* in the eastern part of the Great Ocean with the chronometric longitude and the distances measured between the moon and the sun. Huge errors in longitude caused by the strength of the currents make navigation in this area as extended as it is costly: errors multiply dur-

1. See my map of the route from Acapulco to Mexico (*Atlas du Mexique*, plate v).

ing crossings of 2,000 leagues, and nowhere is the use of chronometers and the method of calculating lunar distances more indispensable than in such a vast sea basin. For this reason, in the past few years even the least-skilled navigators have begun to realize how extremely useful astronomical observations can be. I met Spanish merchants in Lima who had purchased chronometers for 6,000 or 8,000 francs, intending to carry them on newly built ships. I was even pleased to learn that many British and Anglo-American ships that sail round Cape Horn to hunt for whales and to visit the northwest coast of the Americas are provided with chronometers. IV.99

The crossing from Acapulco to Lima is often trying and even longer than sailing from Lima to Europe; it can be done in *winter* by sailing north to 28° or 30° southern latitude before approaching the Chilean coastline; one is occasionally forced to steer south-southwest beyond Juan Fernández Island. Such a passage *por altura*, first made in 1540 by ▾ Diego de Ocampo under the viceroy of Mexico, Antonio de Mendoza, normally takes three to four months, but only a few years ago the ship *Neptune*, which sails for the Guayaquil trade, took seven months to go from the Mexican coastline to the port of Callao.

Thanks to the terral [land breeze] that blows in summer from December to May, one can sail northward from Punta Pariña[1] (latitude 4°35′ south; longitude 83°45′) to Lima. This route is called the *navigación por el meridiano*, because instead of sailing away from the coastline 300 or 400 leagues to the west, one tries to change longitude as little as possible. The land wind is very cool at night in Peru between Paita and Callao; in Mexico between Sonsonate and Acapulco; and generally on most of the coastline in the Torrid Zone; it is variable from the southeast to southeast one-quarter east; inversely, between Cabo Blanco and Guayaquil, the night wind blows landward from the sea. Navigators know how to take advantage of this circumstance as soon as they have landed at Punta Pariña: they tack to the south-southwest in the open sea for eighteen hours by day; when the cool land wind arrives at night, they set sail for the coast for six more hours, beating to windward with full sail because of the currents. In this *navigation on the meridian*, one must not be farther than sixty to seventy leagues from land. A Portuguese navigator has recently proven that the tacking method can even be used in winter,[2] if the ship responds well enough to the rudder. Furthermore, this method offers the great advantage of shortening the route: by adopting it, one avoids the tempests that prevail in the months IV.100

1. See [Humboldt and] Oltmanns, *Recueil d'observations astronomiques*, vol. II, p. 430.
2. *Moralero Derotero de la mar del Sur* (a very valuable unpublished document).

of August, September, and October between 28° and 33° southern latitude. I felt it was important to record here these detailed notes on navigation in the eastern part of the Great Ocean, not only because they are of interest to trade in the New Continent but especially because they prove a principle that should strongly influence all political strategies: to wit, that nature IV.101 has set enormous obstacles in the way of maritime connections between the peoples of Peru and Mexico. In fact, these two colonies, which by their position are rather near to one another, consider each other almost foreign as they do the inhabitants of the United States and Europe.

The oldest and most important branch of trade in Acapulco is the exchange of merchandise from India and China for precious metals from Mexico. Restricted to a single galleon, this exchange is very simple and, although I have been in the places where the most famous *fair* on Earth is held, I have little to add to the information that others have already been provided.[1]

The galleon, which usually weighs 1,200 or 1,500 tons and is commanded by an officer of the Royal Navy, sets sail from Manila in mid-July or at the beginning of August, when the monsoon from the southwest is in full swing. It is loaded with muslin, printed calicoes, coarse cotton shirts, raw silk, silk stockings from China, gold and silver plate made by Chinese in Canton or Manila, spices and seasonings. The voyage is made either through the San Bernardo Strait or around Cabo Bojador, which is the northernmost IV.102 tip of Luzon Island; it used to take five to six months, but since the art of navigation has been perfected, the crossing from Manila to Acapulco takes only three or four months. Winds from the northwest and southwest prevail in the Great Ocean, as they generally do in all seas, beyond the natural boundaries of the trade winds, north and south of the twenty-eighth and thirtieth parallels. Blowing in the opposite direction from the trade winds, these winds may be considered atmospheric *counter-currents*. During my stay in Peru, British ships, truly excellent masted ships, sailed from the Cape of Good Hope to Valparaiso in Chile in ninety days, although they had to travel almost two-thirds of the circumference of the globe from west to east, thanks to the southwest winds. In the northern hemisphere, the northwest wind speeds the crossing from the coastline of Canada to Europe as well as the passage from eastern Asia to the western coastline of Mexico.

1. Anson, *Voyage [around the World]*, book II, chap. X, pp. 63–73. Le Gentil [de la Galaisière, *Voyage dans les mers de l'Inde*, vol.] II, p. 216. Raynal, [*Histoire philosophique et politique*, vol.] II, p. 90. De Guignes, [*Voyage à Péking*, vol.] III, p. 407. Renouard de Ste.-Croix, [*Voyage commercial et politique aux Indes Orientales*, vol.] II, p. 357.

Formerly, the galleon sailed above 35° latitude to land in New California near the tall Santa Lucia Mountains, which rise east of the Santa Barbara Channel. For the past twenty years, ships have landed much farther south, since, having learned of the island of Guadalupe (latitude 28°53′), navigators set their helm southeast, thereby avoiding the dangers of the reef called *Abreojos* and the two *cliffs of Los Alisos*. Most inconveniently on such a long crossing, there is not a single stopover point for the galleon between Manila and the island of Guadalupe and the California coasts. One would IV.103 have hoped for the discovery of some other archipelago, north of the Sandwich Islands and between the Old and the New Continent, that could provide some refreshment and a safe mooring.

By law, the value of the merchandise on the galleon should not exceed half a million piasters, but it is usually one and a half to two million piasters. After the merchants of Manila, the ecclesiastic companies have the greatest share of this lucrative trade; these companies invest almost one-third of their capital, and this use of their money is inappropriately called *dar a corresponder*. As soon as the news arrives in Mexico City that the galleon has been sited off the coastline, the Chilpanzingo and Acapulco roads are filled with travelers; all the merchants hurry to be the first to negotiate over the supercargo arriving from Manila. A few powerful trading houses in Mexico City usually band together to buy all the merchandise, and it has occured that the entire cargo is already sold before news of the galleon's arrival reaches Veracruz. Purchases are made almost without opening the packages, and although the merchants from Manila are accused in Acapulco of *trampas de la China*, or *Chinese trickery*, we must admit that the trade between two countries, separated by 3,000 miles, is conducted in rather good faith and IV.104 perhaps with even more fairness than trade between some nations of civilized Europe that have never had any connection with Chinese merchants.

While the merchandise from the East Indies is in transit from Acapulco to the capital of Mexico, thence to be distributed throughout the Kingdom of New Spain, the bars of silver and the piasters that will form the return cargo are sent down from the interior to the coast. The galleon usually departs in February or March: it sails almost in ballast, since its cargo on the voyage from Acapulco to Manila consists only of silver (*plata*), a very small amount of cochineal from Oaxaca, cacao from Guayaquil and Caracas, wine, oil, and Spanish woolens. On average, the amount of precious metals exported to the Philippine Islands is one million, often 1,300,000 piasters. There are usually a considerable number of passengers, which sometimes increases because of the colonies of friars that Spain and Mexico send to the Philippines. In the year

1804, the galleon transported seventy-five of them, which prompts the Mexicans to say that upon its return, the *Nao de China* is laden with *plata y frailes.*

The voyage from Acapulco to Manila, the longest that can be made in the equinoctial region of the ocean, is accomplished with the help of the trade winds: it is almost three times the length of the crossing from the coastline of IV.105 Africa to the Antilles. As we have seen above, the galleon sails first southward, taking advantage of the northwest winds that prevail on the northern coast of Mexico. Once it has reached the parallel of Manila, it runs westward full sail, in consistently calm seas, with fine cool breezes from the area between east and east-northeast.[1] Nothing disturbs the heavens' tranquility in these regions, other than the occasional brief squall that is felt when it is at its peak. The pilot ▼ Don Francisco Maurelli was therefore bold enough to cross the entire Great Ocean, a length of almost 3,000 nautical leagues, in his bridged longboat (*lancia de navio*); this longboat, called the Sonora, was dispatched from San Blas to bring news to Manila of the most recent rupture between Spain and England. The launch is preserved in the port of Cavite, as should have been done at Timor with the ship on which the unfortunate Captain Bligh made his memorable voyage from the Society Islands to the Moluccas.

Passage from Manila to the coastline of Mexico is as long and difficult as the crossing from Acapulco to the Philippine Islands is short and pleasIV.106 ant, generally lasting from only fifty to sixty days. In the past few years, the galleon sometimes puts in at the Sandwich Islands to take on supplies and water, if the local priests have not *placed a taboo* on the ship's watering place. Since the crossing is not long and the chiefs of these islands are not always favorably disposed toward whites, this stopover, rarely necessary, is often dangerous. As the galleon progresses farther west, the breezes become fresher but also more inconsistent, and one begins to feel strong gusts of wind. The galleon puts in at the island of Guahan, or Guam, where the governor of the Marianas resides in the town of Agana.[2] It has been correctly observed that this island is the only point in the vast expanse of the South Sea, strewn with countless islands, where there is a European style town, a church, and a fort. Furthermore, this delectable country, which nature has endowed with the greatest variety of products, is one of the numerous possessions that the Spanish court has never been able to exploit. The fanatical monks and the disgusting greed of the governors conspired earlier to depopulate this archi-

1. Farther north, especially between 20° and the tropic of Cancer, the trade winds are less regular in the Great Ocean than in the Atlantic Ocean.

2. ▼ Surville, *Nouveau voyage à la mer du sud*, p. 176.

pelago. The commander of the Agana fort is one of those individuals in the king of Spain's employ who can exercise arbitrary power with the most impunity. He communicates with Europe and the Philippine Islands only once a year, and if the *nao* is intercepted or lost in a storm at sea, he can remain for several years in complete isolation. Although Agana is located 4,000 leagues east of Madrid, as the crow flies, it is confirmed that upon seeing the galleon arrive in two consecutive years, the governor of Guahan expressed the wish that *he lived on an island that was not so close to Spain*, so that he would be less susceptible to its ministers' surveillance. IV.107

Besides the *situado,* that is, the silver with which the troops' salaries and royal officers' commissions are to be paid, the galleon brings woolens, fabric, and hats to the colony in the Marianas (*islas de los Ladrones*) to clothe the few whites who live in this archipelago. The governor provides fresh supplies for the galleon, especially pork and beef. Horned animals have propagated remarkably on these islands, where there is a fine breed of cattle that are completely white with black ears. ▼ Commodore Byron[1] insists that he saw *guanacos* like the ones in Peru on the island of Saypan [Saipan], north of Tinian, with its low mountains; this observation should be verified by naturalists. Since the Spanish have introduced neither *llamas* nor *guanacos* or *alpacas* either to Mexico or the kingdom of New Granada, it seems probable that some of them came there from a group of islands near Asia.[2]

In addition to the galleon from Acapulco, a ship is also sent occasionally from Manila to Lima. This navigation, one of the longest and most difficult, is ordinarily made via the same northern route as the crossing from the Philippine Islands to the California coast. After sighting the coastline of Mexico, the galleon headed for Lima sails southward as far as 28° or 30° southern latitude, where the southwest wind prevails. When it is free of the yoke that the monopoly of the Philippines Company forces upon it, Peru will be able to trade freely with the East Indies, and a route that leads south of New Holland over seas where one is sure to find favorable winds may be preferable on the return voyage from Canton to Lima. IV.108

A few years before my stay in Lima, Don José Arosbide had sailed the galleon *El Filippino* via a direct west-east route from Manila to Callao in ninety days. Aided by the mild variable winds that blow in the vicinity of the South Sea Islands, especially at night, the ship had sailed up between the sixth and tenth southern parallels, against the *rotation current*. Fear

1. Hawkesworth's *Compilation*, vol. I, p. 121.
2. Marchand, *Voyage*, vol. I, p. 436.

of falling into the hands of English corsairs led him to choose such an extraordinary route counter to the direction of the trade winds. Forgetting that chance had played an important role in the success of a voyage during which calm spells were interrupted by squalls from the south and south-

IV.109 southwest,[1] Mr. Arosbide wanted to attempt the route from west to east a second time. After struggling against the trade winds for a long time, he had to sail to higher latitudes and follow the former method of navigation. The lack of provisions compelled him to put in at the port of San Blas, where he died of excessive fatigue and disappointment.

It has been asked how Spanish ships since the sixteenth century have been able to cross the Great Ocean from the western coastline of the New Continent to the Philippine Islands without discovering the islands strewn over that vast sea basin. This quandary is easily resolved when we consider that few voyages were made from Lima to Manila and that the archipelagos that are known through the work of ▾ Wallis, Bougainville, and Cook are almost all found between the equator and the tropic of Capricorn. For the 300 years, the navigators of the Acapulco galleon have always had the good sense to sail on the same parallel when crossing from the Mexican coasts to the Philippine Islands: it appeared to them to be all the more indispensable to follow this route since they assumed they would encounter shallows and reefs as soon as they drifted off this course to the north or to

IV.110 the south. In a period when the use of lunar distances and timekeepers was unknown to navigators, one attempted to correct the estimated longitude through the observation of magnetic variation. It was long ago observed that the variation at the San Bernardino Strait was almost zero, and as early as 1585 ▾ Juan Jayme had navigated with Francisco Gali from Manila to Acapulco to test an instrument of his own invention capable of determining magnetic needle declinations.[2] This method of correcting estimates could have been of some interest at a time when a pilot was often nearly 8° or 10° off in his longitude estimates. Very precise measurements made in our own

1. A learned navigator, Mr. de Fleurieu, has already correctly observed that it is not rare to see south-southwest and even northwest winds prevail for several days in the equinoctial region, especially at 15° and 18° southern latitude and along 114° and 118° western longitude. (Marchant, *Voyage*, vol. II, p. 269.)

2. [Espinosa y Tello, *Relacion del viage hecho por las goletas Sútil y Mexicana*], p. 46. [La Pérouse,] *Voyage de Lapérouse*, vol. II, p. 306. In December 1803, I found the magnetic variation in Mexico City (latitude 19°25′45″ north, western longitude 101°25′) to be 8°8′ east, and in the South Sea, at 13°50′ north latitude and 106°26′ longitude, it was 6°54′.

time have proven that changes in magnetic declination are extremely slow in that area, even approaching the San Bernardino Strait.

Furthermore, one should not be surprised that galleons carrying cargos valued at six to seven million francs would prefer not to abandon their prescribed route. True discovery expeditions can only be made at a government's expense: it is an undeniable fact that during the reigns of Charles V, Phillip II, and Phillip II, the viceroys of Mexico and Peru encouraged few undertakings that did honor to the Spanish name. In 1542 *Cabrillo* visited the coastline of New California or New Albion as far as 37° latitude. Drifting northward on his return voyage from China to the Mexican coasts, *Gali* discovered the mountains of New Cornwall in 1582, covered with perpetual snow, and located at 57°30′ north. *Sebastián Vizcaíno*'s expedition discovered the coastline between Cabo San Sebastian and Cabo Mendocino. As early as 1542, *Gaetano* had found scattered islands near the Sandwich Islands group; there is no doubt that even this latter group was known to the Spanish over a century before Cook's voyages, since the island of La Mesa, shown on an old map of the Acapulco galleon, is identical to Owyhee Island [Big Island, Hawai'i] where the tall mountain called *the Table* or *Moana-Roa* [Mauna Loa] is found.[1] In 1595, Mendaña, accompanied by Quirós,[2] discovered the group of islands known as the Marquesas de Mendoza or Mendaña Islands that include San Pedro [Moho Tani] or O-Nateya, Santa Cristina [Tahuata] or Wahitaho, Dominica, or Hiva 'Oa, and Magdalena [Fatu Hiva]. We also owe our knowledge of the Santa Cruz de Mendaña Islands, which Carteret called the Queen Charlotte Islands, to the same intrepid navigators, as well as Quirós's archipelago of Espíritu Santo,[3] which are Bougainville's New Cyclades and Cook's New Hebrides, the archipelago of Mendaña's Salomon Islands, which Surville[4] called the Arsacides, the Dezena Islands (Maitea), Pelegrino (Wallis' Scylly Island), and most likely also O-Taïti [Tahiti] (Quirós's Sagittaria), all three of which are all part of the group of the Society Islands. Is it fair to say that the Spanish crossed the Great Ocean without discovering any new land,

IV.111

IV.112

1. Marchand, *Voyage*, vol. I, p. 416.

2. Alvaro Mendaña de Neyra and Pedro Hernández de Quirós. See the *Succesos de las islas Filippinas* (Mexico City, 1699, chap. VI. *Hechos de* ▼ *Don Garcia Hurtado de Mendoza, marques de Cañete, virey del Perü, los escribó el doctor Don Cristobal Suarez de Figueroa,* p. 238. After ▼ Mendaña's death, his wife, Doña Isabella Baretos [Barreto], famous for her strong spirit and extraordinary courage, took command of the expedition, which ended in 1596.

3. Fleurieu, *Découverte des Français dans le sud-est de la Nouvelle-Guinée*, p. 85.

4. ▼ Shortland's New Georgia (Marchand, *Voyage*, vol. IV, p. 63).

when we recall the plethora of discoveries that we have just cited,[1] which were made at a time when the arts of navigation and astronomy were so far from IV.113 the degree of perfection to which they have risen today? The names of Vizcaíno, Mendaña, Quirós, and Sarmiento [de Gamboa] certainly deserve to be placed next to those of the most famous navigators of the eighteenth century.

We have already observed above that the Sandwich Islands archipelago provides a respite for ships sailing from Acapulco or the northwest coastline of the Americas to the Philippines or China, just as the islands of the Marquis of Mendoza or the Society Islands provide excellent mooring and a great store of provisions for ships that have rounded Cape Horn to hunt for pelts in Nootka and Norfolk Bay. Despite these advantages, the inhabitants of Mexico, who are interested in trade with Asia, would prefer it if the Sandwich Islands were not on the route from Acapulco to Manila; they fear that a European power will establish settlements there or that the islanders, who are by nature lively and entrepreneurial, will instigate piracy in these seas. It is true that the *treaty of Karakakooa*, in which ▾ Tamaahmaah [Kamehameha], the King of Owyhee, *freely* and *voluntarily* ceded his empire to the King of Great Britain in 1794, has had no more long-lasting effects than the numerous treaties concluded among the civilized people of Europe. Constantly warring among each other, the chiefs give preference to the nation that supplies them with the most firearms and ammunition. Shortly thereafter, these arms IV.114 are turned against the very same people who had unwittingly supplied them. Many Europeans, most of them disloyal subjects who deserted British or Anglo-American ships, have settled among the islanders.

With their assistance, an enterprising European power could easily succeed in taking over the Sandwich Islands and settling a colony there. The islanders are excellent sailors: several of them have already embarked on European ships and visited the United States, on the northwest coast of the Americas and in China; they have endeavored to build schooners and even armed vessels, with which they plan to make distant expeditions. Northwest sea currents bring them enormous pine-tree trunks from the northern coast of the American continent. All these circumstances will be of great assistance in establishing a colony in this archipelago. More than all the other is-

1. To this list of Spanish discoveries in the South Sea I might have added those made by Garcia Jofre de Losisa [Loyasa] (*Viage [Viaje] al estrecho de Magellanos*, p. 206), Grijalva, ▾ Gallego, Juan Fernández, Luís Vaz de Torres, and Saavedra Cerón who were the first to sight the northern coastline of New Guinea. See the fine map of the southern part of the South Sea, drawn according to Mr. Dalrymple's scholarly research.

landers in the Great Ocean, the Sandwich Islands indigenous peoples have profited from their connections with Europeans. Their minds have been expanded and new needs have been instilled in them; in the past twenty years, they have made notable advances toward the social state that is very inappropriately known by the name of civilization.[1] Such progress, which would be very slow if the islanders were left to their own devices, will become accelerated under European domination, and it is possible that these people will become as feared one day in the Great Ocean as the corsairs of the Bermudas, those in the Bahamas, and the Barbary pirates are dreaded in the Atlantic Ocean and the Mediterranean. A squadron stationed in the Bay of Karakakooa and cruising to the south and to the east might make itself formidable to ships sailing to the Philippine Islands or China, either from Acapulco and San Blas or from the northwest coast of the Americas. IV.115

Smuggling on the western coast of New Spain is less pronounced than what takes place between Campeche, the mouth of the Río Huasacualco, rebaptized port *Bourbon*, Veracruz, and Tampico. Following the coastline from southeast to northwest, one finds the following ports: Tehuantepec, Los Angeles, Acapulco, Zihuatanejo, Zacatula, Colima,[2] Gualán, Navidad, Puerto Escondido, Jalisco, Chiametla, Mazatlan, Santa Maria Aorne, Santa Cruz de Mayo, Guaymas, Puerto de la Paz (or the port of the Marques del Valle),[3] Monterrey, San Francisco, and Puerto de Bodega. This long list of ports, of which most provide excellent mooring, justifies what we have said above about the visible contrast between the eastern and western coasts of Mexico. IV.116 Strong currents, frequent monsoons, and winter storms make smuggling very difficult. From the coasts of Guatemala to the Sea of Cortés, crossings are so long and arduous that in 1791 the corvettes commanded by Malaspina—two excellent sailing ships—took fifty-eight days to go from Realejo to Acapulco: in the same year, the merchant ship *La Galga*, advantaged by currents and winds, sighted the Azores sixty days after leaving the port of Lima: the first passage is 300 nautical leagues, and the second is 4,500 leagues.

1. Due to the effects of this supposed civilization, the inhabitants of O-Taïti [Tahiti], now accustomed to European-made tools and fabrics, are gradually forgetting how to make tools from stone and bones and neglect paper mulberry cultivation. See Mr. Vancouver's very learned observations on the state of these islanders since their frequent connections with Europeans. (*Voyage autour du monde [Voyage of Discovery to the North Pacific Ocean, and Round the World in the Years 1791–95]*, vol. I, p. 179.)

2. [Lorenzana,] *Cartas de Hernán Cortés*, p. 348.

3. See chap. VIII, vol. II, p. 259 [p. 440 in this edition].

The ports of Acapulco, San Blas, Monterrey, and San Francisco offer the best locations for whale fishing and trading in otter skins, which are found everywhere between 28° and 60° northern latitude. We have already discussed these commodities in chapter ten in our discussion of the marine animals who inhabit the coastline of the Great Ocean. To reach the areas inhabited by the saricovian otters, a composite of river and sea otters, Anglo-Americans must travel all the way around the New Continent, from 40° or 43° degrees north latitude, they ascend as far as 58° and 60° south. After rounding Cape Horn, they sail up the South Sea as far as the same northern latitudes from which they set sail. During my brief stay in the United States in 1804, there were fifteen to twenty American ships,[1] most of them belonging to shipbuilders from Nan-

IV.117

tucket and Boston, on the northwest coastline. After exchanging their pelts in Canton and Macao for tea, raw silk, and nankeen cloth, these ships sail around the globe, returning by way of the Cape of Good Hope. It takes the Mexican Spaniards, whose possessions extend to 38° north, twenty days to travel to these coasts, which European nations can only reach after a six- or seven-month journey. On the seaboard of New California, especially around Monterrey, one finds the superb *abalone shell* whose mother-of-pearl interior is of the most brilliant luster; the inhabitants of the islands of Quadra and New Cornwall prize them as much as the *Haliothis iris* and *Haliothis australis* from New Zealand.[2] Otherwise, trade with Chile provides copper from Coquimbo, which is sought after by the indigenous peoples on the northwest coastline. After the colonists in Russian America, no other nation is more advantageously placed for trading in otter skins than the Mexican Spaniards.

These skins, whose color and quality varies depending on the season

IV.118

and the animal's age and sex, are jade black; it is so highly valued in China that before 1780, the price of an otter skin ranged from forty to sixty and even from 100 to 120 piasters for the highest-quality skins. Since then, however, imports have far exceeded trade demand, and the value of these skins has decreased so much that in 1790, the finest Nootka skin was sold in Canton for fifteen piasters. At that time, the Chinese government sometimes forbade the importation of these skins through its southern ports: this ban was only temporary. According to the list of imports in Canton from

1. In the year 1792 there were only seven. Vancouver, [*Voyage autour du monde*, vol.] III, p. 519.

2. [Espinosa y Tello, *Relacion del*] *viage* [. . . *para reconocer el*] *estrecho de Fuca*, vol. CXLVIII, pp. 121 and 161. Lapérouse, *Voyage*, vol. II, pp. 276–82; vol. IV, p. 276.

1804 to 1806, 34,144 pieces[1] of otter skin were imported over three years, of which nearly five-sixths arrived on Anglo-American ships. During this period, the average price of a skin was from eighteen to twenty-five piasters.[2] This information demonstrates that profit from the fur trade has diminished enormously since Lieutenant King's and ▼ Captain Hanna's stay in China; one also recognizes the inflated calculations of some writers on political economy, who assume that the forty-four million pounds of tea that Europeans consume would be paid for in large part by the furs from the northwest coastline of the Americas. It appears that the Canton and Macao markets are amply provided annually with 30,000 to 35,000 otter skins, and the total value of this importation is only 6,000 piasters. The price of pelts in China will certainly decrease even further if the Americans of the United States profit from the knowledge they have acquired thanks to Captain Lewis's expedition and open a direct trade route between Hudson Bay, Canada, and the mouth of the Columbia River. IV.119

When the account of Cook's third voyage revealed to Europe the advantages of trading in sea otter skins, the Spanish also made a few feeble attempts to engage in this trade. A commissioner was sent to Monterrey in 1786 to collect all of the otter skins from the presidios and missions of New California: it was thought that up to 20,000 skins could be collected in this way. At first the government reserved exclusive rights to the fur trade: but seeing that such a measure was odious, it gave permission to a few merchants in Mexico to send loads of skins to the Philippines. The ship owners took almost no profit because the Spanish government imposed exorbitant taxes on this fledgling branch of its national industry, because merchants in Manila were intermediaries in handling the furs, and because this speculation was begun only when the price of furs had already dropped considerably. What an enormous profit this trade would have been for the Mexicans IV.120

1.

Imports in	1804	11,176 pieces
	1805	22,180
	1806	780
		34,144

According to the trade tables for Russia published by the Count of Romanzoff, China received various types of pelts from sea and land animals via Kyatkha, for a value in an average year of 1,450,000 rubles.

2. Compare ▼ Coxe, *Russian Discoveries*, p. 13, and Dixon, *Voyage Round the World*, p. 316, with Renouard de Sainte-Croix, *Voyage commercial*, vol. III, p. 152.

if at the time of Pérez, Heceta, and Quadra's[1] expeditions in 1774, 1775, and 1779, the court of Madrid had established trading posts in the harbor of Nootka (*Puerto de San Lorenzo*), in the port of Bucareli, or on Hinchinbrook Island, in those northern regions where otters have the finest coats, shinier and thicker than those found south of the forty-eighth parallel? At that time, the hunters from Kamchatka were still the only masters of the fur trade on the northwest coastline of the New Continent.[2]

IV.121 In presenting the trade tables of Acapulco and Veracruz, I have had to IV.122 restrict myself to exported and imported items that have been *registered*, in other words, for which entry and exit duties imposed by Spanish law have been paid: these duties (*derechos reales*) are paid in the Americas, according

1. See chap. VIII, vol. II, p. 301 [p. 461 in this edition].

2. When the Russians had conquered Siberia, the Kamchatka Peninsula was the terminus of their fur-gathering expeditions. They crossed over from Kamchatka to the Aleutian Islands. Several small ships were built in Okhotsk; armed with depraved men, they inflicted acts of cruelty upon the Aleutian population for a long time. Under the direction of ▼ Chelikoff [Shelikov], a group of merchants who behaved more humanely and consistently was formed in Irkutsk. Emperor Paul proclaimed himself the protector of this company on the northwest coastline. Anyone who was not a member of the company was forbidden to trade in the Aleutian Islands. Emperor Alexander expanded the privilege of the company, whose office is in Petersburg, from the Bering Strait to 54° northern latitude. One share of its stock costs five hundred rubles, and the government appoints its directors. The principal or main trading post was first located on the island of Kodiak, but since the sea otters became rarer and rarer in the Aleutian Islands, it became necessary to make expeditions farther southeast and hunt for these animals in the King George archipelago. When the indigenous peoples saw that they had been deprived of trade with the inhabitants of the United States, who traded more valuable things with them, they fell upon the Russians and massacred many of them. Realizing both the dangers and the great advantages of these connections with King George's Sound, the company resolved to transfer the main post from Kodiak to the Norfolk Sound. Not wanting to seize the local people's villages, ▼ Mr. Baranoff, who was then governor, carried out this project cautiously and humanely, and even left them the hill on which the new fortress has now been built. The locals took advantage of Mr. Baranoff's absence to massacre the Russians and built themselves a small fortress in which they put the cannons that they had taken from the Russians and others purchased from foreign boats. Recently arrived from Kronstadt, Mr. Baranoff retook the position with the assistance of a Russian vessel. The fortress has forty cannons; it also contains a governor's house, shops, and barracks. Since that time, the locals have retreated, and have remained neither at peace nor openly at war with the Russians. The local people's skin is white in Sitka and throughout the Prince of Wales archipelago. Novo-Arkhangelsk, the capital of all the Russian colonies on the coastline of the Americas, is located at 57°3′ latitude and 224°22′ longitude east of Greenwich. One hopes that ▼ Mr. de Schabelsky, a learned traveler who sailed on the sloop *Apollon* in October 1821, equipped with excellent physics instruments, will shed some light on these little known regions of Russian America.

to the regulations[1] of 1778 and 1782, by which the price of all merchandise that may be introduced into the colonies, from leather and printed fabric to chemical equipment and astronomical instruments, has been fixed in a rather arbitrary manner. It is because of this assumed value that each item must pay a duty set at a fixed percentage.

Royal duties are distinguished from *municipal duties* in the Spanish colonies: this distinction is made in all ports, from Coquimbo to Monterrey. The *puertos mayores* pay both kinds of tax at the same time, but only the municipal duties are demanded in the *puertos menores*. Furthermore, the customs system is far from uniform in the various parts of the Americas. The *alcabala*, which is paid when merchandise enters but not when it leaves, is 2 percent in Cartagena of the Indies [de Índias], 3 percent in Guayaquil, 4 percent in Veracruz and Caracas, and 6 percent in Lima. The *almojarifazgo* on Spanish products upon entry is generally 3 percent; 7 percent is demanded for foreign merchandise; and the departure *almojarifazgo* is from 2 to 3 percent. Among the municipal duties, there is the *derecho de consulado*, from one-half to 1 percent; the *derecho del fiel executor* and the *derecho del cabildo*. Upon entry into the Spanish colonies, the customs office demands a tax of 9 and one-half percent on *free goods* or agricultural products and Spanish manufactured goods; 12 and one-half percent on *tax-paying goods* or products from foreign soil manufactured in Spain; and 7 percent on *foreign goods*. It must be noted that a tax of 22 percent has already been paid on these latter goods before they entered American ports, 7 percent as they left Spain, and 15 percent upon first entering Spain. For details on the customs system, I can direct the reader to the informative work that Mr. de Pons has provided on the statistics of the province of Caracas.[2] As a commercial agent, this writer has found himself in the most favorable circumstances for studying all aspects of taxation, trade tariffs, and Spanish customs offices.

IV.123

The poor condition of the eastern coastline, the lack of ports, the difficulty of landing and fear of *damages to the ship* make illegal trade more difficult in Mexico than on the coastline of Terra Firma. Smuggling is done almost exclusively through the ports of Veracruz and Campeche: small craft

1. [Spain,] *Arancel general de los derechos reales de aduanas de los años* 1778 *y* 1782. *Calendario mercantil de España y Indias* [*Correo Mercantil de España y sus Indias*], 1804. [▼ Cladera,] *Espíritu de los mejores diarios*, 1789, no. 170 [Ugartiria, "*Carta*,"] p. 953; no. 172; p. 987 [Ugartiria, "Carta segunda,"]; no. 173, p. 1013 [Consulado de Barcelona, "Informe sobre el comercio de América"].

2. [De Pons,] *Voyage à [la partie orientale de] la Terre-Ferme*, vol. II, pp. 357, 360, and 441; vol. III, p. 11.

sent out from these ports are used to collect merchandise from Jamaica and IV.124 maintain what are called *telegraphic connections* in Veracruz. In wartime, frigates blocking the harbor are often seen unloading contraband on the small island of Sacrificios. In general, trade in the colonies is extremely active during naval wars, since that is when countries enjoy the advantages of independence to a certain extent. As long as connections with the metropole remain uninterrupted, the government is forced to slacken its restrictions and occasionally allow trade with neutral parties. Since the customs officials do not scrutinize documents too closely, smuggling goes unchallenged; if it accounts for four to five million piasters annually in peacetime, this figure certainly reaches six or seven million piasters in wartime. During the most recent rupture with Great Britain from 1796 to 1801, the metropole was able to bring in only 2,604,000 piasters[1] of national and foreign merchandise. The shops in Mexico, however, were glutted with Indian muslins and products from British factories.

For the past half-century, the ministry in Madrid has regularly de- IV.125 manded annual reports, either from the viceroys, or the highest finance council [junta], or the provincial intendants, on ways to decrease smuggling. In 1803 it tried a more direct approach, by contacting the *Consulado* of Veracruz, composed of the most powerful merchants in the city. It is quite understandable that none of these reports provided a solution to a problem that is as much of interest to public morale as it is to the treasury. Despite the *coast guard* and a multitude of customs officers maintained at considerable expense, and despite the extreme harshness of the penal code, illegal trade will undoubtedly continue as long as a complete transformation of the customs system does not diminish the lure of illicit profit. Duties are now so high that they increase the price of foreign merchandise imported by Spanish ships by 35 to 40 percent.

After presenting the importance of domestic and foreign trade in Mexico, the state of its roads and ports, possibilities for canals, the hindrances to navigation presented by currents and monsoons in the South Sea, all based on information collected locally, we must still consider briefly the *annual increase of national wealth*. We shall not endeavor here to retrace the history of trade in the Americas, from the time when it was confined to galleon from Portobelo and the fleet in Veracruz to the happier period when King IV.126 Charles III largely removed the obstacles that had encumbered it for three

1. *Reflexiones acerca del comercio de Veracruz y de la influencia que ha tenido la guerra* (a very interesting unpublished paper by ▼ *Don José Donato de Austria*).

centuries. Mr. Bourgoing has discussed these matters with the wisdom and clarity that characterized the work in which he first introduced modern Spain to Europe.[1] Without repeating what has already been sufficiently discussed by several authors of political economy, we shall follow the path that we have heretofore taken, collecting facts and leading the reader, thanks to this information, to general conclusions.

When one reflects on the state of the colonies before the reign of King Charles III and the odious monopoly that Seville and Cádiz have exercised for centuries, it should come as no surprise that the famous regulation of October 12, 1778, was called the *free trade* edict. In matters of trade as well as politics, the word freedom expresses merely a relative idea concerning the passage from the oppression under which the colonists suffered at the time of *galleons*, *account ledgers*, and *fleets* to the state of affairs when fourteen ports were opened almost simultaneously to products from the Americas, analogous to the passage from the most arbitrary despotism to a freedom that is authorized by law. It is true that without completely adopting the *economists'* theory, one might be tempted to think that the homeland and the colonies would have profited at the same time, if the *free trade* law had been followed by the abolishment of a *duties tariff* that was contrary to American agriculture and industry. But should one have expected Spain to be the first to extricate itself from a colonial system that has been adopted for so long by the most enlightened nation of Europe, despite experiences that were most detrimental to individual happiness and civic harmony?

IV.127

At the time when all trade in New Spain was carried out by *registered ships* assembled in a fleet that arrived every three or four years from Cádiz to Veracruz, purchases and sales were in the hands of eight or ten trading houses in Mexico City that held an exclusive monopoly. At that time there was a fair (*feria*) in Jalapa, and the provisioning of a vast empire was handled like supplying a besieged fort: there was almost no competition, and the price of iron, steel, and all things that were indispensable to mining was increased at will. The famous traveler Don Antonio Ulloa commanded the last fleet that arrived in Veracruz in January 1778. The following table shows the value of the merchandise exported by this fleet compared to the value of exports from Veracruz during the four years 1787, 1788, 1789, and 1790, which are included in the period designated as a time of *free trade*.

1. Bourgoing, *Tableau de l'Espagne moderne*, Fourth ed., vol. II, chaps VII, VIII, and IX; pp. 188–96. Laborde, *Itinéraire descriptif de l'Espagne*, vol. IV, pp. 373–84. [Demeunier,] *Encyclop[édie] method[ique], Économie politique*, vol. II, pp. 319–24.

IV.128 *Exports from New Spain via Veracruz, in the days of the Fleets and during the Period of Free Trade*

NAME OF MERCHANDISE	TOTAL EXPORTS FOR THE YEARS 1787, 1788, 1789, 1790		EXPORTS CARRIED BY THE FLEET ULLOA COMMANDED IN 1778		DIFFERENCE IN FAVOR OF FREE TRADE FROM 1787 TO 1790	
	Quantity	Value in Strong Piasters	Quantity	Value in Strong Piasters	Difference in Quantity	Difference of Value in Piasters
Cochenille (first qual.)	91,346 arr.	7,764,469	26,400 arr.	2,243,203	64,946 arr.	5,521,266
Cochenille (second qual.)	7,979 arr.	159,470	1,052 arr.	21,049	6,921 arr.	138,421
Cochenille (as powder)			14 arr.	222		
Vanilla	1,103,295 pieces	49,647	367,765 p.	16,549	735,530 p.	33,089
Medications			732 p.	2,690		
Roucou [Annatto]			95 p.	380		
Sugar			78 p.	159		
Cacao	471 zurrones	37,536	157 zurr.	12,512	314 zurr.	25,024
Cotton	83,769 arr.	83,796	173 arr.	173	83,596 arr.	83,596
Tanned hides	52,539 pieces	105,078	1,313 p.	2,642	51,226 p.	102,436
Cordovan	145,140 doz.	1,886,820	56 doz	734	145,083 doz.	1,886,086
Dry sausage	200 doz.	50	1,000 doz.	250		
Indigo	6,386 arr.	199,562	5,422 arr.	169,459	964 arr.	30,103

Campeche wood	88,393 quint.	110,491		88,393 quint.	110,491
Tabasco chili	18,832 quint.	131,829		18,832 quint.	131,829
Cow horns	693 doz.	693		693 doz.	693
Cow skins	70 doz.	105		70 doz.	105
Sheepskins	103, 057 doz.	618,345		103,057 doz.	618,345
Buffalo skins	57 doz.	570		57 doz.	570
Bear skins	43 doz.	172		43 doz.	172
Deer skins	94 doz.	282		94 doz.	282
Goat skins	59,000 doz.	44,250		59,000 doz.	44,250
Goat skins	200 doz.	112		200 doz.	112
Pouches (leather)	7,224 doz.	25,284		72,024 doz.	25,284
Misc. tanned hides	21,130 doz.	176,130		21,130 doz.	176,130
Total		11,394,664	2,470,022		8,928,293

Since Don Antonio Ulloa's fleet was laden with Mexican agricultural products from 1774 to 1778, the preceding table shows the strong influence that *free trade* had on the development of industry. The value of recorded exports in an average year before 1778 was 617,000 piasters; during the period that began in 1787 and ended in 1790, recorded exports rose to 2,840,000 piasters.

IV.129

Although the fleet in the year 1778 was the last to arrive in New Spain, that country only fully enjoyed the privilege granted by the regulation of October 12, 1778, from the year 1786 onward, when many trading houses were established in Veracruz, where they have prospered. Merchants living in towns in the hinterland who used to supply themselves with European merchandise in Mexico City became accustomed to going directly to Veracruz to do their buying (*para emplear*). The change of trade venue was contrary to the interests of the inhabitants of the capital; but the growth of all branches of public revenue that has been observed since the year 1778 amply proves that what was detrimental to some individuals has been beneficial to national prosperity. The three following tables were compiled to give the most complete illustration of this important truth.

IV.130

Tableau I

Gross Yield from the National Revenue of New Spain

BEFORE FREE TRADE		AFTER FREE TRADE	
Years	**Value in Piasters**	**Years**	**Value in Piasters**
1765	6,130,314	1778	15,277,054
1766	7,841,457	1779	15,544,574
1767	8,130,147	1780	15,010,974
1768	8,622,145	1781	18,091,639
1769	8,465,432	1782	19,594,490
1770	9,694,583	1783	19,579,718
1771	9,560,740	1784	19,605,574
1772	10,805,532	1785	18,770,056
1773	12,216,117	1786	16,826,1,16
1774	11,116,638	1787	17,983,448
1775	11,845,130	1788	18,573,561
1776	12,588,292	1789	19,044,840
1777	14,118,759	1790	19,400,213
Total	131,135,286	Total	233,302,557

Total effect of free trade on gross revenue, during thirty years: 102,167,204 piasters

Tableau II

IV.131

A. Value of Precious Metals Sent on Behalf of the Crown from Veracruz to Spain

BEFORE FREE TRADE		AFTER FREE TRADE	
Years	Value in Piasters	Years	Value in Piasters
1766	90,387	1779	6,795
1767	2,923	1780	3,096,696
1768	623,855	1781	
1769		1782	
1770	1,858,784	1783	691,756
1771	922,306	1784	2,473,866
1772		1785	2,980,332
1773	3,114,046	1786	3,544,489
1774		1787	3,920,680
1775	1,903,649	1788	3,605,719
1776	1,724,907	1789	3,612,623
1777	2,542,086	1790	2,152,961
1778	2,244,129	1791	3,496,065
Total	15, 027,072	Total	29,581,982

Effect of free trade on net revenue realized in Spain: 14,554,910 piasters

IV.132

B. Amount of Piasters Sent from Veracruz to Cádiz and to the Antilles on Behalf of the Crown

DESTINATION	BEFORE FREE TRADE, 1766–1778	AFTER FREE TRADE, 1779–1791	TOTAL EXPORTS ACCORDING TO THE PUBLIC TREASURY
Spain	15,027,072	29,581,982	44,609,054
Antilles	36,259,508	78,846,695	115,106,203
Total	51,286,580	108,428,677	159,715,257

IV.133 *C. Export of Precious Metals from Veracruz to Havana, Puerto Rico, and Louisiana, as Much on Behalf of the Crown (as Situados[1]) as by Individuals*

YEARS	VALUE IN PIASTERS BEFORE FREE TRADE DECLARATION		YEARS	VALUE IN PIASTERS AFTER FREE TRADE DECLARATION	
	Crown	Individuals		Crown	Individuals
1766	2,393,309	437,256	1779	5,463,220	449,193
1767	2,038,937	858,925	1780	6,401,804	159,404
1768	2,391,969	832,216	1781	7,961,168	120,71 4
1769	2,628,613	626,175	1782	9,563,619	138,054
1770	1,667,102	923,815	1783	9,894,072	238,054
1771	2,774,053	320,113	1784	3,561,887	1,231,786
1772	2,809,054	141,948	1785	6,385,034	640,990
1773	2,641,028	340,620	1786	4,643,228	454,076
1774	3,115,206	792,686	1787	5,082,057	508,667
1775	3,089,043	625,895	1788	4,966,481	512,389
1776	3,300,927	423,599	1789	5,611,364	494,561
1777	3,681,746	701,007	1790	4,292,250	266,604
1778	3,728,521	521,822	1791	5,020,511	566,741
Total	36,259,508	7,546,077	Total	78,846,695	5,781,233

[1] The term *situados para las islas* refers to funds sent to Havana, Louisiana, Puerto Rico, and sometime Caracas, to subsidize the administrative expenses of these colonies and troops' wages.

RESULTS			
Piasters Exported from Veracruz to the Spanish Colonies	1766–1778	1779–1791	Difference
Crown and Individuals	43,805,585	84,627,928	40,822,343

IV.134 Tableau III

Amount of Piasters Exported from Veracruz to Spain and the Spanish Colonies, as Much on Behalf of the Crown as by Individuals

DESTINATION	BEFORE FREE TRADE DECLARATION, 1766–1778	AFTER FREE TRADE DECLARATION, 1779–1791
Spain, Crown, from Table II (A)	15,027,072	29,581,982

Havana, Puerto Rico, and Louisiana, Crown, from Table II (C)	36,259,508	78, 846,695
Spain and the Antilles, Individuals	103,873,984	115,623,348
Total	155,160,564	224,052,025

Let us now compare the annual income from the mines of New Spain with the country's loss of currency through its unfavorable trade balance. Having digested the figures that we have recently received related to exports from Veracruz and Acapulco, we shall now be able to resolve the question of whether precious metals are accumulated in a region that contains the most abundant silver mines in the known world.

Several papers presented to the court of Madrid suggest that in peacetime, before the year 1796, the trade balance in Veracruz (not including illegal trade) was as shown in the following table: IV.135

IV.135

IMPORTS	
Imports from Spain	11,100,000 piasters
Imports from Spanish America	1,300,000 piasters
	12,400,000
EXPORTS	
Of Mexican agricultural products	3,400,000 piasters
Of precious metals	9,000,000 piasters
	12,400,000

IV.136

This balance shows a state of exports that was apparently unfavorable for the Kingdom of New Spain. If one accounts for the specie exported on behalf of merchants in the preceding table, then there is no reason not to include the amount of piasters sent annually on behalf of the government, either to Europe or the Spanish colonies, which in an average year amounts to eight or nine million piasters. We have seen above that from 1779 to 1791, the exportation of gold and silver from Mexico via the port of Veracruz on behalf of both the crown and private individuals was more than 280 million piasters, which means eighteen and a half million in an average year. IV.136

In general, according to the tables presented above, the export of precious metals from the port of Veracruz from 1766 to 1791, was	379,000,000 piasters
For the same period, the amount of precious metals extracted from Mexican mines was	460,000,000
Difference	81,000,000

These data suggest that in a twenty-five-year period, the annual accumulation of currency did not exceed one million piasters, for although the consumption of luxury items prior to 1778 was considerably less than it is now, it would be difficult not to estimate the value of smuggling at two and half million piasters, a large part of which is paid for in coin of the realm.

The state of trade in New Spain has changed greatly in the past twelve or fifteen years. The amount of foreign merchandise that illegal trade brings in on the eastern and western coastline of Mexico has increased not in volume but in intrinsic value. It is not the case that more ships are engaged in the *smuggling trade* with Jamaica, but rather that the imported items have changed IV.137 with the growth of luxury and national wealth. Mexico now requires finer fabrics and a much greater quantity of muslins, gauzes, silks, wines, and liquors than it did prior to 1791. Although contraband is valued at four or five million piasters each year, one should not conclude that an equivalent amount[1] of *non-recorded* piasters flows back to Asia and to the English Antilles: some of these illegal imports are exchanged for Mexican or Peruvian agricultural products; others are discharged either in the Americas, Cádiz, Málaga, or Barcelona.

If in the past fifteen years, increased luxury has made Mexico more dependent on Europe and Asia, the product of the mines has also increased significantly. According to information provided by the Consulado, the value of imports to Veracruz, calculated only according to the customs offices' ledgers, was eleven million piasters before 1791; it is now more than fourteen million piasters in an average year. In the ten years before 1791, the annual average output of the mines of New Spain[2] was 19,300,000 piasters, whereas from 1791 to 1801, their annual output was twenty-three million piasters. In the latter period, local plants were exceptionally prosperous. But since IV.138 then, the Indian lower class and people of color also dress better, advances in Mexican manufacturing has not had a noticeable impact on the import of

1. See chap. XI, vol. III, p. 397 [p. 233 in this edition].
2. *Idem*, p. 302 [p. II.170 in this edition].

woolens from Europe, Indian calicoes, and other foreign-made fabrics. Agricultural production has grown proportionally faster than the manufacturing industry. We have seen above the zeal with which the inhabitants of Mexico have taken up growing sugar cane. The amount of sugar exported from Veracruz is already six million kilograms, and in a few years the value of this commodity will equal that of cochineal in the intendancy of Oaxaca.

Viewed as a whole, the data that I have collected on trade in Acapulco and Veracruz from a single perspective suggest that at the beginning of the nineteenth century,

Imports of products and foreign merchandise in the Kingdom of New Spain, including smuggling on the eastern and western coastline, equal *twenty million* piasters.
Exports from New Spain of local agricultural and manufactured goods equal *six million piasters.*

The mines now produce *twenty-three million piasters* in gold and silver, of which eight to nine million are exported on behalf of the crown, either to Spain or other Spanish colonies; consequently, if one subtracts from the re- IV.139 maining *fifteen million* piasters, *fourteen million* to account for the surplus of imports over exports, the result is barely one million piasters. The national wealth, or rather, the currency level of Mexico therefore increases annually.

Based on precise data, this calculation explains why the country with the richest and most consistently productive mines does not have a large amount of currency, and why the cost of labor remains low there. Enormous sums are amassed in the hands of a few individuals,[1] but it is the local people's extreme poverty that strikes the Europeans who travel throughout the countryside and towns in the interior of Mexico. I am tempted to believe that of the ninety-one million piasters[2] that we have assumed exist as currency among the thirteen to fourteen million inhabitants of the Spanish colonies in continental America, nearly fifty-five or sixty million are in Mexico. Although the population of this kingdom is not entirely in a ratio of one to two in proportion to the population of the other continental colonies, the ratio of its national wealth to that of the other colonies is almost two to three. The estimate of sixty million piasters is only ten piasters per capita, but this amount must appear rather large, when one reflects that the estimate for Spain is seven piasters per capita, and fourteen IV.140

1. See chap. VII, vol. I, p. 437 [p. 281 in this edition].
2. See chap. XI, vol. III, p. 425 [p. II.251 in this edition].

in France. In the *capitanía general* of Caracas it was reckoned in 1801 that the currency circulating among a population of seven to eight million inhabitants was only three million piasters[1]; but what a difference there is also between an empire rich in mines like Mexico and another completely lacking mines, whose export products barely equal the value of its imports! Many writers on political economy assume that the ratio of the currency of a country to its gross revenue is four to one. Now the revenue of the Kingdom of New Spain is sixteen million piasters, discounting the government's profit from the mines. According to this figure, the mass of currency would be sixty-four million, which differs very little from our first estimate. We have seen above that the Spanish ministry did not always have accurate ideas of the national wealth of Mexico. Concerned, as it was, with redeeming the *vales* or *public debt* in 1804, the metropole believed that it could suddenly seize the sum of forty-four and a half million piasters belonging to religious bodies.[2] It was easy to foresee, however, that the landowners into whose hands this money fell, who used it effectively IV.141 to improve their land, would not be able to return it in hard currency, so this budgetary transaction was a complete failure.

It would be difficult to deny that since the war that broke out between Spain and France in 1793, Mexico has occasionally suffered great monetary losses. Besides the *situados*, the revenue of the crown, and individual funds, several million were transferred annually to Europe in free gifts intended to subsidize the cost of a war that the lower class considered a war of religion. Such largesse was not always the result of the enthusiasm provoked by the monks' sermons and the viceroys' proclamations; magistrates' authority often intervened to force communities to make their *voluntary offering* and to stipulate its value. In 1797, long after the peace of Basel, an extraordinary loan was established in Mexico City in the amount of seventeen million piasters: this tremendous sum was sent to Madrid and the *revenue from the royal estate* (*renta de tabaco*), which ordinarily yields three and a half million piasters, was pledged to the Mexican creditors. These facts sufficiently demonstrate that exports of currency through Veracruz and Acapulco sometimes exceed the output of the mints, and that the Spanish administration's recent transactions have contributed to impoverishing Mexico.

IV.142 In fact, this decrease in currency would be acutely felt if, for several consecutive years, the mint in Mexico City were to issue fewer piasters, either because of a decrease in the amount of mercury that the amalgamation fac-

1. De Pons, [*Voyage à la partie orientale de la Terre-ferme*,] vol. I, p. 178; and vol. II, p. 380.
2. See chap. X, vol. III, p. 105 [p. II.58 in this edition].

tories required, or because of poor administration of the mines, which are now producing at their peaks. This is a rather critical position for a population of five or six million that, because of an unfavorable trade balance, might find itself vulnerable to a decrease in its capital of over fourteen million piasters per year, were it ever deprived of its wealth in precious metals. This is because twenty million piasters' worth of foreign merchandise are now imported into Mexico and exchanged for six million piasters' worth of domestic agricultural products and fourteen million piasters in specie, which one may consider as having been drawn from the bowels of the earth.

On the other hand, if the kings of Spain had governed Mexico through princes of their house who resided in the country, or if in the aftermath of those events of which history offers us examples in all ages, the colonies had separated from the metropole, then Mexico would experience an annual currency loss of at least nine million piasters, which would be partly deposited in the royal treasury in Madrid, partly in the provincial counting houses in Havana, Puerto Rico, Pensacola, and Manila as improperly designated *situados*. If national industry is given sufficient latitude, if agriculture and manufacturing are revitalized, then imports will naturally decrease and it will then be easy for Mexicans to pay for the value of foreign merchandise with products originating in their own country. The unimpeded cultivation of wine grapes and olives on the plateau of New Spain; the free distillation of brandies from sugar, rice, and grapes; the exporting of flour, enhanced by the construction of new roads; the enlargement of the sugarcane plantations; iron and mercury mining; and steel production may all someday become sources of more inexhaustible wealth than all the veins of gold and silver combined. Under happier external circumstances, the trade balance could become more favorable to New Spain, without paying the account open between the two continents for centuries exclusively with Mexican piasters.

IV.143

In the present state of trade in Veracruz and Acapulco, the total value of exported agricultural products is barely equal to the value of the sugar produced by the island of Cuba: the value of the latter commodity is 7,520,000 piasters, allowing for the export of only 188,000 crates of sugar containing sixteen arrobas each and estimating the price of a crate of sugar at forty piasters. Imports into Mexico, however, which we reckon to be twenty million piasters in an average year, are an object of the greatest importance to the mercantile class in Europe, which seeks outlets for its own products. We should recall at this juncture 1) that the United States of America, whose exports were valued at 71,957,144 *dollars* in 1802, exported a value of only 19,000,000 *dollars* in 1791; 2) that England, at the time of its greatest trade activity with France in 1790, only

IV.144

imported 5,700,000 piasters' worth of merchandise; and 3) that exports from England to Portugal and Germany in 1800 did not exceed 7,600,000 piasters and 12,400,000 piasters, respectively.[1] These data lend sufficient justification to the efforts that Great Britain has made since the end of the past century to participate in the trade between the European Peninsula and Mexico.

When the ports of Spanish America are ranked according to the importance of their trade, Veracruz and Havana occupy first place: an enormous amount of business was conducted there during the last war, during the brief interval during which the court of Madrid granted neutral ships entry to the colonies. The other ports may be ranked in the following order: Lima, Cartagena de Indias, Buenos Aires, La Guayra, Guayaquil, Puerto Rico, Cumaná, Santa Marta, Panamá, and Portobelo.

IV.145 To enable the reader to judge the *relative trade activity* in the Spanish colonies of the Americas, I shall briefly show the value of the exports and imports in several of the ports that I have just listed. These are only general results that are of interest to political economy and the study of trade relations: all minor details are reserved for the notes accompanying the *Personal Narrative* of my journey to the equinoctial regions.

Veracruz. Imports, fifteen million piasters. Exports (not including precious metals), five million piasters.

Havana. Exports of local products, eight million piasters, of which 31,600,000 kilograms of sugar or 6,320,000 piasters (with one crate of sugar valued at forty piasters); 525,000 kilograms of wax, or 720,000 piasters (one arroba valued at eighteen piasters); 625,000 kilograms of coffee, or 250,000 piasters (one arroba valued at five piasters). Sugar exports, which were almost nonexistent before 1760, totaled 14,600,000 kilograms in 1796. In 1802, the sugar crop was so abundant that exports rose to 40,880,000 kilograms. As a result, this branch of trade has almost tripled in ten years. From 1799 to 1803, revenue from the royal customs office in Havana rose on average to 2,047,000 piasters; in 1802, it was in excess of 2,400,000 piasters. IV.146 Total trade activity in Havana was twenty million piasters.

Lima. Imports valued at five million piasters. Exports (including precious metals): seven million piasters.

Cartagena de Índias, including the small nearby ports of Río Hacha, Santa Marta, and Portobelo, with which it has the closest commer-

1. Playfair, *Commercial Atlas*, 1801, plates VIII and X.

cial ties. Exports of local agricultural products, excluding precious metals, 1,200,000 piasters, of which there are 1,500,000 kilograms of cotton, 100,000 kilograms of sugar, 10,000 kilograms of indigo, 400,000 kilograms of Brazilwood, 100,000 kilograms of quinine from New Granada, 1,000 kilograms of Tolu balm, and 6,000 kilograms of ipécacuanha.[1] Imports: four million piasters.

La Guayra, the main port of the province of Caracas. In an average year[2] from 1796 to 1800, exports were valued at 1,600,000 piasters, of which there were 2,985,000 kilograms of cacao, 99,000 kilograms of indigo, 354,000 kilograms of cotton, and 192,000 kilograms of coffee. But from 1789 to 1796, imports were reckoned in an average year[3] to be 2,362,000 piasters; exports of local products were 2,739,000 piasters, of which there were 4,775,000 kilograms of cacao, 386,000 kilograms of indigo, 204,000 kilograms of cotton, 166,000 kilograms of coffee, and 73,000 pieces of leather.

IV.147

Guayaquil. Exports of local products, 550,000 piasters, of which three million kilograms of cacao. Imports, 1,200,000 piasters.

Cumaná (including the small nearby port of Nueva Barcelona). Imports, one million piasters. Exports, 1,200,000 piasters, of which there were 1,100,000 kilograms of cacao, 500,000 kilograms of cotton, 6,000 mules, and 1,200,000 kilograms of *tasajo* or salted meats.

These estimates are based on information that I collected during my journey in the Americas. The *balances* were created according to declarations made at the customs office: smuggling was taken into account only in the trade tables for Cartagena and Cumaná. Collectively, these data will enable us to look briefly at the general trade balance for all of Spanish America. Only by comparing Mexico's trade with that of the other colonies can one judge the political importance of the country, which I have endeavored to present in this work. I begin by first collecting into a single table what the ledgers of the Spanish customs offices tell us about the trade balance of the metropole with its colonies, before and after the famous regulation of 1778.

1. *Raicilla* or ipécacuanha, which reaches Europe via the Spanish ports and smuggling from Jamaica, is the root of the *Psychotria emetica* and not Borero's *Calicocca* or Mutis's *Viola emetica*, as some botanists have suggested. Mr. Bonpland and I have studied the Psychotria as we ascended the Magdalena River near Badillas. Spanish ipécacuanha should not be confused with the Brazilian variety.

2. De Pons, [*Voyage à la partie orientale de la Terre-ferme*,] II, p. 439.

3. See the notes to the first volume of my *Relation historique*.

IV.148

YEARS	VALUE OF EXPORTS FROM THE SPANISH AMERICAS TO SPAIN IN PIASTERS			VALUE OF IMPORTS FROM SPAIN TO THE SPANISH AMERICAS IN PIASTERS		
	Agricultural Products	Precious Metals	Total Exports	Domestic Merchandise	Foreign Merchandise	Total Imports
Average years, from 1748 to 1755	4,955,00	18,060,000	23,015,000	4,039,000	7,076,000	11,115,000
1778	3,728,000	Unknown	Unknown	1,431,000	2,314,000	3,745,000
1784	16,720,000	46,456,000	63,176,000	9,799,000	11,941,000	21,740,000
1785	19,415,000	43,888,000	63,303,000	16,863,000	21,499,000	38,362,000
1788	Unknown	Unknown	40,234,000	7,900,000	7,120,000	15,020,000

In this table,[1] one is struck by the lack of agreement between discrete figures: the years 1778 and 1788 contrast the most with the years immediately preceding them; yet, all of the authors who discuss the beneficial influence of Count de Gálvez's regulation on the development of national industry and the prosperity of the colonies cite these two years, in which trade does not appear to have followed its normal trend. The years 1784 and 1785 offer examples of extraordinary trade activity, because after the peace of Versailles, the surplus of products from the colonies that had accumulated during the war flowed back to Europe all at once. The peace of Amiens recently presented a similar but even more striking phenomenon. In 1802, only the port of Cádiz[2] received colonial

IV.149

1. The results presented in this table for the five years preceding 1753 differ from the results that Raynal presents ([*Histoire philosophique et politique*,] vol. II, book VI), because that well-known author did not take into account the imports and exports from the Spanish Antilles. The balance for the year 1778 is taken from Mr. *Bourgoing*'s tableau of Spain [*Tableau de l'Espagne moderne*,] vol. II, p. 200. For the years 1784 and 1785, see Demeunier, *Encyclopédie méthodique, article Espagne*, p. 322. Imports and exports for the year 1784 are indicated in [Pierre François] Page's work, [*Traité d'économie politique et de commerce des colonies,*] vol. I, pp. 115 and 300. Exports of national merchandise from Spanish ports to the colonies were estimated in 1789 to be 7,220,000 piasters; in 1790, 5,100,000 piasters; in 1791, 5,800,000 piasters; in 1792, 13,500,000 piasters (Laborde, [*Itinéraire descriptif de l'Espagne*,] vol. IV, p. 383).

2. Value of the exports of the Spanish American colonies through the port of Cádiz, from the time of the Peace of Amiens until December 31, 1802.

NAME OF MERCHANDISE		QUANTITIES	VALUE IN STRONG PIASTERS, PRICE AT CÁDIZ
Cotton		34,112 quintals	1,535,040
Indigo		3,892,675 pounds	9,931,687
Sugar		1,029,613 arrobas	4,375,855
Vanilla		11,947,000 pieces	1,075,230
Caracas cacao		33,075 fanegas	1,984,500
Guayaquil cacao		21,532 fanegas	861,280
Coffee		1,799,800 pounds	478,072
Campeche wood			90,380
Cinchona		893,100 pounds	1,786,200
Copper		17,877 quintals	375,417
Hides		339,382 pieces	1,527,219
Cochenille	grana	1,527,219	2,528,007
	granilla	2,528,007	57,447
Tabasco chili		99,875 pounds	16,646

Tallow	3,269 quintals	42,484
Jalap	7,507 arrobas	375,350
Yellow wood (*moralete*)	3,777 quintals	7,554
Sarsaparilla	364 quintals	37,856
Brasilwood (*brasilete*)	1,059 quintals	10,590
Total	Products	27,096,814
	Silver	54,742,033
Total exports of the Spanish colonies to Cádiz		81,838,847

IV.150 productions and precious metals valued at 81,838,847 piasters or 409,000,000 livres tournois; this is equivalent to total English imports[1] in 1790.

The tables that are deceptively referred to as *trade balances* provide useful information only when they give the averages for several years. In this sense, the first result in the preceding table appears preferable to the others: this re- IV.151 sult might even be of great importance to the history of trade in the Americas, if one could be sure that the work done in the customs offices in Cádiz in the account books for the six-year period from 1748 to 1753 was accurate.

The output of the mines that flows annually to Europe, which is listed among the items exported from the colonies, is divisible into three parts: the first is extremely small and belongs to the American colonists settled in Spain; the second, between eight and nine million piasters, goes to the royal treasury as net revenue from all of the colonies in the Americas; the third and largest part pays for the surplus of imports from Europe to the Spanish colonies. Knowing that in 1785 the Americas sent sixty-three million piasters of both precious metals and agricultural products (*en plata y frutos*), to Spain, but received in return a value of only thirty-eight million piasters, one might be tempted to conclude that together the net income of the crown and the revenue of the Spanish families who own property in the New Continent amount to twenty-five million piasters annually. Nothing could be more false, however, than such a conclusion, since the metallic wealth of the colonies serves not only to pay the debt contracted in Spain for merchandise imported from Europe and Asia that has been recorded there, but also to pay English bills of exchange either in Cádiz or Barcelona

1. England's trade with all parts of the world, according to the lists presented to Parliament: imports in 1790 were valued at eighteen million pounds sterling; in 1800, twenty-eight million; exports in 1790 were valued at twenty million pounds sterling; in 1800, thirty-four million.

for whatever smuggling has brought back from Jamaica or Trinidad to the coastline of Mexico, or Caracas and New Granada. IV.152

In general, the Spanish customs offices' ledgers cannot shed any light for us on this significant problem: what is the value of the commodities and merchandise from Europe and Asia that the Spanish colonies require annually, in their present state of civilization? To clarify this discussion, it is more important to know the extent of the Americas' needs than to have accurate knowledge of the active role that the metropole has played until now in supplying its colonies. Furthermore, the phrase *national merchandise*, which is found in all the Spanish trade tables, simply indicates that the merchants succeeded in passing off to customs officials such and such amounts of merchandise as the product of either agriculture or factories on the peninsula. Spanish industries have made considerable advances in the past few years, but it would be an egregious error to use the customs offices' ledgers as the basis for judging the speed of these advances.

To ascertain the approximate value of imports to Spanish America, I have attempted to gather knowledge on site about the state of trade in the principal ports of each province: I have collected information on registered merchandise and on whatever was brought in by smuggling; I have focused IV.153 my attention on the years when, either through free trade with *neutral parties* or through the sales of confiscated smuggled goods, provinces were glutted with commodities from Europe and the East Indies. After discussing with many intelligent merchants the various trade tables that I have presented above, most of which where drawn up under the supervision of the *consulados*, I felt that I could decide on the following figures, which seem the closest to the truth.

The estimates of population that accompany this table [below] are IV.154 based on my own research.[1]

1. I am surprised to see that such an esteemed and otherwise very precise author, Mr. De Pons, has suggested that in 1802 the *Capitanía general* de Caracas included 218,400 blacks [noirs] (*Voyage à la Terre-Ferme*, vol. I, pp. 178 and 241). He finds this figure because at the beginning of his work, he assumes that slaves accounted for three-tenths of the total population, which he reckons to be 728,000 souls. How could Mr. De Pons, who has lived from many years in this beautiful country, assume that there was one black man [Nègre] for every three inhabitants? In 1803, the island of Cuba itself did not have half the number of slaves that this author supposes to exist in the *Capitanía general* of Caracas. I intend to prove elsewhere that in the province of Venezuela, the number of black and mulatto slaves [esclaves noirs et mulâtres] does not exceed one-fourteenth of the total population. It is important to examine this fact in detail because it concerns the welfare and political tranquility of the colonies.

Imports and Exports from the Spanish Colonies of the New Continent

POLITICAL DIVISIONS	EUROPEAN AND ASIAN IMPORTS, INCLUDING CONTRABAND	EXPORTS FROM THE COLONIES		NOTES ON CONSUMPTION
		Value of Agricultural Products	Value of the Product of Gold and Silver Mines	
Captaincy-general of Havana and Puerto Rico	11,000,000	9,000,000		On the island of Cuba: free men, 324,000, of those 234,000 whites. The free people of color [gens de coleur libre] consume more than in Mexico. Less than the Indians.
Viceroyalty of New Spain and Captaincy-general of Guatemala	22,000,000	9,000,000	22,500,000	Total population: 7,800,000. This includes 3,337,000 whites and mixed-blood castes [castes de sang-melé]. The numbers of natives or Indians, who do not consume nearly as much foreign merchandise, reaches two and a half million; whites make up only for 1,100,000.
Viceroyalty of New Granada	5,700,000	2,000,000	3,000,000	Population: 1,800,000. In 1778, an exact census determined 747,641 in the Audiencia of Santa Fe, 531,799 in Quito, for a total of 1,279,440.

Captaincy-general of Caracas	5,500,000	4,000,000		Total population of the seven provinces of Caracas, Maracaibo, Barinas, Coro, Nueva Andalucía, Nueva Barcelona, and Guiana: 900,000; of them, 54,000 slaves.
Viceroyalty of Peru and Captaincy-general of Chile	11,500,000	4,000,000	8,000,000	Population: 1,800,000. In Peru alone, the 1791 census found 130,000 whites; Mixed [métis], who consume much because they enjoy a certain measure of wealth, 240,000. In Chile, there are many whites, but generally life is very simple.
Viceroyalty of Buenos Aires	3,500,000	2,000,000	5,000,000	I have not yet found satisfactory ideas about the population of this viceroyalty, which is rather large in the western provinces known as *provincias de la Sierra*.
Total, in piasters	59,200,000	30,000,000	38,500,000	Total exports of agricultural and mining products: 69,000,000 piasters.

IV.155 The same table indicates that if Asia took no part in trade with the Americas, European manufacturing nations would now have an annual sale of merchandise in the Spanish colonies valued at 310,800,000 livres tournois or 59,200,000 piasters. These enormous imports are balanced by only 160,125,000 livres,[1] or 30,500,000 piasters, the value of the colonial agricultural products. The surplus of imports, which amounts to 157,675,000 IV.156 livres or 28,700,000 piasters, is paid for with the gold and silver extracted from the mines of the Americas. We now know, from what has been developed above, that the value of the precious metals that flow annually from the Americas to Europe, is 38,500,000 piasters or 202,125,000 livres; if one subtracts from this amount the 28,700,000 piasters designated to pay for the surplus of imports over exports, the remainder is 9,800,000 piasters or 51,450,000 livres, which is approximately equivalent to the income of the American landowners living on the peninsula, together with the amount of gold and silver that is deposited annually in the treasury of the king of Spain as the *net revenue of the colonies*. Together, these data suggest the following principle, knowledge of which is very important for political economy, viz., that at the beginning of the nineteenth century, the value of imports in Spanish America was almost equal to the production of the mines, when one deducts the value of colonial agricultural exports, the piasters that flow into the royal treasury in Madrid, and the insignificant sums that colonists residing in Europe withdraw from the Americas.

When, following this principle, one examines the reports of gold and silver imports in Spain and compares them with the output of the mints in the Americas, one can easily see that most authors who discuss Span- IV.157 ish trade have exaggerated both the proceeds of English smuggling and the profit of the Jamaican merchants. In widely circulated works, one can read that prior to 1765, the English earned over twenty million piasters every year from smuggling: adding this sum to the amount of gold and silver that was recorded at the same time in Cádiz as arriving from the colonies, either on behalf of the crown or to pay for the value of Spanish merchandise, one finds a mass of silver far in excess of the actual output of the mines. Despite

1. When comparing the exportation of Spanish and foreign merchandise, estimated according to the account books of the Spanish customs offices, with the importation of the same merchandise, one must bear in mind that the latter exceeds the former, 1) because the merchandise that has arrived in the Americas has previously paid export duties in Spain; 2) because their cost increases due to freight charges, the difference in currency *exchange rates*, and by import duties. Many authors overlook these considerations, and by collecting figures that are not comparable, they arrive at contradictory results.

the smuggling that occurs on the coast of Caracas since the English have been the masters of the islands of Trinidad and Curaçao, it appears that the illegal importation of merchandise in all of Spanish America during the recent years of peace has not surpassed one-quarter of all imports.

It remains for us at the end of this chapter to discuss the epidemic that prevails on the eastern coasts of New Spain for a large part of the year and poses obstacles not only to trade with Europe but also to internal connections between the seaboard and the Anahuac plateau. The port of Veracruz is considered the center of the yellow fever (*vómito prieto* or *negro*). Thousands of Europeans who land on the coasts of Mexico during the sweltering heat perish as victims of this brutal epidemic. Some ships prefer to land in Veracruz at the beginning of winter, when the *los nortes* storms begin to rage, rather than risk losing most of their crew in summer to the ravages of the *vómito* and submitting to a long quarantine upon their return to Europe. These circumstances often have a noticeable influence on the provisioning of Mexico and the cost of merchandise. The scourge of yellow fever has more serious consequences on domestic trade: when connections between Jalapa and Veracruz are broken, the mines lack iron, steel, and mercury. We have seen above that trade between the provinces is carried out by mule caravans: but both the mule drivers and the merchants who inhabit the cold and temperate regions of the interior of New Spain are afraid to descend to the coastline as long as the *vómito* prevails in Veracruz.

IV.158

As trade has grown in this port and as Mexico has felt the need for a more active connection with Europe, the disadvantages that arise from the insalubrious air on the seaboard have also been felt more seriously. The epidemics that prevailed in 1801 and 1802 gave rise to a political issue that had not been debated with the same intensity in 1762 or in previous periods, when the ravages of yellow fever were even more violent. Reports were presented to the government to decide whether it would be better to raze the town of Veracruz and force its inhabitants to move to Jalapa or to some other place on the Cordillera or else to attempt new ways of disinfecting the port. The latter approach seemed preferable, since the fortifications had cost over fifty million piasters, and the port, however bad it may be, was the only one on the eastern coastline capable of providing shelter for warships. Two factions have formed in the country, one that seeks the destruction of Veracruz, and the other its enlargement. Although the government has long appeared to favor the former faction, it is highly likely that this great trial, which involves the property of 16,000 individuals and the fate of a great number of wealthy and therefore powerful families, will be alternately suspended and taken up again without

IV.159

ever being finished. During my stay in Veracruz, I saw the *cabildo* undertake construction of a new theater, while in Mexico City the viceroy's assessor was drawing up a long *informe* to prove that it was necessary to destroy the town, since it was the breeding ground of plague.

We have just seen that in New Spain as in the United States, yellow fever not only attacks the health of the inhabitants but also undermines their financial security, either because it causes their domestic trade to stagnate or because it blocks their exchange of products with other countries. Consequently, anything that has to do with this scourge is as much of interest to the statesman as to the observant naturalist. But the same insalubrity of the IV.160 coastline, which impedes trade, also facilitates the military defense of the country against invasion by a European enemy. To complete this political panorama of New Spain, we must conclude by examining the nature of the disease that makes living in Veracruz so scary to the inhabitants of the cold and temperate regions. I shall not go into the details of a nosological description of the *vómito prieto* here: I have reserved several observations that I gathered during my stay in the two hemispheres for the Personal Narrative of my journey [*Relation historique*] and shall limit myself here to indicating the most salient facts, carefully distinguishing between the indisputable results of observation and whatever belongs to physiological guesswork.

Typhus, which the Spanish call the black vomit (*vómito prieto*), has long been prevalent between the mouth of the Río Antiguo and the present port of Veracruz. The Abbot Clavijero[1] and other writers confirm that this disease first appeared in 1725. We have no idea upon what authority such an assertion (which is contrary to the stories preserved among the inhabitants of Veracruz) is based: there is no old document that provides information about the first occurrence of this scourge, since throughout the warm part of equinoctial America, where termites and other destructive insects pro- IV.161 liferate, it is infinitely rare to find documents that are fifty or sixty years old. Furthermore, it is believed in both Mexico City and Veracruz that the former town [of Veracruz], which is now no more than a village, known as *la Antigua*, was abandoned at the end of the sixteenth century[2] because of the diseases that were already decimating Europeans there.

Long before Cortés's arrival, an epidemic that the locals called *matlazahuatl* prevailed almost periodically in New Spain; some authors confused it with the *vómito* or yellow fever. This plague is probably the same

1. [Clavijero,] *Storia [antica] di Messico*, vol. I, p. 117.
2. One of Alzate's letters in the *Voyage de Chappe*, p. 55.

one that forced the Toltecs to continue their migrations southward in the eleventh century: it ravaged the Mexican population in 1545, 1576, 1736, 1737, 1761, and 1762. But as we have already indicated above,[1] it had two characteristics that distinguished it fundamentally from the *vómito* in Veracruz: it almost exclusively attacked the indigenous peoples or the copper-skinned race and raged in the hinterland of the country on the central plateau, at 1,200 or 1,300 toises above sea level. It is true that in 1761, the Indians in the Valley of Mexico who perished by the thousands as victims of the *matlazahuatl* vomited blood through the nose and mouth, but such hematemeses are frequent in the tropics and accompany bilious ataxic fevers: they have also been observed in the epidemic disease that was rampant throughout South America in 1759 from Potosí and Oruro to Quito and Popayán. According to Ulloa's[2] incomplete description, it was a *typhus* proper to the high regions of the Cordillera. Doctors in the United States, who share the opinion that yellow fever originated in that region, believe to have identified this disease in the *plagues* that prevailed in 1535 and 1612[3] among the redskins in Canada and New England. According to what little we know of the *matlazahuatl* among the Mexicans, one might be led to believe that from the earliest times in the two Americas, the copper-skinned race is subject to a disease which, in its complexity, is related in several ways to the yellow fever in Veracruz and Philadelphia, but which is fundamentally different from it in the ease with which it spreads in a cold zone where the thermometer stays at ten or twelve degrees Centigrade during the day.

IV.162

It is certain that the *vómito* that is endemic to Veracruz, Cartagena de Indias, and Havana is the same disease as the yellow fever that has relentlessly devastated the inhabitants of the United States since 1793. A very small number of doctors have raised doubts about this common identity, which is generally acknowledged by specialists who have visited not only the island of Cuba but also Veracruz and the coastline of the United States, as well as by those who have attentively studied ▾ Mr. Makittrich's, Rush's, Valentini's [Valentin?], and [Ruiz de] Luzuriaga's excellent nosological descriptions.[4] We shall not decide whether yellow fever is identifiable in Hip-

IV.163

1. See chap. V, vol. I, p. 333 [p. 224 in this edition].

2. [Ulloa,] *Noticias Americanas*, p. 200.

3. *Stubbins Ffirth, [A Treatise] on Malignant Fever*, 1804, p. 12. ▾ Gookin [in *Historical Collections of the Indians of New England*] relates the remarkable fact that during the *plague* that prevailed in 1612 among the Pawkunnawhutts near New Plymouth, the skin of the sick Indians was yellowish [p. 148].

4. Aréjula, *[Breve descripción] de la fiebre amarilla [padecida] en Cádiz*, vol. I, p. 143.

pocrates's *causus*, which, like many bilious, remittent fevers, is followed by the vomiting of black matter. But we believe that yellow fever has sporadically emerged on both continents ever since men born in a cold zone have exposed themselves to miasma-infected air in the low regions of the Torrid Zone. Wherever stimulating causes and the sensitivity of internal organs are synonymous, diseases that arise from a disorder of vital functions must take the same form.

It should not be surprising that at a time when connections between the Old and New Continent were so infrequent, and when only a very small number of Europeans traveled annually to the Antilles, that a fever attacking only the non-acclimated would be of such little interest to European doctors. Mortality in the sixteenth and seventeenth centuries must have IV.164 been lower 1) because at that time, the equinoctial regions of the Americas were visited only by Spanish and Portuguese, two nations of southern Europe whose constitution was less likely to react to the drastic effects of an excessively hot climate than the English, Danish, and other inhabitants of northern Europe who now travel to the Antilles; 2) because the first colonists on the islands of Cuba, Jamaica, and Haiti did not gather in towns as densely populated as those that have since been built; 3) because when continental America was discovered, the Spanish were less drawn by trade toward the seaboard, which is usually warm and humid, and they preferred to settle in the hinterland on high plateaus where they found a temperature similar to their homeland. In fact, at the beginning of the conquest, the ports of Panamá and Nombre de Dios[1] were the only ones where foreigners gathered in large numbers at certain times of the year, but also from 1535 onward, Europeans were as afraid to stay in Panamá[2] as they are now reluctant to reside in Veracruz, Omos, or Portocabello. The information reported by ▼ Sydenham and other excellent observers IV.165 offers incontrovertible proof that under certain circumstances, the germs of new diseases cannot be developed,[3] but there is nothing to prove that the yellow fever has not existed for centuries in the equinoctial regions. The moment when a disease is first described, because it has done great damage over a short period of time, must not be confused with the time of its first appearance.

1. Nombre de Dios, located east of Portobelo, was abandoned in 1584.

2. Pedro de Cieza [de León, *La chronica del Peru*,] chap. 2, p. 5.

3. For a description of a throat infection that has prevailed in Tahiti since the arrival of a Spanish vessel, see Vancouver, [*Voyage de découvertes*,] vol. I, p. 175.

The oldest description of yellow fever is by the Portuguese doctor ▼João Ferreira de Rosa;[1] he observed the epidemic that raged in Olinda, Brazil, from 1687 to 1694, shortly after a Portuguese army had conquered Pernambuco. We also know with certainty that in the year 1691, yellow fever appeared on the island of Barbados, where it was called *kendal's fever*, with no proof whatsoever that this disease was brought by ships that came from Pernambuco. In speaking of the *chapetonadas* or fevers to which Europeans are exposed upon arriving in the West Indies, Ulloa[2] relates that according to the locals, the *vómito prieto* was unknown in Santa Marta and Cartagena prior to 1729 and 1730, and in Guayaquil before 1740. The first epidemic in Santa Marta was described by ▼Juan José de Gastelbondo,[3] a Spanish physician. Since that time, yellow fever has raged several times outside the Antilles and Spanish America, in Senegal, the United States,[4] Málaga, Cádiz,[5] Leghorn [Livorno], and even on the island of Menorca,[6] according to ▼ Cleghorn's excellent work. We felt obliged to relate these facts, many of which, in general, are not sufficiently well-known, because they shed some light on the nature and cause of this brutal disease. Furthermore, the theory that the epidemics that have devastated North America almost annually are fundamentally different from those that have surfaced in Veracruz for centuries, and that yellow fever was brought in from the coastline of Africa to Grenada and from there to Philadelphia is as devoid of a basis in fact as the once highly accredited hypothesis that a squadron from Siam had introduced the *vómito* into the Americas.[7]

IV.166

In all climes, people appear to find some consolation in the notion that a disease considered pestilential is of foreign origin. Since malignant fevers are easily born among large crews crowded onto unsanitary ships, the onset of an epidemic often dates from the arrival of a squadron. So instead of attributing the sickness to the close air on ships with no ventilation or to the effect of a sweltering, insalubrious climate on newly disembarked sailors,

IV.167

1. [Ferreira da Rosa,] *Tratado da constituição pestilencial de Pernambuco, por João Fereyra da Rosa, em Lisboa*, 1694.

2. [Ulloa,] *Voyage*, vol. I, pp. 41 and 149.

3. [Ruiz de] Luzuriaga, *De la calentura biliosa* [in *Colección de las disertaciones físico-médicas*,] vol. I, p. 7.

4. In 1741, 1747, and 1762.

5. In Cádiz in 1731, 1733, 1734, 1744, 1746, and 1764; in Málaga in 1741.

6. From 1744 to 1749 (▼ Tommasini, *Sulla febre del Livorno del 1804*, p. 65).

7. ▼ Labat, *[Nouveau] voyage aux isles [de l'Amerique,]* vol. I, p. 73. On the plague in Boullam in Africa, see Chisholm, *[An Essay on the Malignant] Pestilential Fever*, p. 61; and Miller, *Histoire de la [maladie maligne appelée] fièvre [jaune . . .] de New-York*, p. 61; Volney, *Tableau [de climat et] du sol de l'Amérique*, vol. II, p. 334.

people claim that it was brought from a nearby port where the squadron or convoy landed during its navigation from Europe to the Americas. This is why one often hears in Mexico City that the warship that brought such and such a viceroy to Veracruz introduced the yellow fever that had stopped raging there for several years; this is why, during the season of extreme heat, the ports of Havana, Veracruz, and the United States accuse one another of spreading the germs of the contagion from one to the other. The same goes for yellow fever as for deadly *typhus*, known as the Oriental plague, which the inhabitants of Egypt attribute to the arrival of Greek ships, whereas in Greece and Constantinople, the same plague is considered to have come from Rosetta or Alexandria.[1]

▼ Pringle, Lind, and other distinguished physicians consider our bil- IV.168 ious complaints in summer and autumn as the first stage[2] of yellow fever. A slight analogy is evident in the pernicious and intermittent fevers that prevail in Italy, which ▼ Lancisi, Torti, and more recently the famous Frank,[3] in his *Traité de nosographie générale,* have described. The deaths of people with almost all the pathognomic signs of yellow fever, jaundice, vomiting, and hemorrhages have been visibly confirmed from time to time in the Roman Campagna. Despite these relationships, which are not accidental, one may consider yellow fever, wherever it resembles an epidemic disease, as a *sui generis* typhus that is related to both gastric and ataxo-adynamic fevers.[4] We shall therefore distinguish between stationary bilious fevers and the pernicious intermittent fevers that prevail on the banks of the Orinoco, on the coastline that extends from Cumana to Cabo Codera, in the valley IV.169 of the Río de la Magdalena, in Acapulco, and in a large number of other humid, unhealthy places that we have visited, from the *vómito prieto* or yellow fever that ravages the Antilles, New Orleans, and Veracruz.

The *vómito prieto* has not appeared until now on the western coastline of New Spain. The inhabitants of the seaboard stretching from the mouth of the Río Papagallo through Zacatula and Colima as far as San Blas are

1. ▼ Pugnet, *[Mémoires] sur les fièvres [de mauvais caractère] du Levant et des Antilles*, pp. 97 and 331.

2. Lind, *Sur les maladies des Européens dans les pays chauds*, p. 14. ▼ Berthe, *Précis historique de la maladie qui a régné en Andalousie en 1800*, p. 17.

3. Petrus [Peter] Frank, *De curandis hominum morbis*, vol. I, p. 150. The apparent analogy between *cholera morbus*, bilious fever, and gastro-adynamic fever has been very intelligently demonstrated in ▼ Mr. Pinel's admirable work, *Nosographie philosophique* (3rd ed.), vol. I, pp. 46 and 55.

4. [Pinel,] *Nosographie,* vol. I, p. 139, 152, and p. 209. Mr. Frank refers to yellow fever as the *febris gastrico-nervosa*.

subject to gastric fevers that degenerate often into adynamic fevers; one might say that a bilious constitution prevails almost continuously on the arid, burning plains interspersed with small marshes that often harbor crocodiles.[1]

Bilious fevers and *chorera morbus* are very frequent in Acapulco, and the Mexicans who come down from the plateau to buy goods when the *galión* arrives all too often fall prey to them. We have already described the position of the town, whose unfortunate inhabitants, oppressed by earthquakes and hurricanes, breathe a burning hot air, filled with insects and tainted by putrid fumes: for much of the year, they see the sun only through a cloud of greenish vapors that do not affect the hygrometer placed in the low regions of the atmosphere. Comparing the plans of the two ports that I have presented[2] IV.170 in the Atlas of New Spain, one can easily surmise that the heat must be even more stifling, and the air more stagnant in Acapulco than Veracruz. In the first of these two places, just as in Guayra and Santa Cruz in Tenerife, houses are built against a rock wall that heats the air by reflection. The basin of the port is so surrounded by mountains that in order to provide some access to the sea wind during the sweltering summer, colonel ▼ Don José Barreiro, the *castellano* or governor of the fortress of Acapulco, has made a mountain cut to the northwest; the bold project, which is locally called the *Abra* of San Nicolás, has proved useful. Forced to spend several nights in the open air during my stay in Acapulco to make astronomical observations, I constantly felt a slight breeze that entered through the breach at San Nicolas two or three hours before sunrise, when the temperature of the sea was very different from the temperature on land. This breeze is all the more healthful since the atmosphere in Acapulco is infected by the miasmas that rise from the marsh called the *Ciénaga del castillo*, located east of the town: the stagnant marsh waters disappear every year, causing countless numbers of small thoracic fish with a mucilaginous skin to die; the Indians call these fish *popoyote* IV.171 or *axolotl*,[3] although the true *axolotl* in the lakes of Mexico City (▼ Shaw's Sires pisciformis) is essentially different from it, being merely the larva of a large salamander, according to Mr. Cuvier. As they rot in heaps, these fish spread through the surrounding air their putrid smell, which one rightly

1. *Crocodilus aquitis*. Cuvier.

2. Plates IX and XVIII.

3. The axolotl in Acapulco shares only the color of its skin with the same fish in the Valley of Mexico: it is a scaly, olive brown fish with small yellow and blue speckles and two dorsal fins.

considers as the main cause of the bilious-putrid fevers that prevail on this coastline. Lime furnaces, where great masses of madrepore are calcinated, are placed between the *ciénaga* and the town. Despite ▼ Mr. Mitchill's[1] spe-
IV.172 cious theories on nitrogen oxide, Acapulco is one of the most insalubrious places on the New Continent. Perhaps even if this port were to receive ships from Chile and the northwest coastline of the Americas instead of being frequented by ships from Manila, Guayaquil, and other places located in the Torrid Zone, and if the town were visited by more Europeans and inhabitants of the Mexican plateau at the same time, the bilious fevers would soon degenerate into yellow fever there, and the germ of this disease would develop even more drastically in Acapulco than in Veracruz.

On the eastern coastline of Mexico, the north winds freshen the air so that the thermometer drops to seventeen degrees Centigrade. At the end of the month of February, I have seen it remain below twenty-one degrees day after day, whereas in the same period, when the air was calm, the thermometer in Acapulco reads twenty-eight or thirty degrees. The latitude of the latter port is 3° farther south than the latitude of Veracruz: the high Cordilleras of Mexico protect it from the currents of cold air that flow down
IV.173 from Canada toward the coastline of Tabasco. The air temperature there in summer during the day is almost constantly between thirty and thirty-six degrees on the Centigrade thermometer.

I have observed that on the entire coastline the temperature of the sea greatly influences the temperature of the neighboring continent: now, the

1. According to this author, nitrogen oxide, considered to be the cause of malignant and intermittent fevers, is absorbed by limestone; for this reason, the most salubrious parts of France, England, and Sicily have limestone in their soil. (*American Medical Repos[itory]*, V, II, p. 46.) The influence of rock on the great aerial ocean and the physical constitution of men reminds us of the fantastical notions of the ▼ Abbot Giraud- Soulavie, according to whom "basalts and amygdaloids increase the electrical charge in the atmosphere and affect the inhabitants' spirits, making them fickle, revolutionary, and inclined to abandon the religion of their forefathers." Whatever theory one may form of the miasmas that lead to the air's insalubrity, it appears unlikely that, according to the present state of our knowledge of chemistry, tertiary or quaternary combinations of phosphorus, hydrogen, nitrogen, and sulfur can be absorbed by limestone, especially by carbonate of lime. The political influence of Mr. Mitchill's theories is such, however, in a country where the magistrate's intelligence is rightly admired, that finding myself quarantined in Delaware upon arriving from the Antilles to Philadelphia, I saw public health officers seriously engaged in having the opening of the hatchway painted with limestone, so that the septon or miasma of yellow fever from Havana, which was assumed to exist on our ship, would affix itself on a band of limestone three decimeters wide. Should we be surprised that our Spanish sailors believed that there was something magical in this supposed method of disinfection?

heat of the sea does not vary only according to latitude but also according to the number of shallows and the swiftness of the currents that flow from different climes. On the coastline of Peru at 8° and 12° southern latitude, I found the surface temperature of the South Sea between fifteen and sixteen degrees Centigrade, whereas beyond the current that flows forcefully from the Strait of Magellan toward Cabo Pariña, the temperature of the equinoctial Great Ocean is between twenty-five and twenty-six degrees; consequently, in 1802 the thermometer in Lima dropped to thirteen and a half degrees in the months of July and August, and orange trees hardly grow there. Similarly, in the port of Veracruz, I have observed that the warmth of the sea in February 1804 was only twenty to twenty-two degrees, while on approach to Acapulco, I had found temperatures in March 1803 between twenty-eight and twenty-nine degrees. This set of circumstances increases the intensity of the climate on the western coastline: the heat is more constant in Acapulco and Veracruz, and it is believable that if yellow fever ever begins to prevail in the first of these ports, it will last throughout the year, as it does on the island IV.174 of Trinidad, in Saint Lucia, Guayra, and wherever the average temperatures[1] in different months vary only from two to three degrees.

In the low regions of Mexico as in Europe, the sudden inhibition of sweat is one of the main occasional causes of gastric or bilious fevers, especially *cholera morbus*, which is identifiable by frightening symptoms. The climate of Acapulco, whose climate is uniform at different times of the year, inhibits perspiration because of the extraordinary coolness that prevails for a few hours before sunrise. On the coastline, the non-acclimated run great risks when traveling lightly clothed by night or sleep in the open air. In Cumaná and other places in the equinoctial Americas, the air temperature decreases toward sunrise only from one to two degrees Centigrade: during the day, the thermometer stays at twenty-eight or twenty-nine degrees, and at night at twenty-three or twenty-four degrees. In Acapulco I found the heat of the air during the day to be twenty-nine or thirty degrees; during the night, it remains at twenty-six degrees, but from three o'clock in the morn- IV.175 ing almost until sunrise, it drops suddenly to seventeen or eighteen degrees. This change makes the strongest impression on the organs. Furthermore,

1. The difference between average temperatures in the coldest and the hottest month are 25.5° in Sweden at 63°50′ latitude; 23.2° in Germany at 50°5′; 21.4° in France at 48°50′; 20.6° in Italy at 41°54′; and 2.7° in South America at 10°27′. See my comparative tables in the additions to the *Chimie de Thomson* (Mr. Riffault's translation [of Thomas Thompson's *A System of Chemistry*]), vol. I, p. 106.

nowhere else in the tropics have I felt such a great freshness during the last part of the night: it is as if one goes suddenly from summer to autumn, and the sun has scarcely risen, when one already begins to complain about the heat. In a climate where one's health depends mainly on skin functions and the organs are affected by the slightest changes in temperature,[1] a cooling of the air by ten to twelve degrees inhibits perspiration in a way that is very dangerous for non-acclimated Europeans.

It has been wrongly claimed that the *vómito* had never before prevailed in any part of the Southern Hemisphere, and people have attributed the cause of this phenomenon to the cold that they believed to be proper to that hemisphere. I shall have occasion to demonstrate elsewhere how much the temperature differences in the countries located north and south of the equator have been exaggerated. The temperate part of South America has IV.176 the climate of a peninsula that narrows to the south: summers are cooler there and winters more severe than in the lands[2] that widen to the north on the same latitude in the northern hemisphere. The average temperature in Buenos Aires barely differs from that of Cádiz, and the influence of ice, which certainly accumulates more at the North Pole than at the South Pole, is hardly felt below 48° southern latitude. We have seen above that it is precisely in the southern hemisphere, in Olinda, Brazil, that yellow fever has struck down a large number of Europeans. The same disease prevailed in Guayaquil in 1740 and during the early years of this century in Montevideo, a port that is so well-known, moreover, for the healthiness of its climate.

For approximately the past fifty years, the *vómito* has practically disappeared from the Great Ocean coasts, except in the city of Panamá. In that port, as in Callao,[3] the onset of great epidemics is most often marked by the arrival of a few ships from Chile, not because that country, one of the most salubrious and most pleasant on Earth, could spread a disease that does not exist there, but because its inhabitants, transplanted to the Torrid Zone, experience the drastic effects of excessively warm air contaminated by a mixture of putrid fumes. IV.177 The town of Panamá is located on a dry strip of land

1. The air temperature in Guayaquil remains so uniformly between twenty-nine and thirty-two degrees Centigrade that the inhabitants complain about the cold whenever the thermometer suddenly drops to twenty-three or twenty-four degrees. Such phenomena are quite remarkable when considered from a physiological point of view: they prove that the sensitivity of the organs increases through the uniformity and prolonged action of *habitual stimuli*.

2. See chap. VIII, vol. II, p. 326 [p. 474 in this edition].

3. ▼ Leblond, *Observations sur la fièvre jaune*, p. 204.

devoid of vegetation, but when the tide is out, it exposes a large expanse of land covered with rockweed, sea lettuce, and jellyfish. These heaps of sea plants and gelatinous mollusks are left behind on the beach, exposed to the heat of the sun. The air is infected by the decomposition of so much organic matter, and the miasmas that have practically no effect on the indigenous peoples' organs have a powerful effect on individuals born in the cold regions of Europe or the two Americas.

The causes of the unhealthiness of the air are very different on the two coasts of the isthmus. In Panamá, where the *vómito* is endemic and the tides are very strong, the beach is considered to be the source of the infection. In Portobelo, where remittent bilious fevers prevail, and where the tides are hardly noticeable, putrid fumes arise from the strength of the vegetation itself. Only a few years ago, the forests that cover the interior of the isthmus extended as far as the gates of the town, and bands of monkeys entered the gardens of Portobelo to pick fruit there. The healthiness of the air has improved considerably since and the excellent administrator, ▼ Don Vicente Emparán, has had the surrounding woods cut down.

The position of Veracruz is more analogous to Panamá and Cartagena de Índias than to the position of Portobelo and Omos. The forests that cover IV.178 the eastern slope of the Cordillera scarcely reach the Encero farm: sparser woods begin there, composed of Mimosa cornigera, Varronia, and Capparis breynia, gradually disappearing within five or six leagues from the sea coast. The area surrounding Veracruz is terribly dry: arriving on the Jalapa road, one finds a few coconut palms near La Antigua that decorate the gardens of that village; they are the last large trees found in the desert. The excessive heat that prevails in Veracruz is increased by the hills of moving sand (*meganos*) created by the impetuous north winds that surround the south and southwest sides of the town. These cone-shaped dunes are up to fifteen meters high: strongly heated in proportion to their mass, they retain throughout the night the temperature that they have acquired during the day. A Centigrade thermometer plunged into the sand in the month of July will rise to forty-eighth or fifty degrees from the progressive accumulation of heat; but the same instrument in the open air and in the shade will remain at thirty degrees. The *meganos* may be considered ovens that heat the ambient air: they are active not only because they *radiate* caloric heat in all directions, but also because, due to their grouping, they block the free circulation of air. The same cause that created them can also easily destroy them: the dunes change IV.179 place every year, as one can observe especially in the part of the desert called the *Meganos de Cathalina*, *Meganos del Coyle*, and *Ventorillos*.

But unfortunately for those inhabitants of Veracruz who have not acclimated, the sandy plains that surround the town, far from being completely dry, are interspersed with marshy terrain where rainwater collects and filters down through the dunes. ▾ Mr. [Pérez y] Comoto, Mr. Jiménez, Mr. Moziño, and other intelligent physicians who have examined the causes of the unhealthiness of Veracruz before me, consider these reservoirs of muddy, still water as so many sources of infection. I shall list here only those marshes known as the *Ciénaga Boticaria*, behind the powder warehouse, the *Laguna de la Hormiga*, the *Espartal*, the *Cienega de Arjona*, and the *Tembladera* marsh, located between the *Rebentón* road and the *Callejones de Aguas-Largas*. At the foot of the dunes, there are only small shrubs of Desmanthus, Euphorbia tithymaloïdes, Capraria biflora, Jatropha with cotton tree leaves, and a few Ipomoca whose stems and flowers barely emerge from the arid sand that covers them: wherever this sand is bathed by the marsh water that overflows in the rainy season, vegetation becomes more vigorous. The Rhizophora mangle, the Coccoloba, Porthos, IV.180 Arum, and other plants that grow well in moist soil saturated with saline particles form scattered mounds. These low, marshy places are even more threatening because they are not always covered by water. A layer of dead leaves mixed with fruit, roots, the larvae of aquatic insects, and other debris from animal matter begins fermenting as it is gradually heated by the rays of the burning sun. I shall present elsewhere the experiments that I conducted during my stay in Cumaná on the action exerted by mangrove roots on the ambient air when they are slightly moistened and remain exposed to light. These experiments will shed some light on the remarkable phenomenon long ago observed in both Indies, namely, that of all the places with vigorous manchineel and mangrove tree growth, the unhealthiest places are those where the roots of the trees are not always covered with water. Rotting vegetal matter is generally more dangerous in the tropics because of the considerable number of astringent plants whose bark and roots contain large amounts of animal matter combined with tannin.[1]

If it is indisputable that causes exist for the unhealthiness of the air in IV.181 the terrain around Veracruz, it is equally undeniable that others also exist within the town itself. The population of Veracruz is too large for the small plot of land that the town occupies: 16,000 inhabitants are confined within a

1. [Fourcroy and] Vauquelin on the tannate of gelatine and albumin. ["Mémoire sur l'existence d'une combinaison de Tannin et d'une matière animale dans quelques Végétaux"], *Annales du Muséum*, vol. XV, p. 77.

space of 500,000 square meters, since Veracruz is a semi-circle whose radius is barely 600 meters long. As most of the houses have only one story above the ground floor, a large number of people of the lower class often live in the same dwelling. The streets are wide and straight, the longest running from northwest to southeast; the shortest streets, or cross streets, run southwest to northeast. But since the town is surrounded by a high wall, there is hardly any air circulation. The breeze that blows softly during the summer from the southeast and east-southeast is only felt on the terraces of the houses, and the inhabitants, whom the north wind often prevents from crossing the street in winter, breathe stagnant, burning air in the hot season.

Foreigners who frequent Veracruz have greatly exaggerated[1] the slovenliness of its inhabitants. For some time, the police have taken steps to ensure the salubrity of the air. Veracruz is already less dirty than many towns in southern Europe, but since it is frequented by thousands of non-acclimated Europeans, located under a burning sun, and surrounded by small marshes whose fumes infect the surrounding air, it will only see a decrease in the drastic effects of its epidemics after the police have persisted in their efforts for many, many years. IV.182

On the coastline of Mexico, one notices a close connection between the spreading of diseases and variations in atmospheric temperature. There are only two recognizable seasons in Veracruz, the season of the northern storms (*los nortes*) from the autumn to the spring equinox, and the season of the breezes or southeast winds (*brisas*) that blow fairly regularly from March to September. January is the coldest month of the year, because it is the farthest from the two periods when the sun passes through the zenith of Veracruz.[2] In general the *vómito* begins to rage in this town only when the average monthly temperature reaches twenty-four degrees on the Centigrade thermometer; in December, January, and February, the heat stays below this level. Consequently, it is extremely rare that yellow fever does not completely disappear in that season, during which one often feels a noticeable chill. The great heat begins in the month of March and with it the scourge of the epidemic. Although May is hotter than September and October, the *vómito* does the greatest damage in those two months, since it takes all epidemics a certain amount of time to develop the germ of the disease IV.183

1. ▼ Thorne, ["An Account of the Situation and Diseases of La Vera Cruz,"] in the American *Med[ical] Rep[ository]*, vol. XXX [III], p. 46. [Ruiz de] Luzuriaga, *De la calentura biliosa*, vol. I, p. 65 (a translation of Benjamin Rush's work, enhanced by Mr. Luzuriaga's observations).

2. May 16 and July 27.

completely, and the rains that last from June to September certainly affect the creation of the miasmas that form around Veracruz.

The arrival and the end of the rainy season are the most dreaded times in the tropics, because excessive humidity interrupts, almost as much as a great drought does, the rotting of the vegetal and animal substances that accumulate in marshy areas. Over 1,870 millimeters of rainwater fall in Veracruz each year: in 1803, in the month of July alone, an exacting observer, Mr. de Costansó, a colonel in the army corps of engineers, collected over 380 millimeters of rainwater, which is only one-third less than what was collected in London during an entire year. The reason why more heat is not accumulated in the air during the second rather than the first passage of the sun through the zenith of Veracruz must be sought in the evaporation of this rainwater. Europeans, afraid of falling prey to the *vómito* epidemic, consider as very good years the ones when the north wind blows strongly

IV.184

until March and those when it is already felt in the month of September. To ascertain the effect of temperature on the spread of yellow fever, I examined with great care tables containing over 21,000 observations that the captain of the port, Don Bernardo de Orta, made there during the fourteen years preceding 1803. That tireless observer's thermometers have been compared to the ones that I used in the course of my expedition.

In the following table, I present average monthly temperatures, derived from Mr. Orta's meteorological tables: I have added the number of the sick who died of yellow fever in the San Sebastián hospital in 1803. I would have liked to know the state of the other hospitals, especially the one maintained by the monks of *San Juan de Dios*. Scholars living in Veracruz will one day complete the framework that I have only sketched here; I have only indicated the individuals about whom no doubt remained as to the nature of their illness, because of the frequent vomiting of black matter. Since the combinations of foreigners in 1803 were the same at different times of the year, the number of patients indicates adequately the progress of the *vómito*. The same table also presents the climatic variations between Mexico City and Paris,[1] whose average temperature contrasts singularly

IV.185

with temperatures on the eastern coastline of New Spain. In Rome, Na-

1. The average temperature of Mexico City is based on Mr. Alzate's observations. (*Observaciones meteorológicas de los últimos nueve meses del año* 1769, *Mexico*, 1770.) Since the observations made within the walls of the city of Paris indicate a temperature slightly higher than the one that corresponds to 48°50′ latitude, we have chosen the figures presented in the *[Troisième] calendrier [météreologique] de Montmorenci*, which ▼ Mr. Cotte calculated for the years 1765–1808 (*Journal de Physique*, 1809, p. 382).

IV.185 *Meteorological and Nosographic Table of Veracruz (Latitude 19°11′52″) on the Centigrade Thermometer*

DIVISION OF THE YEAR		AVERAGE TEMPERATURE VERACRUZ	PROGRESSION OF YELLOW FEVER (HOSPITAL OF ST. SEBASTIÁN)		NOTES	AVERAGE TEMPERATURE	
			Arrivals	Deaths		In Mexico City	In Paris
North winds	January	21.7°	7	1	At La Guairá, in Cumana, on the parallel of Veracruz with the Antilles, and everywhere where the North winds do not blow, the monthly average temperature is never below 25°.	I do not doubt the av. temp. In Jan., the therm. dropped to five or six degrees and even lower.	1.2°
	February	22.6°	6	2			3.4
	March	23.3°	19	5			8.0
IV.186 Breeze, average temperature above 24°. Yellow fever season	April	25.7°	20	4	Sometimes the north winds blow again.	18.6	10.5
	May	27.6°	73	11	First passage of the Sun across the zenith of Veracruz.	18.8	14.1
	June	27.5°	49	6	Beginning of the rainy season.	16.9	18.0
	July	27.5°	51	11	Second passage of the Sun across the zenith of Veracruz.	17.0	19.4

(*continued*)

DIVISION OF THE YEAR		AVERAGE TEMPERATURE VERACRUZ	PROGRESSION OF YELLOW FEVER (HOSPITAL OF ST. SEBASTIÁN)		NOTES	AVERAGE TEMPERATURE	
			Arrivals	Deaths		In Mexico City	In Paris
	September	27.4°	68	8	End of the rainy season.	15.8	16.4
	October	26.2°	19	3	Sometimes the north winds alternate with the breeze.	16.4	12
North winds	November	24.0°	9	2	These two months are so dry that in 1803 the amount of rainwater did not exceed fourteen millimeters, while on August 18 and September 15, more than seventy millimeters of rain fell within twenty-four hours.	14.4	6.5
	December	21.1°	3	0		13.7	3.8

The average temperature of Veracruz is 25.4°; that of Mexico City 17°; that of Paris 11.3°.

ples, Cádiz, Seville, and Málaga, the average heat in the month of August exceeds twenty-four degrees and differs as a result very little from the heat in Veracruz.

I would have added the thermometric movement in Philadelphia to this table, as well as the number of individuals who died of yellow fever there in each month, had I been able to obtain observations capable of giving the average temperature in the different months of the year 1803. In temperate climes, the results drawn from the highest and lowest points that the thermometer reached at specific times tell us nothing about average temperatures. This very simple and old observation appears to have escaped several physicians who have debated the question of whether the recent epidemics in Spain were caused by the heat waves that might be considered extraordinary in southern Europe. It has been affirmed in many works that the year 1790 was two degrees warmer than the years 1799 and 1800 because in the latter two years, the thermometer in Cádiz rose only to 28° and 30.5°, while in 1790 it rose to 32°. The ▾ Knight Chacón's fine meteorological observations, published by Mr. Aréjula, will shed the most light on this important subject, if someone goes to the trouble of deducing monthly averages from them. Medicine will find support from physics only when precise methods for examining the influences of the heat, humidity, and electrical tension of the air on the spread of diseases have been adopted.

IV.187

We have just outlined the typical spread of yellow fever in Veracruz, having seen that the epidemic stops raging when the average monthly temperature drops below twenty-four degrees with the advent of the northern tempests.[1] Human phenomena are certainly subject to immutable laws, but we know so little about all the conditions under which a disorder introduces itself into our bodily functions that to us a sequence of pathological phenomena appears to exhibit the strangest irregularities. When the *vómito* begins violently during the summer in Veracruz, one sees it prevail all winter long: the drop in temperature weakens the disease, but cannot succeed in extinguishing it completely. The year 1803, when mortality was so low, offers a striking example of this kind. One can see in the table that we have provided above that some individuals were attacked by the *vómito* every month; but also, during the winter of 1803, Veracruz still suffered from the epidemic that had raged with extraordinary force the previous summer.

IV.188

1. Susceptibility to heat and the influence of temperature on the bodily organs depend on the degree of *habitual excitation*, since the same air that feels cold in Veracruz might still be favorable to the development of an epidemic in the temperate zone.

Since the *vómito* was not very prevalent during the summer of 1803, the disease disappeared completely at the beginning of 1804. When Mr. Bonpland and I descended from Jalapa to Veracruz toward the end of February, IV.189 no one in the town was sick with yellow fever. A few days later, in a season when the wind was still blowing fiercely and the thermometer did not rise to nineteen degrees, Mr. Comoto escorted us to the bedside of a dying man in the San Sebastián hospital. He was a mule driver, a very hale Mexican Mestizo, who came from the plateau of Perote and had been attacked by the *vómito* while crossing the plain that separates Antigua from Veracruz.

Fortunately, cases of the disease appearing sporadically in winter are very rare; a true epidemic develops in Veracruz only when the summer heat is first felt and the thermometer rises often above twenty-four degrees. The fever advances in a similar fashion in the United States: in fact, ▼ Mr. Carey[1] has observed that the weeks when the temperature was the highest in Philadelphia have not always been those when the mortality rate was the highest; but this observation only proves that the effects of temperature and atmospheric humidity on the production of miasmas and the state of irritability of the bodily organs are not always instantaneous. I am far from considering extreme heat as the single true cause of the *vómito*, but how can one deny that in those places where the disease is endemic, there is a close correlation between the state of the atmosphere and the spread of the epidemic?

IV.190 It is undeniable that the *vómito* is not contagious in Veracruz. In most countries, people consider as contagious diseases that actually lack this trait, but in Mexico, there is no popular belief that prevents the non-acclimated foreigner from approaching the sickbed of those attacked by the *vómito*. No fact can be cited that makes it probable that either the touch or the breath of a dying person could be dangerous for the non-acclimated who care for the sick. On the continent of equinoctial America, yellow fever is no more contagious than intermittent fevers are in Europe.

According to information that I was able to obtain during a long stay in the Americas, and according to Mr. Mackittrick's, ▼ Mr. Walker's, Mr. Rush's, Mr. Valentin's, and Mr. Miller's observations and those of almost all the physicians who have practiced in both the Antilles and the United States, I am inclined to believe that this disease is not naturally contagious either in the temperate zone[2] or in the equinoctial regions of the New Continent: I say "nat-

1. Carey, *Description of the Malignant Fever of Philadelphia*, 1794, p. 38.

2. See the two excellent papers by ▼ Mr. *Stubbins Ffirth* of New Jersey and Mr. *Edward Miller* of New York on the noncontagious character of yellow fever in the United States.

urally" since it is not contrary to the similarity exhibited by other pathological phenomena that a disease that is not fundamentally contagious may—under the specific influence of climate and the season, through the accumulation of infected persons, and through their individual disposition—assume a contagious character. Apparently, these exceptions, which are infinitely rare in the Torrid Zone,[1] are more prevalent in the temperate zone. In Spain, where in 1803 over 47,000—and in 1804 over 64,000—individuals perished of yellow fever, "this disease was contagious, but only in the places where it raged, since numerous facts, observed especially in Málaga, Alicante,[2] and Cartagena, have proven that infected persons had not spread the disease in the villages where they retired, although the climate there was the same as in the contaminated villages." This theory resulted from the observations made by the enlightened commission[3] sent by the French government to Spain in 1805 to study the development of the epidemic there. IV.191

Looking successively at the equinoctial regions of the Americas, the United States, and the parts of Europe where yellow fever has wreaked its devastation, one sees that despite the regularity of the temperature over several months in summer in zones far removed from one another, the disease assumes a different character. In the tropics, its noncontagious character is almost universally recognized. In the United States, this aspect of the disease has been strongly contested by the school of medicine at the University of Philadelphia, as well as by ▼ Mr. Wistar, Mr. Blane, Mr. Cathral, and other distinguished doctors. Finally, moving northeast in Spain, we find that yellow fever is undoubtedly contagious, as proven by the examples of those persons who escaped it through self-isolation, although they were surrounded by the disease. IV.192

Near Veracruz, the *Encero* farm, whose elevation I found to be 928 meters above sea level, is the highest limit of the *vómito*. We have already seen that the Mexican oaks do not appear at lower elevations than this, as they cannot grow in the heat necessary for the yellow fever germ to foster. Individuals born

1. Fiedler, *Über das gelbe Fieber nach eigenen Beobachtungen*, p. 137. Pugnet, [*Mémoires sur les fièvres de mauvais caractère du Levant et des Antilles*,] p. 393.

2. ▼ Bally, *Opinion sur la contagion de la fièvre jaune*, 1810, p. 40.

3. Mr. Duméril, Bally, and ▼ Nysten. Furthermore, it has in no way been confirmed that yellow fever was introduced into Spain by the *polacra* Jupiter sent from Veracruz or the corvette Le Dauphin, built in Baltimore, on which the intendant of Havana, ▼ Don Pablo Valiente, and the physician Don José Caro embarked. (Aréjula, [*Breve descripción de la fiebre amarilla*,] p. 251.) Three distinguished doctors from Cádiz, ▼ Mr. Ameller, Mr. Delon, and Mr. Gonzálcs, believe that yellow fever developed spontaneously in Spain itself: a disease can be contagious without being imported.

and raised in Veracruz are not susceptible to this disease, and the same goes for the inhabitants of Havana who do not leave their country. But merchants who were born on the island of Cuba and have lived there for many years may be attacked by the *vómito prieto* when their business requires them to visit the port of Veracruz in the months of August and September, when the epidemic IV.193 rages most fiercely. Similarly, Spanish-American indigenous peoples of Veracruz have been seen falling victim to the *vómito* in Havana, Jamaica, and the United States. These facts are certainly remarkable, if we consider them in relation to changes in the sensitivity of the internal organs. Despite the great similarity of climate between Veracruz and the island of Cuba, the inhabitant of the Mexican coasts, resistant to the miasmas spawned in the air of his native country, succumbs to the irritating and *pathogenic* causes that act on him in Jamaica or Havana. It is likely that the gaseous emissions that produce the same diseases on the same parallel are virtually the same; however, a slight difference is enough to throw the vital functions into disorder and determine the specific series of phenomena that characterizes yellow fever. As I have demonstrated in a long series of experiments,[1] in which galvanic excitation serves to measure the state of sensitivity of the organs, this is how chemical agents stimulate the nerves: not only by their own inherent properties but also by the order in which they are applied. In the Torrid Zone, where the barometric pressure and air temperature remain virtually the same all year long, IV.194 and where electrical tides, wind direction, and all other meteorological variations are unswervingly uniform, the human organs, which from birth have grown accustomed to the same impressions in one's native climate, become sensitive to the slightest changes in the surrounding atmosphere. It is because of this extreme sensitivity that the inhabitant of Havana who is transported to Veracruz while the *vómito* is raging most cruelly there sometimes runs the same risk as those who are not acclimated:[2] I say "sometimes" because in general there are few examples of colonists born in the Antilles who are attacked by yellow fever in Veracruz, the United States, or Cádiz as there are of blacks[3] [nègres] who succumb to this disease.

1. [Humboldt,] *Expériences sur l'irritation de la fièvre musculaire et nerveuse* (in German), vol. II, p. 147. The second volume of this work, which was published after my departure from Europe, has not been translated into French.

2. Mr. Pugnet, (*[Mémoirs] sur les fièvres de mauvais caractère*, p. 346) has made the same observation concerning the indigenous peoples of Saint Lucia who visit the nearly islands.

3. [Ruiz de] Luzuriaga, [*Relación de la calentura biliosa*,] vol. I, p. 133. Mr. Blane and Mr. Carey refer to fifteen black men and women [nègres et negresses] who died of yellow fever on the island of Barbados and in Philadelphia.

Furthermore, it is a very striking phenomenon that in the equinoctial regions, in Veracruz, Havana, and Portocabello, the indigenous peoples need not fear the scourge of yellow fever, whereas in the temperate zone, in the United States and Spain, the indigenous peoples are as exposed to it as the foreigners are. Should we not look for the cause of this difference in the uniformity of impressions experienced by the organs of the inhabitant of the tropics, who is surrounded by an atmosphere where the temperature and electric tension vary only slightly? Perhaps the mixture of putrid fumes emanating from soil that is constantly warmed by the rays of the sun and covered with organic debris is always the same. The inhabitant of Philadelphia experiences a winter like the one in Prussia followed by a summer whose heat rivals that of Naples, and despite the extreme *flexibility* that one observes in northern peoples' [physical] organization, these same inhabitants can never succeed, so to speak, in acclimating to their native country. IV.195

When they descend from the Encero to Plan del Río and from there to Antigua and the port of Veracruz, whites and mestizos [le blancs et les métis] living on the interior plateau of Mexico, where the temperature is between sixteen and seventeen degrees and where the thermometer drops occasionally to below freezing, contract the *vómito* more easily than do Europeans or inhabitants of the United States who arrive there by sea. Crossing by degrees into southern latitudes, the latter are prepared little by little for the extreme heat they will feel upon landing: the Spanish-Americans, on the other hand, move briskly from one clime to another when they travel from the temperate to the Torrid Zone in the space of a few hours. Mortality is especially high among two classes of men whose customs and way of life are very different, viz.: the mule drivers (*arrieros*) who are exposed to extraordinary fatigue as they descend with their beasts of burden over twisting roads like those over the Saint Gotthard Pass, and the soldier recruits who are sent to fill out the garrison in Veracruz. IV.196

Of late, every imaginable care has been lavished on these unfortunate young men born on the Mexican plateau in Guanajuato, Toluca, or Puebla, but they have not yet been successfully protected from the effects of the harmful miasmas on the coast. They have been left in Jalapa for several weeks to acclimate them gradually to a higher temperature; they have descended to Veracruz by night on horseback, so as not to expose them to the sun as they cross the arid plains of La Antigua; and they have been lodged in well-aired apartments in Veracruz—but it has never been observed that these men were struck with yellow fever less swiftly and violently than the soldiers for whom these precautions were not taken. A few years ago, by an extraordinary set of circumstances, out of 300 Mexican soldiers, all between the ages of eighteen and thirty-five, 272 perished in three months. Consequently, as I was leaving Mexico, the government was finally planning to entrust the defense of the town and the castle of San Juan de Ulúa to companies of blacks [nègres] and acclimated men of color [hommes de couleur].

In the season when the *vómito* is raging violently, even the shortest stay IV.197 in Veracruz or in the atmosphere surrounding the town is long enough for non-acclimated persons to contract the disease. The inhabitants of Mexico City who intend to sail to Europe but fear the insalubrity of the coasts stay ordinarily in Jalapa until the moment that their ship is about to depart: they travel during the coolness of night and cross Veracruz in a litter before embarking in the launch that awaits them at the pier; such precautions are sometimes useless, and it happens that these same persons are the only passengers who succumb to the *vómito* during the first few days of the crossing. One might accept in this case that the disease was contracted aboard the ship, which, having moored in the port of Veracruz, contained harmful miasmas, but the swiftness of the infection can be more definitively proven by the frequent examples of well-to-do Europeans who die from the *vómito*, despite having found, upon their arrival at the Veracruz pier, litters ready and waiting to carry them to Perote. At first glance, these facts appear to support the theory that yellow fever is contagious in all zones. But how are we to imagine that a disease is communicable over long distances,[1] when it is certainly not contagious through immediate contact in Veracruz?[2] Is it IV.198 not more acceptable that the atmosphere in Veracruz contains putrid emissions that introduce disorder into the vital functions when they are inhaled for even the shortest period of time?

During their stay in Veracruz, most recently disembarked Europeans feel the first symptoms of the *vómito*, which begins with pain in the lumbar regions, the yellowing of the connective tissue, and signs of congestion near the head. For many individuals, the disease appears only when they have already reached Jalapa or the mountains of La Pileta, in the region of pine and oak trees at 1,600 or 1,800 meters above sea level. Those who have lived in Jalapa for a long time believe that they can recognize the germ of the disease in the features of the travelers who ascend from the coast to the interior plateau, although the latter may not have noticed it themselves. Low spirits and fear increase the predisposition of the organs to receive the impact of the

1. *Contagium in distans.*
2. *Contagium per intimum contactum.*

miasmas, and these same causes make the beginning of yellow fever more violent when patients are unwisely advised[1] of the danger of their situation.

We have just seen that people born in Veracruz are not exposed to contracting the *vómito* in their native country; in this respect, they have a great advantage over the inhabitants of the United States, who are susceptible to the insalubrity of their own climate. Another advantage of the Torrid Zone is that Europeans and all individuals in general who were born in temperate countries are not attacked twice by yellow fever. In the Antilles, only a few rare examples of a second attack have been observed, and such examples are very common in the United States. But in Veracruz, once a person has been attacked by the disease, he or she has no fear of subsequent epidemics. Women who disembark on the coast of Mexico or descend from the central plateau are less at risk than men. This prerogative of the fair sex appears only in the temperate zone. In 1800, 1,577 women died in Cádiz versus 5,810 men, and in Seville, 3,672 women versus 11,013 men. It was long-believed that individuals suffering from gout, intermittent fevers, or syphilitic diseases did not contract the *vómito*, but this theory goes against several facts observed in Veracruz. Furthermore, what is experienced there has been observed in most epidemics:[2] namely, that as long as yellow fever rages violently, other *intercurrent* diseases become noticeably rarer.

IV.199

IV.200

Examples of individuals who died thirty to forty hours after the first invasion of the *vómito* are rarer in the Torrid Zone than in temperate regions. In Spain, infected individuals have been seen to move from a healthy state to death in six or seven hours.[3] In this case, the disease reveals all its simplicity and appears to act only on the nervous system. The excitation of this system is followed by complete prostration; the vital principle is extinguished with

1. In this respect, I can cite a characteristic that is all the more unusual since it describes both the phlegm and sangfroid of the copper-skinned race. During my stay in Mexico City, a friend of mine had only been in Veracruz a short while upon his first voyage from Europe to the Americas: he arrived in Jalapa with no idea of the danger in which he would soon be. "You will have the *vómito* by evening," an Indian barber told him as he was lathering his face, "the soap dries as soon as I apply it; this sign never lies, and I have been shaving the *chapetones* who pass through this town on their way to Mexico City for twenty years; three out of five of them die." This death sentence had a deep effect on the traveler's mind, and it was useless for him to tell the Indian how exaggerated his calculation was and that very warm skin is not a sure sign of infection; the barber persisted in his prognosis. In fact, the disease appeared a few hours later, and the traveler, who was already underway to Perote, had to be carried to Jalapa, where he nearly succumbed to the violence of the *vómito*.

2. ▼ Schnurrer, *Materialien zu einer allgemeinem Naturlehre der Epidemien und Contagien*, 1810, p. 40; this work contains valuable material for *pathological zoonomy*.

3. Berthe, [*Précis historique de la maladie qui a régné dans l'Andalousie*,] p. 79.

frightening rapidity. At that point bilious complications cannot manifest themselves and the patient dies, hemorrhaging badly but without his or her IV.201 skin yellowing[1] or vomiting the matter known as black bile. In Veracruz the yellow fever generally lasts more than six to seven days; this is enough time for the irritation of the digestive system to mask, so to speak, the true character of the adynamic fever.

Since the *vómito* in the equinoctial region attacks only individuals born in cold countries and never the locals, the mortality rate in Veracruz is lower than one would think, considering the heat of the climate and the extreme sensitivity of the organs that results from this heat. Large epidemics have only struck down about 1,500 persons within the city annually. In my possession are tables that indicate the state of the hospitals over the last fifteen years, but since these tables do not specifically refer to patients who died from the *vómito*, they tell us almost nothing about medical advances toward decreasing the number of victims.

The mortality rate is excessively high in the hospital administered by the monks of Saint John of God (*San Juan de Dios*): from 1786 to 1802, 27,922 patients entered the hospital, of which 5,657, or more than one-fifth, died. The number of deaths must be interpreted as even higher, since the *vómito* IV.202 did not prevail from 1786 to 1794, and over one-third of all the patients who entered the hospital were affected by intermittent fevers or other non-epidemic illnesses. The mortality rate at the hospital of *Our Lady of Loreto* was much lower. From 1793 to 1802, 2,820 individuals entered the hospital, of which 389, or one-seventh, died. *San Sebastián* is the best hospital in Veracruz, administered at merchants' expense (*Hospital del consulado*) and directed by a doctor[2] who has acquired a well-deserved reputation for his knowledge, fairness, and diligent activity. Here [in the table below] is the state of that small establishment in 1803.

IV.203 According to this table, the average mortality rate was one-seventh or 14 percent. The *vómito* alone carried off only 16 percent; furthermore, it must be observed that over one-third of those who died had been taken to the hospital when the disease had already made alarming progress. In general, according to the trade tables published by the *Consulado*, only 959 people died in Veracruz in 1803, either from various illnesses or old age. Assuming the

1. Mr. Rush observed that during the epidemic of 1793 in Philadelphia, the skin of those who were in the best health, and even that of blacks [nègres], had turned yellow and their pulse was extremely fast.

2. Don Florencio Pérez y Comoto.

MONTH	ARRIVALS			DISCHARGED			DIED		
	Yellow Fever	Other Illnesses	Total	Yellow Fever	Other Illnesses	Total	Yellow Fever	Other Illnesses	Total
January	7		7	6		6	1		1
February	6		6	4		4	2		2
March	19		19	14		14	5		5
April	20	21	41	17	18	35	4	2	6
May	73	30	103	62	30	92	11		11
June	49	4	53	43	3	46	6	1	7
July	51	4	55	40	3	43	11	1	12
August	94	4	98	78	4	82	16		16
September	68	4	72	60	4	64	8		8
October	29	22	51	26	20	46	3	2	5
November	9	17	26	7	15	22	2	2	4
December	3	19	22	3	16	19		1	1
Total	428	125	553	360	113	473	69	9	78

population to be between sixteen and seventeen thousand souls, one finds the total mortality rate to be 6 percent; now, of the 959 deaths, fewer than half were caused by the *vómito*. As a result, the proportion of the total number of deaths to that of *acclimated* inhabitants is approximately one to thirty, which confirms the theory,[1] very widespread in the country, that individuals accustomed since childhood to the great heat of the Mexican coasts and to the miasmas in the atmosphere reach a ripe old age. In 1803 the hospitals of Veracruz admitted 4,371 patients, of which 3,671 left having recovered. As a result, the death rate was only 12 percent, although, as we have seen from the state of the San Sebastián hospital, there were always some patients who succumbed to yellow fever, even when the north winds cooled the air.

IV.204 We have heretofore given detailed information on the ravages of the *vómito* within the walls of Veracruz in a year when the epidemic raged less violently than usual. But many Mexican mule drivers, sailors, and young people (*polizones*) who embark from the ports of Spain to seek their fortune in Mexico fall victim to the *vómito* in the village of La Antigua, on the Muerto farm, in La Rinconada, Cerro Gordo, even in Jalapa, when the invasion of the disease is too sudden for them to be transported to hospitals in Veracruz, or when they do not sense the attack until they are ascending the Cordillera. Mortality is extremely high, especially when several warships and a large number of merchant ships arrive in port at the same time. In some years the number of deaths within the town and in the surrounding area rises to 1,800 or 2,000. This loss is even more grievous since it falls on a class of hard-working men of robust constitutions, who are almost all in the prime of life. The result of the sad experiences at the hospital run by the IV.205 monks of *San Juan de Dios*[2] in the last fifteen years is that wherever patients crowded into small quarters are not treated attentively, mortality rises to 30 or 35 percent during great epidemics, whereas in those places where good care can be provided, and where the physician varies treatment according to the different forms in which the disease appears in a given season, mortality does not exceed 12 or 15 percent. The lists of the *consulado*'s hospital, directed by Mr. Comoto, provided this figure, which certainly seems rather

1. See chap. IV, vol. I, p. 309 [p. 206 in this edition].

2. In 1804 this hospital was about to be abolished and replaced by another one that was to be called the *House of Benevolence* (*Casa de beneficiencia*). Throughout Spanish America, enlightened men complain of the healing methods used by the monks of *San Juan de Dios*. The task that this congregation has set itself is among the noblest: I could cite many examples of these monks' selflessness and bravery, but at the patient's beside, charity is no substitute for the lack of professional capability.

small when compared with the recent ravages of the yellow fever in Spain.[1] But in comparing these facts, we must bear in mind that the disease does not rage every year nor does it afflict everyone with the same violence. To obtain precise results on the ratio of deaths to infected persons, one would have to distinguish between the different degrees of *exacerbation* that the *vómito* reaches as it develops progressively. According to ▼ Russel[l], even the plague appears occasionally in Aleppo under such benign atmospheric influences that many of the afflicted are not confined to their beds during the course of the epidemic. IV.206

On the outskirts of Veracruz, the *vómito* has only been experienced in the interior of the country within ten leagues of the coastline. Since the ground level rises rapidly as one travels west, and since the elevation of the ground affects the air temperature, New Spain cannot help us answer the question of whether yellow fever develops in places that are far from the sea. Mr. Volney,[2] an excellent observer, reports that an epidemic disease with great similarities to yellow fever raged east of the Alleghany Mountains in the marshland around Fort Miami, near Lake Erie; Mr. Ellicott has made similar observations on the banks of the Ohio, but we must remember that bilious remittent fevers sometimes assume the adynamic character of yellow fever. In Spain as in the United States, the epidemic has followed the seaboard and the course of the great rivers: there is some doubt as to whether it raged in Córdoba, but IV.207

1. The following table may be used to judge the average mortality rate observed in Spain during the epidemics of 1800, 1801, and 1804; it is based on information that I owe to Mr. Duméril's gracious kindness.

YEARS	CITIES	SICK	DEATHS	AVERAGE MORTALITY
1800	Cádiz	48.520	9,977	20 percent
	Seville	76,000	20,000	26
	Xeres	30,000	12,000	40
1801	Seville	4,100	660	60
1804	Alicante	9,000	2,472	27
	Xeres	5,000	2,000	40

Mr. Aréjula informs us that nineteen of every one hundred patients died in Seville in 1800; twenty-six in Alicante in 1804; nearly forty in Málaga in 1803, and over sixty in 1804. He claims that Spanish doctors should be proud of having cured three-fifths of the patients who were already vomiting black matter. ([*Breve descripción*] *de la fiebre* [*amarilla*,] pp. 148 and 433–44.) In the case of a great exacerbation of the disease, that famous practitioner's claim would suggest a mortality rate of 40 percent.

2. [Volney,] *Tableau [du climat et] du sol d'Amérique*, vol. II, p. 310.

is appears certain that it ravaged Carlota, five leagues south of Córdoba, a very clean town located on a high hill and exposed to the most salubrious winds.[1]

Brown's system did not arouse as much enthusiasm in Edinburgh, Milan, or Vienna as it did in Mexico. Scholars able to observe impartially both the good and the bad effects of the *stimulant method* believe in general that American medicine has benefited from this revolution. The overuse of bleedings, purgatives, and all debilitating remedies was extremely prevalent in the Spanish and French colonies. Such abuse not only increased mortality rates among patients but was also harmful to newly disembarked Europeans who were bled despite being in perfect health: the prophylactic treatment predisposed this group to the disease.[2] Should we be surprised that despite

IV.208

its inaccuracy and its deceptive simplicity, Brown's method was effective in a country where an adynamic fever was treated as an inflammatory fever, where they feared to administer quinquina, opium, and ether, where, even during the worst prostrations, they patiently awaited crises, prescribing saltpeter, mallow water, and infusions of Scoparia dulcis? Reading Brown's work inspired Spanish and Mexican doctors to become involved in studying the causes and forms of diseases: ideas already put forth long ago by Sydenham, the ▼ Leyden school, Stoll, and Frank were accepted in the Americas. The reform that is now attributed to Brown's system is the result of the reawakening of the observant mind and the general progress of enlightenment.

Although the first sign of the *vómito* is a sthenic diathesis, the bleedings that Rush so vociferously recommended, and which Mexican doctors frequently used during the great epidemic of 1762, are considered dangerous in Veracruz. In the tropics, the passage from synoche to typhus, and from the inflammatory state to a languid state, is so rapid that the loss of blood, which is inaccurately described as being in dissolution, accelerates the general prostration of strength. In the first stage of the *vómito*, purgatives, ice water, and the use of ices and other debilitating remedies is preferred. To

IV.209

use the language of the Edinburgh school, when indirect debility is felt, the most energetic excitants are used, beginning with strong doses and gradually decreasing the *power* of the stimulants. Mr. Comoto has had great success administering hourly doses of over one hundred drops of sulfuric ether and from sixty to seventy drops of opium tincture. This treatment is mark-

1. Berthe, [*Précis historique de la maladie*,] p. 16. Carlota is twenty-six miles from the sea, as the crow flies.

2. Pinel, [*Nosographie philosophique*,] vol. I, p. 207. Gilbert, *Maladies de Saint-Domingue* [*Histoire médicale de l'armée française, a Saint-Domingue*,] p. 91.

edly different from the folk remedy that consists not of relieving vital forces through the use of excitants but rather of administering warm, mucilaginous drinks, infusions of tamarind, and fomentations [hot packs] to calm the irritation of the abdominal system.

The experiments on the use of cinchona in treating yellow fever, which were conducted in Veracruz until 1804, were unsuccessful,[1] although that bark has often had the most salutary effect in the Antilles and Spain.[2] It is possible that these different reactions result from the variety of forms that the disease assumes, depending on whether remission is more or less pronounced or whether gastric symptoms predominate over the adynamic symptoms. Mercurial preparations, especially calomel [mercury chloride] or muriate of sweet mercury, blended with jalap, have often been used in Veracruz, but these remedies, which were highly praised in Philadelphia and Jamaica and were prescribed for ataxic fevers by Spanish physicians as early as the sixteenth century,[3] have been widely abandoned by Mexican doctors. They were more satisfied with the results of olive oil massages, the usefulness of which was recognized by Mr. Jiménez in Havana, Don Juan de Arias in Cartagena de Índias,[4] and especially by my friend ▾ Mr. Keutsch, a distinguished doctor on the island of Saint Croix who has collected many interesting observations of yellow fever in the Antilles. For some time now in Veracruz, ice, pineapple juice (*jugo de piña*), and an infusion of *palo mulato*, a vegetable in the genus amyria, have been regarded as specific remedies for the *vómito*, but extensive and disappointing trials have gradually discredited these remedies, even among the Mexican people. Although they must be ranked among the best prophylactic measures, they cannot form the basis of a curative treatment.

IV.210

Since excessive heat increases the action of the bilious system, the use of ice can only be very beneficial in the Torrid Zone. Relays have been set up to transport ice as quickly as possible on muleback, from the slope of the Orizaba volcano to the port of Veracruz. The road traveled by the *snow*

IV.211

1. According to Mr. Rush and ▾ Mr. Woodhouse's observations, they were no more successful in Philadelphia during the epidemic of 1797. [Ruiz de] Luzuriaga, [*Relación de la calentura biliosa*,] vol. II, p. 218.

2. Pugnet, [*Mémoires sur les fièvres de mauvais caractère du Levant et des Antilles*,] p. 367. [Juan Manuel de] Aréjula, [*Breve descripción de la fiebre amarilla*,] pp. 151 and 209. ▾ Mr. Chisholm and Mr. Seaman preferred to use *Cortex Angusturæ* (the bark of Bonplandia trifoliata) rather than cinchona.

3. ▾ Luis Lobera de Ávila, *Vergel de sanidad*, 1530. Andrés de Laguna, *Sobre la cura . . . de la pestilencia*, 1566, Francisco Franco, *[Libro] de las enfermedades contagiosas*, 1569.

4. [Ruiz de] Luzuriaga, [*Relación de la calentura biliosa*,] vol. II, p. 218.

post[1] (*posta de nieve*) is twenty-eight leagues long. The Indians select pieces of snow mixed with clumps of bonded hailstones. Following an ancient custom, they wrap these blocks with dry grass, and sometimes even with ashes, two substances that are known to be poor conductors of heat. Although the mules bearing snow from Orizaba rush at a full trot to Veracruz, over half of the snow melts on the road, since the atmospheric temperature in summer is constantly between twenty-nine and thirty degrees on the Centigrade thermometer. Despite these obstacles, the inhabitants of the coast can have sherbets and iced water every day. Such an advantage, which does not exist in the Antilles, Cartagena, and Panamá, is infinitely valuable for a town that is normally frequented by people born in Europe and on the central plateau of New Spain.

Although in Veracruz yellow fever is not contagious by direct contact and though it is in no way likely that it was ever introduced from the outside,[2] IV.212 it is nonetheless certain that it only appears in specific periods, yet no one to this day has discovered which atmospheric modifications in the Torrid Zone have given rise to these periodic changes. Regrettably, the history of epidemics has not been charted back further than half a century. The main military hospital of Veracruz was established in December 1764, but no document preserved in the archives of that hospital mentions the diseases that preceded the vómito of 1762. That epidemic, which began under the viceroy the Marquis de Croix, continued to rage until 1775 when, after the streets were paved, a few feeble attempts at sanitization were made, which had the result of reducing the town's dirtiness. The inhabitants imagined at first that the pavement would exacerbate the air's insalubrity, by increasing the unbearable heat that prevails within the city walls because of the reverberation of the sun's rays. But when they noticed that the vómito had not reappeared from 1776 to 1794, they believed that the pavement had saved them from it forever, forgetting that the ponds of stagnant water located south and east of the town, which had always been considered the main seat of the harmful miasmas in Veracruz, continued to release putrid fumes into IV.213 the atmosphere. It is a very remarkable fact that in the eight years preceding 1774, there was not a single example of the *vómito*, although the frequency

1. See plate IX of my Mexican atlas.

2. "Veracruz received the germ of this cruel disease neither from Siam, Africa, the Antilles, Cartagena de Indias, nor the United States; the germ was forged (*engendrado*) in its own territory; it is constantly found there but develops only under the influence of certain climatic conditions." Comoto in his *Informe al prior del Consulado de la Vera Cruz, del mes de Junio,* 1803 (*unpublished paper.*)

of Europeans and Mexicans from the interior was exceptionally high, unacclimated sailors indulged themselves in the same excesses that are still held against them, and the town was dirtier than it has been since 1800.

The merciless epidemic that appeared in 1794 can be traced back to the arrival of three warships, the vessels *El Mino*, the frigate *Venus*, and the hooker *Santa Vibiana*, which had stopped in Puerto Rico. Since these ships were carrying several young, non-acclimated sailors, the *vómito* began with extreme violence in Veracruz. From 1794 to 1804, the disease reappeared every year, as soon as the north winds stopped blowing. We therefore see that from 1787 to 1794 the royal military hospital[1] had admitted only 16,835 patients, whereas from 1795 to 1802, their number rose to IV.214

1. This hospital admits all infected persons who arrive by sea. There were, in

	TREATED	DECEASED
1792	2,887	71
1793	2,907	77
1794	4,195	453
1795	3,596	421
1796	3,181	176
1797	4,727	478
1798	5,186	195
1799	14,672	891
1800	9,294	505
1801	7,120	226
1802	5,242	441

Before the onset of the epidemic in 1794, the mortality rate was only between 2 and 2.5 percent. It is now between 6 and 7 percent and would be even higher if this hospital, like all military hospitals, did not admit so many sailors who are not seriously ill. In general, in the public hospitals of Paris, fourteen to eighteen out of every one hundred patients die. But one must bear in mind that these hospitals admit a large number of patients who are near death or at a very advanced age. [*Précis de l'état actual des hospitaux civils de Paris; exposé des*] *travaux [et observations] du bureau central d'admission*, 1809, p. 5.

The State of Hospitals in Veracruz in 1806

NAME OF HOSPITAL	SICK	DEATHS	AVERAGE MORTALITY
San Carlos	6,382	85	1 and one-third percent
San Sebastián	2,010	281	11 and forty-nine one-hundreths
Loreto (for women)	281	49	17 and forty-four one-hundreths

57,213. Mortality was especially high in 1799, when the viceroy, the Marquis de Branciforte, fearing that the English would disembark on the eastern coastline, had troops quartered in Arroyo Moreno, a very unhealthy site two and a half leagues from Veracruz.

IV.215

It must be observed that during the period that preceded the epidemic of 1794, yellow fever did not cease to rage in Havana and in the other Antilles islands with which the Veracruz merchants maintained constant trade relations: several hundred ships arrived annually from these infected places without being placed in quarantine, and the *vómito* never appeared among the Europeans in Veracruz. I have examined the monthly temperature for the year 1794 in Mr. Orta's meteorological records: far from being higher, the temperature was actually lower than in preceding years, as the following table demonstrates.

IV.216

Heat and humidity can have two very different effects on the development of epidemics: they can support the production of miasmas or simply increase the irritability of the organs and act as predisposing causes. The facts that we presented earlier demonstrate that temperature has an undeniable influence on the spread of the *vómito* in Veracruz, but there is evidence to prove that when the disease has stopped raging for several years, a very hot and humid summer is enough to revive it: heat, therefore, is not the only factor that produces what is very vaguely called a *bilious constitution*. Despite the yellow color that the skin of infected persons takes, it is highly unlikely that bile enters the blood[1] and that the liver and the portal venous system play a

IV.217

In Mexico City in 1805, 18,398 patients entered the twelve hospitals, of whom 1,773 died. The mortality rate was, therefore, 9.6 percent. In Puebla it was 15.7 percent, since of the 6,566 patients who entered the hospital of San Pedro in 1806, 1,032 died.

In 1806, the total number of deceased in Veracruz, including the hospitals, was 663. Yet according to Mr. Don José Maria Quirós's evaluation, the population of the town in that period comprised 35,520 souls, to wit: the normal population, 30,000; sailors and seafarers, 3,640; the mule drivers required to husband 49,139 mules and other beasts of burden that transport goods from Perote and Orizaba to Veracruz, 7,370; foreigners, travelers, and militia, 4,500 individuals. As a result, the average mortality rate during a period when the *vómito* was not raging was only 1.5 percent. In 1805, it rose to 2.3 percent; the number of dead was 1,049, and the total population was 36,230 souls. It is true that this population includes at most 5,000 children from one to ten years old, and that the mortality rate is quite insignificant in so far as most of the inhabitants are young, healthy men who are used to fatigue and changes in climate; however, all of the considerations and calculations that we have just presented give sufficient proof that in the years when yellow fever did not wreak havoc, the port of Veracruz was no more pernicious to health than most other seaside towns located in the Torrid Zone.

1. Human bile is high in albumin: out of 1,100 parts, forty-two parts are albumin, fifty-eight parts resin, yellow matter, soda, and salt, and 1,000 parts are water. Thénard, [Mémoire sur la Bile] in *Mémoires [de la physique et de la chimie de la Societe] d'Arcueil*, vol. I, p. 57.

Average Temperature in Veracruz (Centigrade Thermometer)

MONTH	NO VÓMITO PRIETO		VÓMITO PRIETO EPIDEMICS	
	1792	1793	1794	1795
January	21.5	20.8	20.6	20.7
February	21.5	22.3	22.8	21.0
March	23.7	22.8	22.6	22.5
April	24.2	26.1	25.3	24.0
May	27.3	27.9	25.3	26.3
June	28.5	27.8	27.5	27.2
July	27.5	26.9	27.8	27.7
August	28.3	28.1	28.3	27.8
September	27.5	28.1	27.1	26.1
October	26.3	25.5	26.1	25.0
November	24.7	24.4	23.0	24.3
December	21.9	22.1	21.7	21.0
Average temperature of the year	25.2	22.2	24.8	24.5

major role in yellow fever, as has been assumed. Black matter, found in the *vómito prieto*, is only mildly comparable to bile: it resembles coffee grounds, and I have observed that it sometimes leaves indelible marks on bed linen and walls. When it is gently heated, it gives off hydrogen sulfide. According to Mr. Ffirth's experiments,[1] it contain no albumin, but a resinous, oily sub- IV.218 stance, phosphates, and muriates of lime and soda. By dissecting cadavers

1. According to the exacting experiments of Mr. Thénard, there is no bile in the blood of those who are attacked by icterus. ▼ Mr. Magendie, who has enriched physiology with his ingenious experiments on the action of poisons, has observed that an ordinary-sized dog will die if more than seven grams of bile are injected into his blood. In this case, the serum does not turn yellow and the conjunctiva of the animal remains white. Immediately after the injection, the bile in the blood cannot be recognized by taste, although smaller amounts of bile make a large amount of water taste bitter. ▼ Mr. Autenrieth has observed that human blood serum turns yellow in patients who show no bilious complications. ([*Handbuch der empirischen menschlichen*] *Physiologie*, vol. II, p. 91. ▼ Grimaud, *Second mémoire sur la nutrition*, p. 78.) It is also known that the skin of the elderly turns yellow when bruised and wherever there is extravasated blood.

in which the pylorus was totally obstructed, the same anatomist has proven that the substance from the *vómito* does not come from the hepatic canals but flows into the stomach through the arteries that spread through the mucous membrane: he confirms—and this is a remarkable assertion—that after death, black matter is still found in these same vessels.[1]

Some physicians in New Spain believe that *vómito* epidemics, like those of smallpox, are periodic in the Torrid Zone and that the auspicious time is not far away when Europeans will be able to disembark on the coasts of Veracruz with no more risk than in Tampico, Coro, Cumaná, and wherever the climate is exceedingly warm but very salubrious. If this hope becomes a reality, it will be of the utmost importance to examine carefully the atmospheric modifications, any possible changes to the surface of the earth, and the draining of the marshes—in a word, all the phenomena that coincide

IV.219 with the end of the epidemic. I should not be surprised, however, if this research yielded no positive results. ▾ Mr. Thénard and Mr. Dupuytren's excellent experiments have demonstrated that when mixed with the air of the atmosphere, extremely small amounts of hydrogen sulfide are enough to produce asphyxiation.[2] A great number of causes, the most powerful of which escape our perception, can modify vital phenomena.[3] Diseases arise wherever organized substances, impregnated with certain degree of humidity and warmed by the sun, come into contact with the atmospheric air. We can surmise some of the conditions under which gaseous emissions called miasmas are formed. Intermittent fevers can no longer be attributed to the hydrogen accumulated in warm, humid places, nor can ataxic fevers be attributed to ammoniacal emissions, nor inflammatory diseases to an in-

IV.220 crease of oxygen in the atmospheric air. The new chemistry, to which we owe so many positive truths, has also taught us that we are ignorant of many things that we have long been complacent enough to think that we knew with certainty.

Whatever we may not know about the nature of miasmas, which may be ternary or quaternary combinations, it is no less certain that the insalubrity of the air in Veracruz would decrease noticeably if the marshes surrounding the town could be drained, if potable water were provided for the inhabit-

1. Stubbins Ffirth, [*A Treatise on Malignant Fever*,] pp. 37 and 47.

2. A dog can be asphyxiated in air containing two-thousandths of hydrogen sulfide. [Depuytren, "Notice sur quatre asphyxiés," pp. 148–49.]

3. Gay-Lussac and Humboldt, *Expériences sur les principes constituans de l'atmosphère*, pp. 25 and 28.

ants, if hospitals and cemeteries were moved away from them,[1] if patients' wards, churches, and especially ship's quarters were frequently fumigated with oxygenated muriatic acid, and, finally, if the town walls, which force the population to crowd into a small area and prevent the air from circulating without impeding smuggling, were torn down.

If, on the other hand, the government were to go to the extreme of destroying a town that has cost so many millions to build, and force merchants to set up shop in Jalapa, mortality in Veracruz would not decrease as much as one would think at first. It is true that the black mule drivers [muletiers nègres] or indigenous peoples of the coast could haul merchandise as far as the Encero farm, which is the *upper limit of the vómito*, and that the inhabitants of Querétaro and Puebla would not need to go down to the port to do their shopping, but the people of the sea, among whom the *vómito* does the most terrible damage, would still be forced to remain in the port. The persons who would be forced to reside in Jalapa would be precisely those who are used to the climate in Veracruz because commercial affairs have long established them on the coastline. We shall not examine here the extreme difficulty with which business that accounts for a capital of two hundred fifty million livres tournois annually could be conducted at such a great distance from the port and warehouses, since the beautiful town of Jalapa, where one enjoys perpetual spring weather, is over twenty leagues from the sea. If Veracruz were destroyed and a market were set up in Jalapa, business would once again fall into the hands of a few Mexican families, who would become immensely rich: the small merchant would not be able to subsidize the expenses demanded by frequent journeys from Jalapa to Veracruz and the upkeep of two establishments, one in the mountains and the other on the coast.

IV.221

Enlightened persons have expounded to the viceroy on the inconveniences that would follow the destruction of Veracruz, but they have also proposed both closing the port during the months when the extreme heat prevails and allowing ships to enter only in winter, when Europeans run almost no risk of contracting yellow fever. This appears to be a very wise measure, even if one considers only the danger to which the seafarers who have already arrived in port are exposed, but one must also bear in mind that the same north winds that cool the atmosphere and smother the germ

IV.222

1. In 1804, the richest merchants in the town thought they had overcome the prejudices of the lower class through their own example by formally declaring that neither they nor their families would be buried within the city walls.

of infection also make navigation in the Gulf of Mexico very dangerous. If all the ships that enter the port of Veracruz each year arrived in winter, shipwrecks would be highly frequent on the coastline of the Americas as well as in Europe. The result of these considerations is that before taking recourse to such extraordinary measures, all appropriate ways of reducing the town's insalubrity should be tried, since the preservation of the town is connected not only to the private happiness of its citizens but also to the civic prosperity of New Spain.

BOOK VI

IV.223

State Revenue—Military Defense

CHAPTER XIII

The Current Revenue of the Kingdom of New Spain—Its Gradual Increase since the Beginning of the Eighteenth Century—Sources of Public Revenue

The object of our research has heretofore been to identify the principal sources of public wealth: it remains for us, at the end of this work, to examine the revenue of the state, which is intended to provide for administrative expenses, magistrates' salaries, and the military defense of the country. According to ancient Spanish laws, each viceroyalty is governed not as a possession of the crown but as a single province distant from the homeland. All the institutions that together form a European government are found in the Spanish colonies, which might be compared to a system of confederated states, were the colonists not deprived of many important rights in their trade relations with the Old World. The result of this is that a table of the public revenue of New Spain may be drawn up, as is done for the revenue of Ireland or Norway, which are governed in the name of the kings of England and Denmark. Most of these provinces, which are not designated on the Peninsula as colonies but rather as kingdoms (*reinos*), contribute no net revenue to the treasury of the king of Spain. Everywhere except Peru and Mexico, the duties and taxes imposed are absorbed by the costs of internal administration. I shall not go on at length here about the vices of this administration: they are the same that one sees in European Spain, against which national as well as foreign writers on political economy have raised their voices since the beginning of the eighteenth century.

IV.224

The revenue of New Spain[1] may be valued at twenty million piasters, of which six million are sent to Europe, to the royal tax office. The extraordinary

1. *Producto de las rentas reales del reino.*

increase in public revenue that has been observed since the beginning of the eighteenth century evidences (as does the increase in tithes[1] of which we have IV.225 already spoken) the advancement of the population, the increased commercial activity, and the growth of national wealth. According to the ledgers preserved in the archives of the viceroy and the chamber of accounts (*Tribunal mayor de cuentas*), state revenue was:

In	1712	3,068,400 piasters[1]
	1763	5,705,876
	1764	5,901,706
	1765	6,141,981
	1766	6,538,941
	1767	6,561,316
	Total from 1763 to 1767	30,849,820
	Five-month average	6,169,164
From 1767 to 1769, annual average: 8,000,000		
From 1773 to 1776, annual average: 12,000,000		
From 1777 to 1779, annual average: 14,500,000		
In	1780	15,010,974 piasters
	1781	18,091,639
	1782	18,594,492
	1783	19,579,718
	1784	19,605,574
	Total from 1780 to 1784	90,882,397
	Five-month average	18,176,479
In	1785	18,770,000 piasters
	1789	19,044,000
	1792	19,521,698
	1802	20,200,000

[1] The table of revenue from 1763 to 1784 comes from an unpublished paper written at the *Tribunal de minería* of Mexico City in 1785 to prove the impact of the progress of the mines on state revenue. Production from 1785, 1789, and 1792 has already been published in the *Viajero universal*, XXVII, p. 117. Also see Pinkerton, *Nouv[elle] géogr[aphie]* (English edition), vol. III, p. 167.

1. See chap. X, vol. III, p. 104 [p. II.58 in this edition].

The average of the five years from 1780 to 1784 differs from the corresponding average for the period from 1763 to 1767 by 12,006,515 piasters, or more than two-thirds. The decrease in the price of mercury from eighty-two to sixty-two piasters per quintal, the regulation of free trade, the establishment of the intendancies, the introduction of the tobacco monopoly and several other governmental measures that we have described above, may be considered the causes of this increase in revenue. IV.226

The following are the main revenue streams of New Spain:

1. Revenue from the output of the gold and silver mines, *five and a half million piasters*,[1] to wit:

Taxes[1] the mine owners (*derecho de oro y plata*) paid into the royal coffers by mass (in demi-quintals), by set rate, and by monetary value, in cash and/or tribute [by rights of mint and seigneuriage], in 1795	3,516,000 piasters
Net profit from the sale of mercury, in 1790[2]	536,000
Net profit in cash[3]	1,500,000

[1] See chap. XI, vol. III, p. 339.
[2] *Id.*, chap. XI, vol. III, p. 294.
[3] *Id.*, chap. XII, vol. IV, p. 26.

In 1793, the mint of Mexico City, together with the *separation* house, produced 1,754,993 piasters; expenses were 385,568 piasters, and the net profit from minting coins was 1,369,425 piasters. Duties on gold and silver have increased in the past forty years, as has the amount of precious metals extracted from Mexican mines: from 1763 to 1767, the latter amounted to 58,192,316 piasters or, in an average year, 11,638,463 piasters; from 1781 to 1785, after the decrease in the price of mercury, the establishment of a high council of mines, and the regulation of free trade, the value of mining production was 101,245,573 piasters, or in an average year, 20,249,114 piasters. In 1790, the duty on gold was 19,382 piasters; the duty on silver was 2,021,238 piasters. The net profit of the mint of Mexico City is now nearly six times greater than the mint in Lima. IV.227

2. Income from the tobacco industry,[2] *four to four and a half million piasters*. In 1802, the value of the tobacco purchased from growers in Orizaba and Córdoba was 594,000 piasters; the value of the tobacco sold on

1. *Renta del producto y beneficio metálico*.
2. *Producto del real estanco del tabaco*, chap. X, vol. III, p. 51 and chap. XII, vol. IV, p. 11.

behalf of the crown was 7,687,000 piasters. Production costs in the same year amounted to 1,285,000 piasters. Yet, since administrative expenses, or the salaries of the employees of the tobacco monopoly, exceed the sum of 794,000 piasters, net revenue was only 4,092,000 piasters.

IV.228 From these precise data, which are taken from a table presented above, in chapter twelve, one can see that this branch of public administration is so corrupt that the salaries of its employees consume 19 percent of net revenue. According to a royal cédula decreed during Gálvez's ministry, this sum must be sent to Spain: it is the *líquido remisible a la Peninsula*, which must remain intact and which the viceroys may not use on any pretext for the country's domestic needs. The main royal factory of Seville processes mainly Brazilian tobacco, although the Spanish Río Negro, the island of Cuba, the province of Cumaná, and many other provinces in the Americas could provide the most aromatic tobaccos. Its output, valued at four million piasters, is more or less equal to the net revenue of the tobacco monopoly in Mexico: both surpass the revenue of the Swedish crown, but it is not so much the high duty as the way that it is collected that people find offensive. Of all the reforms that the colonial financial administration has proposed, the most longed-for are the abolition of the tobacco monopoly and the elimination of the Indians' tribute.

3. The net revenue from the alcabalas, nearly *three million piasters.* According to an average from 1788 to 1792, the gross product of this branch of duties was 3,259,504 piasters. Subtracting 371,148 piasters for collection expenses and salaries leaves a net revenue of 2,888,356 piasters. Trade activity has increased so much in the past forty years that the yield from the alcabala from 1765 to 1777 was 19,844,053 piasters, whereas from 1778 to 1790, it was 34,218,463 piasters. Consequently, the customs office of Mexico barely earned 6,661,900 piasters from 1766 to 1778, it but took in over 9,462,014 piasters from 1779 to 1791. In 1799, the revenue from the *alcabala* was only 2,407,000 piasters, but since that time it has greatly increased. Collection fees subsidized by the people amount to 13 percent of this income. Since the indigenous peoples do not pay the alcabala, we may assume that this tax is equivalent to an annual capitation tax of one and two-thirds piasters for whites and mixed castes [castes mixtes].

IV.229

4. The net product of the capitation tax on the Indians,[1] *1,300,000 piasters.* The increase in the proceeds of tributes demonstrates a phenomenon that is little-known in Europe but is one of the most comforting for

1. *Tributos.* See chap. VI, vol. I, p. 344 and 392 [pp. 231 and 255 in this edition].

humanity, namely, the advancement of the Indian population. From 1788 to 1792, the capitation tax on the indigenous peoples was in an average year 1,057,715 piasters; yet collection expenses and salaries amounted to 55,770 piasters, to which must be added pensions paid to the descendants of Montezuma and to some of the *conquistadores*, amounts earmarked for the maintenance of the viceroy's spearmen (*alabarderos*) and other expenses, 102,624 piasters. Subtracting these 158,394 piasters from the gross product of the tributes, one finds a net product (*líquido*) of 899,321 piasters, whereas in 1746, it was only 650,000 piasters. The capitation tax on the Indians from 1765 to 1777 was entered in the registers as 10,444,483 piasters, and from 1778 to 1790, as 11,506,602 piasters. Collection expenses for this type of duty do not exceed 6 percent of the net product.

IV.230

5. The net product from the duty on pulque, *eight hundred thousand piasters*. This duty on the indigenous peoples' wine, which is the fermented juice of the agave,[1] produced in an average year a net product of 761,131 piasters in the towns of Mexico City, Toluca, and La Puebla de los Ángeles: in 1799, this net product was 754,000 piasters. Collection expenses for this income was 7 percent of the *líquido*. Making pulque was completely forbidden by the laws of Charles I and Philip III.

6. The net product of the entrance and exit duty on merchandise, collected under the name of the *almojarifazco*, *half a million piasters*.

7. The product of the sale of papal indulgences or bulls from the cruzada, *two hundred seventy thousand piasters*.

8. The net product of the post,[2] *two hundred fifty thousand piasters*. This product was 1,006,054 piasters from 1765 to 1777, and 2,420,426 piasters from 1778 to 1790, an increase that demonstrated the progress of both civilization and trade.

IV.231

9. The product of the sale of gunpowder,[3] *one hundred fifty thousand piasters*; from 1788 to 1792, in an average year, 144,636 piasters.

10. The net product of the taxes levied on the privileges of the clergy, called the mesada and the media anata, *one hundred thousand piasters*.

11. The net product of the sale of playing cards,[4] *one hundred twenty thousand piasters*.

1. See chap. IX, vol. II, p. 493 [p. 569 in this edition].

2. *Renta de Correos*.

3. *Líquido del real estanco de la polvera*. See chap. XII, vol. IV, p. 15 [p. II.298 in this edition].

4. *Estanco de naypes*.

12. The product of the duty on stamps (papel sellado) *eighty thousand piasters*; from 1788 to 1792 in an average year, 60,756.

13. The net product of the cockfight farm,[1] *forty-five thousand piasters.*

14. The net product of the snow farm, *thirty thousand piasters.* Were there not countries in Europe where people pay a tax to enjoy sunlight, one might be surprised to know that in the Americas, the layer of snow that covers the high chain of the Andes is considered the property of the king of Spain. The poor Indian who reaches the summit of the Cordilleras at his own risk cannot collect snow or sell it in the nearby towns without paying IV.232 a duty to the government. This bizarre custom of considering the sale of ice and snow as a right of the crown also existed, however, in France at the beginning of the seventeenth century, and the snow depot was suspended in Paris only because the size of the tax led to such a precipitous decline in the custom of cooling drinks that the court preferred to declare free trade in ice and snow. In Mexico City and Veracruz, where the summits of Popocatépetl and the Peak of Orizaba provide snow for making sherbets, the *estanco de la nieve* was introduced only in 1719.

We have already compared the total revenue of New Spain in different periods in the eighteenth century; let us now extend this comparison to the different branches of duties listed in Villaseñor's statistical work, published in Mexico City in 1746. For each article, we shall see undeniable proof [in the table below] of the advancement of the population and public prosperity.

In this table we have shown only the duties whose tariff has not been IV.234 increased since 1746: at that time, the monopoly on the sale of tobacco had not yet been introduced and instead of twenty-three million piasters, metal production amounted to only ten million piasters. In the edition of his history of the Americas published in 1788, Robertson estimates the total revenue of Mexico at four million piasters, whereas in that period it was actually over eighteen million. Europe was so ignorant then of the state of affairs in the Spanish colonies that in speaking of the finances of Peru,[2] that learned and famous historian was forced to draw on an unpublished paper written in 1614.

In ancien régime France in 1784, Mr. Necker[3] estimated contributions at twenty-three livres thirteen sous or four and one-half piasters per head for men and women of all ages. Accounting for 5,837,000 inhabitants of

1. *Estanco de los juegos de gallos.*
2. Robertson, [*The History of America*,] vol. IV, p. 352, note XXXIII.
3. Necker, *De l'administration des finances*, vol. I, p. 221.

Comparative Tableau of the Revenue of New Spain IV.233

SOURCE OF PUBLIC REVENUE	IN 1746 PIASTERS	IN 1803 PIASTERS
Taxes on mining products	700,000	3,516,000
Mint	357,500	1,500,000
Alcabala	721,875	3,200,000
Almojarifazgo	373,333	500,000
Tributes or head tax for Indians	650,000	1,200,000
Cruzada	150,000	270,000
▼ Media anata	49,000	100,000
Tax on pulque or agave juice	161,000	800,000
Tax on playing cards	70,000	120,000
Stamps	41,000	80,000
Snow and ice sales	15,522	26,000
Gunpowder sales	71,550	145,000
Cockfights	21,100	45,000

New Spain and twenty million piasters of revenue, one finds per capita income for all ages, sexes, and races to be three and four-tenths piasters. Peru, which currently has only one million inhabitants, and three and a half million piasters of revenue, presents approximately the same result. Since the Indians who are subject to the capitation do not pay any alcabala and do not use any tobacco, calculations of this sort, which are of little use even for Europe, do not apply to the Americas. Furthermore, it is not the total amount of duties but rather their distribution and the manner of collection IV.235 that causes misery among the people. To reach a certain degree of accuracy in such inherently vague estimates, in calculating the charges born by the inhabitants of New Spain, one should refrain from including in their entirety both the duties on gold and silver and the profit from the mint, which taken together account for over one-quarter of the country's total revenue. We shall not engage here in such unsatisfactory discussions, but shall instead hasten to complete the financial table of Mexico by discussing collection costs and state expenses in the following chapter.

IV.236

CHAPTER XIV

Collection Costs—Public Expenses—Situados—The Net Product Deposited in the Royal Treasury in Madrid—The State of the Militiary—National Defense—Final Summary

In examining the various branches of state revenue, we have indicated the collection costs generated by partial receipts. These costs vary in all countries depending on the nature of the tax or duty levied. Thanks to Mr. Necker's[1] research, we know that in France before 1784, collection costs accounted for 10 and four-fifths percent of all the taxes levied on the people, whereas it cost over 15 percent to collect the duties on consumption alone. To a certain extent, these ratios enable us to evaluate the degree of economy present in the administration of finances. The following table, drawn up on the basis of official documents, presents a troublesome result: IV.237 it proves that the inhabitants of New Spain bear a collective burden that exceeds the net revenue of the state by one-seventh. We shall first present this table in the exact state that it was sent by the viceroy Count de Revillagigedo to the ministry in Madrid; we shall then discuss the results that may be gleaned from it.

The figures in this table refer to an average of the five years before 1789. In that period, the revenue of New Spain did not yet exceed eighteen mil- IV.238 lion piasters. The first class of taxes includes over half of the total income; collection costs were 12 and nine-tenths percent of the net product. The second class contains the branches [of industry] that are exclusive to a particular monopoly, such as the royal tobacco monopoly and the sale of mercury and cards for the profit of the crown. For that portion of public receipts, this table presents results that appear inaccurate: for monopoly

1. Necker, [*De l'administration des finances,*] vol. I, pp. 93 and 188.

REVENUE CATEGORIES (*RAMOS DE REAL HACIENDA*)	GROSS REVENUE IN PIASTERS	COLLECTION AND MANAGEMENT FEES IN PIASTERS	NET REVENUE IN PIASTERS
First class, i.e., *de masa común*: Alcabala, Indian tributes, duties on gold and silver	10,747,878	1,395,862	9,352,016
Second class, i.e., *de masa remisible a l'España*: revenue from the tobacco monopoly, from the sales of playing cards and mercury	6,899,830	3,080,303	3,819,527
Third class, i.e., *des destinos particulares*: cruzada, tithes, medias anatas, salaries, and other taxes levied on the clergy	530,425	13,806	516,621
Agenos, revenue from the municipal assets and from charitable work under government management	1,897,128	1,700,956	196,172
Total:	20,075,261	6,190,927	13,884,336

and administration expenses, it lists the sum of 44 and six-tenths percent. The people responsible for creating this table of Mexican finances have probably confused employees' salaries with manufacturing costs and other expenses of which I am unaware. We have explored in the greatest detail above everything that has to do with the tobacco monopoly; we have seen that employees' salaries absorb less than 800,000 piasters out of a gross product of over seven and a half million piasters. If one adds to employee salaries a number of management costs disguised under the vague name of administrative expenses, collection costs may be estimated at 25 percent. The economy introduced into the collection of taxes on the clergy contrasts markedly with the horrible misappropriation that takes place in the control of common property. I would be tempted to believe that in Mexico collection fees represent in general between 16 and 18 percent of gross receipts: the prodigious number of employees, the extreme sloth of those who fill IV.239

the highest positions, and a highly complicated financial administration all make tax collection as slow and difficult as it is onerous for the Mexican people.

According to the table of finances drawn up at the request of the Count de Revillagigedo, the expenses of the state, taken on the average over the years 1784–1789, were as follows:

USES OF STATE REVENUE	PIASTERS
Situados [salaries] sent to the colonies in the Americas and in Asia	3,011,664
Line infantry	1,339,458
Militia	169,140
Upkeep of *presidios* or military posts	1,053,106
Room and board for enslaved laborers	47,268
Arsenal and dockyard of the port of San Blas	93,004
Administration of justice	124,294
Administration of finance	508,388
Pensions and other expenses assigned to the *masa común*	496,913
Missions in California and the American northwest coast	42,494
Various expenses related to defense, to the warships stationed at Veracruz, etc.	1,000,000
TOTAL	7,886,329
Or, the revenue of the three classes of taxation from the previous table	13,884,336
REMAINING, king's revenue to flow to the metropole	5,998,007

IV.240 During the administration of the last viceroy, Don José de Yturigarray, a new financial table was created at the beginning of 1803, the overall results of which differ only slightly from those for the year 1790. Here are the details of this *budget*, in which the distribution of the various items of public expense leave much to be desired in terms of order and clarity.

IV.242 To give a clearer picture of the financial situation of Mexico, after the budget for the year 1803, I shall [then] present the table of state expenses as I classified them in a paper that I drafted in Spanish during my stay in Mexico City and that the viceroy forwarded to the ministry in Madrid in 1804.

Budget of the Public Revenue of New Spain for 1803

USES OF REVENUE	PIASTERS	PIASTERS
Revenues amounted to		20,000,000
EXPENSES *I. Administrative Expenses*		
Sueldos de hacienda, appointment of viceroy, of the commandant-general of the *provincias internas*, of intendants, of secretaries assigned to the different officials, pensions for retired governors (*jubilados*)	2,000,000	
Shipping costs of goods from province to province and to Spain	750,000	(5,250,000)
Purchases of raw materials for state-owned tobacco, gun powder, and saltpeter (*para especies estancadas*)	1,200,000	
Manufacturing costs associated with the mint and tobacco and gunpowder	1,300,000	
Remaining net income (*líquido*)		14,750,000
EXPENSES *II. Expenses Related to the* Masa Común		
Defense, ships, launches, gunboats, line infantry, militia, *presidios*, and enslaved labor	3,000,000	
Fortifications, arsenal, and dockyards in San Blas, warehouses, use of gunpowder for troop exercises	800,000	(4,650,000)
Appointments of courts of law (*audiencias*), missionaries, damages	250,000	
Retirees with pensions	200,000	
Hospitals and repairs to royal factories	400,000	
Remaining net income (*líquido*)		10,100,000

(*continued*)

USES OF REVENUE	PIASTERS	PIASTERS
III. Shipments to the Metropole and to the Colonies (Cargas Ultramarinas)		
Net income from government-owned tobacco	3,500,000	
Net income from playing card sales (*naypes*)	120,000	
Pensions assigned to *ramos de vacantes*, of which, in turn, a third is sent to Mont-de-Piete established for the Madrid military	60,000	
Mercury purchases from Germany	500,000	(7,780,000)
Income from the *medias anatas* and other taxes levied on the clergy, used for the cannon factory in Ximena.	100,000	
Situados in the Asian and American colonies	3,500,000	
Remaining in the Mexican treasury at the end of the year		2,320,000
Amounts sent to the royal treasury in Madrid:	3,620,000	(5,940,000)
From government-owned tobacco and playing cards	2,320,000	
Surplus (*sobrante*) in the coffers of Mexico		

The revenue of New Spain, valued at twenty million piasters, is absorbed:

I. By expenses incurred in the interior of the kingdom, which amount to *ten and a half million piasters*;
II. By the silver sent annually (*situados*) to other Spanish colonies, which amount to *three and a half million piasters*;
III. By the silver deposited in the treasury of the King of Spain in Madrid as the net product of the colony, which amount to *six million piasters*.

I. The expenses of the administration of the interior, covered by income from the *masa común*, are divided in the following manner [see table below].

People in Europe have in general very exaggerated ideas about the power and wealth of the viceroys in Spanish America: this power and wealth exists only when the head of state is supported by a large faction at court and when, sacrificing his honor to a disgusting self-indulgence, he abuses the privileges that are granted him by law. The salaries of the viceroys of New Granada and Buenos Aires are only forty thousand piasters per year: the viceroys of

1. War expenses		4,000,000 piasters
To wit:		
Line infantry (*tropa reglada*)	1,800,000	
Militia	350,000	
Presidios	1,200,000	
Upkeep of the fortress of Perote	200,000	
Navy, dockyards of San Blas, port arsenals	450,000	
	4,000,000	
In 1792, war expenses were as follows: for line infantry, 1,507,000 piasters; for militias, 292,000 piasters; for the upkeep of the presidios, 1,079,000 piasters.		
2. Appointment of the viceroy, intendants, and employees of the finance administration	2,000,000	
3. Costs of administering justice: *audiencias,salas del crimen, juzgados de penas de camara, juzgado de bienes de defuntos, juzgado de Indios*	300,000	
4. Prisons, work houses, hospitals	400,000	
5. Pensions	250,000	
6. Administrative costs, advances on government-owned tobacco, manufacturing expenses in the royal factories; purchase of raw materials; repairs of public buildings	3,550,000	
	10,500,000	

IV.243

Peru and New Spain receive sixty thousand. In Mexico City a viceroy is surrounded by families whose income is three or four times greater than his own. His house is appointed like that of the king of Spain, and he cannot leave his palace without being preceded by his mounted guards; he is served IV.244

by pages and in Mexico City he is only permitted to dine with his wife and family. Such refined etiquette becomes a means of saving money, and a viceroy who wishes to emerge from his isolation and enjoy the company of others must reside in the country for a while, either in San Agustín de las Cuevas, or Chapultepec, or Tacubaya. A few of the viceroys of New Spain have had an increased salary: instead of sixty thousand piasters, the Chevalier de Croix, Mr. Antonio Bucareli, and the Marquis de Branciforte, received an annual income of 80,000 piasters. But this favor of the court was not extended to the successors of the three viceroys whom we have just mentioned.

A governor who renounces all refined sensibility and comes to the Americas to enrich his family will find ways of attaining his goal by favoring the richest individuals in the country through the distribution of positions, mercury *allotments*, and the privileges granted in wartime to engage in free trade with the colonies of neutral powers. For some years now the ministry in Madrid has found it useful to make even the most minor appointments in the colonies; however, the viceroy's recommendation is still of great importance to anyone who solicits [favors], especially in the case of a military appointment or a noble title (*título de Castilla*), which the Spanish-Americans are generally more keen IV.245 to have than are the Spanish-Europeans. It is true that a viceroy may not make trade regulations, but he can *interpret* the orders of the court: he can open a port to neutral parties by informing the king of the *urgent circumstances* that have determined this action; he can protest a reiterated order, collect papers and *informes*, and, if he is rich, clever, and supported in the Americas by a courageous assessor and, in Madrid, by powerful friends, then he can govern arbitrarily without fear of the *residencia*, that is, the account that all governors who have had appointments in the colonies must make of their administration.

There have been viceroys who were so sure of their impunity that they extorted nearly eight million livres tournois in just a few years, but there have been others, I am happy to report, who, far from increasing their fortunes by unlawful means, have expressed a high-minded and generous disinterest. Among the latter, the Mexican people will long and gratefully invoke the Count de Revillagigedo and the Chevalier de Asanza, two statesmen equally distinguished for their personal and public virtues, whose administration would have been even more beneficial if their external position had allowed them to pursue freely the career that they had planned.

II. Three and a half million piasters, nearly one-sixth of the total revenue of Mexico, are sent annually to other Spanish colonies, as indispensable IV.246 aid to their internal administration. These *situados*, according to the averages taken for the years 1788 to 1792, were distributed as follows:

1. Island of Cuba		1,826,000 piasters
a)	*Atención de tierra*, aid from the interior government of the island. Details: 146,000 piasters for Santiago de Cuba and 290,000 for Havana	436,000
b)	*Atención marítima*, Navy expenses. Details: 700,000 piasters for the port and the dockyards of Havana; 40,000 piasters for the ships stationed off the Mosquito Coast.	740,000
c)	Upkeep of Havana's fortifications	150,000
d)	For the purchase of tobacco from the island of Cuba shipped to Spain	500,000
2. Florida		151,000
3. Puerto Rico		377,000
4. Philippines		250,000
5. Louisiana		557,000
6. Island of Trinidad		200,000
7. Spanish portion of Saint-Domingue		274,000
		3,635,000

Although Spain has lost Louisiana and the islands of Trinidad and Saint-Domingue since this table was drawn up, the *situados* have not decreased by 1,031,000 piasters, as one might suspect. The administration of the Philippine Islands, Cuba, and Puerto Rico was so costly during the last war, especially during the residency of the squadron commanded by ▼ Admirals Álava and Ariztizábal, that the amount sent to the eastern and western colonies was never less than three million piasters. One may be surprised to note that Havana requires the assistance of 1,400,000 piasters when one recalls that the recipients of the *royal duties* pay over two million piasters per year to the colonial treasury. Although the indigenous peoples' *tribute* in the Philippine Islands amounts to 573,000 piasters,

IV.247

and the income of the state-owned tobacco monopoly amounts to 600,000 piasters, in recent times the royal treasury in Manila has experienced a constant need for a *situado* of 500,000 piasters.

III. The net revenue (*sobrante, líquido remisible*) that the metropole takes from Mexico was barely one million piasters before the tobacco monopoly was introduced: it is now five or six million piasters, depending on whether the other colonies require greater or lesser *situados*. This *líquido* or *sobrante* comprises the net product of the tobacco and gunpowder monopolies, which is almost always three and a half million piasters, and the variable surplus of the *masa común*. I must observe that in the Spanish colonies, there is almost no money left in the treasury after the annual accounts

IV.248

have been closed. Those who govern are not unaware that the surest way to maintain their credit at court and preserve their position is to send as much money as possible to the royal treasury in Madrid.

Since most of the population of New Spain is concentrated in the five intendancies of Mexico City, Guanajuato, Puebla, Valladolid, and Guadalajara, these provinces bear most of the expenses of the State: the *provincias internas* may be considered colonies of Mexico itself, but far from providing funds to the national treasury in the capital, these colonies are a burden to it. Taking the average of the five years preceding 1793, the *income* of the provincial tax office (*caja real*) de Guanajuato were [as in table below].

IV.249

Using the table of expenses for Guanajuato, one can form an idea of the financial situation in the twelve other intendancies that comprise the Kingdom of New Spain. The income of Valladolid is now 773,000 piasters; this estimate is probably more accurate than the one for the revenue of the intendancy of Guanajuato, which appears a bit too low.

The profit that the Spanish tax office takes from Mexico accounts for over two-thirds of the net product of the Spanish colonies in the Americas and Asia. Most writers on political economy who have discussed the finances of the peninsula, the depreciation of the *vales* [notes], and the bank of Saint Charles, have based their calculation on the most inaccurate basis, by exaggerating the treasures that the court of Madrid receives annually from its American possessions: even in the most prolific years, these treasures have not exceeded the sum of nine million piasters. When one recalls that, since 1784, in European Spain, the regular expenses of the state have amounted to between thirty-five and forty million piasters, one can see that the money that the colonies deposit in the coffers of Madrid is only one-fifth of all revenue. It would be easy to prove that, were Mexico governed wisely;

Duty[1] from gold, silver, and alcabala	850,000 piasters
Income from tobacco, gunpowder, and stamps	312,000
Total	1,162,000
Annual *expenses*:	
For appointments of intendants	6,000 piasters
For appointments of assessors	1,500
For costs of the administration of the treasury	7,800
For costs of the testing of gold and silver	5,600
For costs of the collection of the alcabala and duties on pulque	8,000
For the appointment of guards (*risguardo*)	10,700
Total	39,600

[1] Here we are concerned with duty only, since the amount of silver that enters the provincial tax office of Guanajuato is more than six to seven million piasters, the entire income from the mines flowing through that conduit to the mint in Mexico City.

if it opened its ports to all friendly nations; if it hosted Chinese and Malaysian colonists to populate its western coastline between Acapulco and Colima; if it increased its plantations of cotton, coffee, and sugarcane; finally, if it established a proper balance between agriculture, mining, and the manufacturing industry, in a few years Mexico alone could provide the tax office of Spain with a net profit twice the size of what Spanish America as a whole currently provides. IV.250

Here then is a general tableau of the financial state of the colonies, in relation to the net revenue that the homeland takes directly from them:

The royal treasury of Madrid receives from the viceroyalty of *New Spain* five to six million piasters each year;

From the viceroyalty of *Peru*, at most one million piasters;

From the viceroyalty of *Buenos Aires*, six to seven hundred thousand piasters;

From the viceroyalty of *New Granada*, four to five hundred thousand piasters.

In the *capitanías generales* of *Caracas*, *Chile*, *Guatemala*, the island of *Cuba* and *Puerto Rico*, income is absorbed by administrative expenses: the same goes for the Philippine and Canary Islands.

: :

In an average year, the colonies therefore provide the tax office of Spain with the combined sum of only 8,200,000 piasters. Considering the colonies as distant provinces, one finds that the revenue of the European part of IV.251 the Spanish monarchy is nearly equal to that of the American part.

The Finances of the Spanish Monarchy in 1804

EUROPE. The Peninsula: Gross revenue, thirty-five million piasters. Total income in 1784 was 685,000,000 reales de vellon; in 1788: 616,295,000 reales, according to ▼ [López de] Lerena's report. Population: 10,400,000 inhabitants; surface area: 25,000 square leagues.

THE AMERICAS: The research that I was able to do on the financial state of the colonies suggested that the gross revenue of all Spanish America could be valued at 36,000,000 piasters. The population of Spanish America is approximately 15,000,000 inhabitants; its surface area is 468,000 square leagues. The colonies whose gross revenue can be given with some certainty are the following:

Viceroyalty of New Spain, 20,000,000 piasters.
Viceroyalty of Peru, 4,000,000 piasters.
Viceroyalty of New Granada, 3,800,000 piasters.
Capitanía general of Caracas, 1,800,000 piasters.
Capitanía general of Havana, the island of Cuba, excluding the Floridas, 2,300,000 piasters. The annual *situado* from Mexico is not included in this calculation.

IV.252 ASIA. The Philippine Islands. Gross revenue, excluding the *situado* from Acapulco: 1,700,000 piasters. Population, including only the subdued Indians who dwell on the island of Luzon and on the Visayan Islands: 1,900,000 inhabitants; surface area, 14,640 square leagues.

AFRICA. The Canary Islands, annexed to Andalusia: Gross revenue, including the income from the tobacco monopoly but excluding assistance from Spain: approximately 240,000 piasters. Population: 180,000 inhabitants; surface area: 421 square leagues.

Of the *thirty-eight million piasters* representing the gross revenue of the *Spanish colonies in the Americas, Asia, and Africa*, the profit from the minting of coins and from duties imposed on the product of the gold and silver mines may account for eight and a half million, the revenue from the tobacco monopoly for nine million, and the income from the alcabala; the almojarifazgo; the Indians' tribute money; the sale of gunpowder, spirits, and playing cards; and other taxes on consumption for twenty and a half million. The interior administration of the colonies absorbs thirty-one million piasters per year; and as we have already seen, nearly eight million[1] are deposited in the royal treasury in Madrid. It is known that this sum, together with the thirty-five million piasters that the tax office receives from Spanish America, has long fallen short of compensating for the civil and military expenses of the homeland. Spain's national debt has been successively more than 120 million piasters,[2] and the annual *deficit* has been even larger since trade and industry have been blocked by naval wars. Furthermore, when one compares gross revenue with the state of the population, as we have shown above, one is easily persuaded that the charges imposed on the inhabitants of the colonies are one-third less than the expenses born by those who live on the Peninsula. IV.253

During the great catastrophe that caused Great Britain to lose nearly all of its continental possessions in the Americas, many political writers examined the direct influence that the separation of the Spanish colonies would have had on the finances of the court in Madrid. The explanations that we have just given regarding the general financial situation of Spain in 1804 enable us to provide data that lead to a solution to this serious problem. If all of Spanish America had declared its independence at the time of the Inca Tupac-Amaru's revolt,[3] that event would have had many repercussions IV.254

1. In the account of the general treasury of Spain for the year 1791 that I obtained in the Americas, which gives 800,488,687 reales de vellon, one estimates the revenue of the Indies at 142,456,768 reales or 7,122,838 piasters.

2. In 1805, there were 1,750 million reales de vellon in *vales* or royal bonds. The national debt of Spain is by no means tremendous, when one thinks of the immense resources of this monarchy, which encompasses the most beautiful parts of the globe in both hemispheres. Before the Revolution, France's national debt was 1,100 million piasters; at present, that of Great Britain probably exceeds 2,821 million piasters. In 1796, the total amount of the *notes of assignation* circulated in France was 45,578,000,000 francs or 8,681 million piasters. But when they lost their value, one hundred francs in notes of assignation were worth three sous, six deniers in currency. According to ▼ Mr. Ramel, 6,254 piasters' worth of notes that were not withdrawn remained in circulation. As for *money orders* and *rescriptions*, 4,800 piasters were issued. These amounts must appear even greater since we have already shown that there are fewer than 1,637 million piasters in Europe, and the total amount of gold and silver extracted from the mines of the Americas since 1492 is less than 5,706 million piasters.

3. See chap. VI, vol. I, p. 405 [p. 262 in this edition].

simultaneously: 1) it would have deprived the royal tax office in Madrid of an annual income of eight to nine million piasters of net revenue (*líquido remisible*) of the colonies; 2) it would have significantly decreased trade on the Peninsula, because Spanish-Americans, freed from the monopoly enforced by the Peninsula for three centuries, would have sought out the supplies and foreign merchandise they needed directly from countries not subject to Spain; 3) this change of direction in trade for the colonies would have provoked a decrease in the duties levied by the customs houses on the Peninsula, estimated at four or five million piasters; 4) the separation of the colonies would have ruined many industries in Spain that are supported IV.255 only by the exclusive rights of sale that they have in the Americas, since in their present state, they cannot withstand competition from goods from India, France, or Great Britain. These effects, which would have been acutely felt in the early years, would have been gradually compensated by the advantages that arise from the concentration of moral and physical energy, the need for better agriculture, and the natural balance established by nations united by blood ties and which exchange products that centuries-old habits have turned into necessities. But we would deviate from our main topic if we returned to a discussion that was elaborated in many works of political economy at the time of the Peace of Versailles.

Comparing the surface area, population, and revenue of Spanish America with the surface area, population, and revenue of the English possessions in India, we find the following results:

Comparative Tableau for 1804

	SPANISH AMERICAS	BRITISH POSSESSIONS IN ASIA[1]
Understood in common square leagues of twenty-five to an equatorial degree	460,000	48,300
Population	15,000,000	32,000,000
Gross revenue in piasters	38,000,000	43,000,000
Net revenue in piasters	8,000,000	3,400,000

[1] Territory over which the British company has acquired sovereignty, not including its allies and tributaries, such as the Nizam, the princes of Oude, Carnasic, Mysore, Cochin, and Travancore. According to Mr. Playfair, whom I have followed in the table published in vol. II, p. 13, their population is only twenty-three and a half million.

This table suggests that New Spain, whose population is less than six million, provides twice as much to the tax office of the king of Spain as Great Britain derives from its fine possessions in India, which have a population five times larger. It would be egregiously wrong, however, if in comparing gross revenue[1] to the number of inhabitants, one were to conclude from this comparison that the Hindus have fewer burdens than the Americans. We must not forget that the cost of a day's labor in Mexico is five times greater than in Bengal or, to use a word immortalized by a famous man,[2] in Hindustan, where the same amount of money *commands* five times more labor than in the Americas. IV.256

Looking closely at the budget of state expenses, it is surprising to see that in New Spain, which has hardly any other neighbors to fear than few warlike Indian tribes, the military defense of the country consumes almost one-quarter of total revenue. It is true that the number of troops of the line is only 9,000 or 10,000; but if we add to it the militias, called *provinciales* and IV.257

I. General Tableau of the Army in 1804

NAME OF THE CORPS			MEN
I. Line infantry (*tropas veteranas*)			9,919
1. In Mexico proper		6,225	
2. In the *provincias internas* administered by the viceroy of Mexico		595	
3. In the *provincias internas* administered by the commanders-general		3,099	
Total		9,919	
II. Militias (*cuerpos de milicias*)			22,277
1. Provincial militias (*provinciales*)		21,218	
Of the viceroyalty	18,631		
Of the *provincias internas*	2,587		
2. City militias (*urbanas*)		1,059	
TOTAL, in times of peace (not including the Yucatán Peninsula and Guatemala)			32,196

1. *Revenue of British India* (*in the Year 1801*): 9,742,937 pounds sterling; charges: 8,961,180; net revenue: 781,757. Playfair, *Stat[istical] Breviary*, p. 59.

2. Adam Smith, [*The Wealth of Nations*], vol. II, pp. 25, 33, and 64.

IV.258 *II. Detailed Tableau Representing the Divisions of the Line Troops*

NAME OF THE CORPS		MEN
A. Line troops across Mexico proper		6,225
a. Infantry	5,260	
Viceregal guard created in 1568 (*alabaderos*)		25
Four regiments: *Fixo de la Coronoa, Nueva, Espana, Mexico [City], and Puebla.* The latter three were formed in 1788 and 1789. All are composed of fourteen companies of 979 men each.		3,916
Battalion of Veracruz, consisting of five companies, created in 1793		502
Artillery corps, consisting of three companies, each 125 men strong		375
Corps of Engineers, eight officers		
Volontarios de Cataluña, two companies, formed in 1762		160
Company of Acapulco, formed in 1773		77
Company of the Presidio of the Isla de Carmen, created in 1773		100
Company of San Blas, created in 1788		105
b. Cavalry	965	
Four squadrons of *dragones de España*, created in 1764		461
Four squadrons of *dragones de Mexico*, created in 1765		461
Dragoons of the Presidio of Carmen		43
B. Line troops stationed in the part of the *provincias internas* administered by the viceroy of Mexico (*compañías presidiales y volantes*)		595
a. In old and New California: Presidio of Nuestra Señora de Loreto, founded in 1720		47
IV.259 Presidio of San Carlos de Monterey, founded in 1770		61
Presidio of San Diego, founded in 1770		39
Presidio of San Francisco, founded in 1772		58
Presidio of the canal of Santa Barbara, founded in 1780		65
b. In the new kingdom of León		
Military post (*presidio*) of San Juan Bautista de la Punta de Lampazos, established in 1781		100
c. In the province (*colonía*) of New Santander		
Three companies of *volantes*, formed in 1783		225
C. Line troops across the part of the *provincias internas* administered by two commanders-general		3,099
TOTAL line troops		9,919

III. Detailed Tableau Representing the Divisions of the Military

NAME OF THE CORPS		MEN	
A. Provincial militias (*milicias provinciales*)		21,218	
a. From the viceroyalty of Mexico	18,631		
1. Infantry	7,249		
Seven regiments: *Mexico, Puebla, Tlaxcala, Cordoba, Orizaba, Xalapa, Toluca, Valladolid, and Celaya*, of two battalions or ten companies each, created in 1788. Each regiment has 825 men in peacetime and 1,350 in times of war		5,775	IV.260
Three battalions: *Guanajuato, Oaxaca, and Guadalajara*, of five companies each, 412 men strong during peace times and 675 during times of war		1,236	
Two companies of men of color (*pardos y morenos*) in Veracruz, each 119 men strong		238	
2. Cavalry	4,592		
Eight regiments of dragoons: *Santiago de Querétaro, Príncipe, Puebla, San Luis, San Carlos, La Reyna, Nueva Galicia, and Michoacán*, created in 1788. Each regiment has four squadrons, and is 361 men strong during peacetime and 617 in times of war		2,888	
Six squadrons of lancers in Veracruz, created in 1767		384	
Six squadrons along the borders (*cuerpos fixos de frontera en el interior del reyno*), 1,320; to wit:			
Four companies in *Sierra Gorda*, formed in 1740		240	
Nine companies in *San Luis Colotlan*, created in 1780		720	
Six companies in *New Santander*, created in 1792		360	
3. Mixed troops of white and colored infantrymen and lancers (*compañías fixas de blancos y pardos*), along the eastern and western coastlines, created in 1793. Total strength; 6,790.			
Northern divisions (Atlantic coastline), twenty-nine companies			IV.261
First division	400		
Second division	670		
Third division	760		
Fourth division	500		
Ten companies in Tabasco	910		
Southern divisions (Pacific coastlines), thirty-four companies			
First division	680		

(*continued*)

NAME OF THE CORPS		MEN
Second division	1,140	
Third division	300	
Fourth division	1,030	
Fifth division	400	
From the *provincias internas*, fourteen squadrons or forty-eight companies		2,587
B. City militias (*milicias urbanas*)		1,059
Regiment of Mexican trade, ten companies, created in 1693		702
Battalion of Puebla trade, four companies, created in 1739		228
Cavalry squadron of Mexico, created in 1787		129
TOTAL militias, in peacetime		22,277

urbanas, we find an army of 32,200 men, spread over an area of land 600 leagues long. We shall examine a few of these *statuses* here, which the court of Madrid have requested annually since the viceroys Counts de Gálvez, de Revillagigedo, and Branciforte have enlarged the militia corps. The [tables above] show in the greatest detail the heterogeneous elements that comprise the military status of Mexico and that of the *provincias internas*.

Not included in these tables is the corps of invalids, created in 1774 and forming two companies, nor the troops stationed in the intendancy of Mérida, commanded by the captain-general of the Yucatán peninsula. I was unable to obtain the status of the military forces on this peninsula. IV.262 There are eight infantry companies (*tropas veteranas*) in Campeche and at the small fort of San Felipe de Bacalar: the defense of Mérida is entrusted to militia composed of whites and colored men.

The Mexican army cavalry is extremely large: it forms nearly half of the total force. In 1804 there were [as follows in the table below].

Estimating the force of the Mexican army at 32,000 men, one must observe that the number of disciplined troops is scarcely eight or ten thousand: among the latter, there are three or four thousand who are very experienced IV.263 militarily: this is the cavalry quartered in the *presidios* of Sonora, Nueva Vizcaya, and Nueva Galicia. We have already observed[1] that the inhabitants of the *provincias internas* live in a state of constant war with the nomadic

1. See chap. VIII, vol. II, pp. 232 and 252 [pp. 425 and 436 in this edition].

Infantry			16,200 men
1. Line troops		5,200	
2. Militias		11,000	
Cavalry			16,000
1. Line troops		4,700	
a. In Mexico	1,000		
b. In the *provincias internas*	3,700		
2. Militias		11,300	
a. In the interior of Mexico	4,700		
b. On the coasts	4,000		
c. In the *provincias internas*	2,600		
TOTAL[1]			32,200

[1] An inventory of troops preserved in the archives of the viceroyalty, which is similar enough to the *Guia de forasteros* published in Mexico City by *Don Mariano de Zuñiga y Ontiveros* (pp. 153–79), gives 32,934. Also compare [Estala], *Viagero universal*, XXVII, p. 320, and Pinkerton, *Modern Geography*, p. 162, where a larger estimate has been adopted.

Indian tribes known as the Apaches, Comanches, Mimbreños, Yatas, Chichimecs, and Taouaiazes. The *presidios* or military posts have been established to protect the colonists from attacks by these Indians who, armed with bows and arrows, are mounted on Spanish-bred horses. Since the end of the sixteenth century, when Juan de Oñate created the first settlements in New Mexico, horses have multiplied to such an extent on the savannas that extend east and west of Santa Fe, toward the Missouri and the Río Gila, that the indigenous peoples have not only become accustomed to eating horse meat when there is no buffalo but also use them as mounts for their warlike forays. Just as several African peoples grow corn with no knowledge of how that plant reached them, the horse has now been domesticated north of the sources of the Missouri among the Indian tribes who, prior to Captain Clark's expedition, had never had any communication with whites. Fortunately for the colonists of Sonora and New Mexico, the use of firearms, so common among the savages of eastern Canada, has not spread among the Indians living near the North River [Río del Norte]. IV.264

The Mexican troops at the *presidios* are subject to constant fatigue: the soldiers who compose them are all indigenous peoples of northern Mexico: they are very tall and robust mountain people, as accustomed to cold winter weather as they are to the heat of the summer sun. Always at arms, they spend their life on horseback, marching for eight to ten days across desert steppes, carrying with them no other supplies than corn flour, which they mix with water when they find a spring or pond on the road. Enlightened officers have assured me that it would be difficult to find troops in Europe that moved more lightly, fought more spontaneously, and were more accustomed to privations than the cavalry of the *presidios*. If this cavalry cannot always prevent Indian incursions, it is because the Indians are enemies who skillfully take advantage of the slightest unevenness in the terrain and have become accustomed over the centuries to all the stratagems of petty warfare.

The provincial militia of New Spain, whose force is over twenty thousand strong, is better armed than the Peruvian militia, who, for want of rifles, train in part with wooden muskets. It was not the military spirit of the nation, but rather the vanity of a small number of families, the heads of IV.265 which aspire to the rank of colonel and brigadier, that encouraged the formation of militia in the Spanish colonies. The distribution of commissions and military ranks has become a fertile source of revenue, not so much for the tax office as for administrators who have considerable influence over ministers. The rage for titles, which always characterizes either the beginning or the decline of a civilization, has made this traffic in titles highly lucrative. Traveling through the chain of the Andes, one is surprised to find all the merchants on mountain ridges and in small provincial towns transformed into military colonels, captains, and sergeant majors. Since the rank of colonel brings with it the *tratamiento* or title of lordship,[1] which is constantly repeated in friendly conversation, one may imagine that it is the one that contributes the most to domestic bliss and for which the Creoles make the most extraordinary financial sacrifices. One sometimes sees these militia officers in full uniform, decorated with the royal order of Charles III, gravely seated in their shops, busying themselves with the slightest details of selling their wares: a singular mixture of ostentation and simplicity of demeanor that astonishes the European traveler.

Until the time of the independence of the United States of North America, the Spanish government had not dreamed of increasing the num- IV.266 ber of troops in its colonies. The first colonists who settled on the New Continent were soldiers; the first generations knew no more honorable and lucrative profession than bearing arms. Military zeal led the Spanish to display the dynamic character that rivals the most brilliant exploits in the history of the crusades. Once the subjected native patiently bore the yoke

1. *La Señoria* [seigneurialism], V.S. commonly called the *ussiá*.

placed upon him, and the colonists, who unperturbedly possessed the treasures of Peru and Mexico, were no longer distracted by the allure of new conquests, the warlike spirit was imperceptibly lost. Since then, they have preferred peaceful farm life to the clamor of armies: rich soil, the abundance of foodstuffs, and the beautiful climate all contributed to milder manners, and the same countries that in the first half of the sixteenth century presented only the agonizing spectacle of wars and pillage enjoyed two and a half centuries of peace under Spanish domination.

The domestic tranquility of Mexico has rarely been disturbed since the year 1596 when, under the viceroy, the count of Monterrey, the Castilians' power was consolidated from the Yucatán peninsula and the Gulf of Tehuantepec to the source of the Río del Norte and the coastline of New California. There were Indian uprisings in 1601, 1609, 1624, and 1692: during the last one, the indigenous peoples burned the viceroy's palace, the town hall, and public prisons, and the viceroy, Count de Gálvez,[1] found safety only under the wing of the Franciscan monks. Despite these events, caused by a dearth of food, the court of Madrid did not see fit to increase its military forces in New Spain. In those days, when the Spanish-Mexicans and European-Spanish were even more closely tied, the distrust of the metropole was directed only against the Indians and the mestizos: the number of white Creoles was so small that for that very reason, they preferred generally to ally themselves with the Europeans. The peace that reigned in the Spanish colonies must be attributed to the state of affairs when, following the death of Charles II, foreign princes disputed the ownership of Spain. The Mexicans, governed at that time first by a descendant of Montezuma and then by an archbishop of Michoacan, remained the tranquil spectators of the great struggle that ensued between the houses of France and Austria. The colonies patiently awaited the fate of the metropole, and the successors of Philip V only began to fear their spirit of independence, which had appeared as early as 1643 in New England[2] when a large confederation of free states was formed in North America.

IV.267

The fears of the court increased further when a few years before the peace of Versailles, Gabriel Condorcanqui, the son of the cacique of Tongasuca, better known by the name Tupac-Amaru, incited the indigenous peoples of Peru to rebel so as to reestablish in Cuzco the ancient empire of the Incas. This civil war, during which the Indians engaged in the most

IV.268

1. *Gaspar de Sandoval, conde de Gálvez.*

2. Robertson, [*The History of America*,] book X (vol. IV, p. 307).

horrible acts of cruelty, lasted nearly two years; and if the Spanish had lost the battle in the province of Tinta, Tupac-Amaru's bold enterprise would have had dire consequences, not only for the interests of the metropole but probably also for the existence of all the whites settled on the plateaus of the Cordilleras and in the nearby valleys. As extraordinary as this event was, its causes were in no way linked to the movements to which the progress of civilization and the desire for an independent government had given rise in the British colonies. Isolated from the rest of the world and trading only with the ports in the metropole, at that time Peru and Mexico took no part in the ideas that stirred the inhabitants of New England.

Over the past twenty years, the Spanish and Portuguese settlements on the New Continent have undergone considerable changes in their moral and political position: the need for education and learning has begun to be felt as the population and prosperity have increased. The freedom to trade with neutral parties, which the court of Madrid, yielding to imperious circumstances, granted from time to time to the island of Cuba, the coasts of Caracas, and the ports of Veracruz and Montevideo, put the colonists in contact with Anglo-Americans, Frenchmen, Englishmen, and Danes: these settlers formed clearer ideas about Spain's position vis-à-vis that of other European powers, and the young peoples of the Americas, relinquishing some of their national prejudices, have formed a marked predilection for nations whose culture is more advanced than that of the Spanish-Europeans. Under these circumstances, it is not surprising that the political movements that have taken place in Europe since 1789 have excited the keenest interest among peoples who have long aspired to rights, the deprival of which is both an obstacle to public prosperity and a motive for resentment against the fatherland.

IV.269

Far from dispelling the settlers' agitation, this attitude led the viceroys and governors in some provinces to take measures that exacerbated the colonists' dissatisfaction. Seeds of revolution were perceived in any association whose goal was to spread knowledge: printing presses could not be set up in towns of forty to fifty thousand inhabitants, and peaceful citizens who retired to the country where they secretly read the works of Montesquieu, Robertson, or Rousseau were suspected of revolutionary ideas. When war broke out between Spain and France, unfortunate Frenchmen who had settled in Mexico for twenty to thirty years were dragged into prisons. One of them, fearing the revival of the barbarous spectacle of the *auto-da-fé*, took his own life in the prisons of the inquisition: his body was burned in Quemadero square. In the same period, the government believed that it had

IV.270

discovered a conspiracy in Santa Fé [de Bogotá], the capital of the kingdom of New Granada: individuals who had obtained French newspapers through the trade with the island of Santo Domingo were put in shackles; young people sixteen years of age were sentenced to be tortured in order to extract secrets that were entirely unknown to them.

Amidst these troubles, respectable magistrates and, people like to repeat, even Europeans lifted their voices against these unjust and violent acts: they professed to the court that a politics of mistrust only agitated people's minds, and that it was not by force and by increasing the number of troops composed of indigenous peoples that they would be able to bind closer ties over time between the colonies and the Spanish Peninsula, but by governing fairly, improving social institutions, and granting the settlers' justified demands. Such constructive advice was not taken: the colonial regime underwent no reform, and in 1796, in a country where the advancement of knowledge had been enhanced by frequent communication with the United States and the foreign colonies in the Antilles, a large revolutionary movement nearly eradicated Spanish domination in a single blow. A rich merchant in Caracas, ▾ Don José España, and an officer in the corps of engineers, Don Manuel Gual, a resident of Guaira, developed a bold project for making the province of Venezuela independent and uniting to it the provinces of new Andalusia, Nueva Barcelona, Maracaibo, Coro, Varinas, and Guiana under the name of the United States of South America.[1] Mr. De Pons's *Voyage à la partie orientale de la Terre-ferme*[2] describes the outcome of this failed revolution. The confederates were arrested before the general uprising could take place. As he was led to the gallows, España saw his death approaching with the courage of a man destined to carry out great projects; Wal [Gual] died on the island of Trinidad, where he found asylum but no assistance.

IV.271

Despite the easygoing nature and extreme docility of the people in the Spanish colonies, and despite the special situation of the inhabitants—who, spread over a vast expanse of country, enjoy the individual liberty that always comes from being very isolated—political disturbances would have been more frequent since the peace of Versailles, and especially since 1789, if the mutual hatred of the castes and the fear that a large number of blacks [noirs] and Indians inspires in whites and all free men had not put a stop to the consequences of popular discontent. As we indicated at the beginning

IV.272

1. *Las siete provincias unidas de la América meridional.*
2. Vol. I, pp. 228 and 233.

of this work,[1] these motives became even stronger after the events that took place in Saint-Domingue, and they have undoubtedly contributed more to preserving the peace in the Spanish colonies than have rigorous measures and the formation of militias, which number over forty thousand men in Peru, and twenty-four thousand on the island of Cuba.[2] The increase

1. See [above] vol. I, p. 198 [p. 144 in this edition].

2. In this note I have collected the information that I have received on the number of troops distributed throughout the Spanish colonies. During my last stay in Havana in the spring of 1804, there were under arms on the island of Cuba:

I. Regimented Militias: Infantry	
In Havana	1,442 men
In the city of Puerto del Principe	742
II. Regimented Militias: Cavalry	
In Havana and in its jurisdiction	517
III. Rural Militias, Non-Regimented (milicias rurales)	
East of Havana and in Matanzas	7,995
West of Havana	5,688
In the suburbs (*extra muros*) of Havana	1,368
In the jurisdiction of four cities (las quarto villas)	2,640
In the jurisdiction of Puerto del Principe	1,728
In the jurisdiction of Santiago de Cuba	2,412
Total military strength	24,511

It appears certain that the island of Cuba could muster in its defense an army corps of 36,000 white men [blancs] between the ages of sixteen to forty-five. (See chap. VII, vol. I, p. 420 [p. 272 in this edition].) The armed forces on the island of Cuba are much stronger than those in the *capitanía general* of Caracas, which, in the provinces of Venezuela, Nueva Andalusia, or Cumana, Maracaibo, Guiana, and Varinas, is only 11,900 men, among whom there are not even 2,500 Europeans. In Peru in 1794 there were:

Line troops	12,000 men
Militia, one-quarter of those cavalry	49,000
TOTAL	61,000

This list is taken from the court almanac, or the *Guía política de Lima*, published [by the Sociedad Académica de Amantes de Lima] by order of the viceroy. We have already seen above that a part of this militia, armed with wooden muskets, is not especially formidable. According to the official documents in my possession, in 1795 there were 3,600 regular troops in the kingdom of New Granada, stationed in Santa Fé de Bogotá, Cartagena de

in armed forces reflects the growing mistrust of the homeland even more strongly, since there were no troops of the line on the coasts of Caracas prior to 1768, and in the kingdom of Santa Fé, the government has recognized no need for militia for over two and a half centuries. There was a draft only in 1781, when the introduction of the tobacco monopoly and the duty on spirits excited popular uprisings. IV.273 IV.274

In the present state of things, the external defense of New Spain can have no other goal than to protect the country from an attempted invasion by a naval power. Dry savannas that resemble the steppes of Tartary separate the *provincias internas* from the territory of the United States. Only recently have the inhabitants of Louisiana reached the town of Santa Fé in New Mexico via the Missouri and the Río de la Plata. It is true that the Arkansas and the Red River of Natchitoches, which flow into the Mississippi, originate in the mountains near Taos. But it is so difficult to ascend these rivers because of the swiftness of the current, that the northern provinces of Mexico are as underexposed to attack by this route as the United States and New Granada are via the Ohio or the Río de la Magdalena.

Above 32° northern latitude, the nature of the soil and the expanse of deserts that border on New Mexico offer the inhabitants strong security against attack by a foreign enemy. Farther south, between the Río del Norte and the Mississippi, there are several lines of rivers on the same frontier: in this part of the country, settlers in Louisiana are the closest to the Mexican colonists, since only sixty leagues separate Fort Clayborn, in Natchitoches county, from the Mexican *presidio* of Nacogdoches. The terrain near the coasts is swampy in this part of the intendancy of Potosí. The ground rises only to the north and northeast, and in the middle of the plains that connect the basin of the North River to the Mississippi basin, the Río Colorado of Texas appears to present the most advantageous military position. This point is even more remarkable since toward the end of the seventeenth century, Mr. de Salle had founded the first French colony in Louisiana between the mouth of the Colorado and the small port of Galveston. It would be useless for us to elaborate here on the defense of the borders in the *provincias internas*: the principles of wisdom and moderation that inspire the IV.275

Índias, Santa Marta on the Isthmus of Panama, in Popayán and Quito, and 8,400 militiamen. According to Mr. de Sainte-Croix, there are 5,500 infantrymen and 12,200 militiamen in the Philippine Islands. Summarizing all the information that I have collected on the Spanish colonies in the Americas, a total population of fourteen to fifteen million inhabitants appears to include 3,000,000 whites, 300,000 Europeans, and at most 26,000 European troopers.

government of the United States lead us to hope that an amicable arrangement will soon determine the boundaries between two peoples who both occupy much more land that they can cultivate.

The petty warfare that the troops stationed at the *presidios*[1] wage relentlessly with the nomadic Indians is as burdensome to the public treasury as it is contrary to the advancement of the indigenous civilization. Since I have not traveled in the *provincias internas*, I shall not comment on the possibility of a general pacification. In Mexico City one often hears that for the settlers' security, one should not suppress but rather eliminate the tribes of savages who roam around the Bolsón de Mapimi and north of Nueva Vizcaya. Fortunately, the government has never embraced such barbarous advice, and history teaches us that such measures are not necessary. In the seventeenth century, the Apaches and Chichimecs pushed their incursions beyond Zacatecas, toward Guanajuato and Villa de Leon. Since civilization has increased in these regions, the nomadic Indians have progressively moved away. One may hope that as the population and public prosperity begin to increase in the *provincias internas*, these warrior hordes will withdraw first beyond the Gila, then west of the Río Colorado, which empties into the Sea of Cortés, and finally into the northern desert regions that border on the mountains of New California. This province, of which only the seaboard is inhabited, is still 600 leagues from American Russia and 200 leagues from the mouth of the Río Columbia, where the inhabitants of the United States intend to create a settlement. The defense of the ports of San

1. The military posts (*presidios*) in Mexico are as follows:

1) Intendancy of Durango:

Conchos, Yanos, Callo, San Buenaventura, Carisal, San Eleazario, Norte or *Las Juntas, Principe, San Carlos, Cerro Cordo, Pasage*, Namiquipa, Coyame, *Mapimis, Huejoquillo*, Julimes, San Geronimo, Santa Eulalia, Batopilas, Loreto, Guainopa, Cosiquiriachi, Topago, San Juaquin, Higuera, San Juan, Tababueto, Reyes, Coneto, *Texami*, Sianuri, Ynde, Oro, Tablas, Cuneza, Panuco, Avino.

2) Intendancy of Sonora:

Bavispe, Buenavista, Pitic, Bacuachi, *Tubson, Fronteras, Santa Cruz, Altar*, Rosario.

3) New Mexico:

Santa Fé, Paso del Norte.

4) The Californias:

San Diego, Santa Barbara, *Monterey, San Francisco.*

5) Intendancy of San Luis Potosí:

Nacogdoches, *Espíritu Santo, Bejar, Coahuila*, San Juan Bautista de Río Grande, Aquaverde, Bavia.

The *presidios* that have the strongest garrisons are printed in italics. None of these posts have more than 140 troops.

Francisco, Monterey, and San Diego is entrusted to a corps of only 200 men, and there are no more than three cannons in San Francisco; however, these forces have been sufficient for forty years in seas that are trafficked only by merchant ships participating in the fur trade.

As for Mexico proper, or that part of the kingdom located in the Torrid Zone, one need only glance at the Atlas[1] that accompanies this work, especially the geological sections, to be convinced that there is hardly a country in the world whose military defense is more favored by the configuration of its terrain. Narrow, twisting roads, like those over the Saint Gotthard and most of the Alpine passes, lead from the coastline to the interior plateau, where the population, civilization, and the wealth of the country are concentrated. The slope of the Cordillera is steeper on the Veracruz road than IV.278 on the Acapulco road, and although the currents of the Great Ocean and several meteorological causes make the western coastline less approachable than the eastern, Mexico may be considered as being more naturally fortified on the Atlantic Ocean side than on the side facing Asia. To protect the country from invasion, however, one can rely only on internal resources, since the state of the ports[2] located on the coastline washed by the Antillean Sea is not conducive to maintaining naval forces.

The forces that the court of Spain intended for the defense of Veracruz have always been stationed in Havana; this port, which offers many fine fortifications, has always been considered the military port of Mexico. An enemy squadron can anchor only at the foot of the castle of San Juan de Ulúa, which rises like a rock in the middle of the sea. This famous fort has no other water than what is in its cisterns, which have recently been improved, because they tended to crack from the shaking caused by the discharge of artillery. Professionals, however, believe that the fort of Ulúa is capable of resisting long enough for the unhealthiness of the climate to take its toll on assailants and for land forces to descend from the central plateau. IV.279

The coastline is low to the north and south of Veracruz, and the mouths of the rivers, blocked by sandbars, are only approachable by longboats. The coastal service was organized fifteen years ago, when the fear of an invasion by sea prompted large-scale troop mobilizations near Orizaba and, for the first time in two and a half centuries, Mexico assumed a warlike stance. At that time it was acknowledged that multiple outposts and signals, flat-bottomed ships loaded with large cannon, and a light cavalry capable

1. Plates 3, 5, 9, 12, 13, and 14.
2. See chap. III, vol. I, p. 286 [p. 191 in this edition].

of moving rapidly to the threatened points, were the most useful and least costly defense system.

Once the enemy has landed, he can head his march toward the plateau, either via Jalapa and Perote, skirting around the north side of the Cofre Mountain, or by ascending the Cordilleras via Córdoba, south of the Orizaba volcano. For the most part, these roads present the same difficulties as those that must be overcome when ascending from La Guaira to Caracas, for Honda to Santa Fé, or from Guayaquil to the beautiful Quito valley. The small fort that bears the pompous name of the Fortress of San Carlos of Perote is on the Jalapa road, at the entrance to the Puebla plateau; annual

IV.280 maintenance costs for the public treasury are over one million francs. The most secure means of blocking the enemy from the road he might take, or at least to hinder his advance, would have been to fortify the gorges themselves for the military defense of the pass.

The ease with which a very small number of well-distributed troops can block access to the plateau is so widely known in the country that the government did not think it necessary to submit to the criticism of those who, opposed to laying out the Jalapa road, attempted to demonstrate the danger that it would create for the military defense of New Spain. It felt that such considerations were intended to paralyze whatever might be done for public prosperity, and that this mountain people, enriched by their agriculture, mines, and trade, needed a viable connection with the coastline: the more populated the coastline, the more resistant it will be to a foreign enemy.

: :

In this work, I have drawn the political tableau of New Spain; I have discussed the astronomical material that served to determine the position and expanse of this vast empire; I have considered the configuration of the

IV.281 land, its geological constitution, the temperature, and the type of vegetation; I have studied the population of the country, its customs, the state of agriculture and mining, and the development of manufacturing and trade; I have attempted to present the revenues of the state and its means of external defense; let us now summarize what we have stated about the present state of Mexico.

Physical appearance. In the center of the country, a broad chain of mountains runs first southeast to northwest, and then from south to north beyond the thirtieth parallel: broad plateaus extend along the ridge of these mountains before gradually descending toward the temperate zone. In the Torrid Zone, their absolute elevation is 2,300 to 2,400 meters. The slope of the Cordilleras is covered with dense forests, while the central plateau is almost uniformly dry and devoid of vegetation: the highest peaks, many of which rise above the line of perpetual snow, are crowned with oak and pine trees. In the equinoctial region, the different climes are arranged as if in layers, one atop another: between 15° and 22° latitude, the average temperature of the seaboard, which is humid and unhealthy for those born in cold countries, is between twenty-five and twenty-seven degrees Centigrade. The temperature on the central plateau, which is famous for the salubrity of its air, is between sixteen and seventeen degrees. Rains are infrequent in the interior, and the most populated part of the country lacks navigable rivers.

Territory. One hundred eighteen thousand square leagues, two-thirds of which are in the temperate zone. Because of the high altitude of its plateaus, the third that lies in the Torrid Zone enjoys a temperature similar to spring weather in southern Italy and Spain. IV.282

Population. Five million eight hundred forty thousand inhabitants, of which two and a half million are indigenous peoples of the copper race, one million Spanish-Mexicans, seventy thousand Spanish-Europeans, and almost no black slaves [Nègres esclaves]. The population is concentrated on the central plateau. The clergy accounts for only fourteen thousand individuals. The population of the capital is 135,000 souls.

Agriculture. Bananas, manioc, corn, cereal grains, and potatoes are the base of the people's nutrition. The cereals grown in the Torrid Zone, wherever the ground is 1,200 or 1,300 meters in elevation, produce twenty-four grains for every one planted. The maguey (agave) may be considered the indigenous peoples' vine. There has been rapid progress recently in sugarcane cultivation: Veracruz exports between 5,500,000 kilograms annually, or 1,300,000 piasters' worth of Mexican sugar. Cotton of the finest quality is harvested on the western coastline. The cultivation of the cacao tree and indigo is equally neglected. Vanilla from the forests of Quilate produces an annual harvest of 900,000 [kilograms?]. Tobacco is carefully grown in the districts of Orizaba and Córdoba; wax is abundant in the Yucatán; the cochineal harvest in Oaxaca is 400,000 kilograms. Horned cattle have multiplied enormously in the *provincias internas* and on the eastern slopes between Panuco and Huasacualco. Tithes to the clergy, whose value indicates the growth of territorial products, have increased by two-fifths in the past ten years. IV.283

Mining. Annual output of gold: 1,600 kilograms; of silver: 537,000 kilograms: in all, twenty-three million piasters or nearly half the value of the precious metals extracted annually from mines in the two Americas. From

1690 to 1803, the mint of Mexico City produced 1,353 million piasters, and from the discovery of New Spain to the beginning of the nineteenth century, probably 2.28 billion piasters or nearly two-fifths of all the gold and silver that has flowed in that interval of time from the New Continent to the Old. Three mining districts—Guanajuato, Zacatecas, and Catorce—which together form a central group located between 21° and 24° latitude, produce nearly half of all the gold and silver that is extracted annually from the mines of New Spain. In an average year, the Guanajuato vein, richer than the mineral depository in Potosí, alone produces 130,000 kilograms of silver, or one-sixth of all the silver that the Americas place annually into circulation. In the past forty years, the Valenciana mine, whose annual operating expenses exceed four and a half million francs, has never failed to provide its proprietors with an annual net profit of over three million francs: this profit has sometimes risen to six million; in the space of a few months, it yielded twenty million to the Fagoaga family in Sombrerete. The output of the mines in Mexican has tripled in fifty-two years and sextupled in one hundred years; it will increase even more as the country becomes more populated and as knowledge contributes to progress. Far from being opposed to agriculture, mining has favored the clearing of the land for planting in the most uninhabited regions. The wealth of Mexican mines consists more in the abundance rather than in the intrinsic worth of the silver ore, whose value on average is only .002 (or three or four ounces per quintal of one hundred pounds). The ratio of the amount of extracted mercury ore to the amount produced by smelting is three and one-half to one. The amalgamation process that is used is time-consuming and wastes a large amount of mercury: in New Spain as a whole, this amounts to a loss of 700,000 kilograms each year. One assumes that the Mexican Cordilleras will one day provide the mercury, iron, copper, and lead that are necessary for the interior of the country.

IV.284

Manufacturing. Value of the annual production of the manufacturing industry: seven to eight million piasters. Leather, cloth, and calico manufacturing has been somewhat on the rise since the end of the past century.

IV.285

Trade. Imports of foreign products and merchandise: twenty million piasters; exports of agricultural products and from the manufacturing industry of New Spain: six million piasters. In gold and silver, the mines produce twenty-three million, of which eight to nine are exported on the king's behalf: if one therefore subtracts fourteen million from the remaining fifteen million piasters to compensate for the surplus of imports over exports, one finds that the annual increase in currency level of Mexico is barely one million piasters.

Revenue. Gross revenue is twenty million piasters, of which 5,500,000 come from the production of the gold and silver mines, four million from the tobacco monopoly, three million from the alcabala, 1,300,000 from the Indians' capitation tax, and 800,000 from the duty on pulque or fermented sugar from the agave plant.

Military defense. It consumes one-fourth of total revenue. The Mexican army is 30,000 men strong, of whom scarcely one-third are infantrymen and over two-thirds militia. The ongoing petty warfare with the nomadic Indians in the *provincias internas* and the maintenance of the presidios or military posts are items of considerable expense. The condition of the eastern coastline and the configuration of the terrain facilitate the defense of the country against an attempted invasion by a naval power.

These are the principal results to which I have been led. May this work, which I began in the capital of New Spain, be useful to those who are called IV.286 upon to safeguard public prosperity; above all, may it infuse them with this important truth: that the well-being of whites is intimately linked to that of the copper race, and that there can be no lasting happiness in both Americas until this race, humiliated but not degraded by long oppression, can participate in all the advantages that come from the advancement of civilization and the improvement of social order.

END OF THE SIXTH AND FINAL BOOK

Notes

IV.287

Note A (Volume I, Page 190)

This information is derived from Don José de Moraleda's unpublished papers, preserved in the archives of the viceroyalty of Lima and referenced in the second chapter, vol. I, p. 240. Although their latitude is 8° farther south than Caylin Island, I have made no mention of the Falkland Islands because there is no stable settlement, properly speaking, in the Falkland group. Two corvettes, commanded by officers of the Royal Navy, bring prisoners annually from Montevideo to the port of La Soledad: these unfortunate men are allowed to build huts, but since the viceroy of Buenos Aires does not dare send women to the presidio of the Falklands, following the orders of the court of Madrid, this military post cannot be ranked at the same level as the ones in New California, which are surrounded by farms and villages.

The archipelago of the Huaytecas and Chonos, which extends from 44°20′ to 45°46′ southern latitude, is only a mass of granite rocks covered with dense forests. The Indians from Chiloé, known as the Guayhuenes and Payos, visit these reef islands periodically: they have placed cattle on the islands of Tequehuen, Ayaupa, Menchan, and Yquilao. On the nearby continent, the coastline that extends to the south from Fort Maullín, is inhabited by the Juncos Indians, who are an independent tribe.

IV.288

Note A *Bis*[1] (Volume I, Page 241)

To feed its free inhabitants and slaves, the island of Cuba requires a great amount of provisions, especially salted meat (*tasajo*), from the coastline of Caracas. When Spain is at war with England, navigation from Cumaná, Nueva Barcelona, and La Guaira to Havana is very dangerous, because it is necessary to sail around Cabo San Antonio. Enemy privateers cruise near the Cayman Islands between Cabo Catoche and Cabo San Antonio, especially in the Tortugas. This group of reef islands is located west of the tip of eastern Florida, and ships without chronometers or other suitable means of determining longitude are forced to sight the Tortugas so as to direct their course from there to the port of Havana, where currents constantly make for rough seas. To avoid most of these dangers, there is a useful plan to set up an interior connection on the island of Cuba between the northern and southern coastline, or to put it incorrectly as the indigenous peoples do, to connect the South and North Seas. A navigable canal for flatboats will be opened over a length of eighteen leagues, from the Gulf of Batabanó to the Bay of Havana, across the beautiful plains in the Los Güines district. This canal, which requires only a few sluices, will also serve to fertilize the country by irrigation: salted meat, cacao, indigo, and other products of IV.289 Terra-Firme will arrive by this route to Havana. The crossing from Nueva Barcelona to Batabanó is not only very short and fairly safe in wartime but also presents the advantage of exposing ships less to the dangers of shallows and storms than the normal navigation around Cabo San Antonio and via the old Bahama Channel.

Note B (Volume I, Page 305)

To give an example of the method by which the parish priests of Mexico have compiled the excerpts that I have used to estimate the surplus of births, I present here detailed tables for Singuilucan and Dolores, two villages in the Torrid Zone inhabited only by Indians, that enjoy a climate that is extremely beneficial to human health. The significant increase in population that these tables indicate is surprising.

1. This footnote was mistakenly cited on p. [I.] 241 as the *second* note.

I.	BIRTHS	I.	DEATHS	II.	BIRTHS	II.	DEATHS
1750–1759	62	1750–1759	18	1760–1769	91	1760–1769	18
	41		4		75		35
	72		5		53		59
	65		22		72		17
	74		16		72		28
	69		10		87		44
	70		10		79		30
	77		13		101		13
	96		13		79		18
	78		19		81		29
	692		130		790		291

III.	BIRTHS	III.	DEATHS	IV.	BIRTHS	IV.	DEATHS
1770–1779	87	1770–1779	19	1780–1789	67	1780–1789	21
	76		21		101		29
	78		37		82		36
	52		33		70		22
	76		21		94		68
	71		25		100		55
	81		32		89		64
	102		35		60		60
	95		31		101		40
	87		43		86		77
	805		297		860		472

V.	BIRTHS	V.	DEATHS
1790–1799	81	1790–1799	47
	105		59
	120		58
	119		59
	127		51
	105		52
	103		51
	126		94
	118		102
	128		52
	1,132		625

BIRTHS IN 1800 AND 1801	DEATHS IN 1800 AND 1801
131	57
150	79
281	135

Births in fifty-one years 4,650

Deaths 1,950

Surplus of births 2,610

I.	BIRTHS	I.	DEATHS	II.	BIRTHS	II.	DEATHS
1750–1760	526	1750–1760	77	1760–1770	1,075	1760–1770	317
	532		137		1,146		315
	1,006		171		1,137		694
	1,009		179		786		1,565
	1,003		160		1,495		187
	842		186		1,054		219
	883		173		1,166		340
	1,027		303		1,407		420
	1,021		250		1,177		349
	1,071		262		1,240		283
	8,920		1,898		11,682		4,689

III.	BIRTHS	III.	DEATHS	IV.	BIRTHS	IV.	DEATHS
1770–1780	1,292	1770–1780	281	1780–1790	1,287	1780–1790	2,580
	1,252		203		1,401		313
	1,099		166		1,271		562
	1,118		242		1,644		471
	1,202		362		1,469		588
	1,421		221		1,095		741
	1,304		255		798		2,663
	1,322		381		850		369
	1,459		391		1,329		315
	1,352		515		1,102		307
	12,821		3,017		12,246		8,909

V.	BIRTHS	V.	DEATHS
1790–1800	656	1790–1800	300
	1,070		318
	1,297		515
	1,331		371
	1,074		313
	1,149		275
	1,482		502
	1,492		650
	1,368		968
	1,567		394
	12,486		4,606

BIRTHS IN 1801 AND 1802	DEATHS IN 1801 AND 1802
1,455	556
1,648	448
3,103	1,004
Births in fifty-two years 61,258	
Deaths 24,123	
Surplus of births 37,235	

IV.292

Note C (Volume II, Page 78)

The following tables present the details of the census taken in Mexico City in 1790 by order of the viceroy, Count de Revillagigedo.

Census of the Population of Mexico City, Compiled in September 1820

NAMES OF MAYORS, (CENSUS OFFICIALS)	DISTRICT	MEN	WOMEN	DISTRICT	MEN	WOMEN	TOTAL
José Juan Fagoaga	1	4,511	6,145	2	2,806	3,948	17,510
Tomas Ferán	3	1,043	2,530	4	1,627	1,951	7,696
Franc. Acipuene	5	4,159	4,571	6	3,030	4,054	16,714
José Barcena	7	2,025	3,114	8	1,162	1,026	7,312
Luis Madrid	9	1,820	2,304	10	1,740	2,185	8,049
Joaquín Cortina	11	4,608	4,248	12	1,761	2,147	12,764
Pedro Jove	13	1,819	2,627	14	4,492	6,526	15,464
Antonio Ycaza	15	1,345	1,945	16	1,438	1,923	6,651
José Ysita	17	2,795	3,394	18	254	1,060	8,203
Angel Puyade	19	3,414	3,626	20	1,488	1,452	9,980
José Palacios	21	1,202	1,801	22	1,634	1,931	6,568
Alatto Palacio	23	1,327	2,011	24	814	1,493	5,645
Marq. of Quardiola	25	2,271	2,634	26	2,509	2,014	10,228
Marquis of Salvato	27	380	355	28	1,362	1,527	3,624
Marshal of Castilla	29	1,469	1,931	30	1,201	1,681	6,282
Count Penazco	31	2,904	3,519	32	1,423	2,116	9,962
José Echave		1,640	1,742				3,382
Gov. of San Juan		6,290	6,507				12,797
Totals		46,622	54,904		29,386	37,934	168,846[1]

[1] In this total of 168,846, there are 76,008 men and 92,838 women; there are thus 16,830 more women than men.

IV.293 *Census of the Population of Mexico City in 1790*

I.

RELIGIOUS ORDERS (MALE)	NUMBER OF CONVENTS	PRIESTS AND CHORISTERS	NOVICES	LAY BROTHERS	DONADOS	SERVANTS	CHILDREN	TOTAL
Santo Domingo	1	60	9	4	1	40	0	114
Porta Coelli (Casa de estudios), *idem.*	1	22	0	0	1	6	0	29
S. Francisco, Observantes	1	91	8	25	9	28	0	161
Santiago Tlatelolco (Casa de estudios), *idem.*	1	33	0	2	1	6	0	42
S. Fernando (Colegio de Misioneros), *idem.*	1	45	0	19	6	1	0	71
S. Cosme (Recolección), *idem.*	1	16	4	10	5	35	0	70
S. Diego (Descalzos), *idem.*	1	45	0	6	16	16	0	83
S. Agustín (Calzados)	1	71	11	2	4	9	0	97
S. Pablo (Casa de estudios), *id.*	1	18	0	0	0	6	0	24
S. Tomás (Hospicio de Misioneros), *idem.*	1	3	0	2	0	5	5	10
S. Nicolás (Hospício de Descalzos), *idem.*	1	4	0	1	0	8	0	13

El Carmen (Descalzos)	1	40	0	7	2	15	5	68
La Merced (Calzados)	1	62	9	4	0	13	0	88
Belén de Mercenario (Casa de estudios), *idem*.	1	24	0	2	0	2	0	28
S. Camillo (Agonizantes)	1	7	0	3	1	7	0	18
S. Juan de Dios (Hospitalarios)	1	5	8	23	2	15	0	53
S. Lázaro, *idem*.	1	2	0	0	2	6	0	10
S. Hipólito (Hospitalarios)	1	2	6	19	3	0	0	30
Espíritu Santo (Hipólitos)	1	1	0	4	1	4	0	10
Belemitas (Hospitalarios)	1	2	1	36	4	9	15	69
S. Felipe Neri (Congr. del Oratorio)	1	14	1	3	0	15	0	33
Monserrate (Benitos)	1	3	0	0	0	4	0	7
S. Antonio Abad (regular canons)	1	3	0	3	2	5	0	13
Total	23	573	59	175	60	255	19	1,141

II.

RELIGIOUS ORDERS (FEMALE)	NUMBER OF CONVENTS	INITIATES	NOVICES	CHILDREN	DOMESTIC SERVANTS		CHAPLAINS		TOTAL
					COMMUNAL	INDIVIDUAL	LAY PERSONS	RELIGIOUS	
Concepción	1	77	1	22	20	78	3	0	201
Regina, *idem.*	1	63	2	9	16	65	2	0	157
Balvarena, *idem.*	1	38	1	14	14	47	2	0	116
Jesús María, *idem.*	1	60	2	0	20	62	3	0	147
Encarnación, *idem.*	1	65	2	7	16	67	2	0	147
Santa Iñés, *idem.*	1	25	1	11	8	26	1	0	159
San Joseph de Gracia, *idem.*	1	40	1	6	9	41	2	0	99
San Bernardo, *idem.*	1	44	0	11	14	44	2	0	115
San Gerónimo (Gerónimas), *idem.*	1	58	6	11	12	68	2	0	157
San Lorenzo, *idem.*	1	37	3	10	14	47	2	0	113

Santa Teresa la antigua (Carmelitas Desc.)	1	21	0	0	0	0	2	0	23
Santa Teresa la nueva, *idem*.	1	17	1	0	0	0	1	0	19
San Felipe de Jesús (Capuchinas)	1	34	2	0	0	0	1	0	37
Santa Brígida	1	30	1	0	0	0	1	0	34
Enseñanza	1	69	3	0	0	0	3	0	75
Santa Catalina de Sena (Dominicas)	1	46	3	28	15	49	0	2	143
Santa Clara	1	60	0	16	16	45	0	3	140
San Juan de la Penitencia (Claras)	1	39	2	10	16	41	0	2	110
Santa Isabel, *idem*.	1	37	3	10	21	52	0	2	125
Corpus Christi or Capuch. (Indians)	1	28	1	0	0	0	0	2	31
Totals	20	888	35	165	211	732	31	11	2073

IV.295 *III. Lay Persons*

AGE	UNMARRIED INDIVIDUALS		MARRIED INDIVIDUALS		WIDOWERS	WIDOWS	TOTAL	
0–7 years	Males	Females	Males	Females			Males	Females
7–16 years	8,559	9,823	0	0	0	0	8,559	9,823
16–25 years	7,458	9,099	71	325	104	149	7,663	9,573
25–40 years	4,819	5608	3,350	5836	228	986	8,397	12,440
40–50 years	2,508	3,237	9,097	9,695	804	4,189	12,409	17,121
50 and above	720	728	2,086	1,112	917	2,613	3,723	4,453
	24,999	29,478	17,739	19,112	2,740	10,692	45,478	59,282
	54,477		36,851		13,432		104,760	
	104,760							

IV.296 *IV. Castes*

CASTES	0–7 YEARS		7–16 YEARS		16–25 YEARS		25–40 YEARS		40–50 YEARS		50 AND ABOVE		TOTAL	
	Male	Fem.	Male	Fem.	Male	Fem.	Male	Fem.	Male	Fem.	Male	Fem.	Male	Fem.
Europeans	5	2	40	11	330	81	714	65	612	33	417	25	2,118	217
Spaniards	3,949	4,085	3,606	4,704	4,050	6,018	5,600	8,551	2,366	3,314	1,767	2,316	21,338	29,033
Indians	1,862	1,896	2,171	2,587	2,111	3,204	3,351	4,523	939	1,170	798	991	11,232	14,371
Mulattos	936	1,240	403	560	514	621	721	944	191	425	193	346	2,958	4,136
Other castes	1,807	2,600	1,413	1,711	1,392	2,516	2,023	3,038	649	930	548	730	7,832	11,525
	8,559	9,823	7,633	9,573	8,397	12,440	12,409	17,121	4,757	5,872	3,723	4,453	44,478	59,282
	18,382		17,206		20,837		29,530		10,629		8,176		104,760	

Total: 104,760

IV.297

V.

SCHOOLS (COLLEGES) FOR MEN	PRECEPTORS		STUDENTS (COLLEGIATE)		DOMESTICS	TOTAL
	Lay Priests	Clergy	Secular	Clergy		
Colegio Major de Santos	0	0	6	0	10	16
Seminario	13	0	261	20	24	318
San Ildefonso	8	0	213	23	56	300
San Juan de Latrán	7	0	59	6	15	87
San Ramón	3	0	15	0	8	26
Santiago Tlaltelolco	0	1	4	2	5	12
San Gregorio (Indians)	0	3	23	0	0	26
Total	1	0	38	8	4	51

VI.

SCHOOLS FOR GIRLS	FEMALE INSTRUCTORS		STUDENTS	CHAPLAINS	DOMESTICS	TOTAL
	Religious	Secular				
Jesús María	6	0	125	1	1	133
La Enseñanza	10	0	60	0	4	74
Las Niñas	0	0	33	2	6	41
San Ignacio (or Vizcaína)	0	4	266	2	0	272
Belén (or de Mochas)	0	8	235	2	0	245
Guadalupe (female Indians)	0	4	40	0	8	52
Total	16	16	759	7	19	817

IV.298 *VII. Hospitals*

NAME OF HOSPITAL	CHAPLAINS		EMPLOYEES	DOMESTICS	SICK PEOPLE		MENTAL PATIENTS		PHYSICIANS	TOTAL
	Secular	Religious			Male	Female	Male	Female		
Real de Indios	4	0	2	33	100	63	0	0	3	205
Gen. Hospital of San Andrés	6	0	17	82	337	136	0	0	8	586
San Juan de Dios	0	2	0	8	44	56	0	0	2	112
Espíritu Santo	0	1	0	5	22	0	0	0	1	29
La Tercera Orden de San Francisco	0	1	3	14	4	11	0	0	2	35
Convalescencia among the Belemitas	0	0	0	6	45	0	0	0	1	52
Mentally ill clergy (of the SS Trin.)	3	0	2	7	0	0	19	0	1	32
Mentally ill of San Hipol.	0	2	0	8	0	0	90	0	1	101

Mentally ill of Casa del Salvador	0	1	3	4	0	0	0	53	0	61
Incurables of San Lázaro	0	2	2	5	41	22	0	0	1	73
Idem. of San Antonio Abad	0	1	3	3	8	9	0	0	0	24
Jesús Nazar. of Estado del Valle	2	0	2	10	12	6	0	0	4	36
Total	17	8	34	185	613	303	109	53	24	1,346

WELFARE HOUSES	CHAPLAINS	EMPLOYEES	DOMESTICS	ORPHANED CHILDREN		MEN	WOMEN	TOTAL
				Male	Female			
Foundlings	1	4	5	118	95	0	0	223
Poorhouses	2	2	24	113	56	312	429	938
La Misericordia (married women)	1	2	2	0	0	0	4	9
Total	4	8	31	231	151	312	433	1,170

IV.299 *VIII.*

PRISONS	MALE	FEMALE	CHAPLAINS	EMPLOYEES	DOMESTICS	TOTAL
Of the court	195	24	0	1	2	222
Of the town	75	35	0	1	3	114
By decree	286	16	2	3	12	319
Of the Inquisition	0	0	1	3	1	5
Of the Archbishop	30	3	1	7	6	47
La Magdalena de Recogidas	0	88	1	3	5	97
Of Indians	15	3	0	0	0	18
TOTAL	601	169	5	18	29	822

IX. Inhabitants of Mexico City by Occupation

Prebends	26	Students under religious jurisdiction (*de capa*)	368
Priests	16	Students under military jurisdiction	510
Clerics	34	Finance employees	311
Secular priests	517	Slatterns	63
Inquisition officials	33	Employees of the *Acordada*	177
Crusade officials	5	Laborers	97
Title holders (*títulos de Castillo*)	44	Miners	40
Knights of Royal Orders	38	Businessmen	1,384
Doctors	204	Artisans	8,157
Lawyers	171	Day laborers	7,430
Physicians	51	Individuals subject to tithes	9,086
Surgeons and barbers	227		
Manufacturers	1,474		

X. SUMMARY

IV.300

Secular state				104,760
	Individuals living in convents and colleges			
		Male	3,484	
				6,530
		Female	3,046	
Priests			748	
				1,636
Nuns			888	
Total (not including the military)				112,926

Note D (Volume II, Page 96)

I must add two unpublished works, entitled *Relación de la visita del Desagüe real hecha en* 1764 and *Auto formado en San Cristóbal, en el mes de Enero de 1764, por mando del ilustrísimoseñor Don Domingo de Trespalacios, del supreme consejo y camara de Indias*, respectively, to the material that has helped me draw up the history of hydraulic works in the Valley of Mexico. According to these papers, the engineer Ildefonso Yniesta found a distance of 65,250 *varas* between the shores of Lake Texcoco and the Tula waterfall, whereas Professor Velázquez's trigonometric calculations and direct measurements indicate that this distance is only 62,363 *varas*. This finding, which was used in the map of the Valley of Mexico (*Atlas mexicain*, Plate III), must be considered the most accurate, not only because of the perfection of the instruments used in 1774 but also because of the agreement between the distances that Mr. Velázquez found and those which Mr. Martínez determined in 1611: the latter estimated the distance from IV.301 Lake Texcoco to Vertideros at 35,421 *varas*; Velázquez set it at 35,168; Yniesta's measurement was 38,740 *varas*.

Note E (Volume II, Page 152)

I have discussed elsewhere the striking similarity between the temple of Jupiter Belus and the pyramids of Sakkara with the teocalli, or houses, of the Mexica gods, which were both temples and tombs. See my *Vues des Cordillères et monumens des peuples indigènes de l'Amérique*, pp. 24–40.

Note F[1] (Volume II, Page 277)

The following table describes the state of the missions in New California in 1802. In the census of the Indians, the sexes have been distinguished by the initials *m* and *f*. Both domestic horses and those that roam the savannas are grouped under the heading *horses*: the former are only 2,187 in number. These details on the state of agriculture and civilization on the northwest coastline of the Americas are of great interest since the congress in Washington has resolved to establish colonies at the mouth of the Columbia River. (See above, chap. II, vol. I, p. 206; chap. VIII, vol. II,

1. This note was mistakenly given on p. 277 as letter D.

IV.302–303

VILLAGES OR MISSIONS	BIRTHS	MARRIAGES	DEATHS	TOTAL (INDIANS)	BULLS AND COWS	SHEEP	HORSES	MULES
San Diego	5,952	792	1,283	1,559 (737 m.–822 f.)	6,050	6,000	900	66
San Luis R. de Francia	568	113	104	532 (256 m.–276 f.)	1,400	2,700	226	18
San Juan Capistrano	2,137	491	1,033	1,013 (502 m.–511 f.)	8,710	15,300	660	58
San Gabriel	3,397	746	2,151	1,047 (523 m.–297 f.)	77,500	13,045	1,430	100
San Fernando	748	169	188	614 (317 m.–297 f.)	900	2200	270	43
San Buenaventura	1,669	318	693	938 (436 m.–502 f.)	12,450	5,306	2,085	112
Santa Barbara	2,251	494	989	1,093 (521 m.– 572 f.)	2,100	9,082	627	58
La Purissima Concepción	1,582	356	557	1,028 (457 m.–571 f.)	2,640	5,400	326	44
San Luis Obispo	1,735	467	962	699 (374 m.–325 f.)	5,100	5,300	1,120	100
San Miguel	729	164	163	614 (309 m.–305 f.)	606	3,099	284	28
Soledad	887	218	401	563 (296 m.–267 f.)	1,000	4,000	520	19

(continued)

IV.302–303

VILLAGES OR MISSIONS	BIRTHS	MARRIAGES	DEATHS	TOTAL (INDIANS)	BULLS AND COWS	SHEEP	HORSES	MULES
San Antonio de Padua	2,730	641	1,527	1,052 (568 m.–484 f.)	2,221	5,530	635	37
San Carlos de Monterey	2,418	633	1,496	688 (376 m.–312 f.)	1,200	6,000	875	34
San Juan Bautista	1,079	203	184	958 (530 m.–428 f.)	618	3,800	454	6
Santa Cruz	1,031	306	591	437 (238 m.–199 f.)	1,407	2,915	1,861	88
Santa Clara	4,407	1,010	2,967	1,291 (736 m.–555 f.)	5,000	6,000	6,100	30
San José	857	218	243	622 (327 m.–295 f.)	620	3,500	263	10
San Francisco	2,540	760	1,442	814 (433 m.–381 f.)	8,260	8,000	793	26
Total	33,717	8,009	16,984	15,562 (7,945 m.–7617 f.)	67,782	107,172	19,429	877

pp. 313 and 327 [pp. 468 and 475 in this edition].) It takes eight to ten days to navigate from Monterey to the mouth of the Columbia; the new settlers will be able to bring cows and mules from the missions of New California.

END OF NOTES. IV.304

Supplement

IV.305

Note. In this supplement, the author has included some observations of which he became aware only after the printing of the first three volumes of this edition.

Astronomical Positions

An acute observer, ▼ Lieutenant Glennie [Glenie] of the Royal British Navy, recently plotted Durango at lat[itude] 24°0′55″; Guarisamey, at 24°5′45″. According to the movement of his chronometer, he assumes the longitude of these two places to be 107°8′6″ and 108°25′30″ west of Paris.

Taking the average of his observations, ▼ Don José María Bustamante, who has contributed with tireless zeal to the progress of the astronomical geography and the geology of Mexico, places the following:

Zacatecas	lat. 22°46′3″
Veta-Grande, N. of Zacatecas	22°50′2″

▼ Doctor Culter [Coulter] finds 22°49′53″ for Veta-Grande. According to Mr. Bustamante, the longitude of Zacatecas by lunar distances and a triangulation exercise is 103°13′9″, supposing the longitude of Guanajuato to be 103°14′19″. The same scholar gives 104°11′55″ as the longitude of Veta-Grande. Doctor Culter sets this longitude by lunar distances at 104°17′30″.

We also owe the following measurements to Mr. Bustamante:

Lagos	lat. 21°20′0″	
Aguas Calientes	21°56′55″	
Bolaños	21°50′45″	long. 105°43′
Fresnillo	23°9′0″	104°26′
Plateros	23°14′0″	104°25′
Ramos	22°51′0″	103°40′

IV.306

The latitude and longitude of the three last places are based only on surveys. The longitude of Zacatecas is derived from two series of lunar distances.

The results of ▾ Don Juan de Horbegozo's astronomical observations on the Isthmus of Tehuantepec in 1825.

Confluence of the Sarabia with the Río Coatzacoalco	17°11′46″
Petapa	16°49′30″
San Miguel Chimalapa	16°42′42″
Santa Maria Chimalapa	16°52′31″
Venia de Chicapa	16°35′15″
Zuchitan	16°22′53″
Tehuantepec	16°20′10″
Chibuitan	16°33′54″
San Mateo del Mar	16°12′49″
La Orilla del Mar	16°10′49″
Santa María del Mar	16°13′43″
The coast of the Pacific Ocean close to the above village	16°11′48″
Oaxaca	17°2′33″
Tehuaca	18°26′35″
Orizaba	18°49′50″
Córdoba	18°52′14″
Xalapa	19°30′4″

General Orbegozo determined the following longitudes by means of eclipses of the satellites of Jupiter:

Tehuantepec	3°58′17″
Oaxaca	2°24′37″
Tehuacan	1°51′48″
Orizaba	2°9′7″

I owe the following positions to the extreme kindness of ▾ Mr. Mornay, who barometrically surveyed the country between Veracruz, Mexico City, and Oaxaca and made several astronomical measurements: IV.307

Mitla	lat. 16°55′17″ eastern long. of Oaxaca	22′43″ in arc
Trapiche de Almendaras, near Totolapa	16°37′33″	
Oaxaca	17°2′40″	
San Pedro Nolasco	17°15′47″	19′0″
Mina de Almendaras	16°42′15″	
Mina de Yuyucundo	16°53′36″ western long. of Oaxaca	28′50″
Villa de Elotepec	16°51′30″	34′29″
Mina de San Pablo		
Teovomulco	16°34′ 0″	30′8″

Mr. Glennie's, Bustamante's, Culter's, Orbegozo's, and Mornay's respective latitudes are all based on observations made with a sextant. A careful analysis of my map of Mexico drawn up in 1803 will reveal that no astronomical position was known at that time in the hinterland north of Guanajuato and south of the parallel of Acapulco, the boundary of my own observations.

Measurements of Elevation

Mr. José María Bustamante's barometric observations made between Guanajuato and Bolaños provided him with the following results:

Silao	1,853 meters
Lagos	1,940
Zacatecas	2,490
Buffa, near Zacatecas	2,622
Xerez	2,058
Villa de Colotlán	1,735
Temastian	1,798
Alto de los Guacamayos	1,934

IV.308

Bolaños	947
Buffa, near Bolaños	1,385
Hacunda de Atotonilco	2,191
Huehuetoca	2,286
Tetepango	2,138
Pachuca	2,431
Cerro del Ventoso	2,769
Real del Monte	2,785
Mina de Cabrera	2,620

IV.308 Mr. Bustamante's elevations were calculated according to Laplace's formula, using Mr. Oltmanns' tables.

The barometric results found by General Orbegozo,[1] who found that the connection between the Río Coatzacoalcos (Guasacualco) and the Río Chimalapa via a canal was unfeasible because of the elevation of the terrain between the two rivers.

I. On the Road from the Town of Orizaba to Acayucan	
Orizaba	1,235 meters
Santiago Tuxtla	196
San Andrés Tuxtla	329
Acayucan	137
IV.309 II. On the Isthmus of Tehuantepec	
Confluence of the Sarabía and the Coatzacoalco	45 meters
Guichicaro	265
Petupa	229
Hacienda de la Chivela	241
Hacienda de Tarifa	264
Highest point of the road from Tarifa to San Miguel	358
San Miguel Chimalapa	173
Portillo, in the Cordilleras between San Miguel and Santa María	393

(*continued*)

1. Misled by Mr. Cramer and Mr. Corral's travel journals, I have found that according to Mr. Orbegozo's map of the isthmus, I had found the Río Sarabía to be a tributary flowing from the east, whereas it actually comes from the west. I have published Mr. Orbegozo's map in the *Journal géographique* that is found in Germany under the title *Hertha*.

Rancho de la Cofradía	402
Cime du Cerro Pelado	615
Santa María Chimalapa	286
Río Coatzacoalcos, three leagues to the east of Santa María	160
Venta de Chicapa	55
Zuchitan	30
Tehuantepec	41
III. On the Road from Orizaba to Songolica	
San Andrés Tenejaca	1,167
Portezuelo de Amolapa	1,931
Songolica	1,218
Altura de Tianguetezingo	1,922
Atlanca	1,669
IV. On the Road from Orizaba to Puebla	
Hacienda de Tecamalupa	1,388
Puente Colorada	2,234
Tercera Cumbre	2,503
Tepeaca	2,263
Amozotl	2,331
Puebla	2,150

The elevation above sea level of Lake Nicaragua, which I was not able to provide in vol. I, p. 211 [p. 150 in this edition], was determined in 1781 by order of the court of Madrid. The engineer Mr. Manuel Galisteo found this elevation to be 336 ascending stations and 339 descending ones (*ascensos*, 604 Castilian feet, *descensos*, 470 feet) of 137 feet [each?]. Now, since the lake is eighty-eight feet deep, its bottom is still forty-six Castilian feet above sea level. For this measurement, see the *Rel[ation] hist[orique]*(quarto), vol. III, p. 320. IV.310

Mining Output

In vol. III, p. 179 [p. II.100 in this edition], I have stated the production in gold and silver of the mines in Guanajuato from 1766 to 1803. The ensuing production to 1825 is as follows:

YEARS	BARS	SILVER MARKS	GOLD MARKS
1804	5,734	755,861	2,128
1805	5,510	723,789	2,495
1806	4,716	618,417	2,188
1807	4,417	578,735	2,396
1808	4,685	617,474	1,842
1809	4,737	620,012	2,189
1810	3,896	511,445	1,419
1811	2,067	270,206	550
1812	2,702	357,930	907
1813	2,204	292,211	462
1814	2,568	337,795	708
1815	2,088	275,905	841
1816	2,041	269,711	694
1817	1,580	199,706	523
1818	1,215	155,112	401
1819	1,149	145,362	450
1820	814	100,465	326
1821	600	73,983	298
1822	795	95,057	597
1823	804	96,802	413
1824	901	106,775	517
1825	803	100,193	419

The relative proportion of gold appears to be growing. The year 1791 gave the *maximum* of 767,607 marks of silver (of twelve deniers) and 1,001 marks of gold (of twenty-two carats).

Excerpt from the Will of Hernán Cortés, from the Archives of the Monteleone Family, in Mexico City by Alexander Von Humboldt

IV.313

IN THE NAME OF GOD, AMEN.

To all those who shall read this document, let it be known that in the very noble and loyal city of Seville, on Saturday, the eighteenth day of the month of August, in the year of our Lord and Savior Jesus Christ 1548, García de Huerta, His Majesty's notary, delivered and entrusted to me, Melchor de Portez, notary public of Seville, the original testament that the most illustrious Don Hernán Cortés, Marquis of the Valle of Oaxaca in New Spain, made and presented to me, Melchor de Portez, notary public. Said document was stamped and sealed, and had been written down on Wednesday, the twelfth of October in the preceding year 1547. Due to the death of the said lord and marquis Don Hernán Cortés, this will was unsealed in the presence of said Garcia de Huerta in the village of Canilleca de la Cuesta on December 3 of the same year 1547, following the order of the licensed Master Don Andrés de Jauregui. Having myself petitioned their honors the judges of the Royal Audience of Degrees of the city of Seville that the original of said will be delivered into my own hands, since it had been drawn up in my presence, said Garcia de Huerta was ordered to send me this original will for award of view and degree of review and leave it in my jurisdiction, for which they pronounced the following decree:

IV.314

"We, the judges of the Royal Audience, residing in this town of Seville, in His Majesty's service, order you, Garcia de Huerta, Their Majesties' notary, as soon as you shall have received this order, to deliver to Melchor de Portez, notary public of this town, the marquis *del Valle*'s original will, which was opened in your presence. We order you to do so without exception, in accordance with the sentence pronounced against you in the suit you brought against said Melchor, which was pleaded before us, on the question of which both of you should safeguard said testament. We enjoin you to do this immediately, on pain of arrest by our order. As for the rest, we order you to conform to our sentence.

"Done the sixth day of the month of August 1548.

"*Master* MEDINA, *master* CAMILLA, *master* BALTHAZAR DE SALAZAR *doctor* CARO.

"Written by order of the above and by myself, *Juan Hurtado*, Their Majesties' notary and member of the Audience of judges."

In virtue of which decree, said Garcia Huerta delivered to me the original of said will that said lord Don Hernán del Valle had made, sealed, and stamped in my presence, with the initialed authorization of the said lord and countersigned by me, a notary public, and by the two witnesses present. I then made a copy for my register as well a copy of the authorization IV.315 initialed and given before me, when said will had been closed and signed, exactly as Garcia Huerta had delivered it to me, of which a copy follows here.

In the most noble and loyal city of Seville, on Wednesday, the twelfth day of October in the year of our Lord Jesus Christ 1547, in the house where now repose the ashes of the great and powerful lord Don Hernán Cortés, Marquis del Valle, said house being located in the parish (*colación*) of San Marcos, and in the presence of myself, Melchor Portez, notary public of Seville, and the undersigned witnesses, said lord, the Marquis del Valle, appeared, bodily ill but still with all the faculties that it had pleased our Lord God to bestow upon him, and handed over to me this sealed and stamped document, which he considered to be his will, composed of eleven sheets, including the one bearing his signature and those of Melchor Mojica, his treasurer, and master Infante. Each sheet ends with his signature, as I myself was able to prove, having sealed it with my own hand. It was delivered by the illustrious lord Don Hernán as his only will, which he desired to be fully and properly executed, naming as heirs and testamentary executors those whose names are inscribed in said will, and thereby revoking all previous wills, donations, bequeaths, and codicils, requesting me to nullify them so that only this one would be valid, which act I performed for him, executed on said day and year. And lord Don Hernán signed it in his name.

Present as witnesses were:

Martín de LEDESMA, DIEGO DE PORTEZ, PEDRO DE TREXO, *notaries of Seville*, ANTONIO DE BERGARAS, JUAN PEREZ, *procuror*; DON JUAN DE SAAVEDRA, *grand alguazil of Seville*; JUAN GUTIERREZ TELLO, *son of Francisco Tello, who lives in this city.*

Certified and signed by myself, MELCHOR DE PORTEZ.

In the name of the Most Holy Trinity of the Father, the Son, and the Holy Spirit, three persons in one true God, whom I confess to be my master and savior, in the name of the powerful and most happy Virgin Mary, mother of God, our protector, and to all those who shall see this document, I, IV.316

DON HERNÁN CORTÉS, *Marquis of the Valley of Oaxaca, governor of New Spain and the South Sea*, for His Imperial Majesty Charles the Fifth, King of Spain, my sovereign master and lord, reveal what follows here:

Since I am ill but still in full possession of my reason, such as it has pleased God, Our Lord, to grant me, and awaiting death, since all here below depends on his will, and hoping, when God shall retire me from this world, to have done everything possible to save my soul and ease my conscience, I have written this will that I declare valid, whose execution I order as my final and unique wishes:

1. Firstly, if I should die in Spain, my body shall be placed in the church of the parish to which the house in which I have died belongs, until my successor sees fit to transport my remains to New Spain, which hc must do within the ten years following my death. I should like to be buried in the town of Cuyoacan on the day of my funeral, in the nuns' convent that I order to be established there, which will receive the name of the *Conception of the Franciscan Order*, since I and my descendants will be interred there.

2. If it please God that I should die in Spain, I order that my descendants, or even some of those whom I have named as executors of my will, pay for the expense of my funeral.

3. Besides the priests, beneficiaries, and chaplains of this parish who shall be pallbearers, all monks of all orders in the town or village where I die shall be assembled and accompany the cross as they attend my funeral. According to custom, they shall receive alms, which the executors of my will shall set as they see fit. IV.317

4. That same day fifty beggars shall be dressed in long hooded robes of brown cloth. Bearing lit torches, they will follow my body, and each of them shall receive one *real* after the ceremony.

5. On the day of my funeral, if it takes place before noon, or on the following day, I want as many masses as possible said in all the convents and monasteries near my place of death. Furthermore, in the days that follow, up to five thousand masses total shall be said: one thousand for the souls in purgatory; two thousand for the souls of those who died by me and in my

service during the discovery and conquest of New Spain. The two thousand remaining masses will be said for the souls of those to whom I have some obligation of which I am unaware or which I have forgotten. I order that those commitments that I do recall be fulfilled and paid according to the instructions that I give in this will. As for the cost of the five thousand masses, my heirs will have the usual alms distributed, and I beseech them, in whatever else they may do, to try and minimize expenses, the sole purpose of which is the pomp and vanity of this world, and which would be better used for the salvation of our souls.

6. On the day of my burial, my heirs will present my servants and those of my family with mourning garments. In addition, and for six months after my death, they will be paid the salary that they presently earn or will earn at that time, and they will also be fed the entire time. They will be paid all IV.318 the wages that they are due whenever they wish to take leave of my successor *Don Martín*.

7. When my bones are transported to New Spain to be interred at the convent that will be established based on my orders, I want everything to be done in accordance with the wishes of my wife, *Doña Juana de Zuñiga*, who will pay my funeral expenses, and with those of my successor or any other administrator at that time.

[. . .]

9. I order that the hospital of *Nuestra Señora de la Concepción*, which will be built in Mexico City in accordance with the model found in the chapel of the cathedral and according to the geometer Pedro Vasquez's plan, as well as with the instructions that I have sent this very year (1547) to New Spain, be done so entirely at my expense. For this cost, I designate the sum of the rents from the shops and houses that I possess in that city of Mexico on the square and in the streets of Tacuba and San Francisco, etc. I order that this revenue be specially allotted for these buildings until they are completely finished; my successor may not direct any portion of it to another use. This income shall be placed at the disposal of my successor, who will order its use in his capacity as the founder of the new hospital, for this is my will. When building has been completed in accordance with the aforementioned plan, he will manage the remaining contributions with the rest of the income. As for the internal administration of the hospital, the orders I shall give, which will be written and notarized, will be executed; in their absence, the administrative practice in vigor in the town of Seville at the *Hospital of the Five Wounds*, founded by *Catalina de Rivera*, will be fol-

lowed concerning administrators, chaplains, and other workers or servants at the hospital.

10. In the chapel where the body of Martín Cortés, my father and lord, rests in the monastery of *San Francisco de Medellín*, I order that a mass be henceforth said on the anniversary of his death, and that the ceremonies be conducted according to the instructions that I have given and leave to that effect, which my successors shall thereafter follow. To that effect, I name my son and successor, *Don Martín Cortés*, as protector of that chapel, and those members of my family who shall succeed him; and if he should leave and go to the lands of my majorat, he may name whomever he chooses in his place, with the power to retract or cancel this appointment as he sees fit; the powers of the appointed person will be those of the protector himself. IV.319

11. Ever since merciful God, my Lord, led and guided me in the discovery and conquest of New Spain and all of its dependent provinces, his beneficial hand has bestowed endless favors upon me. I have been victorious against the enemies of the holy Catholic faith, I have pacified and populated this kingdom, which I hope will be useful in the service of God, our Master. To give thanks for all these gifts and also to erase the mistakes that I have forgotten and which might burden my conscience, I order the following buildings to be built:

12. In addition to the hospital of *Our Lady of the Conception* that will be built in Mexico City, a nuns' convent that will be called the *Conception of the Order of Saint Francis* will be built in my town of Coyoacan in New Spain, according to the instructions that I have left, which I recommend be strictly followed. If I do not give all the necessary orders for it, I order that my successor or those who follow him build this convent, summon nuns to it, and bestow upon it the proceeds of an income for this purpose. This convent will be my burial place and that of my successors, as I have said above, and I order that my remains be interred in the large chapel built in the convent church, where no one but my legitimate descendants may be buried for any reason. IV.320

13. A college where theology and canon law are taught will be built in my town of Coyoacan, so that learned people may graduate from it who are capable of leading our churches in New Spain, instructing its inhabitants, and teaching them everything about our holy Catholic faith. As for the number of students, their privileges, and the rules that govern them, the instructions that I leave in this regard, of which I have already spoken, are to be followed. The same applies to the construction, location, shape, and

ordonnance of the building; in the event that I do not leave the necessary orders, my successor or those who follow him will build this college and enforce the regulations and customs that govern the college of *Saint Mary of Jesus*, founded in the town of Seville. Construction costs will be paid with the funds and income designated for this purpose.

14. For the endowment of the aforementioned hospital of *Our Lady of the Conception* that I intend to have built in Mexico City, I have allocated two plots of land, one opposite George Alvarado's house, and the other opposite the home of the treasurer Juan Alonso de Soza, between my home and the *Asequia*, through which one reaches Don Louis Saavedra's house. I have contracted (as one will see more explicitly in the act of endowment, by which I stand completely) to have houses built there and to give from my own holdings for the hospital and its expenses, one hundred thousand maravedís in good currency until these houses are finished. I desire that the act of endowment and the appended notes be followed scrupulously. If instead of these houses, my successor should one day wish to give to the said hospital an income of one hundred thousand maravedís instead of any other possession, he will be free to do so and to allocate this income to whatever fund he chooses, provided it is insured.

IV.321 15. By virtue of this act, I have committed to give to the said hospital near Mexico City land that yields three hundred fanegas of wheat. I order this to be done and allocate to that effect a parcel of land belonging to me in the Coyoacan district between the town and the river that crosses the Chapultepec road, unless impediments require that another piece of land near Chapultepec, where I still have tillable land, be used for this purpose. My successor will choose the most appropriate terrain, and he or his heirs will be able to change this terrain, if they wish, provided that three hundred fanegas of wheat can be harvested from land that is as good as what I have designated. Since some of these pieces of land may not belong to me, I order that they be returned to whomever they belong, if they have more right to them than I, although I am lord of this town. Furthermore, if the owner prefers to be monetarily compensated, I order that the value of this land be transmitted to him, and since I have worked and profited from it, in the belief that I did so freely, the size of the harvests will be estimated and appropriate compensation will be made, so that I am above reproach in this matter. If this land is not sufficient, my successor will give to said hospital whatever will be necessary to fulfill the conditions stated in the act of endowment.

[. . .]

18. Since revenues from land and houses evidently increases in New Spain, just as it does in this kingdom (*Spain*), it follows that the houses and shops that I possess in Mexico City, which are listed above, may produce considerably more than the four thousand ducats that I have allocated in perpetuity (one thousand for the nuns' convent, two thousand for the college, and one thousand for the hospital of the *Conception*), as prescribed in the act of endowment, it is my wish that if the rent from these houses and shops should one day increase, then the surplus should be divided as follows: half for the college, a fourth for the nuns' convent, and the other fourth for the hospital. IV.322

19. [. . .] I recommend that my son Don Martín and his successors ensure to the best of their ability that the profits go to able persons of irreproachable morals who endeavor to teach Christian doctrine to the inhabitants through ongoing exercises, and who visit their parishes and ensure that the obligations that religion imposes on us are fulfilled.

20. I order that my wife, the marquise Doña Juana de Zuñiga, be reimbursed the ten thousand ducats that were her dowry, since I have received and spent them; as they belong to her, it is my wish that they be returned to her without cost from the best and largest of my possessions.

21. Lord Don Pedro Alvarez Osorio, the marquis of Astorga, and I have agreed that his son, Don Alvarez Pérez Osorio, the eldest and heir of his house, will be betrothed to Doña María Cortés, my daughter by my marriage with the marquise Doña Juana de Zuñiga, and that all points of the agreements stipulated in a contract that we made will be observed. It is my wish that the terms of said contract be respected to the letter. Since I have promised one hundred thousand ducats as her dowry, of which the marquis de Astorga has already received twenty thousand pursuant to our conditions, I wish above all that the balance of eighty thousand ducats be paid from my possessions and those of the marquise my wife; the terms of payment stipulated in said contract will be observed regarding the amount that is not immediately turned over. My daughter Doña Maria should consider this dowry as an installment of the inheritance that she will one day receive.

22. Since I must also provide dowries for Doña Catalina and Doña Juana, my legitimate daughters, to fulfill this obligation as best I can, I wish for each of them to receive fifty thousand ducats as an irrevocable gift in their lifetime, and for Melchor de Mojica, my treasurer and secretary, who is present here, to accept this money in their name. My daughters should consider these hundred thousand ducats from my possessions and those of the marquise de Zuñiga as part of the legal inheritance that they will one IV.323

day receive. I order that these hundred thousand ducats be taken from the holdings of the marquise de Zuñiga and those that I leave after my death; and if I should not leave enough to release myself from this commitment, my son and successor, Martín Cortés, or anyone else who comes into possession of my land will make up the balance of this sum. To that effect, fifteen thousand ducats will be withdrawn annually from the revenue of my estates until one hundred thousand ducats have been redeemed, as agreed.

23. I order my successor to pay the sum of one thousand gold ducats annually to Don Martín and Don Luis Cortés, my illegitimate sons, from the revenue of my possessions, or three hundred seventy-five thousand maravedís each for the rest of their lives or until each has more than five hundred thousand maravedís in income. This sum is to be paid to them annually, net and free of charge, and I declare that this income is theirs starting immediately and will be paid to them from the majority of my possessions. I enjoin them to respect my successor, follow his advice, and obey his orders, insofar as honor allows, because they must consider him as the leader and head of the family. Not only do I insist on obedience in this respect, but I also wish that my two sons assist Don Martín in all ways, except in actions against God, the holy Catholic religion, and their legitimate king. Should either of them not extend this indulgence that I demand, and if he has truly IV.324 decided to disobey me, he will have lost all rights to my kindnesses from then on, his allowance will be withdrawn, and he will be considered a stranger to my family.

25. Doña Catalina Pizarro, who lives in Mexico City, my child by Leonore Pizarro, the wife of Juan de Salcedo, will receive the value of the revenue produced by the increase in the number of cows, mares, and sheep that I relinquished to her upon my arrival in Spain; furthermore, the total revenue of the village of Chinantla is to be relinquished to her, with the complete dowry that I gave her for her marriage, all these things having been turned over to Juan Salcedo, the husband of said Leonore Pizarro.

[. . .]

33. I spent a considerable part of the revenue from New Spain and its dependent provinces when I conquered it, restored peace, and reduced it to obedience, in the name of the King of Castile. These advances were made both to support the war in New Spain and external expeditions, for example, to form the body of men sent to Amaluco, whose command I entrusted to Captain Alvaro de Saavedra Geronimo Primo. I sent one troop toward Ibureras, under the command of Captain Pobladores, and another

commanded by Francisco de las Casas. All these troops were sent by order of the emperor, our master, as his royal instructions prove. To absolve his conscience as a most Christian prince, Our Majesty has decreed by a royal ordinance, which has remained in master Juan Altamirano's possession, as well as by his royal council's pronouncement that the expenses incurred during my conquests for said expeditions would be settled with me. I order that all these accounts be reckoned and that the balance be paid, since His Majesty has seen fit to order its restitution; my heir, don Martin, and his successors are to administer this matter.

IV. 325

[. . .]

38. Upon receiving the property in New Spain as a gift, I made every effort to find out what tributes, income, royalties, and contributions the native chiefs received customarily; if I have made any error in this estimation, I order that whatever does not belong to me be returned.

39. As for the native slaves who were taken or purchased there, *the question arose long ago as to whether one can keep them in one's possession with a clear conscience. Since this issue has not yet been resolved, I advise my son, Don Martín, and his successor to spare nothing in attaining precise knowledge of the truth of this matter, for my peace of mind and their own.*

40. Since land in some parts of my estates has been set aside for the planting of kitchen gardens, I order that an attempt be made to find out if this land belonged to any of the local indigenous peoples. If this is the case, the land will be returned to them, with profits equal to whatever their masters reaped from the land, to be paid along with the sum of the rent and tributes that the indigenous peoples may have paid to their masters. The same would apply to a portion of land in the Cuyoacan district that I had ceded to Bernardino del Castillo, my servant, where he built a sugar mill, were that land to be recognized as belonging to one or several other persons.

41. Besides the tribute money collected from my vassals, I have received services from them on many occasions, and they have assisted me with their person and possessions. Since it has not yet been decided if one can legitimately demand these services, I order that after scrupulous investigation, each of those who paid me these services be paid an amount consistent with their value.

IV.326

[. . .]

45. For as long as it may please my cousin, Señora Cicilia Vazquez Altamirano, to live as she now does with the marquise, my wife, or with one of my daughters or my son's wife, she will be free to do so and always enjoy

the same consideration as I have shown her until now: from my possessions, I give her twenty thousand maravedís annually, which are to be paid to her exactly, wherever she may choose to reside.

[...]

49. If María de Torres, a housekeeper presently in the marquise's service, wishes to remain with her or with one of my daughters or my son's wife, she will receive fifteen thousand maravedís annually. If, however, she takes another position, she will be given a lump sum of one hundred thousand maravedís, in compensation for the services she has heretofore performed for us, not excluding the fifteen thousand maravedís that I wish her to receive as long as she remains in service in my household.

[...]

54. To a young woman raised in my household *who is said* to be the daughter of *one* Francisco Barco, who was with me in Tehuantepec, I bequeath the sum of thirty thousand maravedís to assist her in marrying.

[...]

62. Since my son Don Martín Cortés, who must be my heir and successor, will only come of age when he is twenty-five and since he is now only fifteen, it is my wish that he be placed under the surveillance and guardianship of those whom I appoint to that effect, as tutors for all my sons until they have all reached age twenty-five. During this time, my son Don Martín must not leave his mentors' side nor cease to follow their advice, for it is also my wish that his revenue and wealth continually increase until he is of age, in order to provide most expeditiously the means of complying with all the commitments that I contract herein. To that end, for the administration and management of my son Don Martín's possessions, for the guardianship of my legitimate daughters Doña María, Doña Catalina, and Doña Juana, I appoint the most illustrious lords Don Juan Alonzo de Guzmán, duke of Medina, Don Pedro Alvarez Osorio, the marquis de Astorga, and Don Pedro de Arellano, count de Aguillar, and enjoin them to comply with my wish, to assume responsibility for this guardianship, and to care for my children, who are of their blood, bringing them under their protection and fulfilling any obligations they owe to the lords who are their closest relatives as a debt discharged to their family. As a token of my appreciation, and instead of the rights that they would otherwise legitimately receive from the possessions bequeathed to them, I order that for as long as they bear this responsibility, they be given fifty marks of silver annually, which I beg them to accept graciously, for the reasons that I have stated above.—Until he reaches his twentieth year, Don Martín, my son and heir, will receive twelve thousand

IV.327

ducats annually for the upkeep of his household and for his personal use; during that time, the rest of my revenue is intended to fulfill the obligations set forth in this will.—Once my son has reached twenty years of age, he will have full and complete use of my income. Since the towns, villages, sugar mills, mines, and other properties in New Spain that I have leased and that are part of my possessions are far from each other and dispersed in several provinces, and since I am the person most familiar with them, I must delegate those who are the most capable of administering them at the head of each of these properties. Therefore, I ask my son's guardians to approve the appointments that I shall make, which I shall leave in a signed document.— IV.328 This seems to be the appropriate step and will ensure the proper administration of my properties, at the same time, allowing my children's guardians to avoid the burden of appointing themselves as administrators.

63. Finally, I leave Don Martín Cortés, my son, and Doña Juana de Zuñiga, my wife, the successors of my estates and heirs to my household, and after them, the persons mentioned in the institution of my estate, established with the consent of the emperor and King, our master. [. . .]

64. Executed in Seville, on the eleventh day of the month of October in the year of O.L.J.C. [Our Lord Jesus Christ] 1547. [. . .]

Registered in duplicate in the town of Mexico City, the twenty-seventh of January 1771 by myself, DON HERNÁN CORTÉS, assisted by the witnesses DON JOSÉ CALDERÓN, DON Ignacio Sigüenza, DON JOSÉ SANCHEZ, *residents of this town.*

Authentically certified, IGNACIO MIGUEL DE GODOY, royal notary.

Geographical and Physical Atlas of the Kingdom of New Spain

The Geographical and Physical Atlas of New Spain, based on astronomical observations, trigonometrical measurements, and barometrical leveling by ALEXANDER VON HUMBOLDT, 20 folio plates on vellum paper.

As correct as it is magnificently carried out, this Atlas contains the most precise and most exact information about the entire region of the Americas known until today as New Spain.[1] Compiled for use with the *Political Essay*, it can either be bought separately or together with the four volumes of this new edition, revised and enlarged by the author. The four volumes can be purchased with or without the Atlas.

1a. General Map of Kingdom of New Spain from the sixteenth to the thirty-eighth parallel (northern latitude), based on astronomical observations and on the collection of materials in Mexico City at the beginning of the year 1804. By Alexander von Humboldt. Drawn in 1804 Mexico City by the author; perfected by the author himself, Mr. Friesen, Mr. Oltmanns, and Mr. Thuilier in 1809. Engraved by Barrière, with script by L. Aubert Sr. [16.5 × 48 cm]

1b. Contuination of the General Map, southward from the thirtieth parallel.

2. Map of Mexico and border countries situated to the north and the east. Based on the Great Map of New Spain by Mr. Humboldt and

1. See the Reasoned Analysis of the Atlas, or Geographical Introduction, at the beginning of this edition. [This note appeared at the beginning of vol. 4.]

other materials by J. B. Poirson, 1811. Engraved by Barrière, with script by L. Aubert Sr. Can be found in Paris at F. Schoell.

3. Map of the Valley of Mexico and the neighboring mountains, sketched on site in 1804 by Mr. Louis Martin, revised and corrected by Jabbo Oltmanns in 1807, based on the trigonometric measurements by Mr. Joaquín Velásquez and on the astronomical and barometrical observations by Mr. de Humboldt. Designed by G. Grossmann, completed by F. Friesen in Berlin, 1807, and by A. Humboldt in Paris, 1808. Engraved by Barrière, with script by L. Aubert Sr.
4. Watersheds and communications projected between the Great Ocean and the Atlantic Ocean: I. Paix River and Tacoutché Tessé. II. Río del Norte and Río Colorado. III. Río Huallaga and Río Huanuco. IV. Gulf of St. Georges and the Aysen Estuary. V. Río de Huasacualco and Río Chimalapa. VI. Lake Nicaragua. VII. Isthmus of Panama. VIII. Raspadura Ravine and Embarcadero de Naipi. Designed by J. B. Poirson. Engraved by Barrière, with script by L. Aubert Sr.
5. Condensed map of the route from Acapulco to Mexico City, based on astronomical observations and barometric leveling by A. de Humboldt. Designed by A. de Humboldt in Berlin, 1807. Engraved by Barrière, with script by L. Aubert Sr.

6.–8. Map of the route that leads from the Capital of New Spain to Santa Fe of New Mexico, based on journals of Don Pedro de Rivera and in part on astronomical observations by Mr. de Humboldt: Route from Mexico to Durango. Route from Durango to Chihuahua. Route from Chihuahua to Santa Fe. Designed and drafted by F. Friesen in Berlin, 1807. Engraved by Barrière, with script by L. Aubert Sr, the director of said work.

9. Condensed Map of the eastern part of New Spain, from the plateau of Mexico City to the port of Veracruz. Based on the geodesic work by Mr. Miguel Costanzo and Mr. García Conde, following the astronomical observations and the barometric leveling by Mr. de Humboldt. Designed after a sketch by Mr. de Humboldt by F. Friesen in Berlin, 1807. Engraved by Barrière, with script by L. Aubert Sr, the director of said work.
10. Map of the faulty positions of Mexico City, Acapulco, Veracruz, and the Orizaba. Designed by A. de Humboldt in Mexico City, 1804. Engraved by Barrière, with script by L. Aubert Sr.

11. Map of the port of Veracruz, sketched by Mr. Bernardo de Orta, ship's captain in the service of His Catholic Majesty. F. Bauza made it in Madrid (copied and reduced by half by F. Wittich, 1807), after the map published by the Hydrographic Office in Madrid. Engraved by Barrière, with script by L. Aubert Sr, director.
12. Physical tableau of the eastern slope of the plateau of New Spain (route from Mexico to Veracruz via Puebla and Jalapa.) Based on barometric and trigonometric measurements taken in 1804 by Mr. de Humboldt. Designed by A. de Humboldt in Veracruz, 1804. Completed by Wittich and Friesen, 1807. Engraved by Bouquet. Scale and script by Aubert.
13. Physical tableau of the western slope of the Plateau of New Spain (route from Mexico City to Acapulco). Based on barometric measurements taken in 1803 by Mr. de Humboldt. Designed by Wittich after a sketch by Mr. de Humboldt, 1807. Engraved by Bouquet. Scale and script by L. Aubert.
14. Tableau of the central plateau of the mountains of Mexico, between 19° and 21° northern latitude (route from Mexico City to Guanajuato). Based on barometric leveling by Mr. de Humboldt. Sketched by Alex. de Humboldt in Mexico City, 1803. Designed by Raphael Dávalos in Mexico City, 1804 (completed in Berlin, 1807). Engraved by Bouquet. Scale and script by L. Aubert.
15. Profile of the canal of Huehuetoca (Desague Real). Dug to prevent the city of Mexico from flooding. Drawn by F. Friesen in 1808 after plans by Mr. Ignacio Castera and Mr. Louis Martin. Engraved by Bouquet. Script by L. Aubert Sr.
16. Volcanoes of Puebla as seen from Mexico City. Lud. Martin ad nat. del. 1803. Fr. Gmelin perf. Romae 1805. Fr. Arnold sc. Berol. 1807. Printed by Langlois.
17. Peak of Orizaba as seen from the forest of Jalapa. A. de Humboldt ad. nat. prim. del. 1804. Fr. Gmelin perf. Romae 1803. Fr. Arnold sc. Berol. 1807. Printed by Langlois.
18. Map of the port of Acapulco. Drawn by officers of the Royal Navy of His Catholic Majesty, on board the corvettes Descubierta and Atrevida, year 1791. Designed in Madrid by the Hydrographic Office. Engraved by Barrière. Script by L. Aubert.
19. I. Map of the diverse routes on which metal wealth flows from one continent to another. Vol. II. p. 660. Designed by J. B. Poirson after

a sketch by Mr. de Humboldt. Engraved by L. Aubert. II. Output of the mines of the Americas since the discovery. Vol. II. p. 652. III. Amounts of gold and silver extracted from Mexican mines. Vol. II. p. 578. IV. Proportion in which different parts of the Americas produce gold and silver. Vol. II. p. 633. V. Proportion in which different parts of the world produce silver. Vol. II. p. 634.

20. I. Comparative tableau of the territorial extent of the intendancies of New Spain. II. Territory and population of the metropoles and the colonies in 1804.

Indexes

While we have kept all of Alexander von Humboldt's entries and subentries intact, we chose to create three indexes out of his original one for ease of use. Humboldt's own entries appear in lightface, and his subentries have not been alphabetized to conform to today's indexing standards. Where we have added relevant information to his entries, such as dates and cross-references, those additions appear in brackets. New entries we have provided appear in boldface. Page references to volume II of this edition are preceded by "II." Tables (or tableaus) and footnotes are denoted by a "t" or "n," respectively, following the page number.

Index of Names

Abad [y Queipo, Manuel Ignacio de, 1751–1825], vicar-general of the bishopric of Michoacán: information he provided to the author, 67, 283n, 358, 537; his enthusiasm for introducing inoculation, 222; his observations on the Colima volcano, 398.

Abincopa, Gonzalo: discovered the mercury mine of Huancavelica, II.184.

Acerbi, Giuseppe (1773–1846), 511n.

Acipuene, Francisco, II.471t.

Acosta, [José de, 1540–1600]: [327, 529n, 557, 561n; II: 34, 49n, 144n, 209n, 222, 223, 225, 295]; his remarks on the plants of Mexico, 514; how high this author had the fifth [tax] raised, paid to the king from the output of the Potosí mines, II.220–21.

Acosta, [Tomás] Joaquín [Pérez de Guzmán, 1800–1852]: information on platinum he provided for the aut[hor], II.87n.

Acuña [y Bejarano], Juan [Vázquez] de, Marquis of Casa Fuerte [1658–1734]: the only viceroy of Mexico to be born in America, 352n, [II.165n].

Adams, George, Jr. (1750–1795), 60, 276.

Adanson, Michel (1727–1806), 384, 513.

Aguilar Galeote, Martín de (d. 1603), 447, 458.

Aguirre [y Viana, Guillermo de, d. 1811]: member of the audiencia of Mexico City; gave the author La Peña and Crespi's unpublished journal; Pérez's traveling companion, 459.

Ahuizotl, king of Mexico [r. 1486–1502]: built the large teocalli of Tenochititlan, 325; his carelessness caused a flood, 325, 357.

Ainslie, Whitelaw (1767–1837), 520n, 526n.

Alamán, [Lucas, 1792–1853], [161n, II: 112n, 121]; his remarks on population, 476, 481; his opinion of a general tunnel, II.110n.

Álava y Sáenz de Navarrete, Ignacio María (1750–1817), II.441.

Alberni, Pedro de (1747–1802), 436.

Alcalá Galiano, Dionísio. *See* Galiano, Dionísio.

Alcedo y Bexarano, Antonio de (1735–1812), 303.

Alexandro [Guerrero], José [María, fl. 1764–1803]: his work on Lake Nicaragua, 499.

Allen, John (1771–1843), II.266.

Al-Ma'mūn (786–833), 245n.

Almanza, Martín Enríquez de [d. 1583], viceroy of Mexico, [244], 256.

Alonzo Barba, Alvaro (1569–1662), II.145n, 150–51, 206, 207n, 208n, 221, 223n, 225n.

Altamirano de Velasco, Juan Manuel (1733–1793), II.503.

Alva Cortés Ixtlilxochitl, Fernando de (c. 1580–1650), xxiv, 268, 342n.

Álvar Pérez Osorio (c. 1413–1471), II.501.

Alvarado Tezozomoc, Fernando de (c. 1525–c. 1610), xxiv, 342n.

Alvarado [y Contreras], Pedro de [c. 1485–1541]: celebrated leap he took to save his life, 342.

Álvarez Barreiro, Francisco (fl. 1716–1729), 54.

Álvarez Cabral, Pedro [c. 1467–c. 1520]: landed in the Americas, 557.

Álvarez [de Toledo], Juan: ideas he conveyed to Captain Cochrane, 167n; his project for draining the valley of Mexico City, 363; obtains permission to extract gold from the Granada volcano, II.301.

Álvarez Ordoño y Rebin, Baltasar (b. 1768), 35, 115.

Álvarez Osorio, Pedro, second marquis of Astorga (d. 1461), II.501.

Álvarez y Jiménez, Antonio (d. c. 1813), II.203n.

Alzate [y Ramírez], José Antonio [de, 1737–1799]: how he determined the position of Mexico City, 31–32; that of Veracruz, 33–34; [that of San José, 46]; his maps of the archdiocese of Mexico City, [68n], 61–61, 76–77, [90], his map of the surroundings of Mexico City, 77; he sets the position of Picacho, 134; praise of this scholar, 277; his opinion of the elevation of Cuernavaca, 380n, [279, 341n, 344n, 347, 466n; II: 46, 48, 300, 390n, 402n].

Amadas, Philip (c. 1565–c. 1618), 553.

Ameller y Clot, Carlos Francisco (1753–1836), II.407n.

Andréossy, Antoine François (1761–1828), 328, 356, 360, 370n, 378.

Anghiera, Petrus (or Peter) Martyr d' (1457–1526), 508.

Angot des Rotours, Noël-François-Mathieu (1739–1821), II.250n.

Anson, George (1697–1762), II: 251n, 356n.

Antillón [y Marzo], Isidro de [1778–1814]: how he determined the longitude of Mexico City, 29, [57–58]; the one of Veracruz, 33; of Acapulco, 36, 59; of Santa Fe, 52, [89]; [27, 43, 47, 49n, 51–53, 70n, 90, 116].

Anville, Jean Baptiste Bourguignon d' (1697–1782), 17, 27, 32, 34, 37, 86.

Anza [José], Vicente de [fl. 1780s, d. c. 1814]: information he provided, 76; built a large drainage tunnel at Tasco II.81, [II.132].

Apartado, Marquis del. *See* Fagoaga y Villaurrutia, José Francisco.

Arago, Dominique François Jean (1786–1853), 41, 502.

Arcet, Jean d' (1724–1801). *See* Darcet, Jean.

Arciniega, [Claudio, fl. 1580]: his project to protect Mexico City from floods, 358–59.

Arcos, Marquis of (1736–1804). *See* Peñalver y Cárdenas, José Ignacio.

Aréjula, Juan Manuel de (1755–1830), II: 391n, 405, 407n, 415n, 417n.

Arias, Alonzo de: superintendent of the arsenal, responsible for the drainage canal, 361; his opposition to Martínez's project, 372.

Arias, Juan de, II.417.

Arireta, Juan Bautista: owner of the Talenga factory, II.207n.

Aristobulus of Cassandreia (c. 375–301 BCE), 559n.

Ariztizábal, Gabriel de (1743–1805), II.441.
Arnold, Johann Friedrich (1780–1809), 116n, 467,
Arnold, John (1737–1799), 114; II.509.
Arnould, Ambrose Marie (1750–1812), 348, 541; II: 250n, 257, 259n.
Arósbide, Josef: traveled between Manilla and Lima on a direct route, II.259–60.
Arriaga, Antonio de (d. 1780), 262.
Arricivita, [Juan] Domingo [1720–1794]: his report on the College of the Propaganda [Fide] in Querétaro, [50n], 431n.
Arrieta, Juan Bautista: owner of the Talenga manufacture, II.207n.
Arrowsmith, [Aaron, 1750–1823]: how he determined the position of Mexico City, 30; of Veracruz, 34; of Acapulco, 37; his error about the Orizaba volcano, 41–43, 113, [53, 64–66, 68n, 72, 74, 154].
Arteaga [y Bazán], Ignacio [de, 1731–1783]: his expedition to North America, 460.
Asanza, Miguel of, Knight: collected manuscripts about travels in California, 95, 328n; accompanied the Visitador [envoy] Gálvez on his travels in California, 261, 441–42; was arrested, 442; was appointed viceroy of Mexico, 442; praise of his administration, II.440.
Ascásubi y Matheu, Manuel de. *See* Maenza, Marquis of.
Asentzio, Manuel: Velázquez's teacher, 277.
Astorpilco family, II.236.
Atahualpa [Atahuallpa] (Peruvian Inca, [c.1500–1533]), [176, 544; II.249–50]; his name was given to the rooster, II.36.
Atienza, Piedro de: planted the first sugarcane in Mexico, II.4.
Aubuisson de Voisins, Jean François d' (1769–1841), II: 90n, 96n, 117n, 134n.
Austria y Achútegui, José Donato (d. 1806), II.368n.
Autenrieth, Johann Heinrich Ferdinand von (1772–1835), II.421n.
Avendaño y Loyola, Andrés de (1695–1705), 407.
Ávila, Pedro Arias de. *See* Dávila, Pedro Arias.
Axayacatl (king of Mexico [d. 1482]), [268, 341, 342n]: destroyed the kingdom of Tlatelolco, 326.
Axcotla: wealthy Indian family in Cholula, 255.
Ayala, Gabriel de, baptized Indian: author of an unpublished work on the history of Mexico, 342n.
Ayala [y de Aranza], Juan [Manuel de, 1745–1597]: his journey to the northwest of America, 460.
Azanza Navarlaz y Alegría, Miguel José de (1746–1826), 47; II.19.
Azara, Felix de (1746–1821), 544, 547.

Bacon, Baron Francis, Viscount Saint Alban (1561–1626), 278.
Balbi, Adrien (1782–1848), II.266n.
Balboa, Vasco Núñez de (1475–1519), 154.
Baldwin, William (1779–1819), 555.
Bally (Bailly), François-Victor (1775–1866), II.247n,
Balmis [y Berenguer, Francisco Javier de, 1753–1819]: introduced vaccination to the Spanish possessions, 222–23.
Banks, Sir Joseph (1743–1820), 551, 553.
Baños, Count of, viceroy of Mexico, 365.
Baranov (Baranoff), Alexander Andreyevich (1746–1819), II.366n.
Barba [Álvaro] Alonzo [1569–1662]: inventor of heat amalgamation, 150–51, [206, 207n, 208n; II: 145n]; by how much this increased the quantity of silver extracted at the Cerro de Potosí, II: 221, [223n, 225n].

Barcena, José, II.471t.
Barco, Father [Miguel del, 1706–1790]: author of a History of California, 445n; [II.504].
Barlow (Barlowe), Arthur (c.1550–1620), 553n.
Barreiro [Quijano], José [1741–1809]: cut in the mountain he ordered to improve the climate of Acapulco, II.395. [*See also* Quijano, José Barreiro.]
Barreto, Isabel (c. 1567–1612), II.261n.
Barrington, [Daines, c. 1728–1800]: published the navigator Mourelle's journal, 459–60.
Barrow, John, Sir (1764–1848), 174n, 341, 474; II: 259, 260n.
Barry, Roger (1752–1813), 20n.
Barton, Benjamin Smith (1766–1815), 186, 233, 451.
Bastide, Martín de la, 151n.
Baudin, Thomas Nicolas (1750–1803), xi.
Bauzá [y Cañas], Felipe [1764–1834]: [21, 41, 57, 115, 501; II: 209n, 509]; his opinion on the position of Santa Fe, 52–53; on the elevation of Lake Nicaragua, 499.
Becerra de Mendoza, Diego (d. 1533), 45.
Beckford, William (1744–99), 555, 560n; II: 10–11.
Beckmann, Johann (1739–1811), 553; II: 29n, 31, 34n, 37n, 144n.
Belalcázar, Sebastián de (c. 1480–1551), II.35n.
Beltrán, Pedro. *See* Santa Rosa María, Pedro Beltrán de.
Beltrán de Guzmán, Núño (c. 1490–1544), 439n.
Bering, Vitus Jonassen (bap. 1681–1741), 458, 470.
Berlangas, Tomás de [c. 1487–1551]: introduced the banana plant to the Americas, 512, 514.
Bernal de Piñadero, Bernardo (fl. 1663–1677), 433; II.49.
Bernardin de Saint-Pierre, Jacques-Henri (1737–1814), 257.
Berrio de Montalvo, Luis (d. 1643), II.244.
Berthe, Jean-Nicolas (1761–1819), II: 394n, 411n, 416n.
Berthier, [Pierre, 1782–1861]: his documents about the Halsbrücke mine in Saxony, II: 58n, [161, 163].
Berthollet, Claude Louis (1748–1822), II: 31n, 88n.
Berthoud, Pierre-Louis (1754–1813), 276n.
Betancourt y Molina, Agustín de (1758–1824), 324n, 379.
Beudant, François Sulpice (1787–1850), II: 77, 106, 269.
Bieberstein, Friedrich August Marschall von (1768–1826), 551n.
Billings, Joseph (c. 1760–1806), 471, 475.
Blane, Gilbert (1749–1834), II: 407, 408n.
Blas de Iñena, II.300.
Bligh, William (1754–1817), 563; II.358.
Blodget, Samuel (1757–1814), 209, 229n, 294n, 545, 560.
Bochica (Idacanzas), 245n.
Bodega y Quadra, Juan Francisco de la [1744–1794]: 45, 149n, 154, 276; how he determined the position of San Lucas, 47; his map of California, 64; his journey to the north-northwestern coast of the Americas, 460, [462, 466–67, 469, 470, 473; II.364–65].
Bonavenura. *See* Suárez Bonaventura.
Bonilla, Antonio [fl. 1772–1784]: his unpublished report on the Spanish voyages to northwest America, [47n], 457.
Bonilla, Gabriel López de: how he determined the longitude of Mexico City, 30.
Bonne, Rigobert (1727–1794), 34, 37.
Bonneville, Pierre-Frédéric (b. 1768), II.235n.

Bonpland, Aimé Goujaud (1773–1858), xi–xiii, 113, 243n, 249, 250n, 323n, 391, 412n, 450, 489, 506n, 550, 555; II: 23, 31n, 40, 361n, 406.

Boot, Adrian [fl. 1614–1634]: in charge of the hydraulic projects of Mexico City; had the desagüe of Nochistongo stopped, 361.

Borda, Jean-Charles (1733–1799), 35, 277.

Borda Sánchez, José de la. *See* Laborde, Joseph de.

Born, Ignaz Edler von (1742–1791), II: 77, 143, 150, 157.

Boturini Benaducci, Lorenzo (1702–1755), 38, 266, 268, 324n, 339.

Bougainville, Louis de (1729–1811), xxv, II.360–61.

Bouguer, Pierre (1698–1758), 51, 155, 189; II: 224n, 300n.

Bourgoing, Jean-François de, Baron (1748–1811), 303, 483; II: 168, 193n, 250n, 368, 381n.

Boussingault [Jean-Baptiste Joseph Dieudonné, 1802–1887]: his discoveries in the Chocó mines, II.228.

Bowen, Emanuel (c. 1693–1767), 77.

Branciforte, Marquis de, viceroy of Mexico: [62; II: 220, 311, 420, 440, 450]; had a statue of Charles IV erected, 331n. *See also* Grúa Talamanca de Carini y Branciforte, Miguel de la.

Briones, Juan Ignacio (fl. 1793–1803), 381n.

Brisseau Mirbel, Charles-François (1776–1854), 547n.

Brochant de Villiers, André Jean François Marie (1772–1840), II.290.

Brongniart, Alexandre (1778–1847), II: 86n, 90n, 168.

Broughton, William Robert (1762–1821), 148, 467.

Brown, John, M.D. (1735–1788), 349; II.416.

Brown, Robert [1773–1858]: his observations about musa, 520; about corn, 409, [526, 556n, 557, 559n; II.29n].

Browne, Patrick (1720–1790), 521; II.23.

Bruce, James (1730–1794), 173n.

Bruun, Malte Conrad (1775–1826). *See* Malte-Brun, Conrad

Buache de la Neuville, Jean-Nicolas (1741–1825), 466.

Bucareli [y Ursúa, also Bucarely], Antonio [María de], viceroy of Mexico [1717–1779]: [68n, 345n; II.440]; had engineers study the terrain between the Huascualco sandbar and the port of Tehuantepec, II.316–17.

Buch, Christian Leopold, Baron von (1774–1853), 173, 520n; II: 107n, 126.

Buffon, Georges Louis Leclerc, Count of (1707–1788), xxiv; II.33

Bullock, [William, c. 1780–1849]: [7, 113n, 338n]; corrects an error, 446.

Burckhardt, Johann Karl (1773–1825), 19, 48.

Bürg, Johann Tobias (1766–1834), 19, 28, 33, 35, 48, 116.

Burgoa, Francisco (c. 1600–1681), II.317.

Bustamante [y Guerra, José Joaquín, 1759–1825]: [II: 193, 491–92]; his barometric observations, II.489, appendix.

Bustamante y Bustillo, José Alejandro (d. 1750), II.127–28.

Caamaño [Moraleja], Jacinto [b. 1759]: his expedition to the Northwest Coast of America, 469.

Caballero y Góngora, Antonio (1723–1796), II.230n.

Cabrera, Manuel [fl. 1652–1691]: superintendent of the desagüe of Huehuetoca, 365, [367; II.127].

Cabrillo, Juan Rodriguez [c. 1500–1543]: his voyage to New California, [45], 447, 455, 457–58; II.361.

Cadamusto (Cadamosto), Alvise (c 1432–1488), 556, 557.
Cadena, Count of: had the first electrical conductors built, II: 298, [294]. [*See also* Flon y Tejada, Manuel de.]
Cadereyta, Marquis of [1575–1640], viceroy of Mexico: assigned the revenue from a tax on alcohol to the account for the desagüe, 364.
Caldcleugh, Alexander (d. 1858), 520n, 555n; II: 208n, 266, 268n.
Calderón, Francisco [1584–1661]: his plan to dry out the valley of Mexico, 363.
Calle, Juan Díaz de [fl. 1645–1648, d. 1662]: his report to Philip IV, 354n; his research on the inventor of amalgamation, II.144.
Camacho [y Brenes], José [d. 1795]: his journey to New California, 70.
Camargo, Diego Muñoz [c. 1529–1599]: his unpublished work, 342–43, [420].
Campomanes, [Count of]: [202, 207n; II.241n], his calculation of gold and silver imports to Spain, II.232.
Campo Marín, Antonio del, II: 166n, 173.
Candolle, Augustin Pyramus (Pyrame) de (1778–1841), 511; II.44.
Cañizares [Rojas], José de [fl. 1769–1796]: his map of California, 64, [460n].
Carvajal [y Sande, Juan de, c. 1590–16567], oídor in Mexico City: his mineral collection, 334n; [II.150n].
Cardanus (Cardan or Cardano), Girolamo (Geronimo) (1501–1576), II.179.
Carey, Mathew (1760–1839), II: 406, 408n.
Carlos IV, King of Spain (1748–1819), xi, xv, xvii, xxv.
Carteret, Philip (d. 1796), II.361.
Casa Fuerte, Marquis of. *See* Acuña y Bejarano, Juan Vázquez de.
Casas, Bartolomé de las (1484–1566), 397.
Casas, Francisco de las (1461–1536), II.503.
Casasola: his recollections of travels in California, 47n; his manuscript of his recollections of the Spaniards' journeys to the northwest of America, 457.
Cassini de Thury, César-François (Cassini III, 1714–1784), 24, 31–32, 46, 95, 146; II.311n.
Castera, Ignacio de (1777–1811), 108, 354, 367, 370, 377; II.509.
Castilla, Marshal of, II.471t.
Castillo, Bernardino del, II: 5n, 503.
Castillo, Cristóbal del [1526–1606]: baptized Indian, author of unpublished work on the history of Mexico, 342n.
Castillo, Domingo de (fl. 1540s), 32, 37, 47n, 50, 70, 440.
Catherine II of Russia (1729–1796), 475.
Cathral (Cathrall), Isaac (d. 1819), II.407.
Catineau Laroche, Pierre-Marie-Sébastien (1772–1828), 518.
Caulín, Antonio (1719–1802), 303n.
Cavanilles, Antonio Josef (1745–1804), 142n; II.88n.
Cavelier, René-Robert (1643–1687). *See* Sieur de la Salle.
Cavero y Cárdena, Ignacio (c. 1756–1834), II.87n.
Ceán Bermúdez, Juan Agustín (1749–1829), 457n.
Cebrián y Agustín, Pedro, Count of Fuenclara (1687–1752), 202.
Cepeda. *See* Zepeda, Fernando de.
Cerda Sandoval Silva y Mendoza, Gaspar de la, Count of Galve (1653–1697): 353; II.453.
Cerralvo, Marquis of, viceroy of Mexico, 356, [364]. *See also* Pacheco y Osorio, Rodrigo.

Cervantes, Miguel de [sic: Vicente, 1758–1829]: professor of botany in Mexico, [xiv, 250, 462n], 506; II.304; his mineralogical collection, [276], 334n.
Cevallos [y de Bustillo], Ciriaco [1764–c. 1816]: [33, 35, 124t, 457n, 462, 467]; explorer of the coasts of Nuevo Santander, [21, 59], 192.
Chabrol de Volvic, Gilbert Joseph Gaspard (1773–1843), 541; II: 276, 278–79.
Chacón, José María (1749–1833), II.405.
Chappe [d'Auteroche, Jean, 1722–1769]: [101; his determination of the position of Mexico City, 31–33, [96]; of Veracruz, 34 [146]; his voyage in California, [34n, 45–46, 48, 62, 95, 101, 126t], 278, [330, 445; II: 311, 320, 390n].
Chaptal, Jean Antoine Claude, Count of Chanteloup (1756–1832), II.9n.
Charles I, [King of Spain and, as Charles V, Holy Roman Emperor, 1500–1558]: encouraged Cortés to find the *secret of a passage* from America to Asia, 150; supported the cultivation of hemp and flax, II.19–20. *See also* Charles V.
Charles II (1661–1700), II: 440, 453.
Charles III, [King of Naples and Sicily and] King of Spain [1716–1788]: [275, 352; II: 134, 368–69, 452]; improved the lot of the Indians, 254; supported the cultivation of hemp, II.19.
Charles IV, [King of Spain, 1748–1819]: popularized the benefits of vaccination among indigenous peoples, 222ff; his statue in Mexico City, 275, [331, 335, 439; II.180].
Charles V, King of Spain, 112, 142–43, 256, 343, 562n; II: 38, 128, 134–35, 191, 238, 361; Cortés's letters to him: 200, 250, 283, 286n, 321–22, 333, 335, 343, 387n, 439, 530, 533n; II: 22, 49, 65, 69n, 249, 299, 315, 320, 497.
Charpentier, Johann Friedrich Wilhelm Toussaint von (1738–1805), II: 119, 150, 157.
Chassebœuf, Constantin François de, Count of Volney (1757–1820). *See* Volney, Count of.
Chastenet, Antoine Hyacinth Anne de, Count of Puységur (1752–1809), 35, 73.
Chiapa, Bishop of. *See* Casas, Bartolomé de las.
Chimalpahin [Quauhtlehuanitzin, Domingo Francisco de San Antón Muñón, b. 1579]: baptized Indian and author of an unpublished history of Mexico, [xxiv], 352n.
Chirikov, Alexei Iljich (1703–1748), 458, 470.
Chisholm, Colin (c. 1747–1825), II: 393n, 417n.
Chladni, [Ernst Florens Friedrich, 1754–1827]: his report on meteoric iron in Mexico, II.179n.
Chovell, [José Casimiro, 1775–1810]: information he provided the author, 67; his work on the mercury mines of Mexico, II.180.
Churchill, Awnsham (d. 1728), 340n.
Churruca y Elorza, Cosme Damián (1761–1805), 22, 35.
Cia, José: information with which he provided the author, II.19.
Cieza de León, Pedro [c. 1520–1554]: his account of the riches of Potosí, [551n]; II: [35n, 209], 217, [218–19], 220, [221, 392n].
Ciscar y Ciscar, Gabriel de (1759–1829) 16, 116, 354n, 499.
Cisneros, Diego (c. 1740–1812), 201, 344n.
Ciudad Real, Antonio de (1551–1617), 407.

Cladera, Cristóbal (1760–1816), II.366n.
Clairaut, Alexis Claude (1713–1765), 244.
Claret, Charles Pierre. *See* Fleurieu, Charles Pierre Claret, Count of.
Clavijero [Echegaray, Francisco Javier, 1731–1787], abbot, [12n, 235; II.45]: author of a history of Mexico, 142, [200n, 201n, 267, 323, 324n, 326n, 333n, 342n, 344, 347, 391, 430n, 452, 527n, 563; II: 19, 22n, 28, 29n, 31n, 33n, 64n, 249n, 299n, 390n]; his map of Lake of Tezcuco, II.322.
Cleghorn, George (1716–1789), II.393.
Clerc, [Pierre-Antoine, 1770–1843]: engineer-geographer in Paris, 98n.
Clusius, Carolus (Charles de l'Écluse, 1526–1609), 557n, [558].
Cochrane, Charles Stuart (fl. 1822–1825), 167n.
Cocquebert de Montbret, Charles-Étienne (1755–1831), II.165n.
Colebrooke, Henry Thomas (1765–1837), II.68n.
Colmenares, José Ignacio de (1761–1833), 154.
Colnett, James [1753–1806]: his voyage, 71; was arrested in Nootka, 464.
Colón [de Portugal y Castro], Pedro Nuño [d. 1673], Duke of Veraguas: descendant of Christopher Columbus, viceroy of Mexico, 352n.
Columbus, Christopher (1451–1506), xxiii, 320n, 335, 353n, 513, 553, 557–58,
Columbus, Ferdinand (1488–1539), II.31.
Columella, Lucius Iunius Moderatus (fl. 50 CE), 532n.
Comoto, Florencio Pérez y: director of the hospital of the consulado in Veracruz, II: [400, 406, 412n], 414, [416, 418].
Condamine, Charles-Marie de la (1701–1774), 18, 155, 189; II: 224n, 300.
Conde, Diego García [1760–1825]: his trigonometric work on a part of New Spain, 62–63, [93]; was in charge of the construction of a road from Mexico City to Veracruz, II.310; [312, 508].
Condorcanqui: family claiming to be descended from the Inca, 262.
Condorcanqui, Andrés: José-Gabriel's nephew, 262,
Condorcanqui, Diego [Cristóbal, d. 1783]: José-Gabriel's brother; his cruelty, 262–63; his submittal, 263.
Condorcanqui, José-Gabriel [also Tupac Amaru II, c. 1742–1781]: so-called Inca of Peru, 262; the uprising he incited, 262; his ordeal, 262.
Consag, Fernando (1703–1759), 442.
Cook, James [1728–1779]: xxv, 45, 48, 49n, 201, 451, 558; II: 51, 360–61, 365]; not the first navigator to have entered the harbor of Nootka, 459, [460–66, 469, 470, 473].
Cornejo, Juan, II.156.
Coronado, Seb[astián, fl. 1772–1779]: discovered the mines of Catorce, 423; II.125.
Corral, Miguel de [fl. 1746–1794]: engineer; his map of the Rio Huasacualco, 62–63, 82; his research on creating a connection between the two oceans, II.317.
Correia da Serra, José Francisco (1750–1823), 559; II: 223, 266, 270.
Corro Segarro, Juan de [fl. 1670s]: inventor of an amalgamation process, II: 150, [206].
Corso de Leca, Carlos, II: 150, 206.
Corte Real, Gaspar (c. 1450–1501), 458.
Cortés [de Monroy y Pizarro], Hernán [1485–1547]: [xxiii, 9, 38, 47, 51n, 75, 165, 150, 178, 249–50, 309, 319, 324–29, 357, 380n, 387, 414, 416, 439, 454, 513, 532–33, 566; II: 28, 37–38, 49, 390; letters of, 32, 111n, 200, 244; family of, 387; II.32; family archive of, 50, 70; on slavery, 287; his retreat, 320; his residence, 341]; explored the coasts of California,

45; what he called the capital of Mexico, 142–43; title Charles V advised him to take, 142–43; his arrival in Mexico, [199, 247], 268, [446]; title he gave himself, 281; information he gave the emperor about wealth of the clergy, 283; his testament, 286, [379]; II: [5], 495–505; his remorse, 287; what one called him in New Spain, 320n; his description of the valley of Mexico, 321, [322, 333, 561, 563; about gold and silver, II: 22, 64, 69, 238, 244, 249; and mining, II: 81, 131]; his burial monument, 335; account he gave Charles V of the destruction of Tenochtitlan, 336–37, [343–44]; his entry into Tenochtitlan, 341, 530; [and the noche triste, 342]; his exploits led him to make discoveries in the South Sea, 440–41; his voyage to California, 439, [443]; his description of Popocatépetl, II.299–300; his research on a connection between the two oceans, 439–40; II: 315–16, [320, 363]; his voyage in the Pacific Ocean, II.316.

Cortina, Joaquín, II.471t.

Corvera, José Francisco Ibañez de, II.245n.

Costanzó, Miguel [Costansó, 1741–1814]: [29, 40, 416, 442; II; 298, 402]; determined the correct latitudes of Caps San Lucas and Saint-Rose, 46; the position of Santa Fe, 53; his maps of New Spain, 60–63, [67, 89, 93, 320; II.508]; the journal of his voyage in California was confiscated, 442n, [453–54]; his fate, 448, 275.

Cotte, Louis (1740–1815), II.402.

Coulter, Thomas (1793–1843), II.489.

Covens, Jean [Johannes, 1697–1794]: how he determined the position of Veracruz, 43; of Acapulco, 37; [of Mexico City, 31–32].

Coxe, William (1774–1828), II.365.

Cramer, Agustín [d. 1780]: [82]; his map of the Río Huasacualco, 63; his research on a possible connection between the two oceans, II: 317–18, [492].

Crawfurd, John [1783–1868]: [520n; II: 29, 287n]; spread the news of gold production in the archipelago of the Indies, [526n, 560n]; II.271.

Crespi, Juan [1721–1782]: Juan Pérez's travel companion; his unpublished journal, 459n, [460].

Croix, [Carlos Francisco], Marquis of, viceroy of Mexico [1699–1786]: [xix, 120, 61, 261, 447; II: 418, 440]; organized a group of businessmen in Mexico City to finish the work on the desagüe, 366.

Cruz Cano y Olmedilla, Juan de la (1734–1790), 83, 151n, 303n.

Cuervo, José Tienda de [d. 1763]: map of Sonora dedicated to him. 63.

Cuitlahuatzin, King of Mexico, 268, 343n.

Cuvier, [Georges, Baron of, 1769–1832]: [243, 453]; his opinion on the nature of the axolotl, 323, [II.395].

Dalrymple, Alexander (1737–1808), 68n; II.362n.

Damoreau, Estienne, II.216n.

Dampier, William (1651–1715), 152, 155.

Dandolo. *See* Vincenzo, Count Dandolo.

Darcet (D'Arcet), Jean (1724–1801), II.67.

Dávalos, Rafael [b. c. 1782–1810]: worked with the author on drawings of geological profiles 107; [II.509].

Dávila, Pedro Arias (Pedrarias, c. 1468–1530), II.24. *See also* Ávila, Pedro Arias de.

Dávila [or De Ávila Mesura], Damián [fl. 1505–1514]: worked with Enrique Martínez on desagüe of Huehuetoca, 359.

Davy, Sir Humphry (1778–1829), II.152.

Delambre, Jean Baptiste Joseph (1749–1822), 19, 28, 34, 114, 116.

Delareff, Yefstrat Ivanitsch (Evstratii Ivanovich Delarov, c.1740–1806), 473.
Delessert, Jules Paul Benjamin (1773–1847), II.14.
Delisle, Joseph-Nicolas (1688–1768), 50.
Delius, Christoph Traugott (1728–1779), II: 130n, 152n.
Deluc (de Luc), Jean André (1727–1818), 377.
Démeunier, Jean Nicolas (1751–1814), II: 255, 368n, 381n.
Denon, Dominique Vivant (1747–1825), 383, 404.
Desfontaines, René-Louiche (1750–1833), 513n, II.28n.
Desvaux, Auguste Nicaise (1784–1856), 520n.
Díaz, Juan [1736–81], Father: his ascertainment of the junction of the Colorado and Gila rivers, 50, [130t].
Díaz del Castillo, Bernal [1492–1585]: [200, 244, 324, 407, 443; II: 249, 299]; his opinion about Cortés's conduct, 337n, [342].
Díez de Aux y Armendáriz, Lópe, Marquis of Cadereyta (Cadreita, 1575–1640), 364.
Díez de la Calle, Juan (fl. 1645–1648, d. 1662), 354; II.144.
Diodorus Siculus (first century BCE), 157n, 403, 531n.
Dioscorides, Pedanius (c. 40–90 BCE), 532n; II.31.
Dixon, George (1776–1791), 232n, 462, 466; II.365n.
Dobrizhoffer, Martin (1717–1791), 239n.
Donnet, Alexis (1818–1867), 99n.
Don Tadeo [Ortiz]: plumbed the Huasacualco, 492. [*See also* Ortiz y Ayala, Simón Tadeo.]
Douwes, Cornelius (1712–1773), 38, 44, 52.
Doz [y Funes], Vincente [c. 1734–1781]: his voyage in California, [33], 45–46, [126t], 278.
Drake, Sir Francis [c. 1540–1596]: he was not the first to discover New California, 447; he only got as far as Cape Greenville, 458, [460n, 553, 554n, 558].
Dryander, Jonas Carlsson (1748–1810), 553.
Duchêne (Duchesne), Antoine-Nicolas (1747–1827), II.43.
Du Halde, Jean Baptiste (1674–1743), 174.
Duméril, André Marie Constant (1774–1860), 323n; II: 407n, 415n.
Dunal, Michel Félix (1789–1856), 555n.
Duncan, Charles (fl. 1786–1792), 462.
Dundas, Henry, Lord Melville (1742–1811), II.259.
Dupain-Triel, Jean-Louis (1722–1805), 100.
Dupé [Dupaix], Guillaume de [1750–1819]: his research on the pyramid of Papantla, [276n], 412.
Dupont de Nemours, Pierre Samuel (1739–1817), II.267.
Dupuis, Charles François (1742–1809), 265n.
Dupuytren, Guillaume, Baron (1777–1835), II.422.

Ebel, Johann Gottfried (1764–1840), 110; II.104.
Ebeling, Christoph Daniel (1741–1817), 74.
Echeverría [y Godoy, Atanasio, d. c. 1811]: painter in Mexico City, 276.
Elhuyar [y Zubice, also Delhuyar], Fausto de [1755–1833], director of the royal School of Mines in Mexico City [xiv, xxi]; his materials on the locations of mines in Mexico, 15, 67; [II: 62, 64, 69, 94, 165, 166n, 169n, 177n, 192n, 193n, 231]; his merits, 274; sent to the author samples from the mass of an aerolite, 274; his project of a gallery in the mine of Viscaína, II.128–29.

Eliza [y Reventa], Francisco [de, 1759–25]: his expedition to Nootka, 464.
Ellicott, Andrew (1754–1820), 72; II.415.
Emparán [y Orbe], Vicente [1747–1820]: measures he took to improve the climate in Portobelo, II.299.
Enríquez de Almansa (Almanza), Martín (c. 1510–c. 1583), 224, 356; II.166n.
Enríquez de Ribera Manrique, Payo (1622–1684), 365.
Ensenada, Marquis of, II.291. *See also* Somodevilla y Bengoechea, Zenon de.
Erdmann, Johann Friedrich (1778–1846), II.268.
Escalante, [Silvestre Vélez de, c. 1750–1780], Father, [77, 149, 181]: his apostolic sojourns, 446, [468n].
Escalona, Duke of, [and Marquis of Villena]. *See* [López Pacheco, Diego]; Villena, [Marquis of].
Eschwege, [Wilhelm Ludwig], Baron of [1777–1855], [xiv; II: 104n, 234t]: provided the author with information about Brazil, II: 264–68, [270–72].
Escobar, María de: brought the first wheat to Peru, 532.
España, José [María, 1761–1799]: his plan to make the province Venezuela independent, II.455.
Espinosa [y Tello], José [1763–1815]: his astronomical reports, [21, 29, 22, 47n], 70, [71n, 116, 128t, 440t, 447t, 449n, 452n, 454n, 458n, 459, 462, 467, 528n; II: 360n, 364n].
Estala (Estalla), Pedro (1757–1815), 303; II: 256n, 451n.
Eyriès, Benoît Jean Baptiste (1767–1846), 457n.

Fabre de l'Aude, Jean-Claude (1755–1832), II.30.
Fabricius, Johann Albert (1668–1736), II.39.
Fagoaga, José Juan, II.471t.
Fagoaga [Liyzaur], José María [1764–1837]: [14]; map of the valley of Mexico he drew, 77–78; wealth of the Fagoaga family, 280–82, [562; II: 304, 462].
Fagoaga y Villaurrutia, José Francisco, Marquis del Apartado (d. after 1836), II.124.
Faujas de Saint-Fond, Barthélemi (1741–1819), II.185.
Fer, Nicolas de (1646–1720), 61.
Ferán, Tomás, II.471.
Fernández, Juan (1536–1604), II: 355, 361n.
Fernández de Córdoba Zayas, Francisco, Marquis of de San Román (1756–1818), II.302.
Fernández de Navarrete, Pedro (1564–1632), II: 239–40, 243t.
Fernández de Oviedo y Valdés, Gonzalo (1478–1557), 512–13, 521–22, 524–25, 529, 557, 564n, II: 4, 37.
Fernández de Quirós, Pedro (c. 1563–1615), 439, 563, 565; II: 361–62, 420.
Fernández de Velasco, Pedro (fl. sixteenth century), II: 144, 184, 225.
Ferreira de Rosa, Joao (fl. 1680–1695), II: 306, 393n.
Ferrelo, Bartolomé [fl. 1542–1543]: continued Cabrillo's voyage, 457.
Ferrer Maldonado, Lorenzo (c. 1550–1625), 457, 466.
Ferrer [y Cafranga], José Joaquín [1763–1818]: [21, 24]; how he determined the position of Veracruz, 33, [34–35]; of the Cofre de Perote, 40; of Orizaba Peak, 41; [43–45, 52–53, 59, 72, 73n, 94, 114, 123t, 124t, 125t, 126t, 130t, 183n, 191].
Feyjóo [Feijoo, de Sosa, Miguel, 1718–1791]: errors he made in estimating the population of Peru, 201.
Ffirth, Stubbins (1784–1820), II: 391n, 406n, 421, 422n.

Fidalgo, Joaquín Francisco (1758–1820), 22, 82–83, 154.

Fidalgo [y Lopegarcía], Salvador [1756–1803]: his expedition to North America, 464, [465].

Fidler, Peter (1769–1822), 147, 148n, 181, 232.

Fiedler, G., II.407.

Figueroa y Acuña, José Agustín Pardo de, Marquis of Valleumbroso (c. 1695–1747), 232.

Fleurieu, Charles Pierre Claret, Count of (1738–1810), 27, 167, 232, 458n, 459, 460n; II: 360n, 361n.

Flon y Tejada, Manuel de, Count de la Cadena (d. 1811), 386.

Flores, Luis [1637–1653], Father: in charge of the desagüe, 365; [II.125].

Flores [Maldonado, Manuel] Antonio [c. 1722–1799]: [47n]; member of the Vizcaíno's voyage, 458.

Floridablanca, Count of: established posts throughout Spanish America, 140, [202. *See also* Moñino y Redondo, José.]

Font, Pedro [1738–1781]: his determination of the junction of the Colorado and Gila rivers, 50–51; his map of California, 64, [130t, 161, 205, 212]; his overland travel from la Pimeria Alta to Monterrey, 430–32.

Fonte, Bartolomé [de]: his apocryphal voyage to the northeast of America, 457, [469].

Fonte [y Hernández Miravete, Pedro] José de [1777–1839]: [205n]; his opinion on the population of New Spain, 212.

Forbonnais, Francois Véron Duverger de (1722–1800), II: 239n, 252, 254, 261.

Forcada [y la Plaza], Antonio [d. 1818]: his map of New Spain, 62–63.

Forster, Johann Georg Adam (1754–1794), 512–13, 558n, 564n, 565.

Foster, Henry [bap. 1796–1831] and captain Basil Hall: how they determined the position of Acapulco, 36. [*See also* Hall, Basil.]

Fragoso da Motta de Siqueria, Joaquim Pedro (d. 1833), II.149.

Franco, Francisco (c. 1515–after 1569), II.417.

Frank, Johann Peter (1745–1821), II: 394, 416.

Franklin, John (1786–1847), 71–72, 158, 452; II.104n.

Freiesleben, Johann Carl (1774–1846), II: 107n, 126.

Friesen, Friedrich: [54, 87–88]; made maps of the routes in the northern provinces of Mexico, 108; [II.507–9].

Fuca, Juan de [also Apóstolos Valerianos, c. 1531–1602]: his supposed voyage to the northwest of America, [70, 115, 440n, 447, 454n], 457, 467.

Fulton, Robert (1765–1815), 379.

Gaëtan (Gaetano), Juan, 45, 70, 563; II.361.

Gaëte, Duke of. *See* Gaudin, Martin-Michel-Charles.

Gaius Plinius Secundus (23–79 BCE). *See* Pliny.

Galaup, Jean-François de, Count of Lapérouse (1741–c.1788). *See* Lapérouse, Count of.

Gali, Francisco [1539–1591]: [45]; discovered part of the northwestern coast of America, 457–58, [470; II.360–61].

Galiano, Dionísio [Alcalá, 1760–1805]: [21]; how he determined the position of Mexico City, 29–30, [33, 35, 45, 47n, 49n, 240, 499–50, 452n, 457n, 458–459, 464–65]; his expedition to New California, 527; to Nootka, 467–69.

Galisteo, Manuel [fl. 1781]: his levelling work on the coasts of the South Sea, 499; [II.493].

Gall, Franz Joseph (1758–1828), 243.

Gallatin, Abraham Alphonse Albert (1761–1849), xxvi, 229, 546n; II: 17, 55n.

Gallego, Hernando (fl. sixteenth century), II.361n.

Gallesio, Giorgio (1772–1839), 522.

Galve, Count of. *See* Cerda Sandoval Silva y Mendoza, Gaspar de la.

Gálvez [Gallardo y Ortega], Bernardo [Vicente Apolinar de, 1746–1786]: viceroy of Mexico, [322, 352]; accused of having desired independence from Spain, 353.

Gálvez [y Gallardo], José de, [Marquis of Sonora, 1720–1787]: [xix, xxi, 254, 337; II: 29, 165, 167, 176n, 383]; minister of the Indies, 274, [304; II.430]; his voyage to Sonora, [278], 432; in California, [430], 441–43, 446–47; [II.49].

Gálvez y Gallardo, Matías de (1717–1784), 499.

Gama, Antonio de León y [1735–1802]: [21]; how he determined the position of Mexico City, 29, [31, 33, 245n, 267n, 277]; biographical note on this scholar, 279, [342n, 422].

Gamboa, Pedro Sarmiento (c. 1532–1608), 325; II.362.

Gamio, Juan Ignacio: owner of a Peruvian factory using German amalgamation, II.207n.

Gante, Pedro de [1491–1572]: Franciscan monk who believed that he was the biological son of Charles V, 334.

Garcés, Enrique: the method of amalgamation used in the Americas was attributed to him, II: 144, [297].

Garcés, Francisco [Tomás Hermenegildo, 1738–1781]: his map of California, 64; his voyage, 430–32, [437].

Garcés y Eguía, José [d. 1824]: his work on tequesquite salt, 400; [II: 94, 95n, 123n, 144n, 146, 147n, 152, 157, 207n].

García, Gregorio (c. 1560–1627), 267.

García, José (fl. 1760s), 166.

García Conde, Diego (1760–1825), 62–63, 93; II: 310, 312, 508.

García [de la Vera], Pedro: his travels to the cinnabar veins in Guazún, II.183.

García de Tapia, Pedro, 144n.

García Granados, II.32.

García Guerra (1560–1612), 361.

García Hurtado de Mendoza y Manrique, Marquis of Cañete (1535–1609), II.361n.

García Sarmiento de Sotomayor y Luna, Count of Salvatierra (1595–1659), 354; II.144n.

Garcilaso de la Vega, [El Inca, born Gómez Suárez de Figueroa, 1539–1616]: [xxiv]; his remarks on the plants of Mexico, 514, [529n, 532, 551n, 557, 562n, 563; II: 29n, 35–36, 38n, 49n, 144n, 209n, 218n, 219, 222, 239, 249, 314n].

Gardoqui y Arriquibar, Diego María de (1735–1798), 255n.

Garnier, Germain [1754–1821]: [II: 51n, 57n, 257n]; his estimate of the production of the gold and silver mines in Spanish America, II.241.

Gärtner, Joseph (1732–1791), 513.

Gastelbondo, Juan José de: medical doctor who observed the yellow fever in 1729, II.393.

Gatterer, Johann Christoph (1727–1799), 340n, 413.

Gaubil, Antoine (1689–1759), 549.

Gaudin, Martin-Michel-Charles, the First Duke of Gaëte (1756–1841), II.276.

Gay-Lussac, [Louis-Joseph, 1778–1850]: his experiments with amalgamation of silver muriate, II: 152–54, [422n].

Gellert, Christlieb Ehregott (1713–1795), II: 143, 150, 154, 157.

Gelves, Marquis of, Viceroy of Mexico: stopped the canal of Nochistongo, 361. [*See also* Mendoza y Pimentel, Diego Carrillo de.]
Georgi, Johann Gottlieb (Jean-Théophil, 1738–1802), 264n; II.235.
Gerard, John (1545–c.1611), 553–54.
Gerboux, Fr[ançois]: his estimate of the quantity of gold and silver that has flowed back to Europe since 1492, [II: 239n], 242, [243t, 250n, 254n, 155–57, 274n, 279].
Gijón [y León, Miguel de, first Count of Gijón, 1717–1794]: his attempt to establish settlements of European artisans in the province of Quito, II.292.
Gilbert, Ludwig Wilhelm (1769–1824), II.268n.
Gilbert, [Nicolas Pierre, 1751–1814]: his statistical notes on the Yucatán, 405; [II.416n].
Giraud-Soulavie, Jean-Louis (1751–1813), II.396n.
Glenie, [James], lieutenant [1750–1817]: his astronomical observations, II.489.
Gmelin, [Wilhelm] Fr[iedrich, 1745–1821]: famous artist in Rome, 112; [II.509].
Godoy y Álvarez de Faria, Manuel de (1767–1851), 331; II.505.
Goethe, Johann Wolfgang von (1749–1832), xxii.
Gogueneche. *See* Goyeneche, Joaquín de.
Gómara, Francisco López de (1511–1560), 45n, 264, 440, 488, 556, 558; II: 4, 49n, 250, 299, 301n, 316n.
González Carvajal, Ciriaco (fl. 1800–1812), 334n.
González Gutiérrez, Pedro María (1764–1838), II.407n.
Gookin, Daniel (1612–1687), II.391n.
Gordon, Robert Jacob (1743–1795), 174.
Goyeneche, Joaquín de: Biscayan pilot; he was the first one who brought the attention of the Spanish government to the Cúpica bay, 84, [162].
Graham, James Duncan (1799–1865), 56.
Grant, Charles (1746–1823), II.272.
Gray, Robert (1755–1806), 458, 468.
Grijalva, Hernando de [d. 1537]: [45]; discovered Socorro Island in California, 70, 439; II: 316, [361n].
Grijalva, Juan de [c. 1480–1527]: [142, 407]; visited Ulúa Island in 1548, 414.
Grimarest, Pedro: commander-general of the interior provinces, 310.
Grimaud, Jean-Charles-Marguerite-Guillaume de (1750–1789), II.421n.
Grobert, Jacques François Louis (b. 1757), 383–84.
Grúa Talamanca de Carini y Branciforte, Miguel de la (1755–1812). *See* Branciforte, Marquis de.
Guadalajara [y Tello], Diego [de, 1742–1801]: professor of mathematics on Mexico City, author of a table of latitudes, 54, 278.
Gual, Manuel (1759–1800), II.455.
Guatimucin. *See* Quauhtemotzin.
Güemes Pacheco de Padilla, Juan Vicente (1740–1799). *See* Revillagigedo, Count of.
Guerra, García, archbishop of Mexico City, viceroy of New Spain [1560–1612]: assigned the work on the desagüe of Alonzo de Arias, 361.
Guignes, [Joseph] de [1721–1800]: [233, 549; II.356n]; his research on the quantity of silver the English quarried in China, II.258–60.
Gutiérrez [de Toledo, García], baptized Indian: author of unpublished writings on the history of Mexico, 342n.
Guzmán, Núño de. *See* Beltrán de Guzmán, Núño.

Haenke, Thaddeus [Peregrinus Xaverius, 1761–c. 1810]: botanist of the Malaspina expedition, 466.
Haidar 'Ali Khan (1722–1782). *See* Hyder Ali.

Hall, Basil, Captain [1788–1844]: [36f, 130n]; determined the population of San Blas, 48, [116n, 117]. *See [also]* Foster, [Henry].
Hanna, James (d. 1787), II.365.
Haro, Gonzalo López de [d. 1823]: his expedition to Russian settlements in America, 461
Harriot, Thomas (c. 1560–1621), 553–55.
Hassel, Johann Georg Heinrich (1770–1829), II: 255, 268n.
Haüy, René Just (1743–1822), II: 67, 88n.
Hawkesworth, John (1715–1773), II.359n.
Hawkins (Hawkyns), John (1532–1595), 554n, 558.
Hearne, Samuel (1745–1792), 72, 232, 453.
Heceta [y Dudagoitia], Bruno de [1751–1807]: his voyage to the northwest coast of America, 460, [468; II.366].
Heeren, Arnold Hermann Ludwig (1760–1842), 531n.
Hell [Höll], Father [Maximilian, 1720–1792]: [31]; how he determined the position of San José, 46.
Helms, Anton Zacharias (1751–1803), II.209n.
Henry, Maurice (1763–1825), 20.
Henry IV (1553–1610), 244.
Herbert, James Dowling (1791–1833), 182.
Herder, Johann Gottfried von (1744–1803), 340n.
Hernández de Córdova (Córdoba), Francisco (d. 1517), 407.
Hernández de Toledo, Francisco (c. 1515–1582 or 1587), 201, 262, 268, 513, 525n, 527, 529n, 548n, 558n; II: 221, 225.
Herodotus (c. 485–425 BCE), 265, 375n, 531.
Héron de Villefosse, [Antoine-Marie, 1774–1852]: his assessment of the production of the mines in Europe, [118n; II: 150n, 168n, 231, 235, 247n, 254, 266], 269–70.
Herrera, [Manuel Díaz de], [21, 59, 124t]. *See [also]* Cevallos [y de Bustillo, Ciriaco].
Herrera y Tordesillas, Antonio de (1549–1625), 556n, 561; II: 220n, 249n, 250n.
Hervás y Panduro, Lorenzo (1735–1809), 451; II.21.
Hesiod (fl. c. 700 BCE), 264–65.
Hipólito Unanué, José (1755–1833), 232, 288; II: 198n, 200n.
Hodgson, John Anthony (1777–1848), 182n.
Holguín, Garcí (García), 343.
Horbegozo, Juan de, II.490.
Horn, Georg (1620–1670), 233.
Hualca, Diego: discovered the metal vein of Potosí, II.99.
Huari Capca: discovered the mines of Pasco, II.201.
Huddleston, Lawson (1754–1811), 379.
Huehue-Motecuzoma: name of Montezuma I, 320n; embankment he built, 357.
Huescar (r. 1527–1532), II: 36, 250n.
Huichilobos. *See* Huitzilopochtli.
Huitzilopochtli, Mexica deity: [143, 323]; where his temple is located, 321, [334]; the meaning of his name, 323n.
Humboldt, Wilhelm von (1767–1835), 17, 233.
Hurtado de Mendoza, Diego, commander (fl. 1500–1530, d. 1532), 45, 399, 440n.
Hurtado de Mendoza y Manrique, García (1535–1609), II.361.
Hyder Ali, 16. *See also* Haidar 'Ali Khan.

Ibañez de Corvera, José Francisco, 245n.
Idacanzas (Bochica), 245n.
Illiger, Johann Karl (1775–1813), II.40.
Iniesta [Bejarano Durán], Ildefonso [de, 1716–1781]: his estimate of the volume of water in the canal of

Iniesta (cont.)
Huehuetoca at the time of the great floods, 368n, [372n].
Isabella I, Queen of Castile (1451–1504), 253, 256, 558.
Isasbirivil, Mariano [d. c. 1811]: how he determined the position of Veracruz, [24], 33, [35]; of the peak of Orizaba, 41, [154].
Isla, Juan de: collaborator of Enrico Martínez's, 359.
Iturbi, Juan: his expedition in California, 443.
Iturigarray [y Aróstegui], José [Joaquin Vicente de, 1742–1815], viceroy of Mexico, [354n], 372, 376, [578]; II: 116n, 310.
Ixtlilxochitl, [Antonio Pimentel; Fernando Pimentel; and Fernando de Alva Cortés, c. 1580–1650]: baptized Indians, authors of unpublished writings on the history of Mexico, [268], 342n.

Jacquin, Nikolaus Joseph, Freiherr von (1727–1817), 250n, 514n, 522.
James, Edwin (1797–1861), 250n, 514n, 522.
Jancigny, Jean-Baptiste Dubois de (1752–1808), 543.
Jayme (Jaime), Juan, II.360.
Jefferson, Thomas [1743–1826]: [xv, xvii, xxvi]; his determination of the position of Santa Fe, 52; his eulogy, 144, 446n, [451; II: 31n, 33n].
Jenner, Edward (1749–1823), 223.
Jiménez [de Bertadoña], Fortún [d. c. 1534): Grijalva's pilot on his California journey, 439.
João I, King of Portugal (1385–1433), II.267t.
João III, King of Portugal (1502–1557), II.267t.
João IV, King of Portugal (1604–1656), II.267t.
João V, King of Portugal (1689–1750), II.267t.
Jofre de Loaysa (Loaísa), García (1490–1526), II.361.
Jones, William (1746–1794), 263n.
Jove, Pedro, II.471t.
Jovellanos, Gaspar Melchor de (1744–1811), II.59n.
Juan y Santacilia, Jorge (c. 1712–c. 1773), 83, 155, 157, 354.
Juarros y Montúfar, Juan Domingo (1753–1821), 111n, 153n, 180n, 404, 487.
Junker (Juncker), Auguste (1791–1865), II.139n.
Jurine, Louis (1751–1819), II.40.

Kamehameha I (c. 1758–1819), II.362.
Kämpfer, Engelbert (1651–1716), 564.
Karsten, Dietrich Ludwig Gustav (1768–1810), II: 78, 88n, 90.
Keutsch, Johan Mathias Frederik (1775–1815), II.417.
King, Gregory (1648–1712), II.276n.
King, James (1750–1784), 461.
Kino, Father, [50, 445n]. *See also* Kühn, [Father Eusebio].
Kircher, Athanasius (1602–1680), 562n.
Kirwan, Richard [1733–1812]: [II.67n]; what he thought about the elevation of Santa Barbara, II.184n.
Klaproth, [Julius Heinrich von, 1783–1835): [238]; analyzed the aerolite of Durango, 427, [526n; II.78]; and the muriates of silver, II: 89–90, [111n, 152n, 180n].
Knox, John (1720–1790), 447.
Kühn, Father Eusebio [Francisco, bap. 1645–1711]: his voyage in California, 50; presumably the first to prove that this land was not an island, 439, 440, 444.
Kunth, Karl Sigismund (1788–1850), 563n; II.31n.

La Bastide, Martin de, 18, 151n, 189, 224n.
La Condamine, Charles-Marie de (1701–1774), 155; II.300.
Labat, Jean-Baptiste (1664–1738), II.393n.
Labillardière, Jacques-Julien Houtou de (1755–1834), 174.
Laborde, Joseph de [1699–1778]: founder of the church of Tasco, 380; his adventures, II: 131, [168, 368n, 383n]. [*See also* Borda Sánchez, José de la.]
Laborde, [Louis Joseph] Alexandre de [1773–1842], 98, [303n, 345–47, 483n; II.59n].
Lacaille, Nicolas Louis de (1713–1762), 24.
Lacépède, Count of, II.51. *See also* Médard de la Ville-sur-Illon, Bernard Germain Étienne.
Lachaussée [Pierre, c. 1691–c. 1754]: built a hydraulic water machine after plans by M. del Río, II.130.
Lafon, Barthélemy (1769–1820), 72, 423.
Lafora, Nicolás [de, c. 1730–after 1788]: his travelogue, 51–55, [60]; his map of the borders of New Spain, 64, [87, 89–90, 130t].
Laguna, Andrés de la, II.417n.
Laguna [Calderón], Pedro de la [c. 1755–c. 1813]: [54–55, 57, 86]; his partial map of Mexico, 63, [129n]; drew the layout of the ruins of Mitla, 403–4.
Lalande, Joseph-Jérôme Lefrançais de (1732–1807), 31, 73.
Lamarck, Jean Baptiste Pierre Antoine de Monet de (1744–1829), 567.
Lambert, Aylmer Bourke (1761–1842), 507n, 555n.
Lambert, William (d. 1834), 73n.
Lampadius, Wilhelm August Eberhard (1772–1842), II: 144n, 150n.
Lancisi, Giovanni Maria (1654–1720), II.394.
Landívar, Rafael [1731–1793], Mexican poet, 391, 395.
Lángara y Huarte, Juan de (1736–1806), 59, 459.
Langlès, Louis Mathieu (1763–1824), 263n; II.232n.
Lapérouse, Count of: how he determined the position of Monterrey, 48–49n, [180n, 449, 460, 465, 470; II: 360n, 364n. *See also* Galaup, Jean-François de].
Laplace, Pierre Simon de (1749–1827), 26n, 133–34, 180n, 245n, 391n; II: 184n, 492.
Lassaga, Juan Lucas de, II: 191n, 194n.
Lasuén, Fermín Francisco de (1736–1803), 449, 450n, 451, 527.
Latreille, Pierre André (1762–1833), II: 40, 42.
Laugier, André (1770–1832), II.268.
Lavoisier, Antoine-Laurent de (1743–1794), 277, 348–49, 537, 541–43.
Le Gentil de la Galaisière, Guillaume-Joseph-Hyacinthe-Jean-Baptiste (1725–1792), II: 184n, 356n.
Le Monnier (Lemonnier), Pierre Charles (1715–1799), 20n.
Leblond, Jean-Baptiste (1747–1815), II.398n.
Leca, Carlos Corso de: inventor of the *beneficio de hierro* [an iron amalgamation method], II: 150, [206].
Lefebvre, Jean-Laurent, Abbot, 543.
Lemos, Francisco Gil [de Taboada y], viceroy of Peru [1736–1809]: had the southern coast of Chile explored, 166; his census of the inhabitants of Peru, 29; [II: 87n, 230n].
Leibniz, Gottfried Wilhelm (1646–1716), 244.
Lemaur [y de la Muraire, Francisco Angel, c. 1770–c. 1841] and Lemaur [y de la Muraire, Félix Ramón Manuel, 1767–1841]: completed the project of the Güines canal, 200.

Lenoir, Étienne (1744–1832), 276.
León y Gama, Antonio. *See* Gama, [Antonio de León y].
LePère, Jean-Baptiste (1761–1844), 159n, 168.
Lewis, Captain [Meriwether, 1774–1809]: his voyage to the mouth of Columbia River, [149, 182], 446, [458; II.365].
Lichtenstein, Martin Hinrich Carl (1780–1857), 174n.
Lind, James (1716–1794), II.394.
Linné, Carl von (Linneaus, 1707–1778), 463, 522, 558; II: 28, 44.
Lizana [y Beaumont], Francisco Javier de [1750–1815]: archbishop of Mexico City; information he gave the author, 205n.
Llanas, Antonio: discovered the mines of Catorce, II.125.
Loaysa [Loaísa], Father Jerónimo [de, 1498–1575]: archbishop of Lima; his census of the inhabitants of Peru, 200.
Lobera de Ávila, Luis (c. 1480–1551), II.417n.
Löfling, Pehr (1729–1756), 522.
Long, Stephen Harriman (1784–1864), 55–56, 90, 91n, 181, 239n, 248, 435n, 555n.
López, Salcedo y Rodríguez, Diego de (d. 1 547), 153n.
López de Bonilla, Gabriel (c. 1600–1668), 30.
López de Gómara. *See* Gómara, Francisco López de.
López de Haro, Gonzalo. *See* Haro, Gonzalo López de.
López de Lerena, Pedro (1734–1792), II.444.
López [de Vargas, Juan, 1765–1830]: his map of the surrounds of Mexico City, 62, 65, [154].
López de Vargas Machuca, Tomás (1730–1802), 76, 151n.
López de Villalobos, Ruy (1500–1546), 70.
López Pacheco, Diego, Marquis of Villena (1599–1653), 365.
Lorenzana [y Buitrón], Cardinal [Francisco Antonio de, 1722–1804]: [209, 319n]; [his edition of Cortés's history of New Spain, 32n, 111, 112n, 143n, 283, 286n, 309n, 322, 328n, 333, 335n, 336n, 337n, 445n, 561n; II: 22n, 33n, 65n, 69n, 249n, 299n, 300n, 318n, 320n, 363n]; his work on Mexican antiquities, 283; report he presented on the desagüe, 354n; [on Popocatépetl, II.300].
Louis XVII (1785–1795), II.282t.
Loureiro, João de (1710–1791), 559n, 564n.
Lubarsky, [Vasili, 1795–1852]: discovered platinum in the Ural mines, II.268.

Macartney, Lord George (1737–1806), the first Earl Macartney, II: 34, 257–60.
Mackenzie, Alexander (1764–1820), 71–72, 147n-148, 181, 232, 463–64, 468n, 474, 549n; II.190n.
Mackittrich, Jacob, II.406.
Madrid, Luis, II.471n.
Maenza, Marquis of [1804–1876]: his attempt to found a colony of European artisans in Quito, II.292. *See also* Ascásubi y Matheu, Manuel de,
Maestre [y Fuentes], Ignacio [b. 1737]: his work on Lake Nicaragua, 499n
Magendie, François (1783–1855), II.421n.
Magens, Nicolaus (c. 1697–1764), II: 235n, 241, 254n.
Malaspina [Meli Lupi], Alessandro [1754–1810]: [xxv]; how he determined the position of Cape San Lucas, 67; his expedition to the northwestern coast of the Americas, [34–37, 47–50], 465–67; II: [59, 74, 115–17, 126t–128t, 154, 166, 279n,

341n], 363, [459, 469, 474]; his imprisonment, 465; evaluated the mines of the Americas, II.231n.

Maldonado, [Lorenzo] Ferrer [c. 1550–1625]: [25]; his apocryphal journey to the American northwestern coast, 457, [466, 469, 555].

Malintzín (Malinche) (c. 1501–1550), 384n.

Malo, Abel, 99n.

Malpighi, Marcello (1628–1694), 237.

Malte-Brun, [Conrad, 1775–1826]: doubts he raised that the Tacoutché-Tessé and the Columbia River were the same, 148n, [471, 473; II.266].

Malthus, Thomas Robert (1766–1834), 209n.

Manco-Capac (Titu-Manco-Capac), ninth Sapa Inca, 551, 553; II.49.

Mange [Mangi], Juan Mateo [1670–c. 1727]: his unpublished journal of a voyage in California, 445n.

Maniau, Ildefonso, II.178n.

Maniau [y Torquemada], Joaquín [1753–1820]: his unpublished work on New Spain, 255n.

Manzo y Zúñiga, Francisco, [Count of Hervías], archbishop of Mexico City [1587–1656]: his charitableness during the floods from 1629 to 1634, 263.

Maquinna of Nootka (b. c. 1760), 463.

Marcgrave, Georg (1610–1644), II.28.

Marchand, Adrien Joseph, 543.

Marchand de la Ciotat, Étienne (1755–1793), 45, 48–59, 151, 167n, 232n, 240, 458n, 460, 474, 561n; II: 50, 359n, 361n.

Marcos de Niza, Father (c. 1495–1558), 50, 429, 441.

Marín, Campo: his assessment of the mint of Mexico City, II.304.

Marqués: name under which Cortés was known in Mexico, 320n; [II: 363, 496–97]. [*See also* Cortés de Monroy y Pizarro, Hernán; Monteleone, Duke of.]

Márquez, Pedro [José, 1741–1820]: his work on Mexican antiquities, 412n, [413].

Martín, Luis [1772–c. 1808]: Mexican engineer, [77–78, 112]; his measurements of the valley of Mexico, 319; his work on the ruins of the palace of Mitla, 403–4; [II.508–9].

Martínez, Enrico [c. 1555–1632]: [30]; creator of the desagüe of Huehuetoca, 358, [361, 366–74]; the beginnings of his work, 358–59; rebukes against him, 360–62, [364]; drew up the maps of Vizcaíno's travels, 458.

Martínez [Fernández], Estéban José [1742–1789]: Juan Pérez's pilot on his voyage to the northwestern coast of the Americas, 459, 461–64, 467; was put in charge of founding a settlement in Nootka, 461; had James Colnett arrested, 464; [unpublished journal of, 473].

Martínez Gómez, Tomás (c. 1487–1551). *See* Berlangas, Tomás de.

Martínez [López de Vía], Alonso [fl. 1607–26, d. 1626]: collaborator of Enrico Martínez's, 359.

Mascaró, Manuel [Agustín, 1747–after 1809]: his unpublished travelogue, 52; his map of New Spain, 53–55, [57], 60, [77, 87–88, 129t, 130t, 319n]; of the surroundings of Doctor, etc., 62.

Maso [Mazo], José [Antonio] del [d. 1805]: owner of a mercury mine, II: [116n], 181.

Mason, Charles (1728–1786), 36, 50, 71.

Massard, Jean-Baptiste Louis (1772–1810), 412n.

Maurelli [de la Rúa], Francisco [Antonio, 1750–1820]: pilot, author of a journal about the voyage of Heceta, Ayala, and Quadra, 460; [II.358].

Mayer, Christian (1719–1783), 20.
Mayer, Johann Tobias (1723–1762), 50, 54, 95.
Meares, John (c. 1756–1809), 240.
Médard de la Ville-sur-Illon, Bernard Germain Étienne, Count of Lacépède (1756–1825). *See* Lacépède, Count of.
Medina, Bartolomé [c. 1503–1585]: inventor of an amalgamation process used in the mines of the New World, II: 144, [150–51, 156, 206, 244t].
Medina, Salvador de [d. 1769]: his travels in California, [126t], 278.
Memminger, Johann Daniel Georg (1773–c. 1840), II.16.
Mendaña de Neira (Neyra), Álvaro de (1542–1595), 439, 563; II.361–62.
Méndez, Simón: his plan for preventing floods, 362; it was examined again by Velázquez in 1774, 373–74.
Mendoza Melendez, Pedro, II.144n.
Mendoza y Luna, Juan de, viceroy of Mexico [1571–1628]: his achievements, [224], 333, [411].
Mendoza y Pimentel, Diego Carillo de (c. 1557–1636). *See* Gelves, Marquis of.
Mendoza y Ríos, José de (1762–1816), 25; II.348.
Mercati, Michele (1541–1593), II.179n.
Mercator, Gerard (1512–1594), 125, 187.
Michaux, André (1746–1802), 187, 506n; II: 28n, 37.
Mier y Trespalacios, Cosme [Antonio] de [1747–1805]: doyen of the highest court of justice in Mexico City, [354n]; maps he had drawn up of the desagüe of Huehuetoca, 359; as superintendent of the desagüe of Huehuetoca, he had two drainage canals built, 370–71, [376].
Millar, James (1762–1827), 360n.
Miller, Edward (1760–1812), II: 393n, 406.
Mirbel. *See* Brisseau Mirbel, Charles-François.
Mitchill, Samuel Latham (1764–1831), II: 105, 396.
Mociño. *See* Moziño, José Mariano.
Molina, Alonso de (c. 1514–1585), 180n, 526, 527n, 551; II.208.
Moncada, Sancho de [b. 1580]: his estimate of the quantity of gold and silver that had flowed back into Europe since 1492, II: 239, [243t].
Monclova, Count of, viceroy of Mexico, [20]; appointed Father Cabrera superintendent of the desagüe, 365. [*See also* Portocarrero y Lasso de la Vega, Melchor.]
Moñino y Redondo, José, Count of Floridablanca (1728–1808), 140, 202.
Montalvo, [Luís] Berrio de [d. 1643]: his report on the metallurgical processing of minerals, II.144.
Montaño, Francisco [b. 1499]: whether he descended into the crater of Popocatepetl, [112]; II: 300, [341].
Monteleone, Duke of: his wealth, 281; monument he had erected in honor of Cortés, 335; [II.32].
Monterey, [Gaspar de Zúñiga y Acevedo], Count of, viceroy of Mexico [1560–1606]: sent Oñate to New Mexico, 343.
Montesclaros, Marquis of. *See* Mendoza y Luna, Juan de.
Montesquieu. *See* Secondat, Charles-Louis de.
Montés y Pérez, Francisco de (1753–1817), 35.
Montezuma, [142, 200–201, 249, 286n, 320, 329, 341, 352, 357–58, 454, 463, 548; II: 21–22, 29, 37, 39, 49, 64–65, 81, 249, 315, 431, 453]. *See [also]* Motecuzoma.
Montezuma, Antonio: author of an unpublished history of Mexico, 342n.
Montezuma, Pedro, son of Montezuma II, 341n. *See [also]* Tohualicahuatzin.

Montezuma II (Motēuczōmah), King of the Aztec Empire (c. 1466–1520), 199, 252, 286, 342–43.

Montúfar y Fraso, Juan Pío (d. 1761), 258.

Montúfar y Larrea, Carlos (1780–1816), xiii.

Moore, [John] Hamilton [1738–1807]: how he determined the position of Veracruz. 34.

Moquihuix, the last king of Tlatelolco, 326.

Moraleda [y Montero Espinoza], José [Manuel Nicolás de, 1747–1810]: his journeys, 166; II.465.

Mornay, Aristedes Franklin (1779–1855), II.491.

Morse, Jedidiah (1761–1826), 224, 476.

Morse, Sidney Edward (1794–1871), 476.

Mortier, Cornelis (1699–1783), 37.

Mosquera, José María (1752–1829), II.88.

Motecuhzoma Xocoyotzin. *See* Montezuma II.

Motecuzoma: real name of Montezuma, 320; two princes with this name, 320; borders of their empire, 141. [*See also* Montezuma.]

Mothès, Frédéric [Friedrich Gottlob Mothes, 1766–1795]: his report on the mines of Potosí, II.99n.

Motolinia [c. 1490–1569]: author of unpublished history of Mexico, [224, 267], 342n. [*See also* Paredes, Toribio.]

Moziño, [José Mariano, 1757–1820]: his botanical work, 275–76, [451; II: 370, 400]; his voyage to Nootka, 462n, 463.

Muñoz Camargo, Diego. *See* Camargo, Diego Muñoz.

Murphy [Porro], Tomás [b. 1768]: his patriotic views, 222; [II.341].

Murr, Christoph Gottlieb (1733–1811), II.94n.

Murray, Johan Andreas (1740–1791), II.28n.

Mushet, Robert (1782–1828), II.272n.

Mutis [y Bosio], José Celestino [Bruno, 1732–1808]: his great botanical works, 275–76; [II: 29, 380]; discovered the mercury mine of Quindío, II.183.

Nadal, Pedro, [50]; made astronomical observations on the Río de Balzas, 429.

Nairne, Edward (1726–1806), 511.

Narváez, Pánfilo [de, 1470–1528], [224]; his itineraries, 446.

Nava, Pedro de (b. c. 1740), 64.

Navarrete, Pedro Fernández de [1564–1632]: his estimate of the quantity of gold and silver that had flowed back into Europe since 1492, II: 239–40, [243].

Navarro y Noriega, Fernando (d. 1826), 144, 211, 230n, 479.

Navincopa. *See* Abincopa, Gonzalo.

Necker, [Jacques, 1732–1804]: [537, 543]; his estimate of the quantity of gold and silver that had flowed back into Europe since 1492, II: 242, [243t, 250n, 251n, 255, 261, 274, 276n, 279, 281, 432, 434].

Née, Louis [c. 1737–c. 1807]: botanist of the Malaspina expedition, 466.

Nelson, Horatio (1758–1805), II.41.

Newcks, Thomas [c. 1719–1771]: how he determined the position of Mexico City, 30; of Veracruz, 34.

Newton, Isaac (1643–1727), 278.

Netzahualcoyotl, king of Tezcuco, 357.

Nicander, Henrik (1744–1815), 208.

Nicholson, William (1753–1815), II.18n.

Nima Quiché (Nima K'iche,' zenith c. 1450 CE), 404.

Niño, Andrés [1475–c. 1530]: claimed that there is no strait between South and North America, II.319.

Niño, Pedro Alonso, El Negro (1468–c. 1505), II.253t.

Niparajá, deity of the Californians, 444.

Niza, Marcos de [c. 1495–1558], [50]; made astronomical observations on the Río de Balzas, 429; his fabled stories about the city of Cíbola, 441.

Niza [de Santa María], Tadeo de: baptized Indian; author of unpublished work on the history of Mexico, 342n.

Noguera, Fernando María (b. 1761; fl. 1792–1797), 82, 154.

Nordenflycht, [Fürchtegott Leberecht], Baron of [1752–1815]: his geological collection, II: 185, [189].

Norris, Robert (d. 1791), 285n.

Núñez Cabeza de Vaca, Alvar [c. 1490–c. 1557]: his voyages, 446.

Núñez de Balboa, Vasco (c. 1475–1519), 154.

Núñez de la Vega, Francisco (1632–1706), 487–88.

Nuttall, Thomas (1786–1859), 484.

Nysten, Pierre-Hubert (1771–1818), II.407n.

Obregón, [Lorenzo Sánchez de, fl. 1574–1581]: licenciado; his plan to protect Mexico City from floods, 358–59.

Obregón [y Alcocer, Antonio], first Count of Valenciana [1722–1786]: 435; II.112–13.

Obregón [y Barrera, Antonio Francisco, 1773–1833]: information he gave to the author, 67; [II.125].

Ocampo, Diego de, II.355.

Ocaño, Rodríguez de: discovered the mines of Chota, II.202.

Ocio, Manuel del (c.1700–1771), 443.

Olid, Cristóbal de (1488–1524), 333.

Olivarius, Holger de Fine (1758–1838), II.260n.

Olmos, Andrés de [c. 1491–c. 1571]: author of an unpublished history of the conquest of Mexico, 342n.

Oltmanns, Jabbo [1783–1833], [13]; how he determined the position of Veracruz, [23, 28, 30–31, 33], 34, [59]; [of Acapulco, 35; of La Venta de Chalco, 40]; of Cofre de Perote, 41; [of the Huehuetoca canal, 44]; of San José, San Lucas, and San Blas, 47–49, [74]; his works on the geography of Mexico, 69, 72–73, 77–78, 116, [131], 133–34; II: 308, 391n, [355n, 492, 507, 508].

Oñate [y Salazar], Juan de [c. 1551–1626]: his conquest of New Mexico, 434, [455; II.451].

Ontiveros, [Felipe de Zúñiga y, c. 1717–1793]: determined the position of the farm of Pazcuaro, 134.

Orbegozo, Juan de, General: his astronomical observations, II.490–92. [*See also* Horbegozo, Juan de.]

Ordaz, Diego [de, 1480–1532], [112]; whether he descended into the crater of Popocatépetl, 298–300; he explored the river of Huasacualco, II.316.

Ordoñez de Montalvo, Juan José, II: 145n, 150.

Oriani, Barnaba (1752–1832), 114.

Origenes Adamantius (c. 185–254), 265n.

Oropesa: title bestowed upon the family of the Inca Sayri-Tupac, 262.

Orta, Bernardo de [fl. 1788–1806]: captain of the port of Veracruz; his meteorological observations, [97,] 192; II: 402, [420, 509].

Ortega y Medina, Juan Antonio, 579.

Ortiz y Ayala, Simón Tadeo (1788–1833), 492.

Osterwald, Georg Rudolf Daniel (1803–1884), 110.

Oteiza, Juan José de [Oteyza, 1777–1810], [37n, 44]; how he determined the longitude of Durango, [53–54], 60, [88, 130t], 436–27; [his journals,

87]; his maps of the surroundings of Durango, 62; his calculation of the surface of Mexico, 308; of the pyramids of Teotihuacan, 338–39.

Otero, Pedro Luciano [1717–1788]: one of the entrepreneurs of the Valenciana mine, 113.

Oudermeulen, Cornelis van der (1735–1794), II: 237n, 239n, 242.

Ovando [y Cáceres, Nicolás de, 1460–1518]: riches he sent to Europe, II.249.

Oviedo y Valdés, Gonzalo Fernández de (1478–1557), 512–13, 521–22, 524–25, 529, 557, 564n; II: 4n, 37.

Pacheco de Mendoza, Antonio (1495–1552), II: 302, 355.

Pacheco y Osorio, Rodrigo (d. 1652). *See* Cerralvo, Marquis of.

Padilla, Cristóbal de: his plans to drain the valley of Mexico by using the natural chasms of Oculma, 363.

Pagaza[urtundúa], Juan José de [1755–c. 1817]: his maps of New Galicia and New Biscaya, 63–64, [67].

Page, Pierre François (1764–1805), 229, 272; II.20

Pagès, [Pierre Marie Francois de, captain, 1748–1793]: his overland journey from Louisiana to Acapulco, 421.

Palacio, Alatto, II.471t.

Palacios, José, II.471t.

Palafox [y Mendoza], Juan de [1600–1659]: bishop of Puebla and viceroy of Mexico; his instructions concerning the desagüe, 354n, [365].

Pallas, Peter Simon (1741–1811), 475; II.154n.

Pánfilo de Narváez (1470–1528), 224, 446.

Paredes, Toribio. *See* Motolinia.

Parilla, Luis: placed at the head of the establishment of Chalco, II.20.

Parish, Elijah (1762–1825), 224n.

Paul I, Emperor of Russia (1754–1801, r. 1796–1801), 74.

Pausanias (second century CE), 324.

Pauw, Cornelius de (1739–1799), xxiv, 81, 200, 228.

Pavón [y Jiménez], José [Antonio, 1754–1840]: one of the leaders of the botanical expedition of Peru, 275, [551].

Pedro I of Portugal (1320–1367), II.68t.

Pedro II of Portugal (1648–1706), II: 238, 268t.

Peña, Tomás de la: accompanied Juan Pérez on his voyage; his unpublished diary, 459.

Peñalver y Cárdenas, José Ignacio, Marquis de Arcos (1736–1804), II.6n.

Penazco, Count, II.471t.

Pérez, Juan [Antonio, d. 1775]: his voyage to the American Northwest, 459–60, [462ff]; he had been in the natural harbor of Nootka before Cook, 459.

Pérez Gálvez, Antonio, II.145.

Pérez y Comoto, Florencio. *See* Comoto, Florencio Pérez y.

Perronet, Jean-Rodolphe (1708–1794), 168.

Peuchet, Jacques (1758–1830), 208, 290, 346n, 348, 476, 541n, 543, 544n; II: 9n, 10n, 30n, 250n, 255, 261n, 274.

Philip II, King of Spain and King of Castile (1527–1598), 273n, 525; II.49.

Philip III, King of Spain (1578–1621), 344, 361, 414; II.431.

Philip IV, King of France (Philip I of Spain, 1268–1314), II.237.

Philip IV, King of Spain (1605–1665), 342n.

Philip V, King of Spain (1683–1746), 415, 444; II.453.

Phillips, William (1775–1828), II.88n.

Pichardo, José Antonio, Father [1748–1812], [xiv]; monk of San Felipe Neri

Pichardo (cont.)
in Mexico City, [34], 62, 325n, [337n, 342n, 343].
Pignatelli Rubi Corbera y San Climent, Cayetano María, Baron de Llinas, Marquis of (b. c. 1725–after 1788), 64.
Pike, Zebulon Montgomery (1779–1813), 55–56, 65, 68n, 90, 181.
Pimentel, Antonio and Fernando. *See* Ixtlilxochitl.
Pimentel y Sotomayor, José Andrés (d. 1768), 394.
Piñadero, [Bernardo] Bernal de [fl. 1663–1677]: his expedition in California, 443; [II.49].
Pinel, Philippe (1745–1826), II: 394n, 416n.
Pingré, Alexandre Guy (1711–1796),
Pinkerton, [John, 1758–1826]: his division of New Spain, 303; [II: 56n, 428n, 451t].
Pinzón. *See* Yáñez Pinzón, Vicente.
Pizarro, Francisco (c. 1470–1541), II.238.
Plato (c. 428–347 BCE), 265n.
Playfair, William (1759–1823), 119, 140; II: 256n, 257n, 259, 379n, 446t, 447n.
Pliny, 350, 517, 519, 520n, 532n; II: 144n, 151.
Plukenet, Leonard (1642–1706), II.45.
Pococke, Richard (1704–1765), 340, 383.
Poinsett, Joel Roberts (1779–1851), 65n, 477; II.177.
Poirson, Jean Baptiste (1760–1831), 69, 73; II: 508–9.
Polo, Marco (1254–1324), II.31.
Pomar, [Juan Bautista, c. 1527–1602]: baptized Indian, author of an unpublished history of Mexico, [xxiv], 342n.
Pombo, José Ignacio de (1761–1815), 161n; [II.88n].
Pommelles, Knight of, 292. *See also* Sandrier de Mitry, Jean-Christophe.
Ponce: baptized Indian, author of an unpublished history of Mexico, 342n.
Ponce de León, Pedro (c. 1540–1628), 349n.
Pons, François Raymond Joseph de (1751–1812), 303n; II: 367, 377n, 381n, 385n, 455.
Portlock, Nathaniel (c. 1747–1817), 462.
Portocarrero y Lasso de la Vega, Melchor, Count of Montclova (1636–1705), 365.
Pownall, Thomas (1722–1805), 61.
Poyférré, Jean-Marie, Baron de Cère (1768–1858), 99n.
Pozuelo Espinosa, Francisco (1651–1691), 395.
Pringle, John (1707–1782), II.394.
Prony, Gaspar Clair François Marie Riche de, Baron Riche of Prony (1755–1839), 378.
Proust, Joseph Louis (1754–1826), 516; II: 8, 154.
Pugnet, Jean François Xavier (1765–1846), II: 394, 407n, 408n, 417n.
Purdy, John (1773–1843), 114n.
Putsche, Karl Wilhelm Ernst (1765–1834), 555n.
Puyade, Ángel, II.471t.
Puységur, Count of, 35, 73. *See also* Chastenet, Antoine Hyacinth Anne de.

Quadra, Juan [Francisco] de la Bodega y [1744–1794], [45, 154]; how he determined the position of San Lucas, 47; his map of California, 64; his journey to the north-northwestern coast of the Americas, [208n, 276], 460, [462n, 466–67, 470, 473; II: 364–66].
Quardiola, Marquis of, II.471t.
Quartara y Guerrini, Antonio (fl. 1800–1821), 154n.
Quauhtemotzin, the last king of Mexico, [268–69], 337n; the heroic character of this prince, 343.

Quesnay, François (1694–1774), 541.
Quetlabaca, King of Mexico. *See* Cuitlahuatzin.
Quetzalcoatl, [247], 266–67; the Mexica applied his prophecy to the Spanish, II: 64, [249].
Quijano, José [Barreiro]: his tableau of the Valenciana mine, II.395.
Quimper [Benítez del Pino], Manuel [fl. 1789–1819]: his journey to Nootka, 467.
Quiroga, Vasco de [c. 1470–1565]: first bishop of Michoacán, benefactor of the Indians, 397.
Quirós, José María (c. 1750–c. 1824), 439, 563, 565; II: 361–62, 420n.

Raleigh, Walter, Sir (c. 1552–1618), 550, 553, 555n.
Ramel de Nogaret, Jacques Dominique-Vincent (1760–1829), II.445n.
Ramírez de Fuenleal, Sebastián, Bishop (c. 1490–1547), 439n.
Ramond, Louis-François Elisabeth, Baron de Carbonnières (1755–1827), 80; II.185.
Ramsden, Jesse (1735–1800), 276.
Raynal, [Guillaume Thomas François, Abbot, 1711–1796], [xxiv, 37, 81, 118, 160n, 228, 522n; II: 23, 28–31]; his view of the wealth of the mines of New Mexico, II.217–18; his estimate of the amount of silver that had flowed back to Europe since 1492, II: [41, 216–17,] 239–41, [243, 245, 252, 255n, 258n, 272, 356n, 381n].
Redhead, [Joseph James Thomas], the doctor [1767–1847], II.209n.
Redouté, Pierre-Joseph (1759–1840), II.44n.
Regla, Count of [Santa Maria de, 1710–1781]: his wealth, 280; [II.128]; as the proprietor of the mines of Viscaína, II.129–30. [*See also* Romero de Terreros, Pedro.]
Reichenbach, Georg Friedrich von (1771–1826), II.139.
Reinaga, Juan de [b. 1509]: introduced camels to Peru, II.314.
Remesal, Antonio de (1570–1619), 11n, 180n.
Rennell, James (1742–1830), 18, 122n, 68n, 86.
Renouard de Sainte-Croix, Félix, Marquis of (pseud., 1773–1840), II: 258n, 356n, 365n.
Restrepo, [José Manuel, c. 1781–c. 1863], [162]; his assessment of what the panning in Antioquia produced, II.228n.
Revillagigedo, Count of [1740–1799], viceroy of Mexico, [xix, 47n, 61, 82, 147, 202, 204, 261, 328n, 332, 348, 354n, 356, 378; II: 20, 69, 318, 434, 436, 450]; census of the inhabitants of Mexico he ordered, [202–4], 273, [292, 475, 477, 541; II.470]; policing he established for the streets of Mexico City, 334; expedition to Nootka he ordered, 467, [469]; praise of his administration, II.440. *See also* Güemes Pacheco de Padilla, Juan Vicente.
Reynolds, William (1758–1803), 379.
Rheede tot Drakenstein, Hendrik Adriaan van, Lord of Mijdrecht (1636–1691), 513.
Riaño [y de la Bárcena], Juan [Antonio] de [c. 1757–1810]: his eagerness to introduce smallpox vaccination, 222.
Ribera, [Manrique, Payo] Enriques [de, 1622–1684]: archbishop of Mexico City, viceroy of Mexico, 365.
Riepl, Franz Xaver (1790–1857), II.104n.
Riffault des Hêtres, Jean René Denis (c. 1752–1826), II.397n.
Río, Andrés [Manuel del, 1764–1849]: professor of mineralogy at the School of Mines in Mexico City, [xiv], 67, 276, [400n; II: 87, 122n], 130, [190].

Río, Antonio del, Captain [fl. 1786–1789]: his work on Mexican antiquities, [404], 487.

Ríos, Pedro de los (d. c. 1565), 264.

Rittenhouse, David (1732–1796), 78.

Ritter, Carl (1779–1859), 149n.

Rivera [y Villalón], Pedro de [General, c. 1664–1744]: his travelogues, 51–55, [57, 60, 87–90, 129–30t; II.508].

Rivera Bernárdez, José de, second Count of Santiago de La Laguna (fl. 1710–1730), II.145n.

Rivero y Ustariz, Mariano Eduardo de (1798–1857), II.154n.

Rixi, Jodoco: monk who sowed the first European wheat in Quito, 592.

Robertson, [William, 1721–1793], [81, 118, 200n, 227, 332, 414, 520; II: 122, 217, 222n]; his estimate of the quantity of gold and silver that has flowed back into Europe since 1492, II: 242–43, [253n, 432, 454].

Robinson, [William] Davis [b. 1774]: provided information on the rip current of the Río San Juan in Nicaragua, 151, [153n, 156].

Robredo, Antonio [d. c. 1830]: how he determined the longitude of Mexico City, 32, [184n, 511n].

Rochette, Louis Stanislas d'Arcy de la (1731–1802), 68n.

Rodríguez, Diego, Father [1596–1668]: how he determined the position of Mexico City, 30–31.

Rodríguez, Juan José [fl. 1800–1805]: assisted the author in making geological maps, 107n.

Rodríguez Cabrillo, Juan (b. c. 1500–1543), 45, 447, 455, 457, 458; II.361.

Roggeveen, Jacob (1659–1729), 558.

Rojas, José [Antonio, b. 1773]: his experiments with the temperature of the hot waters in San José de Comangillas, 390.

Román, Antonio: his plan for draining the valley of Mexico, 363.

Romanzoff, Count, II: 260, 365n. *See also* Rumyantsev, Nikolai Petrovich.

Romero: wealthy Indian family in Cholula, 255.

Romero de Terreros, José María, Marquis of San Cristóbal (1766–1815), 280n.

Romero de Terreros, Pedro (1710–1781). *See* Regla, Count of Santa María de.

Rousseau Saint-Aignan, Nicolas Auguste Marie de (1770–1858), 471n.

Rovira, Francisco Javier (1740–1823), 354n.

Roxburgh, William (c. 1751–1815), 520n.

Rozier, François, Abbot (1734–1793), 355n.

Rubi, Marquis of. *See* Pignatelli Rubi Corbera y San Climent, Cayetano María.

Rubín de Celis, [Miguel, b. 1746]: found an aerolite near Olumpa, 427.

Rühs, Friedrich (1781–1820), 544.

Ruiz de Luzuriaga, Ignacio María (1763–1822), II: 391, 93n, 401n, 408n, 417n.

Ruiz de Montoya, Pantaleón, II.43.

Ruiz [López], Hipólito [1754–1816]: his journey to Peru, 275, [551].

Rul, Diego [1761–1812]: one of the owners of the Valenciana mine, [122t]; II: 98, [145].

Rumphius (Rumf), Georg Eberhard (1627–1702), 513.

Rumyantsev, Nikolai Petrovich (1754–1826). *See* Romanzoff, Count.

Ruprecht, Anton (1748–1814), II: 143, 157.

Rush, Benjamin (1746–1813), 225; II: 391, 401n, 406, 412n, 416, 417n.

Russell, Alexander (1715–1768), II.415.

Saavedra Cerón, Álvaro de (d. 1529), 361.

Sabine, Edward, Captain (1788–1883), 555n.

Sahagún, [Bernardino de, c. 1499–1590]: author of an unpublished history of Mexico, [267], 342n.

Saint-Aignan. *See* Rousseau Saint-Aignan, Nicolas Auguste Marie de.
Sainte-Croix, [Carloman Louis-François] Félix Renouard, [Marquis of, 1773–1840]: notes he provided on the trade with India and China, II: 258, 283n, [284t, 365n, 457n].
Saint-Hilaire, Augustin François César Prouvençal (c. 1779–1853), II.266.
Salamanca y Humara, Secundino (c. 1768–1839), 467.
Salcedo [y Salcedo], Nemesio [1754–c. 1814]: commander-general of the interior provinces, 310.
Salinas, Marquis of. *See* Velasco I [Velasco y Castilla y Mendoza, Luis de].
Salmerón, Martín [1774–1813]: Mexican giant, 242.
Salvatierra, Count of. *See* García Sarmiento de Sotomayor y Luna.
Salvatierra, Juan María de, Father [1648–1717]: his travels in California, 440, 444; his unpublished map, 445n.
Salvato, Count of, II.471t.
San Cristóbal, Marquis of, 280. [*See also* Romero de Terreros, José María.]
San Miguel [Iglesias y de la Cajiga], Antonio de, Father, 1726–1804]: archbishop of Valladolid; report he presented to the king in favor of the Indians, 256n; aqueduct he had built, 397.
San Román, Marquis of [1756–1818]: director of the mint of Mexico City, II.302. [*See also* Fernández de Córdoba Zayas, Francisco.]
Sánchez de Obregón, Lorenzo (fl. 1574–1581), 358–59.
Sánchez de Tejada, Ignacio (1764–1837), II.227n.
Sandoval, Gaspar de. *See* Cerda Sandoval Silva y Mendoza, Gaspar de la.
Sandoval, Gonzalo de [1497–1528]: conquered the province of Tehuantepec, 333; II.316.
Sandoval y Guzmán, Sebastián: his work on the production of the Potosí mines, II.216–20.
Sandrier de Mitry, Jean-Christophe, Knight of Pommelles (b. 1776), 292.
Santa Rosa María, Pedro Beltrán de (fl. 1705–1757), 407.
Sarmiento: wealthy Indian family in Cholula, 255.
Sarmiento de Gamboa, Pedro (1532–1592), II.362.
Sarmiento de Valladares, José, Count of Moctezuma (1643–1708), 352n.
Sarría, Francisco Xavier de [b. c. 1750]: how he determined the position of Zacatecas, 54–55; [II.145n].
Sarychev, Gavril Andreevich (1763–1830), 471, 473.
Sauer, Martin [fl. 1780s–1790s], 471, 473.
Saussure, Horace Bénédict de (1740–1799), 329n, 377n.
Savage, John (b. 1770), 555n.
Schabelsky. *See* Shabelsky, Akhilles Pavlovich.
Schérer, Jean-Benoit (1741–1824), 233.
Schlegel, Karl Wilhelm Friedrich von (1772–1829), 451.
Schlenev, [Nicolai Alexsejevitch von, General]: discovered the mines of Ural, II: [264], 267, [268n].
Schmidtmeyer, Peter (1772–1829), II.268n.
Schneider, Johann Gottlob Theaenus (1750–1822), II.50.
Schnurrer, Friedrich (1784–1833), II.411n.
Schreber, Johann Christian Daniel von (1739–1810), 242n.
Schwartner, Martin von (1759–1823), II.269n.
Seaman, Valentine (1770–1817), II.417n.
Secondat, Charles-Louis de, Baron of La Brede et de (1689–1755).
Sein, Salvador: professor in Mexico City, II.139.

Selvalegre, Marquis of. *See* Montúfar y Fraso, Juan Pio.
Serra, Junípero [1713–1784]: his travels in California, [64], 448.
Serres, Olivier de (1539–1619), II.38.
Sessé [y Lacasta, Martín de, 1751–1808]: one of the leaders the botanical expedition to New Spain, 275–76, [462n], 506; II.304.
Shabelsky, Akhilles Pavlovich (Achille Schabelski, 1802–1856), II.366n.
Shelikohv, [Grigory Ivanovich, 1747–1795], [473]; founded a trading company in Irkutsk, II.366n.
Shortland, John (1736–1804), II.361n.
Shreve, Henry Miller, Captain (1785–1851), 91.
Sierra, [Miguel] Lamberto [de]: Tesorero of Potosí; his evaluation of royal rights, II.209n.
Sieur de la Salle: his settlement west of the Mississippi gave rise to debates about the borders of Mexico, II.457. [*See also* Cavelier, René-Robert.]
Siguenza [y Góngora], Carlos de [1645–1700], [50]; how he determined the longitude of Mexico City, 30–32; his maps of New Spain, 61–62, 76–77, [134n]; his hypothesis about the age of the pyramids of Teotihuacan, 266, [324n], 339.
Simonde de Sismondi, Jean-Charles-Léonard (1773–1842), II.275.
Sinclair, John, Sir (1754–1835), 555.
Skinner, Joseph (c. 1715–1756), 165n.
Smith, Adam [1723–1790], [118; II: 51n, 57n, 194n]; his estimate of the quantity of gold and silver that has flowed back into Europe since 1492, II: 241, [250n, 251t, 257n, 261n, 327, 447n].
Smith, John, Captain, 555n.
Sobreviela, Manuel (d. 1803), 165n; II.201n.
Sochipiltecatl: wealthy Indian family in Guaxocingo, 255.
Solander, Daniel Carlsson (Daniel Charles, 1733–1782), 558.
Solís [y Miranda, Martín de, fl. 1671–1706]: in charge of the administration of the desagüe, 365.
Solís y Rivadeneyra, Antonio de (1610–1686), 81, 142, 333n, 343n, 561; II.300.
Solórzano [Pereira, Juan de, 1575–1655], [258; II: 144n, 220–21]; his estimate of the amount of gold and silver that has flowed back into Europe since 1492, II: 240, [243].
Somodevilla y Bengoechea, Zenón de, Marquis of La Ensenada (1702–1781), II.291.
Sonneschmidt, [Friedrich Traugott, c. 1763–1824]: his reports on the mines of Mexico, 67; II.143; how he determined the height of the Sierra Nevada, 133–34; discovered meteoric iron in Zacatecas, 427; II.179; seven elevations determined by this traveler, 133; [II: 78, 103n, 121–24, 130, 147, 156, 164].
Soto, Hernando de (c. 1500–1542), II.49.
Staunton, George Leonard, Sir (1737–1801), 341, 462; II: 285n, 286n.
Steininger, Johannes (1794–1874), II.104n.
Stoll, Maximilian (1742–1787), II.416.
Storch, Heinrich Friedrich von (1766–1835), 74.
Strabo (c. 64 BCE–c. 23), 156, 158, 324, 391, 559n.
Suárez Bonaventura: how he determined the longitude of Mexico City, 31.
Suárez de Figueroa, Cristóbal (c. 1571–c. 1644), II.361n.
Surville, Jean François Marie de (1717–1770), II: 358n, 361.
Sydenham, Thomas (1624–1689), II: 392, 416.

Taboada y Lemos, Francisco Gil de (1736–1809), 166, 200; II: 87n, 230n.
Tafalla, Juan (1755–1811), 506n.
Talledo y Rivera, Vicente (b. 1760), 162n.
Talleyrand-Périgord, Charles Maurice, first Prince of Bénévent (1754–1838), 297n.
Tamarón [Romeral, Pedro], bishop of Durango [1695–1768]: his travel manuscript, 424, [424n, 434, 436, 437n].
Tangáxuan II (d. 1530), II.28.
Tanner, Henry Schenck (1786–1858), 55–56, 65n, 68n, 72.
Tarbé des Sablons, Sébastien-André (1762–1838), II.281.
Tardieu, Antoine François (1757–1822), 68n.
Taylor, John (1779–1863), 65n, 579; II.157n.
Tecuanouegue: wealthy Indian family in Los Reyes, 255.
Tecuichpotzin: daughter of Montezuma II; various houses in Mexico City descended from her, [269], 342n.
Tejada, Francisco, II.224.
Tejada, Ignacio [Sánchez de, 1764–1837]: his unpublished writing about the government of viceroy Espeleta, II.227n.
Teoyamiqui, Mexica deity: her hieroglyph-covered statue in Mexico City, 325, 338.
Tepa, Count of, II.42. *See also* Viana, Francisco Leandro de.
Tessier, Henri-Alexandre (1741–1837), 516, 546.
Tetlepanguetzaltzin (d. 1525): last king of Tacuba, hung on Cortés's orders, 337.
Tezozomoc, [Fernando] Alvarado, baptized Indian: author of an unpublished history of Mexico, [xxiv], 342n.
Thénard, [Louis Jacques, 1777–1857], [II.152]; his experiments with suffocation, II: [420n,] 421n, [422].
Theophrastus (c. 370–c. 287 BCE), 532n.
Thiéry de Menonville, Nicolas-Joseph (1739–1780), II: 23n, 28, 41–45.
Thorne, Richard V. W., II.401n.
Thouin, André (1747–1824), 449.
Thunberg, Carl Peter (1743–1828), 549n, 558, 560n; II: 232n, 259n.
Tienda de Cuervo, José de (d. 1763), 63.
Tipu Sultan (c. 1753–1799), 16.
Tiscar, José Antonio de, 154.
Titsingh, Isaac (1745–1812), 546.
Tlacahuepan-Cuexcotzin: Mexica deity, 323n.
Toaldo, Giuseppe (1719–1797), 355n.
Tohualicahuatzin: son of Montezuma II, ancestor of the counts of Montezuma and Tula, 341n.
Toledo, [Francisco de, 1515–1582], viceroy of Peru: his census of the Indians of Peru, 201, [262, 346].
Tolsá, Manuel [1757–1816]: creator of the statue of Charles IV, 275, 331n; and of Cortés's sepulchral monument, 335.
Tonantzin: Mexica goddess, 364n.
Tooke, Thomas (1774–1858), II: 10n, 12, 232n, 266, 276n.
Toro e Ibarra, Francisco José Rodríguez del (1761–1851), 161.
Torquemada, [Juan de], Father [c. 1564–1624], [224, 264, 324n]; undertook the construction of roadways in Mexico City, 334, [342].
Torre, José Ignacio de la (fl. 1797–1821), 33.
Torre, Lorenzo de la: inventor of a particular procedure for amalgamation, II.151.
Torre Barrio y Lima, Felipe de la, II.145n.

Torres [y Guerra], Alonso de [fl. 1782–1793]: his travels, [28t], 70.
Torres [y Rueda, Marcos de, 1591–1649], bishop of Yucatan, viceroy of Mexico, 365.
Torti, Francesco (1658–1741), II.394.
Tovar, [Juan de, c. 1541–1626]: author of unpublished writings on the conquest of Mexico, 342n.
Toze, Eobald (1715–1789), II.239n.
Tralles, Johann Georg (1763–1822), 114; II.88n.
Triesnecker, Franz de Paula (1745–1817), 31n, 49n.
Tuckey, James Kingston (1776–1816), 556n.
Tupac Amaru [1545–1572], so-called Inca of Peru: his rebellion and his death, 262; [II: 445, 453–54].
Tupac Amaru II. *See* Condorcanqui, José-Gabriel.
Turnbull, John (fl. 1799–1813), 201.
Tussac, François Richard de (1751–1837), 519n.
Tzin-teotl: Mexica goddess, 364n.
Tzotzomatzin: predicted to Ahuitzotl the danger to which the new Huitzilopochco aqueduct exposed the capital, 357.

Ugarte, Juan de, Father [1662–1730]: his travels in California, [51,] 440, 444.
Ugarte [y Liaño], Tomás de [1754–1804]: how he determined the position of Veracruz, 33.
Ulloa, Antonio de [1716–1795], [83n, 155; II: 45, 184n, 186, 206, 209, 216–20, 224–25, 333, 391, 393]; governor of Huancavelica, II.186; commanded the last fleet that arrived in Veracruz before the trade with the Americas was declared free, II.369.
Ulloa, Francisco de [fl. 1535–1540; d. 1571]: explored the coasts of California the whole way to the Río Colorado, [51, 71], 440.
Unanué [y Pavón, José] Hipólito [1755–1833]: introduced vaccination to Lima, 223, [288n; II: 198n, 200n].
Urrutia [y Montoya], Carlos de [1750–1825]: his map of part of New Spain, 61.
Uztáriz [y Hermiaga, Gerónimo, 1670–1732]: his estimate of the quantity of precious metals that have flowed back into Europe since 1492, II: 239–40, [242–43, 250n, 252].

Vahl, Martin (1744–1804), 558.
Valdés [y Fernández Bazán], Antonio [1744–1816]: minister of Spain; his attempt to regulate the allocation of mercury, II.166.
Valdés [y Flores, Cayetano, 1767–1835]: his travels in California and to Nootka, [48–49], 449, [459, 465], 467–69; [II.95].
Valencia, [Isidro] Vicente [1776–1811], 67; his description of the mines of Zacatecas, 400; [II.122].
Valencia, Tomás (1752–1819), II.229n.
Valenciana, Count of, 279–80, [378; II.197. *See also* Obregón y Alcocer, Antonio].
Valentin, Louis (1758–1829), II: 391, 406.
Valentini, Michael Bernhard (1657–1729), II.391.
Valera, Blas (1545–1597), II.249n.
Valerianos, Apóstolos (c. 1531–1602). *See* Fuca, Juan de.
Valiente y Bravo, José Pablo (1740–1818), II.407.
Valladares, José Sarmiento de, Count of Montezuma [1643–1708]: descendant of king Montezuma, viceroy of Mexico, 352n.
Valle, Marquis de la: Cortés's title, 281, 320n, [341, 440].

Vallejos, intendant of Cuenca: his work on the mercury mines, II.183.
Valleumbroso, Marquis of, 233. *See also* Figueroa y Acuña, José Agustín Pardo de.
Valmont de Bomare, Jacques-Christophe (1731–1807), 327.
Vancouver, [George, 1757–1798], [26, 33, 45, 48]; his determination of the position of Monterrey, 49–50, [51, 53, 70, 74, 128t, 147–49, 276, 449, 457–58, 460, 462–65, 467–70, 73–74; II: 363n, 364n, 392n].
Varaigne, Filiberto Héctor, II.216n.
Vater, Johann Severin (1771–1826), 233, 451.
Vauquelin, [Louis Nicolas, 1763–1829], 427; II: [67, 89], 400n.
Vazie, William (b. 1756), 360n.
Velasco, Pedro Fernández [de]: introduced amalgamation to Potosí, II: 144, 184, 225.
Velasco [y Castilla y Mendoza], Luis de, [Marquis de Salinas del Río Pisuerga II, 1539–1617], viceroy of Mexico, 355–56; had construction on the desagüe of Huehuetoca begun, 357, [374]; introduced the Sisa tribute, 364; established textile factories in Texcuco, II.294.
Velasco [y Ruiz de Alarcón], Luis, el Viejo or Primero [1511–1564], viceroy of Mexico, 357, [165]; founded the town of Durango, 343.
Velázquez [y Cárdenas de León], Joaquín [1732–1786], director of the highest tribunal of mines in Mexico City: how he determined the longitude of that city, 29, [31, 33, 354]; great levelling and trigonometric work this astronomer performed in 1773 [and 1774], 43–44, [76, 320, 338n, 359]; his travels in California, 45–46, [53, 60, 278–79]; his maps of New Spain, 61–62, [65, 67; II.484]; his triangulations, 77–78, [80]; biographical notes on this famous man, 277; his levelling of the waters of the canal of Huehuetoca, [368], 372–73, [376]; his research for Méndez's plan to save Mexico City from floods, 373; [II: 143, 194n].
Vélez de Escalante, Silvestre. *See* Escalante, Silvestre Vélez de.
Velosa, Gonzalo de: built Mexico's first sugar mills, II.4.
Venegas, Father [Miguel, 1680–1764]: his ideas about California, [64], 444–45.
Veraguas, Duke of. *See* Colón de Portugal y Castro, Pedro Nuño.
Vernaci y Retamal-Villaredo, Juan José (1763–1810), 459, 467.
Vespucci, Amerigo (1451–1512), 239n, 513, 523, 557.
Viana, Francisco Leandro de (1730–1804). *See* Tepa, Count of.
Vicuña, Juan (d. 1768), 166.
Villalobos, Ruy López de [1500–1546]: discovered San Benedicto, 70.
Villalpando, Luis de (d. c. 1552), 407.
Villaseñor y Sánchez, José Antonio de (1703–1759), 202; II.432.
Villavicencio, Manuel Galicia de (1730–c. 1788), 64.
Villena, Marquis of, viceroy of Mexico: put Father Flores in charge of hydraulic projects, 365. [*See also* López Pacheco, Diego.]
Villermé, Louis-René (1782–1863), 211.
Vincent, William (1739–1815), 391n.
Vincenzo, Count Dandolo (1758–1819), 546n.
Vines, Richard (1585–1651), 224n.
Visconti, Ennio Quirino (1751–1818), 113n.
Vitruvius (fl. 70–15 BCE), 372; II.144n.
Vizcaíno, Sebastián [1548–1624]: took possession of the peninsula of California, 443; his travels in New

Vizcaíno (cont.)
California, 447–48, [453]; called the capital of this country Monterey, 455; care with which he measured the coastlines, 458–59; [II.361–62].
Vizlipuzli. *See* Huitzilopochtli.
Volney, Count of, 74, 186, 236, 239–40, 243n, 451; II: 393n, 415. *See also* Chasseboeuf, Constantin François de.

Wactupuran: Californian deity, 444.
Wafer, Lionel (c. 1640–1705), 155.
Wal, Manuel: his plan to bring independence to the province of Venezuela, II.455. [*See also* Gual, Manuel.]
Walker, James, II.190.
Wallis, Samuel (1728–1795), II.360.
Wargentin, Pehr Wilhelm (1717–1783), 29, 46.
Webb, William Spencer (b. 1785), 173.
Werner, [Abraham Gottlob, 1750–1817]: his theory on the origin of veins, [276; II: 76, 105], 108, [109n, 223n].
Wilkins, Charles, Sir (bap. 1749–1836), 487.
Wilkinson, [James, 1752–1825], General, 63, [560n].
Willdenow, Carl Ludwig (1765–1812), 558n; II.27.
Wilson, James, Captain (1760–1814), 201.
Wistar, Caspar (1761–1818), II.407.
Witte, Samuel Simon (1738–1802), 340.
Woodhouse, James (1770–1809), II.417.
Wright, William (1735–1819), 522.
Wurm, Johann Friedrich (1760–1833), 73.

Xocojotzin, or the cadet: nickname for Montezuma II, 320n.

Yáñez Pinzón, Vicente (c. 1492–1509), 513.
Yniesta, Ildefonzo: his measurement of the distance between lake of Tezcuco and the casinde de Tula, II.448. [*See also* Iniesta Bejarano Durán, Ildefonso de.]
Young, Arthur (1741–1820), 185, 541, 454; II: 285n, 286n.
Ysasi, Joaquín de [fl. 1760–1780s], 499.
Ysita, José, II.471n.

Zach, Franz Xaver, Freiherr von (1754–1832), 17, 23n, 24, 43n, 49n, 73n, 114; II.258n.
Zapata [y Mendoza, Juan Buenaventura, c.1600–1689]: baptized Indian, author of an unpublished history of Mexico, 342n.
Zárate [Salmerón], Gerónimo de, Father: was put in charge of road reconstruction in Mexico City, 334.
Zea, Francisco Antonio (1770–1822), 233, 525–26, 529.
Zepeda, Bernabé Antonio de [fl. 1778]: his work on the mines of Catorce, 423; [II.125].
Zepeda, [Fernando de, fl. 1636–1647]: his history of the desagüe, 354n, [359n, 367; II.135].
Zoëga, Johann Georg (1755–1809), 324n, 383, 403n.
Zúñiga y Acevedo, Gaspar de (1560–1606). *See* Monterey, Gaspar de Zúñiga y Acevedo, Count of.
Zúñiga y Ontiveros, Felipe de (c. 1717–93), II.451t.
Zúñiga, Juana de: Cortés's wife who outfitted a flotilla to go in search of her husband, 440; [II: 498, 501–2, 505].
Zurita, [Alonso de, 1512–c. 1585]: author of unpublished work on the conquest of Mexico, 342n.

Subject Index

Academía de los Nobles Artes de Mexico: its influence on national taste, 274; new project, 276n; its palace, 335.
Acatl (the First): to which year this period corresponds, 324n.
Acoclames: uncivilized Indians, 424.
Acolhua, [199]; this people's arrival in Mexico, 232, [235, 464, 467–66, 323, 337–58, 414, 437].
Acordada: prison in Mexico City, 334; [II.483t].
Additions and revisions: the most important ones in this new edition. *Volume I*: the importance of geography, 16–18; the longitude of Veracruz, 34–35; about the position of Cape San Lucas, 47–48; about the position of Nootka, 49; about the position of Santa Fe in New Mexico, 54–56; copies made of Mr. von Humboldt's large map of New Spain, 67–69; about the position of the Isthmus of Panama, 78–84; rivers located between 33° and 42° latitude, 90–92; map of incorrect positions, 95–96; tableau of important political divisions, 145; about the Río Colombia, 148–49; about the isthmus of Nicaragua, 152–53; details about the isthmus of Panama, 160ff; about oceanic canals, 167–71; numerical results relating to the climates of Mexico, 193–95; population, 210–19; chronological tableau of the history of Mexico, 263–268. *Volume II*: revisions of and remarks for the statistical tableau, 475–502; about the banana species, 517, 397, 519f; about corn, 525–26; about harvests, 545–46; about Solanum tuberosum, 545–56. *Volume III*: sugar, II: 7, 9, 17; cotton, II.17–18; cultivation of hemp, II.19–20; indigo, II.32–33; numerical signs among the Mexica, II.67–68; Mr. Acosta's ideas, II.87–88; discovery of platinum, II.88–89; Real de Minas in Guanajuato, II.118–22; about amalgamation, II: 156–62, 164; minting, II: 169–74, 198, 221–22; about the production of the gold mines of Brazil compared with those of the Ural region, II.263–74; about the relative amounts of precious metals for coins and for gold and silver smithing, II.273–82; about the changes in the accumulation

Additions and revisions (cont.) of precious metals in Europe, II.283–87. *Volume IV*: Trade balance of Veracruz, II.341 and a tableau; colonies of Russian America, II.365–66. *Supplement*, 489–94. Excerpt from Hernán Cortés's Testament, II.495–505. (In order not to make this list overly long, only the most substantial additions have been included. Many passages interpolated into the text and new notes, among them a large portion of the first volume, have gone unmentioned.)

Administration, public: expenses it causes, II.463–37.

Aerolites: masses of malleable iron that appear to be meteorites, 427. *See also* Iron, meteoric.

Agave. *See* Maguey.

Agricultural products: amount imported to Mexico in 1801, II.330t.

Agriculture: its condition in New Spain, 505ff; what the word means in the tropics, 507; in the interior of Mexico, 507–8; impact of mining on agricultural prosperity in different various parts of Mexico, 507–8; annual total of agricultural production, II.3–4; obstacles to its improvement, II.59. *See also* Plants.

Agulichan: Russian post, 472.

Ahahuete [Ahuehuete] (*Cupressus disticha* [bald cypress]): renowned for its size, 384.

Alatlauquitepec: mines, 388; [II.75t].

Albaradón de San Lázaro: built by Velasco I, viceroy of Mexico, 357–38, [370].

Alcabalas, indirect taxes, [9]; Indians are exempt, 255, [261, 281; II: 176, 433]; what they are, II.367; annual proceeds from, II: 433–35, [443, 445, 463].

Alcosac: remnants of one of the small pyramids that surround the large teocalli of Cholula, 384.

Almojarifazgo: duty paid on merchandise, II.367; annual proceeds from, II: 431, [433t, 445].

Almonds: amount imported to Veracruz in 1802, II.322t; in 1803, II.330.

Almshouse, in Mexico City, 334.

Altar, Presidio del, 52, 57, 130t; II.458n.

Amalgamation: used in the Mexican mines, [228, 276, 528; II: 63, 85, 93–94, 100t–103t, 112, 116, 118, 129, 132, 138, 142–47, 151–54], 155; costs per 100 quint[als] of minerals, II: [156–57, 160t, 163, 166–68, 180], 207; as used in the mines of Peru, II: [144, 150, 157, 182, 184, 187], 207; and in Potosí, II.225.

America, Russian: [218t, 309, 447, 456, 461–62; II: 364, 366n]; description of this country, 470–75.

America, Spanish: comparison of its area with the Russian empire and the British possessions in Asia, 139–140; its division into nine governments, 140; annual amount of its production of gold and silver, II.196–97; comparison of its area, population, and revenue with those of the British possessions in India, II.446–47.

America, United States of: its population, 143–44.

Americans: significance the Creoles give this name, 270.

Anchovies: amount imported to Veracruz in 1802, II.232t. *See also* Salt fish.

Angangueo: mines, 397; [II: 74, 91].

Animal kingdom of Mexico, products of: horned beasts, II.33; sheep, II.35; turkeys, II.37; guinea fowl, II.37–38; Muscovy ducks, II.37–38; silkworms, II.38; silk moth of the species *bombyx*, II.38–39; bees, II.39–40; cochineal, II.40–41; pearls, II.48–49; *murex* snails and mollusk shells of Monterrey, II.50.

Anise: quantity exported from Mexico in 1803, II.335t.

Annatto received by the king in Mexico, annual proceeds from, II: [334t], 370t.
Antimony, [II: 151–52, 155]; mines that furnish it, II: 179, [200n].
Antipathy between the inhabitants of the plains and those of the plateau of the Cordilleras, II.348.
Antiquities, Toltec, 186–87.
Apaches, [20, 237]; uncivilized Indians, 251; where they live, 418, 424–25, [II: 451, 458].
Apiculture: bee keeping in Mexico, II.39–40.
Aqueducts: [246, 381]; that bring drinking water to Mexico City, 327, [332–33, 338, 397]; Texcuco aqueduct, 333n; Jamapa aqueduct, 416; [in Moran, II.130; in Santa Fe, II.298].
Aristocrats, Indian. *See* Caciques.
Army. *See* Forces, armed.
Arsenic: [II: 151–52, 156, 161t]; mines where it is quarried, II: 179, [185, 187].
Artichokes, [351, 561, 562]; Jerusalem, not cultivated in Mexico, 559.
Asiento de Huantajaya, mine: its yield, II.203.
Asientos de Ibarra, mine, II: [201], 203, [206n, 217, 245t].
Atole: broth made with cornmeal, 528.
Axes, II.266–67.
Axes [ajes, 556–57]. *See [also]* Igname.
Axolotl, reptile: food of the Aztecs, 323, [395].
Aztecs, people of Mexico, [199, 243n, 247, 249, 350, 377, 410, 561, II: 22, 28, 33n, 249]; hypothesis about their origin, 231ff; [their language, 235, 264, 398]; their migrations, 232, [245,] 267, [430,] 213, [437]; their settlement on the islands of Acocolco, 323; in Tenochtitlan, [321n], 323, [357; II.316n]; three stops they made during their migrations, 431; their preference for living in isolation on mountainsides, 510.
Baal-Berith, temple of, 335.
Babylon, ancient monument of, 324, 340, 383.
Bacuachi: presidio, II.458n.
Balms: amount exported from Mexico in 1802, II.326t; in 1803, II: 334t, [380].
Banana: cultivation of, [185, 238, 294, 409, 517n, 518, 512–22, 563, 567], 512; three known species, 513–14; its uses, 516; as an object of trade, [516], 518–19; doubts as to its origin, 514–15, [519n, 526n, 556, 558]. *See [also]* Musa.
Baquetes [pouches?]: impact the free trade had on their export, 371t.
Barenadores: category of miners, 227.
Basanes [full-grained sheep leathers]: impact that free trade had on their export, II.371t.
Bayeta [flannel, cloth]: amount exported from Mexico in 1803, II: [294], 335t.
Bayettes [baize]: quantity exported to Mexico in 1803, II.335.
Beasts of burden: used to transport merchandise, 376, 502; II: 33, 114, 137, 182, 314, 420n; lack of, 386; better than carriages, II.309.
Beer: amount imported to Veracruz in 1802, II.323t; in 1803, II: 331t, [332t; in 1804, II.338t].
Belus in Babylon, temple of, 324, 340, 383; II.484.
Bendavales: periodic winds that prevail across the Great Ocean, II.351.
Berendo [pronghorn]: animal of New California, 442, [453].
Biblioteca Americana, 151n, 154n.
Births, [205–6, II.446–70]; birth-death ratio in Mexico, 203t; relation to the population, 227–28; example of how the births registers are kept in Mexico, II.467.
Blacks, [200n, 214t, 215t, 216t, 217t, 219, 237, 240, 242, 243n, 272, 443, 476, 522, 530, 568; II: 7, 255n, 385n,

Blacks (cont.)
408, 412, 455]; not many of them in Mexico, [213n], 230, [246, 248, 379, 284–85, 288–89, 452n, 455t, 475; II: 6, 410]; law that favors their emancipation, 287.

Bolaños: mines, [15, 21], 399–400; [II: 72, 81–83, 84t, 91, 95n, 126, 130, 136, 143–44].

Bombón: mines, II.201.

Borders of the United States and the United Provinces of Mexico: remain undecided, 484.

Botanical Garden, Mexico City, 274–76, 334, [462n].

Brandy, [348t, 570; II: 26, 251, 312]; amount imported annually to Veracruz, II: 322t, [338t]; in 1802, II.322t; in 1803, II.330t.

Brazil: quantity of sugar it exports, II: 12, [14t]; its mines, II: 243t, 245t; yield of its mines compared to those in the Ural, II.263; the greatest wealth its washes produced from 1752 to 1761, II.264; yield of it mines from 1491 to 1805, II.271–72.

Brazilwood [*Caesalpina echinata*] and Bois jaune [*Ochrosia borbonica*], [407; II.308]; exports from Mexico to Cádiz in 1802, II.383t.

Bread: annual consumption in Mexico, 349.

Breadfruit: unknown in Mexico, 563.

British East India Company, 24; II: 257n, 286–87.

Brown's system: its use on yellow fever, II.416.

Buenavista: presidio, [57, 130, 430]; II.258t.

Buenos Aires, viceroyalty of: yield of its gold and silver mines, II.208; its trade balance, II.387t; net revenue the king derives from this viceroyalty, II.443. [*See also the Toponym Index.*]

Buffalo hides: impact the free trade law has had on their export, II.371t.

Butter, [II.34]; quantity imported to Veracruz in 1802, II.324t; in 1803, II.332t.

Cabildo, Libro del: 1524 manuscript, 337.

Cacao: its varieties, II.21n; how much of it is consumed in Europe, II.21; used as currency in Aztec times, II.22; quantity imported to Veracruz in 1802, II.325t; impact free trade had on its export, II.370t.

Cacao from Acayucan, 410.

Cacao from Caracas: quantity shipped to Europe via Mexico in 1803, II: 325t, 383t.

Cacao from Guayaquil: shipped to Europe via Mexico, II.314; quantity imported annually to Veracruz, II.321–22; quantity exported from Mexico in 1802, II: 326t, 149; in 1803 to Spain, II.334t; to other parts of the Spanish Americas, II.334t.

Cacao from Maracaibo: quantity imported to Mexico in 1803, II.323t; exported, II.334t.

Cacao from Soconuzco: quantity exported from Mexico in 1802, II.326t; in 1803, II.334t.

Cacao from Tabasco: quantity imported to Mexico in 1803, II: 333t, [338t].

Cacao of Guayaquil, [II.55]; shipped to Europe via Mexico, II.314–15; annual amount imported to Veracruz, II.21–22; amount exported to Mexico in 1802, II: 325t, 383t; in 1803 to Spain, II.334t; to the other parts of Spanish America, II.334t.

Cacao tree: its cultivation in Mexico, II: 21–22, [461].

Caciques: indigenous aristocrats, 252; how they humiliate the tributary tribes, 252.

Cactus: different species on which cochineal feeds, II.44–45.
Caledonian Canal, [153], 168–69.
Californians: their divinities, 444.
Calli, second: year to which this period corresponds, 324n.
Camburi: banana species, [513–14], 521.
Camels: introduced to Peru, [161], II.314.
Camotes, [558]. *See [also]* Batatas.
Campeche wood [*Hæmatoxylum campechianum*]: province that provides it, 407–8; annual quantity exported from Veracruz, II.321; in 1802, II: 325t, [326t, 327t], 383t; in 1803, II: 333t, [334t]; about its export from other parts of the Spanish Americas in 1803, II: 335t, [338]; impact free trade had on its export, II.371t.
Canal ordered in 1814, 150n.
Canals: of average size, 168.
Canary Islands: gross income, II.444.
Candles, [55; II: 360n, 459]; quantity imported to Veracruz in 1802, II.324t; in 1803, II.331t.
Cannons, [352, II: 69, 128, 300]; melted down in Manila and transported to Veracruz, II.317.
Cañon de los Virreyes: tunnel of the desagüe of Huehuetoca, 369.
Canteens: quantities imported in Veracruz in 1802, II: 323t, 64; in 1803, II.330–32t.
Canvas, made of linen and hemp: [yearly consumption of, 9; cotton canvas, 338]; not made in Mexico, II.338; annual amount imported to Veracruz, II.321t; amount imported to Veracruz in 1802, II.323t–324t; in 1803, II.331t–332t.
Canvas, painted, made in Mexico, II.294.
Capers: quantity imported to Veracruz in 1802, II.323t; in 1803, II.330t.
Capital the clergy put into real estate, II.58.
Capitation tax: paid by Indians; annual revenue from, II.430.
Capuces: Indian tribe, 388.
Caracas, captaincy-general of: annual trade balance, II.387t; numbers of slaves in, II.385n; gross revenues, II.444t.
Carneros cimarrones [bighorn sheep]: animal that lives in the mountains of California, 442.
Carriages manufactured in Mexico, II.306.
Cartagena de Indias: annual trade balance, II.380.
Casa del Apartado in Mexico City: description of this building, II.304.
Casa del Estado, [322]; situated on the site of Montezuma's palace, 341.
Casas grandes del Río Gila: Aztec antiquity, 432.
Casas grandes of New Viscaya, 431.
Cash in circulation in the New World: discussion of this topic, II.38; amount of its annual accumulation in Mexico, II.375t.
Cassava: manioc bread, 521, [523, 541].
Castes among the inhabitants of Mexico, 230ff; relationship among castes, 295ff.
Castillo de San Carlos, 153.
Cathedral in Mexico City, 27n, 33, 77, 320, 325, 331–32, 334, [341, 498].
Cattle: in Mexico, [224, 381, 411, 426, 453]; II: 34, [359, 461, 465].
Causeways: Mexicaltzingo, 357; Tepejacac, 325; Tlacopan, 325; Iztapalapa, 325; Tlalma, 357. *See also* Albaradón; Cruz del Rey.
Cempohualilhuitl: secular calendar of the Mexica, 413.
Census of the inhabitants of New Spain: the first one, 200–201; details of this

Census of the inhabitants (cont.) process, II.470ff; census recently ordered by the provisional junta, 475ff.
Ceremony, religious: celebrated on the occasion of the introduction of vaccination, 222.
Chairs: quantities imported to Mexico in 1803, II.331t.
Chamalitl [sunflower], *Helianthus annuus*: its cultivation in Mexico, 559.
Chapetones: name for whites born in Europe, 270; [II.411n].
Chapultepec: aqueduct that carried potable water to Mexico City, 332; castle built by viceroy Gálvez, 352; its ruin, 353.
Cheese, [II.34]; amount imported to Veracruz in 1802, II.324t; in 1803, II.332t.
Chicha: alcoholic beverage of the Mexicans, 529–30.
Chichimecs: tribe of uncivilized Indians, 142, [331, 235, 237, 251, 268], 388, 396, 424; their arrival in Mexico, 232; their history, 267.
Chicken: bird unknown in Mexico before the conquest, II.36–37.
Chile: production of its gold and silver mines, II.208.
Chili. *See* Pepper.
China: caste known by this name in Mexico, 298.
China: quantity of gold and silver carried from there to Europe, II.257.
Chocó, province of: gold and silver produced there, II: 228, 245; condition of this province, II.230.
Cholula, pyramid of: its size, 383; compared to the ones in Egypt, to the monument of Belus and the teocallis of Teotihuacan, 383, 486–87.
Chugatch: people in Russian America, 473.
Chukchi: people in Russian America, 472.
Chunu: way of preparing potatoes, 554.
Cicimecs [Chichimecs]: nomad people in the north of Mexico, 142.
Cider: quantity of its importation to Veracruz in 1802, II.323.
Cigar factory: in Querétaro, II.296.
Cigars, II: 29, 30, 329.
Cinnabar, [II: 65, 66n, 179, 181–85, 187–89]. *See [also]* Mercury.
Cinnamon: quantity imported to Veracruz in 1802, II.324t; in 1803, II.332t; [in 1804, II.338].
Civilization, Indian, 246–47; inroads it has made among the whites in Mexico, 273f.
Clayborne: fort in Louisiana, 421–22, [457].
Clergy, Mexican: their numbers, 282; their wealth, 282; amount of capital they have invested in real estate, II.58.
Clerigo Bridge: place where the last Aztec king was captured, 343.
Climate of Mexico: [172], 179–80, 505; II.403t–404t; its impact on agriculture, 519–20, 534.
Cloth, woolen: manufactures of, II.294.
Cloves: quantity imported to Veracruz in 1802, II.324t; in 1803, II.332t; [in 1804, II.338].
Cochineal: cultivation of, 404; II.41ff; amount exported annually from Veracruz, II.321t; amount exported from Mexico in 1802, II: 326t, 383t; to Spain in 1803; [in 1804, II.338]; II.334t; to other parts of the Spanish Americas, II.335t; impact of free trade on its exports, II.370t.
Cock: name the Peruvians gave to this bird to ridicule it, II.36.
Cock fights: tax yields from these fights, II.432.
Cocoyames: uncivilized Indians, 424.

Cod, salted: amount imported to Veracruz in 1802, II.324t; in 1803, II.332t.
Coffee: quantity exported from the island Cuba, II.20; quantity imported to Veracruz in 1802, II: 325t, 383t; in 1803, II.333t; exported from Mexico in 1802, II.326; in 1803, II.333.
Coffee plant: its cultivation in Mexico, II.20–21.
Coining: from the mint in Lima, II.119t; in city of Potosí, II.222; comparison of how much metal was used for coinage and for gold and silver smithing, II.273–83.
Coins, silver: amount exported from Mexico to Spain on behalf of individuals in 1802, II.374; in 1803, II.334t; to other parts of the Spanish Americas in 1802, II.327t; in 1803, II.335t; on behalf of the crown in 1802, II.329.
Colonies: principles according to which modern colonies were established, II.291.
Colonies, Spanish: effects that produced their independence, II.445–46. [*See also* Spanish America.]
Colpa, vitriolic earth: its use in amalgamation, II: [151–52], 304–5.
Comanches: uncivilized Indians, 424; [II.451]; their skill in handling horses, 424.
Comilhuitlaohualliztli: religious calendar of the Mexica, 413.
Commodities, colonial: Mexico's wealth of them, II.3ff.
Communication between the South Sea and the Atlantic Ocean: Map of the points that would make it possible, 82; there are nine of those, 147: (1) between the Peace River and the Tacoutché-Tessé, 147; (2) between the Rio del Norte and Rio Colorado, 149; (3) between the River Huasacualco and the Chimalapa, 150; II.315; See *Huasacualco.* Temporary connections, 150; (4) between Lake Nicaragua and the Gulf of Papagayo, 150–54; the ease of this connection, 168; (5) at the Isthmus of Panama, 154–61; physical conditions that would make this connection possible, 157; political consequences it would have, 160–61; (6) between the Bay of Cúpica and the Rio Naipi, 251; (7) via the ravine of Raspadura, a connection that has existed since 1788, 235; (8) via the Gualaga River, 165; (9) via the Gulf of Saint-George, 165–66; remarks about this communication extracted from the *Relation historique*, 167ff.
Conchuco: yield of its mines, II: 203–4, [244t].
Conductors, electrical: introduced to Mexico, II.298.
Conil, bocas de: freshwater surrounded by saltwater, 407.
Connections between the races in Spanish America, 216t. [*See also* Race.]
Conquer, to: meaning of this word the language of the missionaries, 429n.
Conquistadores: descendants of the first Spaniards to settle in Mexico, 252.
Consulado de la Veracruz: trade summaries it published in 1802, II: 322t, 328f; in 1803, II: 330, 336–37; its makeup and its functions, II.348.
Consulado de Mexico: completed the desagüe of Huehuetoca, 336.
Convent of San Francisco, Mexico City, 334.
Copper: did the Mexica know how to turn it into steel?, II.66–67; copper minting, II.177n.
Copper from Coquimbo [Chile], transported to Europe via Mexico, II.315.

Copper ornaments: amount exported from Mexico, in 1802, II: 327t, 383t; in 1803, II.335t.
Copper plate: amount exported from Mexico to Europe in 1802, II.326t; to the other parts of the Spanish Americas in 1802, II.327t; for the king, 335t.
Cora: ancient people of Mexico, 266.
Cordage: amount imported to Veracruz in 1802, 65; II.323t; in 1803, II.325t. [*See also*] Ropes.
Cordovan leather: impact the free trade law has had on its export, II.370t.
Corks: quantity imported to Veracruz in 1802, II.323t; in 1803, II.331t.
Corn. *See* Maïs.
Corvan, Saint Jerome of: monastery, 256.
Cosecheros: Indians who plant vanilla, II.26.
Costales. *See* Sacks.
Cotton: its cultivation in Mexico, [392, 410–11]; II: 6ff, [461]; amount exported from the United States to Great Britain, II.18; to the rest of Europe, II.18; amount exported from Mexico in 1802, II: 326t, 383t; in 1803, II.335t; impact the free trade law had on its export, II.370t.
Cotton, dyeing of: in Tehuantepec, [497]; II.294.
Cotton canvas: amount exported to the intendancy of Guadalajara, II.294; to the intendancy of Puebla, II.294, imported to Veracruz, II.321; amount imported to Veracruz in 1802, II: 323t, 324t; in 1803, II: 331t, 332t.
Cow horns: impact the free trade law has had on their export, II.371t.
Creoles: whites born in the colonies, 270; hatred between them and the Europeans, 416–17.
Cross: ancient sculptures representing it found in Guatemala, 487.
Cruz del Rey [calzada de la]: causeway that divides the two basins of Lake Zumpango, 354.
Cruzada, Bulles de la [Papal indulgences]: revenue from them, II: 431, [433t, 435t].
Cumin: amount imported to Veracruz in 1802, II.323t.
Currency: mint in Mexico City, 334; [II: 85, 168–69, 173–74, 240, 244, 273, 323n]; its work compared to that of the mints in France, II.280ff; quantity of coins produced annually, II: 302–3, [340, 378]; annual profit it produced for the king, II: 429, [442–43]; [mint in Lima, Peru: II: 198, 199, II.429].

Deaths: in relation to births in Mexico, 205ff; to the population, 207–8; to the sexes, 292–93; tableaux of deaths that the author used to calculate the population, II.290ff.
Deer: species in New California, 453–54.
Defense of the country: general observations, II.445–46; its purpose could not have been to be prepared for an invasion by a maritime power, II.457; Indian wars, II.458; defense of the eastern coastlines, II.459.
Depósito Hidrográfico (Madrid), 24, 28, 43, 59, 69, 83, 154, 158–59, 162n, 165, 468.
Deposits, metalliferous: of Mexico, II.78–79.
Derecho de oro y plata: source of revenue for the Spanish king from the yield of the mines, II.429t.
Derecho de tierra caliente: nature of this tax, II.319.
Desagüe de Huehuetoca: its profile, 108; manuscripts the author used for his comments on this topic, 354n; Martínez started building this canal

in 1607, 358; failings for which he is criticized, 359; new leveling by Alonzo de Arias, 361; Adrian Boot placed in charge of the inspection of the hydraulic work, 361; Martínez resumes the work, 361; he causes a flood and is arrested, 361; Simón Méndez charged with completing the project, 361; projects by Antonio Román, Juan Álvarez de Toledo, Cristóbal de Padilla, and Francisco Calderón, 363–64; Martínez again assigned to take over the project, 364; replaced by Father Luis Flores, 365; Martín Solís then directs the work, 365; the damage he causes, 365; Manuel Cabrera becomes superintendent, 365; sluggishness with which the work proceeds, 365; a group of businessmen in Mexico City takes over, 366; they complete the project in 1789, 366; the author's opinion of this work, 367; the dimensions of the desagüe, 368; costs it caused until 1789, 370; the leveling of its waters, 372n; this canal is one of the reasons for the impoverishment of the indigenous people in the Valley of Mexico, 374.

Deus Redicolus: Etruscan temple, 403.

Dikes [embankments or causeways]: used by the Aztecs to protect the city of Tenochtitlan from floods, 357; this system was abandoned after the flood of 1607, 358.

Dioscorea [alata], [407, 443, 515, 522, 524, 556–57, 558–59]. *See [also]* Igname.

Dioscorea sativa, 558.

Diputaciones de minería: thirty-seven in number among which the mines in Mexico are distributed, II: 69ff, 75, [191–92].

Discoveries by the Spanish on the northwestern coast of the Americas: by Cabrillo, 457; by Gali, 457; by Viscaíno, 458; by Pérez, 459; by Heceta, Ayala, and Quadra, 460; by Quadra and Arteaga, 460; by Martínez and Haro, 461; by Martínez himself, 461; by Elisa and Fidalgo, 464; by Malaspina, 465; by Galiano and Valdés, 467; by Caamaño, 469; by the Spanish in the Great Ocean, II.361.

Divisions, political, of the Spanish Americas, 144.

Divisions of the territory of New Spain, 303ff, 311ff. *See [also]* New Spain.

Dogs: [224, 260, 462n]; used as food, II: 33n, [118n].

Dominico: banana species, 514.

Drawing school in Xalapa, II.301.

Dresden Codex, 487.

Durango, bishopric of: its revenues, 283t.

Durango, intendancy of: its size and population, 423–24; it suffers from attacks by uncivilized Indians, 424; its towns, 426ff; names of its reales de minas, II.73.

Earthenware made in Mexico: II.297; amount imported to Veracruz in 1802, II: 323t, 324t; in 1803, II.331t; amount exported by other parts of the Spanish Americas in 1802, II.327t; in 1803, II.335t.

Encomiendas: type of fiefdoms established for the conquistadores, [204], 253; their abolition, 254.

Entradas: type of war the missionaries waged against the Indios Bravos, 286.

Equestrian statue of King Charles IV, Plaza mayor, Mexico City, 275, 331, 335.

Escorial (Madrid), 346, 464.

Esselen: people who live in New California, 450ff.

Europeans: this word is synonymous with the Spanish in Mexico, 273.

Expeditions, botanical: undertaken by order of the government, 275.
Expenditures, public: annual amount from 1784 to 1789, II.436t; in 1803, II.437f; classification of expenses: (1) domestic administration, II.238–39; (2) situados, II.440–41; (3) net revenue, II.442.
Export of merchandise from Mexico: amount in 1802, II.325t; in 1803, II.333t; in 1804, II.338t; in 1805, II.339; in 1806, II.339, 341; annual amount, II.375t; in all of the Spanish colonies, II.380–81.
Eyder (or Holstein) canal (Schleswig-Holstein, Germany), 160, 168–69.

Fábrica, La: lazaret, 492.
Fabrics: amount of their annual import to Veracruz, II.321.
Famine: why this scourge frequently afflicts Mexico, 225.
Faraones. *See* Apaches.
Fifth payed to the king in Potosí, II.209ff.
Figs: amount imported to Veracruz in 1802, II.323t; in 1803, II.330t.
Finances of Mexico. *See* Expenditures, public; Revenues.
Finances of the Spanish monarchy: amount of all revenue, II.443–44.
Flax: the government prevents its cultivation in Mexico, 565; II.29ff. *See [also]* Hemp.
Floods in the valley of Mexico: 355; times when they occurred, 355–56; flood of 1446, 357; of 1498, 357; of 1553, 357; of 1580, 358; of 1604, 333, [358]; of 1607, 344, [358]; from 1629 to 1634, 362; of 1763, 371; of 1772, 371.
Florida: sum its draws annually as subvention for is administration, II.441t.
Flour: amount exported annually from Veracruz, II.321t; from all of Mexico in 1802, II.327t; in 1803, II.335t.
Food products: amount exported from Mexico to others parts of the Spanish colonies in 1802, II.324t; in 1803, II.333t.
Forces, armed, in New Spain: amount of expenses they cause the state, II.439t; costs relative to revenues, II.447; size of the army in 1804, II.447t; tableau showing the division of line troops, II.448t; the militias, II.449t; number of troops, II.450; weariness of those living in presidios, II.451–52.
Fort Maullín (Chile), 139, 470; II.465.
Fort Miami (near Lake Erie), II.415.
Fort Wayne, 239n.
Fronteras: presidio, 67, 430; II: 195, 458n.
Fruit preserves: amount imported to Mexico in 1803, II.330t.

Gachupines: name for whites born in Europe, 230, 270, [364n].
Gauze: amount imported from Spain to Veracruz in 1802, II.323t.
Geography: varied knowledges it requires, 16.
Gold: veins that carry it, II.85–86; the best in the Americas, II.229. *See also* Mines.
Gold bars exported from Veracruz, II.321.
Gold coins and gold articles: amount exported from Mexico to Europe on behalf of private individuals in 1802, II.326t; in 1803, II.335t; from other parts of the Spanish Americas in 1802, II.327t; in 1803, II.335t.
Gold leaf: amount exported from Mexico in 1802, II.327t; in 1803, II.335t.
Gold panning in the province of Antioquia: their number, II.229; in Chocó, II.230.
Goldsmithing: by the Mexica in Montezuma's day, II.64; observations on the quantity of metal used for

goldsmithing relative to that used for coins, II.273–74; Necker's thoughts on this topic, II.274; goldsmith products fabricated in France in 1809, II.274; in Paris in 1810, II.274; different problems in this field, II.275; gold products compared by numbers, II.276ff; Mr. Lowe's opinion, II.278; its current state, II: 301, 305.

Governments, forms of: among the Indians prior the conquest, 245–44.

Grains of the Old World: unknown in the Americas prior to the arrival of the Spaniards, 531–32; region agreeable to them, 532–33; how much watering their cultivation requires, 535–36; richness of their crops, 536; average yield from wheat, 536–37; quantity of the harvest in New Spain, 539t; comparison of the average yield with that in other countries, 542–43; labor costs, 547; average price of wheat, 547.

Grindstones: amount imported to Veracruz in 1802, II.323t.

Guachichiles: Indian tribe, 388.

Guachinangos: name for the inhabitants of Mexico City who do not originate from there, 284.

Guadalajara, bishopric of: its revenue, II.58.

Guadalajaxa, intendancy of: size of, 397; climate in, 398; agriculture of, 399; mines in, 399; towns in, 399; manufacturies in II.239; its Real de minas, II.83.

Guamanes: Indian tribe, 388.

Guanajuato, intendancy of: number of church members it includes, 282n; size of, [313t], 388; population, [271, 314], 388; agriculture in, 388; mines in, 399; II.70; towns and cities in, 389; its mineral waters, 390; detailed description of its mines, II.97–98; description of its Real de minas, II.70; rock formations and the link between deposits, II.98; revenue of, II.442.

Guayhuenes: Indians who periodically visited the islands of Huaytecas and Chonos, II.465.

Guiane [French Guiana]: quantity of sugar it exports, II.14t.

Gulfstream: current of warm water, 158.

Gunpowder: its production as royal privilege, II.297; the only existing factory, II.298; amount produced, II.298; annual revenue the king receives from this merchandise, II: 431, [437t].

Habilitadores: investors that trade in vanilla and quinquina, II.26.

Halsbrücke: mine near Freiberg [Saxony], [II: 79, 150n]; minerals it contains, II.158; research on this mine published by Mr. Berthier, II: 158, [161]; essential costs of exploiting this mine, II.160; the nature of its minerals, II.161.

Hammocks: amount imported to Veracruz in 1802, II.325t; in 1803, II.333t.

Hams: amount exported to Veracruz in 1802, II.323t; in 1803, II.330t.

Harpoons: amount imported to Veracruz in 1802, II.325t.

Harpsichords made in Mexico, II.306.

Havana, captaincy-general of: its trade balance, II.386; gross revenue, 251; [II: 386t, 387t].

Hazelnuts [filberts]: amount imported into Veracruz in 1802, II.323t; in 1803, II.330t.

Hebrides: islands in Scotland, 26, 565; II.361.

Hemp: the government discourages its cultivation in Mexico, 565; II.18; falseness of this claim, II.19–20.

Hides, tanned, made in the Intendancy of Guadalajara, II.293; quantity exported from Mexico in 1802, II: [321t], 335t, [370t], 383t; impact the free trade law has had on its export, II.370t.
Hides and pelts: amount exported from Mexico in 1803, II.335t; in 1803, II.334t.
History of Mexico: chronological tableau, 263ff; Indians' notions about cosmogony, 264; migrations and history of the Toltec, the Chichimec etc., etc., 266ff; founding of Tenochtitlan, 268; Mexica kings, 268; Cortés's arrival, 268.
Horcasitas: presidio, 431, 433.
Horses, [421, 435, 450t, 453; II: 33–34, 35n, 52, 117, 485t, 486t]; their propagation on the savannahs, [424]; II: 451, [484].
Hospitals, II: 419n, [420ff].
Hot steam baths: liking the Aztecs and indigenous people of New California had for them, 452.
Huamachuco: yield of its mine, II.203.
Huancavelica: mercury mine in Peru, II: [144, 164–65], 183ff, 207, [209n], 225.
Huantajaya, [II: 201, 206n, 217, 245t]; yield of its mine, II.203.
Huayres: kilns in Cerro de Potosí formerly used to extract silver from ore, II.224–25.
Huitzilopochco: waters carried in the canals of Tenochtitlan, 327; danger to which this aqueduct exposes the town, 357.
Hydrographic Office. *See* Depósito Hidrográfico

Igname: its cultivation in Mexico, [523], 557. [*See also* Yam; Dioscorea.]
Importation of merchandise from Europe to Mexico: amount in 1802, II: 322, 383; in 1803, II.33t; in 1804, II.338; in 1805, II.339; in 1806, II.341t; annual amount of, II: 339, 377; to the other Spanish colonies, II.384; amount of fraudulent imports, 156.
Indians, American, copper-colored indigenous people: their number, 224; their migration from the north to the south, 234: their languages, 235; their physiognomy, 235–36; color of their skin, 237; they have beards, 239; their longevity, 240; drunkenness, 240–41; are not prone to deformities, 241; their similarities with the Mongolian race, 242–43; their moral faculties, 243; state of their civilization before the Europeans arrived, 245; comparisons between them and blacks, 246–47; their religion, 247–48; their character, 248; their sadness, 249; their taste for painting and sculpture, 249; for flowers, 249–50; their social condition, 251; division into taxpayers and nobles, 252; humiliations they experienced after the conquest, 252–53; they were submitted to *encomiendas*, 253; their living conditions were improved during the eighteenth century, 235–36; mainly thanks to the creation of intendancies, 254; differences in wealth among them, 255; the impoverishment of the masses, 254; examples of great wealth, 255; they do not pay indirect taxes, 255; they are subject to tributes, 255; other taxes they pay, 255; they are denied many civil rights, 256; their debasement, 257; why the last viceroys did not do anything to improve their lot, 261; political complications caused by the isolation of the Indians, 261–62; II.463; their ideas about cosmogony, 264; their

first settlements and their migrations, 266. [*See also* Peoples, indigenous, of the Americas.]
Indians, civilizing of: 224–25; progress being made by white people in Mexico, 274.
Indians, nomadic. *See* Indios bravos.
Indigo: about its cultivation, II.31–32; its different species, II.32; its export, II.32t; its transport from Guatemala to Europe via Mexico, II.314–15; amount annually exported from Veracruz, II.321t; imported in 1802, II.325t; exported from Mexico in 1802, II.383t; in 1803, II.334t; impact free trade has had on this export, II.370t.
Indios bravos: 250, [425]; war the missionaries and the troops from the presidios fought against them, 285; II: [451], 458.
Instruments, scientific: achromatic telescope, 19, 30, 78, 278, 488; artificial horizon, 20n; barometer, 19, 23, 77, 83, 131, 133–34, 152, 155, 175, 183, 191, 192; II: 126, 184, 209n; chronometer, 17, 19, 23, 28, 34–35, 37–38, 49–50, 54, 83, 87, 116–17, 277, 467; II: 354–55, 466, 489; gnomon, 46, 51, 54, 57, 243; graphometer, 60, 62; hygrometer, 119, 377; II.395; octant, 46; theodolite, 22n, 78; thermometer, 134, 183–84, 194–95, 210, 227, 295, 390, 393–94, 407, 417, 427, 511, 548; II: 5, 18, 47, 108, 153, 202, 391, 396ff, 401–2, 403–4, 405–6, 409, 418, 412; sextant, 19, 33, 46, 51, 60, 83, 277, 393n; II.491; repeating circle, 19–20, 22n, 277; quadrant, 33–34, 46, 54, 62. *See also* Adams; Berthoud; Bird; Borda; Lenoir; Ramsden *in the Index of Names*.
Intendancies: their number, 305ff; differences in their sizes, 313ff; in their populations, 314; in their relative populations, 314–15.
Iraca: Mexican dish, 561.
Iron: use was unknown to the ancient Mexica or was at least not especially valued, II.66; mines where it is found, II.178–79; annual amount imported to Veracruz, II.322t.
Iron, meteoric, found in Mexico, II.179.
Iron, wrought: amount exported to Veracruz in 1802, II.232t; in 1803, II.331t; on behalf of the king, II.336–37.
Iron bars: amount imported to Veracruz in 1802, II.323t; in 1803, II.331t.
Isleños, natives of the Canary Islands: they oversee the plantations, 270.
Istenenetl: ruins of a small Mexica pyramid, 384
Itzli. *See* Obsidian.

Jalap: province where it is produced, 409–10; about its cultivation, II.27–28; amount annually exported from Veracruz, II.321; from all of Mexico in 1802, II.326t in 1803, II: 334t, 383t.
Jarucos. *See* Vaqueros.
Jatropha, [240, 294; II: 349, 400, 407, 443, 513, 517, 521ff]. *See [also]* Yucca.
Jesuits: [their explorations, 440, 444]; their institutions in California, 441; conflicts they experienced, 443; the military submitted to their orders, 443.
Juncos: Indian tribe, II.465.
Justice system: expenses it creates for the state, II.436t.

Kamchatka: initially the terminus of Russian expeditions, II.366n.
Kenaizi: people in Russian America, 473.
Koliugi: people in Russian America, 473–74.
Koniagi: people in Russian America, 473.

Languages, indigenous: 234–35; predominant languages of the new continent, 235.
Lead: mines where it is quarried, II.179; amount exported from Mexico in 1802, II.327t; in 1803, II.335t.
Leoba. *See* Mitla.
Leyden School, II.416.
Lima [Peru]: annual trade balance of, II.380.
Lipan: uncivilized Indians, 251.
Liqueur: amount imported to Veracruz in 1802, II.323t; in 1803, II.331t.
Líquido remisible: net revenue that the Spanish king takes from Mexico, [430, 435], 442; from other colonies, II.446.
Love apple. *See* Tomatl.
Lydian stone: [II: 105, 121]; formed in calcareous strata, II.122ff.
Lye [sodium hydroxide]: how it is found, II.191n; province where it abounds, II.296–97; its characteristics and uses in Xaltocan, II.304.

Maguey (agave): its cultivation in Mexico, [526n], 566; beverage made from it called pulque, 568; importance of this cultivation, 568. *See [also]* Pulque.
Maïs [maize]: its cultivation, 525ff; its extraordinary fecundity, 527; it is the people's main staple, 528; its average price, 528; beverage made from it, 528; amount of its yield, 530.
Majorats: contrary to the progress of agriculture, [411]; II.59.
Manganese: mines where it is quarried, II.179.
Mani [peanut]: its cultivation in Mexico, 559.
Manioc: its cultivation, 520ff; its species, 520–21, 524; it is indigenous to the Americas, 522; usefulness of this production, 523.
Manufacturing: constraints the government places on their establishment, II.293; amount of value of their products in New Spain, II.293; manufacturing of cottons and woolens, II.293; of silks, II.295; of tobacco products, II.296; of soap, II.296–97; of earthenware [ceramics], II.286; hats, II.286; gunpowder, II.296–97; gold and silver smithing, II.301; coins, II.302.
Manuscripts, Aztec, 245n, [326, 570; II.316n].
Maps, geographical, created by the Aztecs, II.315–16.
Maps included in the Mexican Atlas: I. *Condensed map of the kingdom of New Spain*; material used in making this map, 15; projection, 25; scale, 25; principles for naming the seas, 27; thirty-six points determined by the author's observations, 27; discussion of the position of Mexico City, 27–28; of Veracruz, 33–34; of Acapulco, 35–36; of various locations on the road from Mexico City to Acapulco, 38–39; on the road from Mexico City to Veracruz, 40; of various points located between Mexico City, Guanajuato, and Valladolid, 43–44; of California, 45–46; unpublished material that the author consulted, 51–52; maps he used, 59ff; advantage of this map over older maps, 65; how mountains were drawn, 66–67; copies made of this map, 68n. II. *Map of New Spain and countries bordering it to the north and east*: purpose of this map, 69–70. III. *Map of the Valley of Mexico*: material used in drawing it, 75; astronomical observations on which it is based, 77. IV. *Map of watersheds and projected connections between the*

two oceans: description of this map, 82; material used in drawing it, 83. V. *Condensed map of the road from Acapulco to Mexico City*, 85. VI. *Map of the road from Mexico City to Durango*; material used for this map, 86. VII. *Map of the route from Durango to Chihuahua*, 88. VIII. *Map of the route from Chihuahua to Santa Fe in New Mexico*, 89–90. IX. *Condensed of the eastern parts of New Spain, from the plateau of Mexico City to the port of Veracruz*, 93ff. X. *Map of the false geographical positions* attributed (by various geographers) to the ports of Veracruz and Acapulco and to Mexico City, 95ff. XI. *Map of the port of Veracruz and the castle of San Juan de Ulúa*, 97. XII and XIII. *Physical tableau of the eastern and western slopes of the plateau of New Spain*, 98–99; remarks about the work on physical maps, 99; description of these maps, 108. XIV. *Tableau of the central plateau of the mountains of Mexico*, 107. XV. *Profile of the canal of Huehuetoca* (Desagüe Real), 158. XVI. *Volcanoes of La Puebla*, 110ff. XVII. *Map of the port of Acapulco*, 115–16. XIX. *Five maps about mines and their products*, 118ff. XX. *Two comparative tableaux of the territorial areas and the population of New Spain*, 119–20.

Marble, 331.

Marquis de San Miguel's estate (wines), 571.

Matalan: people of New California, 450.

Matlazahuatl: illness that particularly afflicts Indians, 224; II.391.

Maya: language of the Indians of the Yucatán, 407–8.

Mayolias: Indian tribe, 388.

Mechoacan: medicinal root, II.28.

Mecos: uncivilized Indians, 251; war the missionaries waged against them, 285–86; II.424.

Meiya: species of banana, 514.

Mengwe: people of California, 444.

Mercurio Peruano, 165n; II: 173, 198n.

Mercury: amount of this metal needed to extract silver from the mines of Mexico, II.163; quantity of mercury lost during amalgamation, II.163–64; impact of its price on quarrying, II.165; attempt to get it from China, II.167; mines where it is quarried, II.179–80; their different settings, II.180–81; imports of this metal to the Americans will soon stop, II.182; places in the Spanish Americans other than Mexico where it is found, II.182ff; quantity of this metal used in Potosí, II.225; amount annually imported to Veracruz, II.322; amount imported to Mexico in 1802, 70; in 1803, II.329; revenues its sale produces for the king, II.429.

Mesada and media anata: annual yield from this tax for the king, II.431.

Mescaleros. *See* Apaches.

Mestizos and métis: 230ff; their numbers, 288.

Metals, precious: changes in their accumulation in Europe, II.283–84; amount annually exported to the islands of the Philippines, II.357.

Metals used to as currency among the Aztec, II.68.

Métis. *See* Mestizos.

Mexica. *See* Aztecs.

Mexical: pulque brandy, 570.

Mexico, country of: meaning of this word, 143

Mexico, intendancy of: area and population, 318; borders, 319; landscape, 319–20; description of the valley of Mexico, 319; major towns, 379–80; mines, 381; II.71–72.

Mexico, Kingdom of: its population, II.26. *See also* New Spain.

Mexico, Valley of: location, 319; area, 319; roads that cross it, 320; Cortés's description of it, 321; map he had made of it, 322; successive institutions and facilities that the Aztecs created there, 323; description of the teocalli of Tenochtitlan, 324 (*see also* Teocalli); description of Lake Tezcoco, 327; ancient monuments that can be found there, 338; pyramids of San Juan de Teotihuacan, 338; military entrenchment of Xochicalco, 340; *chinampas* [gardens] that float on its lakes, 350–51; thermal springs, 351; Chapultepec Castel, 352; tributary rivers in this valley, 353; Desagüe de Huehuetoca, 355; plan for drying out the valley, 358ff; reasons for the depopulation of this valley, 374–75.

Mexico City: geographical position of, 27; consternation caused there by the solar eclipse in 1803, 32; ancient name of, 143; improved situation there after the connections to the rest of the world were opened, 190–91; its temperature, 194; the different castes into which the population is divided, 230, 288; scientific institutions in this city, 254; number of saragates [vagabonds], 443; proportion of sexes among the population, 456; II.329; this city is not surrounded by waters anymore, 320; causeways that connect it to the mainland, 325–26; the city Cortés had rebuilt is much smaller than Tenochtitlan, 325; why it is far away from any lakes, 326; beauty of this city and its surrounds, 330; cleanliness that prevails there, 332; aqueducts that provide it with drinkable water, 333; roads that lead there, 333; remarkable buildings, 334; ancient monuments, 338; description of Montezuma's palace, 341; ruins of king Axayacatl's palace, 342; bridge called Salto de Álvarado, 342; Clerigo bridge, 343; Cortés rebuilt the city in the same place as Tenochtitlan, 343–44; its population, 344, 476; II.472ff; number of its clergy, 282n, 345t; income of its archbishopric, 346; its inquisition tribunal, 346; births and deaths, 346–47; consumption of its inhabitants, 348; compared to consumption in Paris, 348–49; increased consumption of wine after 1791, 349; great flood between 1629 and 1634, 361–62; plans to relocate the city, 363; why the desagüe of Huehuetoca does not make the city flood proof, 371; its elevation above sea level, 379; its manufactories, II.297; its mint, II.302; its separation house, 304; this city is the main location of domestic trade in New Spain, II.308; details on its population from the 1790 census: (1) monks, II.472ff; (2) nuns, II.274; (3) secular clergy, II.476; (4) castes, II.477; (5) male students, II.478; (6) female students, II.479; (7) hospitals, II.480; (8) prisons, II.482; (9) by type of profession, II.483.

Mexico City, archbishopric of: its revenues. II.58.

Mexico City, cathedral of, 27n, 33, 320, 325, 331–32, 334, 341; II.498.

Mexitli. *See* Teocalli.

Meztli Ytzaqual (House of the Moon): ancient pyramid, 338.

Militias: their numbers, II.450; their divisions, II.449; why there are so numerous, II.463.

Mimbreños. *See* Apaches.

Mineral kingdom of Mexico: products of, II.61–284. *See [also]* Mines.

Minería, Colegio de, II.192.
Minería, Cuerpo de [Corps of Miners] in Mexico City: better terms it provided for mine owners, 281.
Minería, Diputaciones de [Mining Councils], II.69.
Minería, Real Seminario de (Royal School of Mines), 15, 112, 281.
Minería, Tribunal de (Mining Board), 9, 15, 67, 279, 281, 427; II: 62, 140, 166, 169, 176, 428n.
Miners: extent to which they brought their craft to Mexico, II.134–35; freedom of their labor in the Spanish Americas, II.140–41.
Mines: their influence on agriculture in different parts of Mexico, 507ff; on the population, II.62; on their locations, II.63; exploitation under the Aztec kings, II.64; geographical positions of the ones that are now being worked, II.69. Geological tableau of New Spain: rocks, 75 (*see* Rocks); mineral deposits, veins, and beds, II.78; groupings, I.80; formation of veins, gold, and silver; the nature of minerals, II.83; washes, II.85; discovery of platinum, II.88f; average wealth of the minerals, II.93; description of the most metalliferous regions: Guanajuato, II: 97–98, 492; some general views on the age and nature of the formations, II.119–20; Zacatecas, II.122; Catorce, II.124; Pachuca and Real de Monte, II.125; Tasco, II.130–31; craft of the Mexican miner; administration of the mines, II.134–35; amalgamation and smelting, II.147–48; type of amalgamation miners use in the Americas compared to those in Halsbrücke, II.158; influence of the price of mercury on the progress of its mining, II.166–67; quantity of gold and silver extracted from the mines of Mexico, II.170t; cash value of the yield from 1733 to 1792, I.173t; in 1805, II.173; might the annual production be increased?, II.173; did it decrease?, II.117–18; common metals: iron, copper, II.178; tin, II.178; lead, II.179; metals of limited use, II.179; mercury, II.179–80; coal, II.189–90; salt, II.190; lye, II.190; mine legislation; high council, II.191–92; tax the proprietors have to pay, II.193; future prospects, II.196–97; comparisons between of the yield of the Mexican mines and that of mines in the other Spanish colonies, II.197; in Chile, II.208; in Buenos Aires, II.208; and in New Granada, II.227; tableau of the current yield of the mines of the new continent (smuggling not included), II.231; precious metals smuggled from the ports of Veracruz and Acapulco, II.232; from Cartagena and Portobelo, II.233; on the Amazon River, II.233; from Chile, II.233; in the viceroyalty of Buenos Aires, II.233; from Brazil, II.233; tableau of the current yield of the mines of the new continent (smuggling included here), II.233; their annual yield at the beginning of the nineteenth century, II.234; Tableau of the current yield of the mines in Europe, northern Asia, and America, II.236; proportion of gold to silver taken from the Spanish Americas, II.237; research on the quantities of gold and silver that have flows from one continent to the other since 492, 238–39, 256; according to Uztariz, II.238; according to Moncada, Navarete, and Solorzano, II.238; according to Raynal, II.240; according to Adam Smith, II.241; according to Robertson, II.242; according to the author of research on

Mines (cont.)
trade, II.242; according to Necker, II.242; according to Gerboux, II.242; documented amount of gold and silver taken from the mines in the Americas from 1492 to 1803, II.244–45; undocumented gold and silver, II.246–47; total amount of gold and silver taken from the mines in the Americas since 1492, II.247t; how much each of the different colonies have contributed proportionally, II.245; proportion of gold to silver, II.247; amount of gold and silver found at the time of the conquest and looted in its aftermath, II.247–48; quantity of cash in circulation in the New World, II.250; quantity of gold and silver that pass directly to Asia and Africa without reaching Europe, II.250–51; total amount of gold and silver that Europe has received from the New World since 1492, and which resulted in the previous calculations, II.251–52; proportion of the wealth that has flowed back to Europe during different epochs, II.253t; research on the question what became of that wealth, II.254–55; different routes by which the wealth of gold and silver reaches Asia: (1) via the trade of the Levant, Egypt, and the Red Sea, II.256–57; (2) via the East Indies and China, II.257; (3) via the Russian trade, II.257; gold and silver accumulation in Europe, II.261; recent observations, II.262–62; Brazilian mines, their yield compared to that of the mines in the Ural, II.263; amount metals used for coin minting compared to that used for gold and silver smithing, II.273–74; work of the coin mints in France compared to those in Mexico City, II.280; changes of the accumulation of precious metals in Europe, II.283–84.

Mining Council: its creation, II.191.

Minting: from the mint in Lima, II.199t; from the city of Potosí, II.222–23; comparison of the quantity of metal used for coins with that used for gold and silver smithing, II.274–80.

Minting fees that the mine owners pay, II.193.

Miraculous virgins of Guadalupe and the Remedios, 364n.

Missionaires: their occasional hostilities to the Indios bravos, 285.

Mita: law that forced the indigenous people to work in the mines; no longer in force in Mexico, 227.

Mitla, palace of: 267; its ruins, 402ff.

Mixed-bloods: different types found in Mexico, 288ff.

Mixteca: ancient people of Mexico, 266.

Monasteries: why, in the Americas, they have had less influence on the progress of agriculture, II.60. *See [also]* New Spain.

Monetary value of the silver minerals of Mexico, II.93.

Monterey, bishopric of: its revenues, II.58.

Monterey shell, II.50.

Morro: lookout post, 497.

Mortmain of the clergy: whether it conflicts with agricultural advancement, [386]; II.259.

Musk duck: bird indigenous to Mexico, 37

Mulattos, 230, 298.

Mulberry trees: the government put a stop to their cultivation in Mexico, II: 38–39, [363n].

Murex of Mexico, II: 50, [394].

Musa [banana species]: its cultivation, [185], 513ff; different preparations of this fruit, 519; how it is grown on

the island of Cuba, 517n. *See [also]* Plants; Bananas.
Muslin: quantity imported to Mexico in 1803, II: 331t, 332t.

Nahualtecs or Nahuatlacs: their arrival in Mexico, 232, 467.
Nevado: meaning of this word, 382n.
New California, province of: size, 445; also called New Albion, 447; missions and presidios that the Spanish court established there, 447–48; increase in its population, 449; products of its soil, 449–50; peoples that live there, 450; difference between their language and the Aztec tongue, 451; the liking for steam baths, 452; their occupations, 453; their animals, 453; missions that the Spaniards founded in this country, 454–55; II.484–85.
New Spain: its size, 139; it is the most important possession of the Spanish, 141; borders of, 141; this name originally only applied to the Yucatán, 142; it is not synonymous with Anahuac, 142; comparison of its size and population with that of the United States of America, 143; configuration of its coastline, 146; physical tableau of this country, 172: climate, 172; makeup of its mountains, 173; description of their plateaus, 176; of their highest peaks, 179; climate on the coasts, 183; differences among calientes, templadas, and frías regions, 183–84; elevations where metals are founds, 186; navigable rivers, 187; lakes, 187; vegetation, 187; snow lines, 188; summer heat, 189; rainfall, 189; earthquakes and volcanic eruptions, 190; physical advantages of this land, 190; its military dependence on Havana, 192; navigational dangers on its coasts, 192; population, 199, 212–13 (*see [also]* Population); territorial divisions: (1) into ten provinces before the Count of Galvez, 304; (2) into fifteen intendancies and districts, 304; (3) into three regions, 306; (4) into the kingdoms of Mexico and New Galicia, 307; (5) into New Spain proper and interior provinces, 307; comparison of its size and population with those of some other countries, 310t; division of the internal provinces, 310–11; surface area and population relative to territorial divisions, 311–12; differences among their interior provinces relative to size, 313; to populations, 314; their relative populations, 314–15; statistical analysis of New Spain, 316: (1) intendancy of Mexico, 318 (*see also* Mexico); (2) of Puebla, 381; (3) of Guanajuato, 389; (4) of Valladolid, 390; (5) of Gudalaxara, 397; (6) of Zacatecas, 400; (7) of Oaxaca, 401; (8) of Mérida, 405; (9) of Veracruz, 408; (10) of San Luis Potosí, 418; (11) of Durango, 423; (12) of Sonora, 428; (13) province of New Mexico, 433; (14) of Old California, 438; (15) of New California, 445; glance at the coasts of the Great Ocean from the port of San Francisco to the Russian settlements, 456; voyages that led there, 457; present population of New Spain, 478t; number of missions and church members, 479; wealth of the clergy, 482t; convents, 482–83; state of the agriculture in New Spain, 505–31 (*see* Agriculture); state of the mines, II.61 (*see* Mines); state of manufactories, II.292–307 (*see* Manufacturing); trade, II.307 (*see* Trade); Finances, II.427–47 (*see* Revenue); military, II.447 (*see* Forces, armed).

New Spain proper: its size, I.307.
Noche triste: the period known as this in history, [320], 342n.
Nopaleros: planters who breed the cochineal, II.47.
Nortes de hueso Colorado: northern winds in Mexico, I.193.
Northwest Passage. *See* Passage to the northwest of America.
Notes, supplementary, to the statistical analysis, 475.
Numbers system, Mexica: their system compared to many others, II.68n.

Oats: cultivation in Mexico, 548.
Oaxaca, bishopric of: its revenues, 58t.
Oaxaca, intendancy of: names of church members, 282n; its size and climate, 401; monuments in, 402; one raises cochineal there, 404; details about its current population, 477; its cities and mines, II.75
Obrajes: large manufactories for woolen fabrics in Querétaro, II.294.
Observatories, astronomical: Berlin, 20n; Greenwich, 33, 49, 116–17; Léon Island, 27; Mannheim, 20n; Mr. Ferrer's, 33; Paris, 126; Santa Ana, 46, 278; Uranienborg, 30–31.
Obsidian, [322, 339]; found among the ruins of *la Casa grande*, 431; how the Aztec used it, II: 68, [124, 126].
Oca [*Oxalis tuberosa*]: its cultivation in Mexico, 556.
Oceloxochitl. *See* Cacomite.
Octli. *See* Pulque.
Olive trees: [184], 380, 392, 449, 511, 522, 565–66; the government tried to stop their cultivation, II: 19, 291, [379].
Olives: amount imported to Veracruz, in 1802, II.322t; in 1803, II.330t.
Olmecs: ancient people of Mexico, 266, [339n].
Oregano: amount imported to Mexico in 1802, II.323t.
Otomí: nomadic people in the north of Mexico, 142, [201, 226, 231, 234–35, 239], 266, [273]; in the intendancy of Valladolid, 396.
Otter pelts, [454, 456, 459, 463; II: 51, 167]; importance of the trade of this product, II: 363, [364f].
Outflow channels: since 1607, they had replaced the system of dikes to protect Mexico City from floods, 355.

Pames: Indians, 388.
Papagallo: storm, 193–94; [II.353].
Papahua Tlemacuzque, or Teopixqui: Toltec and Aztec priests, 339.
Papantla: pyramid of, [324, 402], 412.
Papas. *See* Potatoes.
Paper, ruled: amount imported to Veracruz in 1802, II.323t; in 1803, II.331t.
Paper, white: not manufactured in Mexico, II.296; amount imported annually to Veracruz, II.322t; amount individuals imported to Veracruz in 1802. II: 323t, 324t; in 1803, II: 331t, 332t; on the behalf of the Crown, 1802, II.331t; in 1803, II.338; [in 1805, II.339; price by year, II.340t].
Paper for cigar factories, II: 296, 336; made of agave (maguey), 245n, 338, 487, 570.
Parallel of great elevations: geological phenomenon, 395.
Parallelism of strata: observed in large areas of the country, 395.
Passage to the northwest of America: problem that occupied the Spanish during the sixteenth century, 458–59.
Pathways. *See* Roads.
Paving stones: amount imported to Mexico in 1803, II.331t.
Payos: Indians who regularly visit the Huaytecas and Chonos islands, II.465.
Peanuts. *See* Mani.

Pearl fishing: in California, [440–41], 443, [497]; II: 48ff, [247].
Pelts: quantity exported from Mexico in 1802, II.326t; in 1803, II.335t; impact the free trade law has had on its export, II.371t. [*See also* Furs; Hides; Skins.]
Pensions paid by the government; their amount, II.439t.
Peoples, indigenous, of the Americas. *See* Acoclames; Acolhua; Apaches; Capuces; Chichimecs; Chugachi; Cochimi; Cocoyames; Comanches; Cora; Guachichiles; Guamanes; Guayhuenes; Esselen; Juncos; Kenaizi; Koliugi; Koniagi; Lipans; Matalans; Mayolias; Mecos; Menquis; Mixteca; Mimbreños; Olmecs; Otomí; Pames; Papago; Payos; Pericues; Pimas; Quirotes; Rumsen; Saragates; Salsen; Samues; Seris; Tepanec; Ugalachmiuti; Vehidi; Yabipais, Xicalanca; Zapoteca.
Pepper, [113, 351, 525, 548; II.57]; quantity imported to Mexico in 1803, II.324t; [exported from Mexico, II: 321t, 326t].
Pericú: people of California, 444.
Peru: population, 210, 238; [II: 387t, 433, 360]; mercury this kingdom produces, II.183; yield of its gold and silver mines, II.198; famous mines of this kingdom, II.199–200; division of this kingdom into provinces and intendancies, II.200; mines of different regions, II.206; amalgamation procedure used there, II.206; trade balance, II.387t; net revenue for the Spanish Crown, II.443; gross revenue of this viceroyalty, II.443–44; its armed forces, II.456.
Piasters: quantity minted in 1790 at the mint of Potosí, II.225.
Pig: animal unknown in Mexico at the time of the conquest, II.35.
Pima: Indian tribe, 430.
Pimienta de Tabasco: province where it is produced, 410; amount annually exported from Veracruz, II.321t; in 1802, II.326t; in 1803, II: 334t, 386t; impact the free trade had on this export, II.371t.
Pintepata: Russian post, 472.
Pit coal: places where it exists, II: 190–91, [201n].
Pita: amount exported from Mexico in 1802, II.327t. *See also* Maguey.
Pitch: amount imported to Mexico in 1803, II.333t.
Plant kingdom of Mexico, products of: different climates in which plants prosper, 185. *See* Plants [and how they are used; Plants of New Spain].
Plants, aromatic: amount exported to Veracruz, in 1802, II.322t; in 1803, II.330t.
Plants and how they are used: *Adansonia digitata*, 384, [402]; *Agave americana*, [240, 487, 565, 567n]; *A. cubensis*, 567; *Amyria*, II.417; *Andromeda*, 463; *Annona cherimolia*, 387; *Arachis hypogea*, 559, [559n]; *Arbutus perotensis* [*madroño*], II.39; *Artocarpus incisa*, 425, 563; *Arum macrorhizon*, 524; asparagus, 563, [556, 563]; *Atriplex*, II.304; *Avena sativa* [oat], 532n, 458, 460; *Bonplandia trifoliata*, 506n; II.417n; *Brassica*, 561; *Bromelia ananas* [pineapple], 562n; II.417; *Cactus cylendricus*, 442; *C. coccinellifer* [prickly pear], II.44; *C. opuntia*, II.44; *C. pereskia*, II.244; *Caesalpinia brasiliensis* [Brazilwood], 408; *Calicocca*, II.380n; *Camburi* [banana], 513–14, [521]; *Cannabis sativa*, II.19; *C. indica*, II.19; *Capsicum annuum*, 559; *C. baccatum*, 559; *C. frutescens*, 559; *Carica*, 371n; *Cecropia peltata*, 506n; *Cedrela odorata*, II: [40],

Plants and how they are used (cont.) 306n, 318–19; *Cestrum mutisii*, II.32; *Cheirostemon planatifolium*, 250; *Chenopodium quinoa*, 176, 554, 559; [II.304]; *Cicer*, 561; *Cinchona*, 506; II.417; *Citrus aurantium*, 564; *C. decumana*, 565n; *C. medica*, 565n; *C. trifoliata*, 564; *Cocolloba uvifera*, II.400; *Cocos nucifera*, [558]; *Coffea arabica*, II.20; *Commiphora madagascarensis*, 506n; *Convolvulus batatas*, 524, 558; *C. chrysorhizus*, 524, 558; *C. jalapa*, 409, 558n; II.27–28; *C. platanifolius*, II.558; *C. edulis*, 558; *Coutarea*, 506n; *Cupressus disticha*, 332, 384, 402; *Cycas circinnalis*, 524; *Danais*, 506n; *Dioscorea alata*, 524, 556ff, 559; *Dracontium polyphyllum*, 524; *Epidendrum vanilla*, 409; II: 22–23, 25–26; *Erythroxylon cocca*, 517; *Exostema*, 506; *Filices arborescentes*, 506; *Garcinia mangostana*, 565n; *Gossypium*, II.293; *Gratiola*, II.304; *Gyrocarpus mexicana*, 250n; *Hæmatoxylon campechianum*, 407; *Helianthus tuberosus*, 559; *H. annuus*, 559; *Hevea*, 506n; *Hordeum tuca*, 526; *H. vulgare* [barley], 532n, 548; *Indigofera tinctoria*, II.32; *Ind. argute*, II.31; *Ind. anil*, II.31; *Ind. disperma*, II.31; *Jatropha manihot*, 349, 513, 398, 524–25; *J. janipha*, 522; *J. carthaginensis*, 522; *Laurus persea*, II.38; *Limonia trifoliata*, 564; *Linea borealis*, 305 [?]; *Liquidambar styraciflua*, [166], 409; [II: 24, 29]; *Lobelia*, 506n; *Medicago sativa*, 160; *Meya* (*Maris meridionalis*), 385 [?]; *Mikania guaco*, II.50 [?]; *Milium nigricans*, 160; *Morus acuminata*, II.38; *M. rubra*, II.38; *Musa paradisiaca*, 185, 513; *M. sapientum*, 514; *M. regia*, 514; *M. mensaria*, 514; *Musænda bracteolata*, 506n; *Myrtus pimenta*, 410; *Nicotiana tabacum*, II.92; *N. rustica*, II.30; *Olea europæa* [olive], 449, 565; *Oryza sativa*, 559ff, 560n; *Oxalis tuberosa*, 556; *Padus capuli* [cherry], 563; *Pæderia*, 506n; *Paspalum purpureum*, 160; *Passiflora*, 387, 562; *Phaseolus* [beans], 448, 561; *Phormium tenax*, 563; *Pinus*, 409, 455, 463; II: 362, 410; *Pinkneya*, 506n; *Pisum sativum*, 561; *Platano arton*, 513ff, 518–19; *Polygonum fagopyrum*, 524; *Portlandia hexandra*, 506n; *Porthos*, II.400; *Prunus avium*, 513; *Psychotria emetica*, II.380n; *Quercus* [oak], 409; II.410; *Rhizophora mangle* [red mangrove], 561; II.400; *Rosa mexicana*, 463; *Saccarum violaceum*, 570; *S. officinarium*, II.16; *Schinus molle*, [332], 548; *Secale magu*, 526; *S. cereale*, 532, 547; *Sesuvium portulacastrum*, II.304; *Smilax salsaparilla*, 410; II.27n; *Solanum tuberosum*, 548, 554n; *S. cari*, 550; *S. lycopersicum*, 559; *Spondias*, 563; *Strychnos pseudoquina*, 506n; *Swietenia febrifuga*, 506n; *S. mahagoni*, II.318; *Tacsonia*, 562; *Theobroma cacao*, 517; II.21–22; *T. bicolor*, II.17; *Tigridia cacomite*, 559; *Tithymaloidei*, 524; *Triphasis aurantiola*, 565; *Triticum compositum*, 527; *T. mechuacanense*, 527; *T. hybernum*, 532n, 537, 548, 551, 560; *T. spelta*, 532n; *Tropæolum esculentum*, 554; *T. peregrinum*, 554n; *Tacca pinnatifida*, 524; *Urceola elastica*, 506n; *Vaccinium*, 463; *Viola emetica*, II.380n; *Vitis vinifera*, [428], 438, 448, 566; *Zea maïs*, 350, 525–26; *Z. curagua*, 526, 529; *Zizania* [wild rice], 543n. *See* Temperature.

Plants of New Spain. I. Plants that serve as nourishment for people:

(1) banana, 512; (2) cassava, [349], 521, [523]; (3) corn, 525; (4) European cereals, 531–32; (5) potato, 548; (6) Oca, 556; (7) igname, 556; (8) batatas [sweet potatoes], 558; (10) tomato, 559; (11) peanuts, 559; (12) pepper, 559; (13) chamalitl, 559; (14) rice, 559; (15) all vegetables and fruit trees from Europe, 561; (16) plants used to prepare drinks: maguey, 566. II. Plants that provide raw materials for manufacturing, II.3: (1) sugarcane, II.4; (2) cotton, II.17; (3) flax and hemp, II.18–19; (4) coffee, II.20; (5) cacao, II.21; (6) vanilla, II.22; (7) sarsaparilla, II.27; (8) jalap, II.27; (9) tobacco, II.28–29; (10) indigo, II.31. *See also* Plants *[in other entries]*.

Platano: banana species, 513–14.

Platinum, [II.87ff]; discovered in Brazil, II.89; not found in Mexico, II.201; nor in the viceroyalty of Buenos Aires, II.208; found on Chocó and in Barbacoa, II.288.

Playing cards: revenue from taxes on this item, II.433.

Plaza del Volador, Mexico City, 362.

Plaza Mayor, Mexico City: 331, 335, 338, 341, 354ff, 362, 387, 389.

Plums [or prunes], [562–63]; amount imported to Mexico in 1803, II.330t.

Poitos: type of Indian slave, 286.

Popayán: quantity of gold minted here from 1788 to 1794, II.227.

Population of New Spain, 140, 199; remarks on a tableau comparing the population of the mother country with those of its colonies, 140; interior regions of the country are more densely populated than the coastline, 199–200; it has increased since the arrival of the Spanish, 200; its stand in 1793, 202–3; it has significantly increased since, 204; relation of births to deaths, 207–8; birth-death ratios in relation to the population, 207; compared to other countries, 208; state of the population in 1803, 210; in 1810, 211; current, 212; Black population of the American continent and its islands, 216–17; population relative to religious beliefs, 217; reasons for its stagnation: (1) smallpox, 221; (2) the matlazahuatl, 224; (3) famines, 225; unnecessary to include among these reasons work in the mines, 226–27; nor yellow fever, 229; population has hardly increased because of new settlers, 229; different castes among the inhabitants, 230: (1) Indians, 230, 263; (2) Whites, 270, 443; (3) Blacks, 284–85; (4) Mixed-bloods, 288ff; number of males compared to females, 291t; statistics on the connection between caste and longevity, 295; influence of the mixing of castes on society, 295; comparison of this population to that of several other countries, 209–10, 476; population according to territorial divisions, 311–12; its disproportion in certain intendancies, 314; relative population of the intendancies, 314; current stand within the confederate states of Mexico, 475–76.

Ports of New Spain: their relative importance, II.380.

Ports planned to replace that of Veracruz, 192.

Positions, geographical: of Mexico City, 121; as determined by astronomical observations, 121–122.

Possessions, Russian, in the Americas: plan the Madrid Court hatched to attack them; 470–71; position of these factories, 471.

Postage stamps: one of the branches of public revenue; their amount, II.431.
Postal services: annual gross income from, II.431.
Potatoes, sweet: their cultivation in Mexico, 558.
Potosí, viceroyality of Buenos Aires: quantity of silver extracted from its mines, II: 99, 209n; royal taxes paid on the silver extracted from the mines of Cerro de Potosí, II.210; sum of this tableau, II.211; yield of its mines, II.215ff; quarrying from 1556 to 1787, II.216; minted coins, II.222t; diminishing of the content of its minerals, II.222; old method for extracting the minerals, II.224; amalgamation is introduced there, II.225; quantity of piasters minted there in 1790, II.225.
Poultry: rare before the conquest. II.35–36.
Poverty of the Mexican people: its reasons, II.377.
Presidios: purpose of their creation, II.451.
Prisons: how much they cost the state, II.439t.
Provinces, interior: their divisions (1) into interior provinces of the viceroyalty and the command, 307; (2) into eastern and western, 310–11; lands they take up, 311t; their surface area and population, 311t; their relative population, 316–17.
Puebla, intendancy of: number of its clergy, 282n; area and population, II.148; size of different castes, 381; number of its towns and villages, 386–87; factories, 386; salt marshes, 386; marble, 386; language of its inhabitants, 386; towns, 186–87; mines, 388; relative population, 388; manufactories, II.292.
Puerto Rico, island: amount it receives annually from Mexico to cover the costs of its administration, II.441t.
Pulque: beverage made from the juice of the agave, [240, 417], 468; how much is consumed in Mexico, 349; yield from the taxes taken on this beverage, II: 431, [433].
Pulque de mahis: alcoholic beverage brewed from corn, 530.
Pumps, II: [112, 128, 130, 138–39,] 201.
Pyramids: Cheops, 383–84; Chephren, 384; Giza, 340, 382–83; Mastaba al Faraun, 486; Menschich-Dahshur, 383; Mycerinus, 384–85; Sakkara, 324, 340, 383ff, 486; II.484; Tonatiuh Yztaqual, 338–39, 383–84. *See also* Cholula; Istenenetl; Papantla; Teotihuacan.
Pyramids of San Juan de Teotihuacan, 338: their height, 338; their age, 339; their interior construction, 339; their height compared to that of the Egyptian pyramids and the ones of Cholula, 384n.

Quarterons: Mexican caste, 289.
Quinquina: amount imported to Veracruz in 1802, II.325t; exported in 1802, II: 326t, 383n; its use against yellow fever was not successful, II.417; natural affinity with Cinchona, 506n. *See also* Plants: [*Cinchona*].
Quinterons: Mexican caste, 289.
Quirotes: people of New California, 448.

Races: distribution of races in the continental and insular Americas, I.214t; race relations in the Spanish Americas, I.216t.
Raisins: amount imported to Veracruz in 1802, II.322t; in 1803, 330t.
Reales de plata: three kinds of currency with this name, II.216n.
Reales del minas: their names, II.69.

Repartimientos, [259; II,165–66]; abolished by Charles III, 254.
Revenue of New Spain: impact the free trade had on its increase, II.370; annual amount, II.428t; branches: (1) revenue from the yield of mines, II.429; (2) from tobacco production, II.429; (3) from alcabalas, II.430; (4) from capitation tax by the Indians, II.430–31; (5) from the tax on pulque, II.431; (6) various other branches, II.431; (7) tax on snow, II.432; comparison between the years 1746 and 1804, II.433t; amount per capita, II.433; costs covered, II.435t; amount of the revenue based on the líquido remisible, II.442. *See also* Expenditures, public.
Rice: its cultivation is neglected in Mexico, 560; quantity imported there in 1803, II.333t.
Ridge on the road from Cruces de Panama, 156.
Río: all the composites of this word must searched under the word with which it is joined. [*See Toponym Index.*]
Roads. *See* Routes.
Rocks, formations of: geological makeup of New Spain; general considerations, 176, 179, 181, 269–67, 494–95; (1) primitive rocks: granite or gneiss, 380, 401; II: 126, 231; micaceous schists, Glimmerschiefer, II: 76, 79, 153, 86; primitive schists, Urthonschiefer, II: 76, 104ff, 131; serpentine, II: 77, 104 (syenite, II; 104, 184–85, 108, 119); (2) transitional rocks: transitional schists, Übergangsthonschiefer, II: 119, 122, 124 (transitional Grünstein, II: 108, 120, 184); transitional porphyry, Übergangsporphyr, II: 77–78, 79, 105, 108, 122–23; greywacke, transitional greystone, II: [78–79], 121–22, 124; transitional limestone, Übergangskalkstein, II: 78, 105, 108, 122–23; (3) secondary rocks: older greystone, II: 105, 122 (clayey schists, Schieferton, coal, II: 78, 180, 189–90); alpine limestone, Alpenkalkstein, II: 78, 105n, 126, 185, 187; Jurassic limestone, Jurakalkstein, II: 78, 107, 126; old gypsum, alter Gips, 221 (rock salt, 331); younger greystone, neuer bunter Sandstein, II.126; new gypsum, neuer Gips, II.126; new limestone or upper limestone, II.185; (4) alluvium rocks, II: 85, 87, 178, 228–29, 233, 245; (5) volcanic rocks, trappean formations, 393, 394n, 400, 412; II: 83, 124, 127, 297; trappean porphyry, Trapp-Porphyr, 181, 392; II: 216–17, 180; basalt, amygdaloidal basalt, Mandelstein, 282, [319], 328, [341, 369, 377], 390, 392–93; II: 108, 124, 126, [146, 209, 310–11, 396n]; obsidian, pearlstone, Perlstein, 341; II: 68, 126.
Rope [also packthread]: amount imported to Veracruz in 1802, II.325; in 1803, II.333t.
Roucou [annatto]: amount exported from Mexico in 1802, II.326t; in 1803, II: 334t, [370t].
Route from Philadelphia to Mexico, 422n; from Pueblo Viejo to Mexico City, 500; details of this route and the places it passes, 500; its convenience, 500.
Routes, trade: most frequented, II.308; their division into longitudinal and transversal, II.309; description of the one from Mexico City to Acapulco, II.309; of the one from Mexico City to Veracruz, II.310.
Rumsen: people of New California, 450ff.
Rye: cultivation in Mexico, [526, 532n, 541], 548.

Sabino del Santa María del Tule: tree famous for the size of its trunk, 402.
Sacks, [II.137]; amount imported to Veracruz in 1802, II.325t; in 1803, II.333t; exported by other parts of the Spanish Americas in 1802, II: 327t, 333t.
Saffron: amount imported to Veracruz in 1802, II.322t; in 1803, II: 330t, [388].
Saint-Charles de Perote: fort, 417; its importance, II.460.
Saint-Domingue, [II: 3–4, 6, 20, 23, 29, 41–42, 44, 248, 416n, 456]; quantity of sugar this island exports, II: 8, 10; support that the Spanish part of the island annually received from Mexico to cover administrative expenses, II.441.
Saint-Martin-du-Canigou: monastery in the Pyrenees, 512.
Salamanca: presidio, 406.
Salaries of viceroys and staff; annual amount, II.349.
Salsen: people of New California, 450.
Salt: lack of salt in New Spain, II.190–91; amount imported to Veracruz in 1802, II.325t; in 1803, II.333t.
Salt fish: amount imported to Veracruz in 1802, II.325t; in 1803, II: 331t, 333t.
Salted meat: amount annually exported from Veracruz, II.321.
Salto de Alvarado: name of a bridge in Mexico City, 342.
Samues: Indian tribe, 388.
San Antonio de Bejar: presidio, 90.
San Buenaventura: presidio, 431, 458n.
San Francisco: presidio, 26.
San Luis Potosí, intendancy of: area, 418; division of its territory, 418–19; description of the land, 419; climate, 420; debate about its borders, 420; II.457; description of the route that leads to Louisiana, 421; its cities, 423.
Sandwich: political observations about this island, II.362.
Santa Fe de Bogotá: quantity of gold minted there from 1789 until 1795, II.226.
Saragates: inhabitants of Mexico without permanent residence, 284, [346].
Sardines: amount imported to Veracruz in 1802, II.323t. *See [also]* Salt fish.
Sarsaparilla: its crop, II.27; amount annually exported from Veracruz, II.321t; from all of Mexico in 1802, II.326t; in 1803, II.334t.
Sausages: amount imported to Veracruz in 1802, II.323t; in 1803, II: 330t, 332t; [II.370t].
School of Mines, Mexico City, 274, 276–77, [279, 334]; II: 89, [130, 137, 140, 176, 191, 192n, 305].
Sciences, natural and exact: progress they have made in Mexico, 275.
Sciences and arts: their state in Mexico, 274; Academy of Mexico City [Academía de los Nobles Artes], 274; their revenue, 275; different courses of study: drawing, 275; botany, 275–76; chemistry, 276; School of Mines, 277; mathematics, 277–78.
Seeds, cotton: quantity exported from Mexico in 1803, II.334t.
Segura de la Frontera. *See* Tepeaca *[in the Toponym index]*.
Seniorage, right to: paid by the mine owners, II.193–94.
Serge, [II.294]; quantity exported from Mexico in 1803, 335t.
Seri: Indian tribe, 430.
Sheep, [348t, 349t, 381, 442, 479; II: 292, 294, 485t, 486t]; breeding of, II.35.
Shoes: amount imported to Veracruz in 1802, II: 325t, [328n].
Sierra, provinces of: gold and silver yield of their mines, II.208.

Silks: manufactured domestically, II: [39], 295; amount annually imported to Veracruz, II.321t; in 1802, II.323t; in 1803, II: 331t, 332t.
Silkworms, [II.295]; introduced by Cortés, II.38.
Silver: amount extracted annually from the mines of Mexico, II.84t; veins that contain it, II.89. *See [also]* Mines.
Silver, finely crafted: in Europe, II.262; amount exported from Mexico to Spain in 1802, II.326t; by other parts of the Spanish Americas, II.327t.
Silver ingots: average amount exported annually from Veracruz, II.321.
Sisas: taxation imposed for the completion of the desagüe of Nochistongo, 364, 366.
Situados: aid Mexico sends annually to the other Spanish colonies; annual amount, II.441.
Skins: quantity exported from Mexico in 1803, II.335t.
Skins of billy goats, deer, sheep and bears: influence the free trade has had on their exportation, II.371t.
Slave trade, 144, 285. *See also* Hawkins, John *in the Index of Names.*
Slaves: their number is almost zero in Mexico, 144, 284, [426, 461]; II: 295, 507; types of Indian slaves found there, [253], 285, [287–88, 374–75, 425]; [African slaves, 211, 226, 258, 272, 284ff, 293–94, 440, 522, 531–32, 557; II: 4, 13, 229–30, 466].
Smallpox: destruction it wreaked in Mexico, 221.
Smallpox vaccination: its advancement in, 221ff.
Smuggling of gold and silver: its main depots, II.232; of the English with the Spanish colonies, II.368; its amount, II: 376, 385–86.
Snow: snowline, 188, [488, 505, 510]; [snow farm, II.432]; how it is transported to Veracruz, II.417–18; taxes on the sale of snow, II.433t.
Soap: places where it is manufactured, II: 293, [296–97]; amount annually exported from Veracruz, II.321t; amount annually imported to Veracruz in 1802, II.323t; exported from Mexico to other parts of the Spanish Americas in 1802, II.327t; in 1803, II.335t.
Sonora, bishopric of: revenues, 282.
Sonora, intendancy of: area, 428; rivers, 429; connections with New Mexico and New California, 430; names of its reales de minas, II.74.
Spaniards: hatred that exists between them and the Creoles, 270–71; their numbers in Mexico, 273.
Sperm whales: hunting of, 161, 169, 171; II: 51ff, 54, 353, 363.
Spoons, silver, that Cook found at the port of Nootka: a passage from Father Crespi's journal solved this mystery, 460.
Starch: amount imported to Veracruz in 1802, II.325t; in 1803, II.333t.
Steel: average annual amount imported to Veracruz, II.322; in 1802, II: 323t, 324t; in 1803, II.331t; [price of, II.340t].
Stockings: amount exported to Mexico in 1803, II: 331t, 332t.
Storms, [153]; description of the ones in Veracruz, 192–93, [463, 548; II: 351, 363, 389, 401, 466].
Straw hats: amount imported to Veracruz in 1802, II.325t; in 1803, II.333t.
Sugar: exported from the port of Veracruz, II: 3ff, 7; sugar of Saint-Domingue, II.4; vezú, II.8; amount consumed in Mexico, II.9; sugar export from different cities in Mexico, II.9–10; countries where sugarcane

Sugar (cont.)
is mainly planted, II.10–11; sugar consumption in Europe, II.12; in France, II.15; quantities of sugar exported from the British Antilles, II.13t; from the other Antilles, II.13–14; from Brazil, II.14t; from Guyana, II.14t; from Louisiana, II.14t; from the French West Indies, II.14t; from Bourbon, II.14t; from Île-de-France, II.14t; sugar consumption in various countries, II: 12, 14–15; export from Asia, II.16; amount annually exported from Veracruz, II.321; amount exported from Mexico to Spain in 1802, II.326t; in 1803, II: 334t, 383t; exports to other parts of the Spanish Americas in 1802, II.327t; in 1803, II.335t. *See also* Plants.

Sugarcane: its cultivation in Mexico, II.3–4; it is cultivated without blacks, II.6; value of sugar at Veracruz, II.7; its yield, II.8.

Sulfur: province from which it comes, II.298.

Surface area of New Spain, according to intendances, 313.

Tableaux included in the *Political Essay*. VOLUME *I.*—Introduction: Comparative t. of the position of the major points from Mexico City to Veracruz, according to Arrowsmith; according to the author, 42; positions of various places in the intendancies of Sonora and Guadalajara, 57; position between Acapulco and Tehuantepec, 58; longitude and latitude of several places in Mexico and bordering countries, 73; corrected longitude of eight points on the western coasts, 74; chain of triangles Mr. Velázquez measured in 1773 from the rock basin to the mountain of Sincoque, 78ff; t. of the geographical positions of New Spain determined by astronomical observations, 121ff; first t. of the most remarkable elevations measured in the interior of New Spain, 184; second t. of elevations, 186; twelve t. of elevations, 131–32.—*Book I.* T. of broad political divisions of the Spanish Americas, 145; comparative altitudes of the Andes, the Alps, etc., 182n. *Book II*: T. of the population of New Spain in 1793, 203; t. of the population of the Mexican confederation, 212–13; t. concerning population, 214ff; comparative t. of the relation between castes and sexes, 291, 293. *VOLUME II.—Book III:* comparative population t., 310; t. of the territorial divisions in New Spain, 311f; consumption in Mexico City and of Paris, 348; comparison of the heights of the three largest pyramids in Egypt and the one of Cholula, 384; comparative t. of several languages of California, 451; t. of the capital that the clergy of five orders owned in 1822, 482; comparative t. of the population of some parts of Europe and New Spain, 485–86; harvest of cereals, 439–40; consumption, 541n. *VOLUME III.—Book IV*: indigo export, II.32; tithes paid to the clergy, II.56; mines of Mexico, II.269; yield of the mines of Guanajuato, II.100ff; yield of the mines of Valenciana, II.115; t. comparing the mines of the Americas and of Europe, II.117–18; silver extracted from the mines of New Spain, II.143; t. of the expenditures of the mines of Freiberg, II.159; impact of the price of mercury on its consumption, II.167; gold and silver extracted from the mines of Mexico,1690–1809, II.170ff; advances in mining, II.174–75; decreases in mining,

II.177; yield of the mines of New Spain, II.199; mining in Yauricocha, II.202; in Hualgayoc, etc., II.204; royal taxes (derechos reales), II.210ff; mining in the Cerro de Potosí, II.222; gold for minting in Santa Fe, etc., II.226; annual yield of its mines, II: 231, 234; in Asia, Europe, and the Americas, II.236; in piasters and in pounds, II.241; calculated by different authors, II.243; registered, 1492–1803, II.244ff; unregistered, II.246; recapitulation, II.246; proportions among different extracted metals, II.247; among those exported to Europe, II.253; tea imported from Canton, II.258. *Volume IV.—Book V*: t. of tobacco production, II.296; handicraft objects of gold declared at the mint in Mexico City, 1798–1802, II.301; coin production in 1798, II.303; trade balance of Veracruz in 1802, II: 337, 374; results, II.328; trade balance in 1803, II.330; results, II.336; comparison, II.337; of the years 1796 to 1820, II.342; exports at the time of free trade, II.370; gross revenues of New Spain, II.372; exports and imports compared, II.382; export from the colonies via Cádiz until December 31, 1802, II.383n; state of the Saint Sebastian Hospital, II.413; epidemics of 1800, 1801, [and] 1804, II.415n; hospitals of Veracruz in 1806, II.419n; average temperature in Veracruz, II.421; t. of the revenue of New Spain, II.433; costs of the recovery, II.435; use of the state revenue, II.436; budget of public revenue for 1803, II.437; t. comparing the revenue of the Spanish and British colonies, II.446; t. of the structure of the army in 1804, II.447ff; birth-death ratios, II.467ff; population of Mexico in 1820, II.471; in 1790, II: 472, 483; population of the villages of New California, II.485–86.

Tallenga: German amalgamation factory. II.207.

Tallow: amount exported from Mexico in 1802, II: 327t, [383].

Tarahumara: indigenous people in the state of Chihuahua, Mexico, 51.

Tarasca: ancient people of Mexico, 266, [396]; Indian tribe, II.28.

Taxes, municipal, on merchandise, II.466–67.

Taxes, royal, on mining, II: 429, [433]; their reduction, II.176; on merchandise, II.366.

Taxes that the mine owners paid, II.193.

Tchinkegriun: Russian post, 472.

Tea: quantity annually exported to Europe, II.258; replaced the export of cash by the British East India Company, II.286.

Techichi: dog eaten by the Mexica, II.33.

Temperature, average: of the tierras calientes, 141, 183; of the tierras frías of Mexico, 184; of the tierras templadas, 184; numerical results concerning the climate of Mexico, 194–95; temperatures of New California, 448–49; of Nootka, 463; of the more northern parts of the Americas, 470; of Havana, 511; of Westro-Botnia, 511n; of Acapulco, II.395; of the water at the surface of the Atlantic Ocean and the South Sea, II.396; of Cumaná, 397; of Guayaquil, 398n; minimum average temperature that the cultivation of sugarcane, bananas, coffee, oranges, olives, and grapes seems to require, 512–513; comparison of the average temperature during different months of the year in Mexico City, Veracruz, and Paris, II.403–4; discussion of the question whether the temperature of the two hemispheres is as different as is typically expected, II.398.

Tenateros: type of miners, 227–28; [II.114]; their muscle powers, II: 137–38, [141–42].
Tenochitlan: old name for Mexico City, 143: its founding, 268; changes of its name over time, 321n; origins of this name, 321n; founding of this city, 323; causeways that connected it with the mainland, 326; it was extended by merging with Tlatelolco, 326; its division into four districts, 327; its destruction by Cortés, 336; average temperatures, II.402; number of its inhabitants, 344; II.471t.
Tenochques: one of the names for the Aztec, 321n.
Teocalli: name of Mexica temples, 243, 324; description of the ones of Tenochtitlan and in Mexico City, 324; materials out of which they were made, 324–25, 339.
Teopan. *See* Teocalli; Xochimilco.
Teotihuacan: pyramids, 338.
Tepanec: ancient peoples of Mexico, 266.
Tepetate: kind of clay, 375.
Tequesquite: Mexica name of sodium carbonate, [329], 375, 400; [II.145]; provinces where its abundant, II.296–97.
Tereros. *See* San Cristóbal.
Testament of Hernán Cortés, II.495ff.
Tezontli: nature of this stone, 325.
Thread: amount imported to Veracruz in 1802, II.323t; in 1803, II.331t.
Ticapampa: German amalgamation factory in Requay, II.207n.
Tierras calientes, 183.
Tierras frías, 184.
Tierras templadas, 184.
Tin: mines where it is found, [II: 77], 178; amount exported from Mexico in 1803, II.335.
Tin plate: amount imported to Veracruz in 1802, II.324t; in 1803, II.335t.
Tithes: amount, II.56; paid by the mine owners, II.193.
Tlacosulpan, Rancho de: on the Huasacualco, 492.
Tlamama: Mexican porters, II.33.
Tlaolli: alcoholic beverage made from corn, 530.
Tlaxcala [Tlaxco]: ancient republic now part of the intendancy of Mexico, 385; the privileges of the inhabitants of this city, 385; population, 387.
Tobacco: its cultivation in the intendancy of Veracruz, 410, 417; in Mexico broadly, II.29; its production as a regal right, II.396; amount produced in the royal factories, II.396; annual sum this production brings to the king, II.429–30.
Toltec: their arrival in Mexico, 266; history of the first Toltec, 266; monuments that remain in Teotihuacan, 339.
Tomatl: its cultivation in Mexico, 559.
Tonalpohualli: Mexica secular calendar, 413.
Tonatiuh Ytzaqual, House of the Sun: ancient pyramid, 338–39, [383, 384n].
Tortoiseshell: amount exported from Veracruz in 1802, II.326t; in 1803, II.333t.
Trade: how it started between the Spanish and certain Indian tribes, 436; obstacles to Mexico's trade, II.307; inland trade: II.307; hampered by the lack of water connections, II.307; routes for inland trade, II.307–8; goods traded inland, II.313; advantages of a connection between the two oceans, II.317; foreign trade: II.320; via Veracruz, II.320; goods exported, II.321; goods imported, II.321–22; amount imported in 1802, II.322–23; in

1803, II.330t; exported in 1802, II.322–23; in 1803, II.333t; Veracruz trade in 1804, II.338t; 1805–1806, II.339t; Mexico's trade balance in 1824, II.339 (tableau); Acapulco trade, II.349; duties to which trade is subjected, II.366–67; smuggling: II.36–68; impact of the free trade edict on it, II.369; annual monetary losses New Spain incurred because of its passive trade, II.375t; classification of ports for this trade according to their importance, II.380; general trading balance of New Spain, II.382t; of all the Spanish Americas, II.386t; measures used to stop yellow fever, II.388ff. *See [also]* Yellow fever.

Trade balance of New Spain, annual: II.377; in general since 1748, II.382. *See [also]* Trade.

Trapiche de Almendaras, II.491.

Trapiches: small manufacturies of woolens in Querétaro, II.294.

Tribute: capitation tax that the Indians pay, II.430–31.

Trinkets made in Mexico, II.306.

Tschugatschi: people of Russian Alaska, 470.

Tshutski (also Chukchi): people in easternmost Siberia, 471n.

Turkey, [348, 528]; animal indigenous to Mexico, II.36–37.

Tzapoteca: ancient people of Mexico, 266.

Ugalachmiuti: people in Russian America, 473–74.

University of Mexico City: its building, 335.

Ural Mountains: comparison between the mines of this area and those in Brazil. II.263; these mines were discovered by Mr. de Schlenew, II.267; their yield, II.267–68.

Vaccine introduced to Mexico, 222; exists naturally in the country, 222.

Vaqueros: peasants, I.495; their way of living, I.495.

Valderas, Compuerta de: sluice in the desagüe of Huehuetoca, 368.

Valladolid, city in the eponymous intendancy: ratio castes to sexes among its inhabitants, 293t.

Valladolid, intendancy of: number of clergy in it, 282n; details about this intendancy, 390ff; its area, 390; climate, 390–91; [physical] upheaval that occurred as a result of a volcanic eruption, 391; relative population, 396; tally of its reales de minas, 397; II.74.

Vanilla: province where it is produced, 409; about its cultivation, II.22; amount exported annually from Veracruz, II.370; amount exported from Mexico in 1802, II: 326t, [383n]; in 1803; II.334t; impact that free trade has had on this export, II.370t.

Vara, Mexican: its relation to a pied de roi, 354n.

Vara castillana: its relation to a toise, 354n.

Vehidi: people of California, 444.

Venados: deer of New California, 453–54.

Veracruz, intendancy of: its size and climate, 408–9; its [agricultural] production, 409–10; its population, 410–11; its mountains, 411–12; its antiquities, 412–13; cities in, 414ff; its mines, 417; its reales de minas, II.69 (*see [also]* Reales de minas).

Veracruz, port of: its geographical position, 33–34; map of, 95; its temperature, 292; description of the town, 414; history of its construction, 414–15; reasons for the unhealthfulness of its climate, 415; II.399; movement of the annual trade of this port, II.321; amount of

Veracruz, port of (cont.)
merchandise imported from Spain in 1802, II.*322*t; in 1803, II.*330*t; of foreign merchandise in 1802, II.*324*t; in 1803, II.*331*t; amount imported in 1802 of merchandise from the Spanish colonies, II.*325*t; in 1803, II.*332*t; imports via Cádiz, II.*383*n; trade balance in 1802, II.*328*t; in 1803, II.*336*t; state of its commerce in 1820, II.*338*; imports in 1820, II.*338–39*; exports, II.*338–39*; trade in 1805, II.*339*; in 1806, II.*339*; annual trade balance, II.382t; this town is a stronghold of yellow fever, II.391; annual rainfall, II.402; average temperature, II: 402–3, 409.
Vermicelli: amount imported to Veracruz in 1802, II.*323*t; in 1803, II.*331*t.
Vertideros, Compuerta de: sluice in the desagüe of Huehuetoca, 368, 375.
Veta Biscaína de Real del Monte: rich vein in the intendancy of Mexico, 381; II.129.
Vezú: juice from sugarcane, II: 8, [11]. *See [also]* Sugar.
Viceroys of Mexico: their power and their income, II.439–40.
Villalta: [real de minas], 405; II.75.
Vinegar: amount imported to Veracruz in 1802, II.322t; in 1803, II.*330*t.
Vineyards: in Parras, 438; protected by the Spanish court, 571.
Virgins, miraculous, of Guadalupe and Remedios, 364n.
Visitadores: what they did, 278n.
Volcanoes of Mexico: 111, 190; of Mexico City; 488.
Vomito prieto (black vomit): has little impact on the state of the population in Mexico, 229. *See [also]* Yellow fever.

Washington, Treaty of: was unable to determine the borders between the United States and the United Provinces of Mexico, 484.
Waters, thermal: in the valley of Tenochtitlan, 351; near Guanajuato, II.108.
Wax: quantity produced by Mexico, II.*39–40*; annual amount imported to Veracruz, II.*322*; annual amount imported to Veracruz in 1802, II.*325*t; exported from Mexico in 1802, II.*327*t; in 1803, II: *331*t, [*333*t, *338–39*, 380, 461].
Whale skin, II.51.
Wheat, European. *See* Cereals.
Wheat: its cultivation in Mexico, 547.
Whites: enjoying all legal rights, 270; inequality the administrators introduced among them, 270; their number, 271; compared to whites in the Antilles, 272; and to other parts of the new continent, 273t; progress of civilization among them, 273; their wealth, 279–80. *See also* Chapetones; Creoles.
Wine: amount annually imported to Veracruz, II.*322*t; amount imported to Veracruz in 1802, II: *322*t, *324*t; in 1803, II.*330*t.
Wood tin [casserite]: mines where it is quarried, II.178.
Woolens: amount imported to Veracruz in 1802, II: *323*t, *324*t; exported to other parts of Spanish America in 1802, II.*327*t; in 1803, II: *331*t, *332*t.

Xicalanca: ancient people of Mexico, 266.
Xicotlan: salt manufactory in the intendancy of Puebla, 386.

Yabipais [Navajo]: Indian tribe, 437.
Yam. *See* Igname.
Yanos: presidio, 424, 431, II.458n.
Yanos: third dwelling place of the Aztecs, 431.
Yellow fever: mainly in Veracruz, II.389; impact of this plague on trade, II.389–90; and on the military defense of the country, II.390;

when it was first observed, II.390; should not be mistaken for the *matlazahuatl*, II.390–91; identical with *vomito Prieto*, II.391; why it had not formerly receive much attention from physicians, II.393; times when this illness was observed, II.392; it is endemic to Veracruz, II.393; it is a sickness *sui generis*, II.394; it did not spread to the western coast of Mexico, II.394; it is not exclusive to the northern hemisphere, II.396; links between this malady with atmospheric temperatures, II.401–2; it is not necessarily contagious, II.406; in the tropics it does not afflict the indigenous peoples, II.407; whites and mestizos in the interior are more often infected than Europeans arriving by boat, II.409; men are more often infected than women, II.409–10; duration of the illness, II.412; average mortality rate of the sick, II.412; limits of this malady in inland regions, II.415; treatment of this illness with the stimulant method, II.416; with quinquina ointment, II.416; with olive oil massages, II.417; with ice and snow, II.417; the illness only occurs periodically, II.418; ways of rendering it less frequent, II.422–23.

Yetl: Mexican name of tobacco, II.29.

Yucatán, bishopric of: its revenues, 282.

Yucca: plant that provides manioc, 520. *See [also]* Manioc.

Yuta [Ute]: Indian tribe, 425.

Yxtacmaztitlan: mine, 388.

Zacatecas, intendancy of: size, 400; climate, 400; geological features, 400; its reales de minas, 401; II.122; their geological characteristics, 401; metals being mined there, 401; II.138.

Zambos, 230, [237], 289; [II.56].

Zambos prietos, 289.

Zapoteca: Mexica people, 402. *See [also]* Tzapoteca.

Zinc: mines where it is quarried, II.179.

Zoology: animals in this work: *Anas moschata* [musk duck], II.37; *Apis mellifica* [bee species], II.40; *Balaena mysticetus* [baleen whale], II.53; [*Bombyx de Fabricius*, II.39]; *Bombyx mori* [silkworm], II: 38, [295]; *B. madroño*, II.39; *Bos taurus* [cattle breed], II.*34; B. americanus*, II.33; *B. moschatus*, II.33; *Camelus huanaco*, II.359; *Canis familiaris*; II.33; *Capra berendo* [wild goat], 442; *Cervus Canadensis* [elk], 453; *C. strongyloceros*, 453; *Coccus cacti* [cochineal], II: 40, 48; *Crax nigra*, II.36; *C. pauxi*, II.36; *C. globicera*, II.36 *Haliotis iris* [abalone], II: [5], 364; *H. australis*, II.364; *Meleagris gallopavo* [wild turkey], II.36; *Melipona fasciata* [bee species], II.40; *Murex* [sea snail], II: 50, [294]; *Mustela lutris* [sea otter], II.364; *Numida meleagris* [helmeted guineafowl], II.37; *Ostrea margaritifera* [pearl oysters], II.49–50; *Ovis aries* [sheep], [442], II.35; *O. ammon*, 442; penelope, II.36; *Phasianus gallus* [pheasant], II.36; *Physeter microcephalus* [sperm whale, II.51; *Psittacus* [parrot species], II.36; *Siren pisciformis*, II.395; *Sus tajassu* [peccary], II.35; *Termes fatalis* [termites], II.360; *Ursus caudivolvula* [bear species], II.36.

Toponym Index

Abinito: mines in Durango, Mexico, 88.
Abra de San Nicolás: cut in the mountain made to create a healthier climate in Acapulco, [380]; II.350.
Abusheer (Abu Sir): northern extension to the Saqqara necropolis, Egypt, 486.
Acabuca ravine, II: 103n, 108n, 121.
Acaguisotla (Acahuizotla): village in Guerrero, Mexico, 39; II.126.
Acahualtzinco: formerly near Ocopipila, Zautla, Mexico, 267.
Acámbaro: village in Guanajuato, Mexico, 106.
Acamiscla: village in Zumpango, Mexico, II.131.
Acaponeta: village in Nayarit, Mexico, 399; II.29.
Acapulco, town and port in the kingdom of New Spain: geographical position, 35; temperature, 154; population, 380; one considers relocating the military facilities of San Blas there, 479; description of the harbor, II.349; its trade with Guayaquil and Lima, II.352; obstacles to knowing the perils of navigation, II.353; trade with Manila, II.356; climate of the harbor and reason for its being protected from yellow fever, II.395.
Acatlán: village in Hidalgo, Mexico, 58.
Acatzitzintlan: now Mexicaltzinco, municipality in the state of Mexico, Mexico, 323.
Acayucan: town [in Veracruz state, Mexico], [410, 492], 495; [II.493t].
Acazonica: Jesuit farm, 415.
Achichintla, 131; II.77n.
Acobamba: village in the kingdom of Quito, now in the Huancavelica region in Peru, II.185.
Acocolco: Aztlán, small islands in Mexico, 323.
Acolhuacan. *See* Culhuacan.
Acoria: village in in the kingdom of Quito, 185.
Actopán: village in New Spain, now a city in central-eastern Mexico, 43, 121t, 132t, 175.
Actopán Mountains, 43.
Actopán Valley, 177.
Acusquilco, 400.
Africa: continent, 18 et passim.
Agibaniach: settlement in Alaska, 473.
Agrippa's Baths: ancient Rome, 350.

Aguacaliente, 74. *See also* Aguas Calientes.
Aguas Calientes: town, 399, [483; II.490].
Aguas Claras: city in the kingdom of New Granada, now in Colombia, 163.
Aguasarco, hills of, 392–93.
Aguilar: river described by Martín de Aguilar, possibly along the Oregon coastline, 441, 458.
Ahahuete: area in Mexico City near Barria de Santiago, 326.
Ahuacatancillo: city in Nayarit, Mexico, II.72.
Ahuxcatlan: now most likely Ahuacatlan, municipality and a municipal seat of Nayarit, Mexico, 399; II.29.
Ajusco (Axusco): lava dome volcano just south of Mexico City, 77, 121, 134.
Alacran: shoals near the Scorpion Reef off Yucatán peninsula, 124; II.326.
Alameda: town in New Mexico, now USA, 437.
Alaska, now USA, 309, 470, 472, 473.
Alberca de Santiago, 538.
Albermarle: bay in North Carolina, 553n.
Albuquerque y Alameda [towns in New Mexico], 437.
Alcohuacan [Mexican name for the town of Tezcuco], 342n.
Alcozauca: town and municipality of Guerrero, Mexico, II.72.
Aleutian Islands: islands belonging to both the U.S. state of Alaska and the Russian federal subject of Kamchatka Krai, 472n; II.366n.
Alexander Archipelago. *See* Prince of Wales Archipelago.
Alexandria canal: now Mahmoudia Canal, sub-canal from the Nile River, Egypt, 168.
Allegheny Mountains: mountains in Pennsylvania, now USA, 545t, 550; II: 37, 74, 186, 199, 234, 240, 296.
Almagre Mountain: mountain in Colorado, now USA, 92.
Almaguer plateau: mountain plateau in the kingdom of New Granada, now in Colombia, 544.
Almolonga: municipality in the Quetzaltenango department of Guatemala, 111n, 180n.
Alps: mountains in central Europe, 66 et passim.
Alpuyeca: town in Morelos, Mexico, 39.
Altamira: port city on the Gulf of Mexico near the southeastern tip of Tamaulipas, Mexico, 19, 20, 304, 305, 308, 489.
Altar: small city, now in the Mexican state of Sonora, 52, 57t, 130t; II.458.
Alto de Cajón (Cerro Cajón): mountain in Guerrero, Mexico, 40.
Alto de Camarón: village in Guerrero, Mexico, 39.
Alto de los Cajones (Los Cajones): village in Guerrero, Mexico, II.126.
Alto de los Guacamayos, II.492t.
Alto del Peregrino (Ocote de Peregrino): village in the municipality of Petatlán, state of Guerrero, Mexico, II.354.
Alvadeliste: mining site, II.74.
Álvarado: city in the Mexican State of Veracruz, 494; its position, 494. Port of: trade balance in 1806, II.341t.
Álvarado: river, 178 [190, 192, 207, 410, 491, 494, 496, 560, 561]. *See also* Papaloapan.
Amalapa: settlement near port of Conchagua, today in El Salvador, 152.
Amaluco, II.502.
Amaquemecan: old settlement in Mexico State, Mexico, 267.
Amatlan: prehispanic village in Oaxaca, 267.
Amaxac: de Guerrero, municipality in Tlaxcala, II.72.
Amazons, river of the: facilitates the smuggling of silver from Peru, II.233.

Amboina: now Ambon in Indonesia, 514n, 515.
Ameca: city in Jalisco, established in 1522, 319.
Amilco: spring near Churubusco, a neighborhood in Mexico City, 333.
Amojaque: Mexico, II.126.
Amour (Amur): river between Russia and China, 472n.
Amoy (Xiamen): city in Fujian province, China, II: 258, 285t.
Amsterdam, Netherlands, 117, 573; II: 237n, 239n, 242.
Anahuac [today Valley of Mexico]: countries that were part of it, 142; its area compared to that of New Spain, 199; population, 200. *See also* Valley of Mexico.
Anahuac Andes, 81. *See also* Rocky Mountains.
Ananca Mountains: stony mountains in the Andes, II.208.
Ancon Mountain: mountains extending from Ancon in Salida inlet to Cape Horn, 155n.
Ancud, Gulf of: in Chile south of Puerto Mont, 139.
Andageda [Andagueda, Colombia]: auriferous river in Chocó, [163, 164]; II.230.
Andalusia, Spain, 411, 438, 449, 525, 547, 558, 566; II.444.
Andes. *See* Cordilleras.
Anegada de Fuera: island near the British Virgin Islands, 125t.
Angasmarca: village in the in the kingdom of Quito, now in province of Santiago de Chuco, Peru, II.200.
Anian: ancient name of the Hudson strait, 458, [466].
Annaberg (Annaberg-Buchholz): village in Saxony, Germany, II: 96, 152.
Anse des Amis: inlet near Nootka island, British Columbia, Canada, 49.
Antequera. *See* Oaxaca.
Antigua: river near Antigua, Guatemala, 125t, 414, 416; II.312.
Antigua: village in Guatemala, [63, 111n; II: 210, 390, 399, 406, 409], 414.
Antigua Guatemala: city in Guatemala, 111, 180n, 191.
Antilles, British: sugar exports from, II.13t.
Antilles, French, Dutch, Danish, Swedish: sugar exports, II.13t. *[See also Subject Index.]*
Antilles, Spanish: sugar exports from, II.13t.
Antilles, Swedish: St Barths, Swedish Lesser Antilles, II.13t.
Antioquia, province of: in the kingdom New Granada, today in Colombia): its gold production, [162, 508, 552; II: 83, 182, 228], 229, [233, 245t, 253t, 254].
Antioquia Mountains: near Antioquia in Colombia, 163; II.288, 282.
Antwerp: port in Belgium, II:16, 35n.
Apán Plains: southern part of the state of Hidalgo, Mexico, II.320.
Apazapa (Apazapan): Veracruz, Mexico, 416.
Apomas: mine in the Fuerte region, II.74.
Aquaverde: military post, II.458n.
Aquitapilco: former mine near Tlaxcala, II.72.
Aragua Valley: in Venezuela west of Caracas, 328, 546n, 558.
Aranjuez: city in Spain, 3, 574.
Arastradero of San Pablo: geological formation, 164.
Araya: peninsula in Venezuela, 443.
Archangel (Arkhangelsk): city in Siberia, 315, 474, 485t.
Arequipa: intendancy and city in Peru, II: 200, 203n, 206n, 241.
Arguello: possibly town in Patagonia that no longer exists, 166n.
Arica: city in Chile, II.203.

Arigangueo: mine in the Tlalpujahua region, II.74.
Arispe [designated capital of the Pimería Alta, today Nogales Arizona], [20, 52, 53, 57, 60, 130t, 181, 429, 430], 432ff, [480].
Ariztizabal Island: British Columbia, Canada, 469.
Arjona, Ciénaga de: 415; II.400.
Arkansas: river in Arkansas, now USA, 56, 65n, 72, 90, 91, 92, 435n, 436, 484; II.457.
Arno: river in Italy, 465.
Aroma: mountain in the kingdom of Quito, II.185.
Arroyo Gavilán: stream in today's New Mexico, USA, 415.
Arsacides. *See* Solomon Islands.
Arthabasai station, 149n.
Ascensión, Bay of: 406, 407, 430.
Asia, 107 et passim: central Asia, 189, 245, 531, 556, 575; II.235; eastern Asia, 167, 233, 240, 458, 526n, 549, 561, 564, 566; II: 4n, 34n, 356; Russian Asia, 472, 484, 485; II.256.
Asia Minor: 66, 513, 566; II.16.
Asientos de Ibarra: district, 399; II.72, 82t, 83.
Astoria: colony at the mouth of the Columbia River, 149n.
Astrakhan, Russia, 486t.
Asturias, kingdom of: Iberian Peninsula, 260, 388; II: 19, 351.
Atacama Desert: desert plateau in Chile, II.200.
Atatlahuca: town in Oaxaca, Mexico, 478t.
Atempa: hot springs, 500.
Atlacuechahuayan: ancient city, II.64n.
Atlanca: village in Veracruz, Mexico, II.493t.
Atlantic Ocean, 38, 53, 63, 82, 105, 140, 146–47, 154, 167, 169, 178, 190, 223, 329, 360, 375; II: 51–52, 314, 317, 320, 352, 358n, 363, 459, 508.
Atlapezco (Atlapexco): town and municipality in Hidalgo, now in Mexico, 500.
Atlixco: village in the intendancy of Puebla, home to a famous cypress, [222], 384, 387, [402, 482t, 488, 540].
Atotonilco el Chico: village in Hidalgo, Mexico, 72t.
Atotonilco el Grande: Indian village, now town and municipalitiy in Hidalgo, Mexico, II: 68, 72.
Atoyaque (Atoyac): municipality in Jalisco, Mexico, II: 17, 313.
Atrato: river in Chocó [kingdom of New Granada, now in northwest Colombia]: the range of the Andes interrupts it between Cupica and the Atrato, 162ff; its mouths functions as a depot for the secret gold and silver trade, II: [230], 233.
Attakapas: county in Louisiana bordering on Mexico, 422.
Atzacualco: one of the districts of Tenochtitlan, 327.
Aude: department in southern France, 486t.
Augustinos, 482t.
Australia, 11.
Autlán (de Navarro): city and municipality in the Costa Sur region of Jalisco, Mexico, 399; II.29.
Auvergne: region in France, 105, 173, 177, 393; II.66n.
Avino: military post, II.458n.
Axapusco (Axapuzco or Axapusco): town and municipality in Mexico State, Mexico, 205, 206t.
Axtlan: also Aztlán, Aztec settlement, 437.
Axuchitlán (Ajuchitlán): city and municipality in Guerrero, Mexico, II: 74t, 180.
Ayahualulco: now municipality in Veracruz, Mexico, II.28.

Ayatitlán: village near Volcán de Colima, Mexico, 307n.
Ayaupa: island, II.465.
Ayoguesco: ancient settlement in Oaxaca, 478t.
Ayotitan: village in Jalisco, Mexico, 304.
Ayssen (Aysén) Estuary: region in Chile, 166.
Azangara (Azángaro): province, Puno Region in Peru, 262; II.208.
Azcapotzalco: one of present-day Mexico City's sixteen boroughs, 268; II.64.
Azogue: village in New Granada): provides mercury, II: 183.
Aztlan: native country of the Toltec, 232–33, [247, 267, 323, 431, 553].

Bab-el-Mandeb, Strait of: between Yemen on the Arabian Peninsula and Djibouti and Eritrea in the Horn of Africa, 157.
Bacalar: river, 406; II.450.
Bacamuchi: town in Sonora, Mexico, II.74t.
Bacatopa: mining site, II.74t.
Bahía: population of, 476.
Bahía de Chetumal: in the western Caribbean Sea on the southern coast of the Yucatán, 406.
Bahía Honda: island in the lower Florida Keys, II.23.
Bajío de la Tembladera: marsh, 415; II.400.
Bajo del Alerta, 125t.
Bajo del Obispo, 125t.
Baltic Sea: northern Europe, 160, 330.
Baltimore: city in Maryland, USA, 73, 579; II.407n.
Banks Island: Canadian Arctic archipelago, 469.
Baños de Jesús: baths in Arequipa, kingdom of Quito, II.183.
Baracoa: municipality and city in Guantánamo, Cuba, II.10.
Baraderas: river in Mexico, also called River of Crocodiles, 406, [408, 410].
Baranca de las Tinajas: Guerrero, Mexico, II.91.
Barbacoas: province in the kingdom of New Granada, now in Colombia, 508; II: 87, 228, 229, 233, 253t.
Barbados: island in the Caribbean, 485; II: 13t, 88, 291, 393, 408n.
Barbary Coast: term used by Europeans to refer to the coastal regions of North Africa inhabited by Berber people, II: 16, 257.
Barcelona: city in Spain, 17, 376, 574t; II.384.
Baroyeca: mining site in Sonora, Mexico, II.74t.
Barquisimeto: city on the Turbio River and capital of Lara in northwestern Venezuela, 163.
Barrio de las Muertas, 477t.
Barrio de Santiago, 113, 326.
Basochuca: town in the municipality of Arizpe in Sonora, Mexico, II.74t.
Batabanó: municipality and town in Mayabeque, Cuba, 72, 166, 531; II.466.
Batabanó, Gulf of: II.466.
Batas: village, 177.
Batopilas: military post, II.458n.
Bavaria, Germany: 173; II.139.
Bavia: military post, II.458n.
Bavispe: presidio, II.458n.
Bay of Tehuantepec, Mexico, 180.
Bayona, bay or gulf of: near present-day Cies islands, Spain, 398, 429; II.50.
Bejar: military post, II.458n.
Bemuchco: village in Mexico, 501t.
Benares: Bengal, India, II.11.
Bení: river in northern Bolivia, 514; II.233.
Berbice: region in British Guiana, II: 12, 14t.
Bering Strait: sea between Sibera and Alaska, 236, 382, 458, 466, 470ff; II: 351, 366n.

Berlin: city in Germany, II: 508, 509.
Bernal Chico: islet in the Gulf of Mexico, 125t.
Bernal Grande: island in the Gulf of Mexico, 125t.
Berne: city and canton in Switzerland, 110, 173.
Betolatia: mining site, II.75t.
Bevara: river, 163.
Bhutan: country in the Himalaya Mountains, 551, 555.
Big Diomede Island. *See* Imaglin.
Big Horn: mountains in today's Wyoming and Montana, now USA, 71t.
Bighorn. *See* James Peak.
Birch Bay: in Whatcom County, Washington, now USA, 149.
Bizaru, 386.
Black Sea: sea bordering Russia, 232; II.308.
Boca Chica: one of the entries to the port of Acapulco, II.350.
Boca de Jagua: community in Colombia, 73t.
Boca de Leones: Nuevo Leon, Mexico, II.71t.
Boca de San Gregorio: part of the desagüe of Huehuetoca, [360], 368t, [373t].
Boca Grande: one of the entries to the port of Acapulco, II.350.
Bocaneme: mining town in the kingdom of New Granada, now in Colombia, II.231.
Bocchetta Pass: pass near Genoa, Italy, II.119.
Bogotá: the ancient Cundinamarca in the kingdom of New Granada, now in Colombia, 99, 161, 162, 176, 182n, 195t, 245, 276, 331, 485; 551, 559, 574t; II: 88, 182, 190, 226t, 227, 250n, 306, 309, 348, 455, 456.
Bolaños: municipality in northern Jalisco, Mexico, 15, 21, 67, 181, 389, 399, 400, 534; II: 72, 81t, 82–83, 84t, 91, 95n, 126, 130, 136, 143t, 196n, 197, 490t, 491, 492t.
Bologna: city in Italy, 54, 391.
Bolsón de Mapimi: mountainous area [internal drainage basin at the center-north of the Mexican Plateau], [63, 88, 183, 418], 424, [428, 436; II.458].
Bombacho: likely Mombacho, volcano in Nicaragua, 150.
Bonavista (Buena Vista): village in Coahuila, seven miles south of Saltillo, Mexico, 342.
Bonny, Bight of: now Bight of Biafra, West Africa, 288.
Borbón: mining site, II.71t.
Bordeaux: city in France, 28, 168, 304, 544, 574.
Bordones: valley and waterfall in Colombia, II.23.
Borysthenes. *See* Dnieper.
Bourbon: island, II: [11], 14, [20n, 363].
Boutet Canal: now possibly Loreauville Canal, 422.
Boveda Real: site of canal trench in New Spain, 368t, 372, 373.
Brazil. *[See Subject Index.]*
Brazo de Miraflores. *See* Gila River.
Bremen: Hanseatic port city in northern Germany, 544; II.16.
Brenta: river in Venice, Italy, 356.
Bridgewater: underground canal near Worsley, UK, 360.
Bristol Bay. *See* Kamischezkaja, Gulf of.
British Antilles. *See* Antilles, British.
Bruselas: abandoned town in Nicaragua, 153n.
Buena Esperanza. *See* Villa de la Purificación.
Buena Guía. *See* Colorado River.
Buenos Aires. *[See Subject Index.]*
Bufa: rock formations, II: 104, 106, 122.
Buga: in the kingdom of New Granada, now municipality in Valle del Cauca, Colombia, II: 35n, 233.

Bukharia: Uzbekistan, 505.
Burgundy canal: France, 168.
Buritica: town and municipality in Antioquia in the kingdom of New Granada, now in Colombia, II: 228n, 229.
Burras: mining village in Guajanuato, Mexico, 389t; II.98, 105, 107.

Cabo Blanco. *See* Cape Orford.
Cabo Bojador, II.356.
Cabo Burica, 141n.
Cabo Catoche: Yucatán Peninsula, Mexico, 25, 158n, 191, 406, 466.
Cabo Codera, II.394.
Cabo Corrientes, 48n, 106, 126n, 161, 479; II: 53, 351.
Cabo de la Galera, Trinidad, 27.
Cabo Engaño. *See* Cape Edgecumbe.
Cabo Finisterre, Galicia, 182.
Cabo Garachine, 163.
Cabo Mendocino, 45, 53, 60, 74t, 128t, 140, 446, 456, 458, 461, 467, 470; II.361.
Cabo Pariña, II.397.
Cabo Pilares, 166.
Cabo San Antonio, Cuba, 73t, 158n, 406, 466.
Cabo San Bartolomé, 473.
Cabo San Francisco, 162; II: 351n, 353.
Cabo San Lucas, 45, 47, 48, 49, 70n, 74, 126t, 467; II.50.
Cabo San Sebastián, 74, 458; II.361.
Cabo Santa Elena, II.154n.
Cabo Tiburón, 163.
Cacanumacán: town in Mexico State, Mexico, 568.
Cadereita: Jiménez: administrative seat of Mexico, 381.
Cádiz: city in Spain, 7, 27, 45, 63, 115, 160, 162, 223, 296, 304, 407, 428, 439, 446, 465, 565; II: 19, 25ff, 48, 235, 241–42, 314–15, 329, 369, 373, 376, 383–84, 388, 391n, 393, 398, 405, 408, 411, 415t.
Cairo canal: canal from Old Cairo to the Red Sea, 168.
Cajamarca: city, plains, province in the kingdom of Quito, now northern highlands of Peru, 176, 544; II: 36, 199, 206, 238, 250, 253t, 348.
Cajamarquillo: in the kingdom of Quito, now a district of the province Ocros, Peru, II.184.
Cajatambo: province and town in the kingdom of Quito, now in Peru, II: 200, 207.
Cajón, El: mining site, II.74t.
Cajurichi: town in the municipality of Ocampo, Chihuahua, Mexico, II.73t.
Calabar: port city in southern Nigeria, 288.
Calcutta: city in India, 168, 348, 557; II: 10, 41, 286.
Caledonian Canal, [153], 168–67.
Caleta: shallows near Cozumel, Mexico, 191.
Callo: military post, II.458n.
Culhuacán: now a Mexico City borough, 51.
Cali: city in the kingdom of New Granada, now in Colombia); 163, 164; II: 228, 233.
Caliche: rock formation, 163; II: 108, 110.
California, New, 26, 45, 64n, 69, 70, 140, 191, 312t, 430, 442, 444ff, 457–58, 469–70, 475, 527f, 539, 540t, 548, 562, 565; II: 33n, 50, 356, 361, 364–65, 448t, 453, 458.
California, Gulf of: geographical positions of some points on its coastlines, [19], 50, [51n, 148–49, 181, 428, 440–41; II: 49–50, 53].
California or Old California: Cortés explored its coastline, 45; Chappe's, Doz's, and Velasquez's travels in this country, 45; population and area, 438–39; history of the discovery of this prov[ince], 439; climate, 440; mountains, 442; pearls, 443; Jesuit

California or Old California (cont.) settlements there, 444; indigenous tribes living there, 444; villages, 445; mines, II.75.
Calimal: river in the kingdom of New Granada, now in Colombia, 164.
Calimaya: town and municipality south of Toluca, now Mexico, 205, 206t, 292.
Callao de Lima, II: 350, 352.
Calpulalpan Mountain: in Tlaxcala, now southeastern Mexico, 177.
Calzada de Iztapalapán: today one of Mexico City's sixteen boroughs, 326.
Calzada de San Pedro de Tlahua: today one of Mexico City's sixteen boroughs, 354.
Camalecón: river in Honduras, 165n.
Campaña de Trujillo: mountainous region in Mexico, II.354
Campeche: port city on the Gulf of Mexico, [124t, 267], 405ff, [492–93, 496; II: 39–40, 320–21, 338, 363, 367, 450]. *See [also]* Wood from Campeche *[in the Subject Index].*
Campeche, Bay of (Campeche Sound): in the southern area of the Gulf of Mexico, 405.
Campedrito: village, 501t.
Canaan: region in the ancient Near East, 335.
Cañada de la Virgen: recently excavated Otomi archaeological site in Guanajuato, Mexico, II.110.
Cañada de Marfil: ravine near Guanajuato, Mexico, II.98.
Canal du Midi: canal in southern France, 328, 356n, 360, 370, 378n, 379.
Cañar: region in the kingdom of Quito, now province in Ecuador, 335.
Canary Islands: islands in the Atlantic, 27, 158n, 161, 270, 275, 402, 442–43, 508, 512ff, 521–22, 564, 566, 571, 574; II: 8, 21, 443–44.
Cancharani Mountain: in the kingdom of Quito, now Puno province, Peru, II.208.
Cancún: island in Mexico, II.336.
Canelas: mining site, II.73.
Canilleca de la Cuesta: village in New Spain, II.495.
Caño de las Animas: canyon or plain in the kingdom of New Granada, now in Colombia, 163.
Cañon de los Virreyes, 368t.
Cantal: department in the Auvergne-Rhône-Alpes region, France, 173.
Canton: region in China, 168, 170–71; II: 19, 167, 233, 258, 284t, 286, 356, 359, 364–65.
Cap Français. *See* Cap Haïtien.
Cap Haïtien. *See* Saint-Domingue.
Cape Chirikov. *See* Cabo San Bartolomé.
Cape Chukotsky, 472.
Cape Douglas, 469.
Cape Elie. *See* Cape Suckling.
Cape Gregory, 469.
Cape Grenville, New Georgia, 458.
Cape Hatteras, 73.
Cape Horn, 160, 165–66, 169–70, 462, 565; II: 51–52, 314, 317, 355, 362, 364.
Cape Malovodnoy, 472.
Cape Mendocino: 45, 53, 60, 74t, 128t, 140, 309, 445, 456, 458, 461, 467, 470; II.361.
Cape Morant, 73.
Cape Morant: easternmost point of mainland Jamaica, 73.
Cape of Good Hope, 160, 169–70, 174, 557; II: 52, 259, 261t, 285t, 317, 356, 364.
Cape Orford, 457, 468n, 469.
Cape Portland, 73t.
Cape Rodney, 472.
Cape San Abad, 71n.
Cape San Lucas (Cabo San Lucas), 45ff, 70n, 71n, 74, 126t, 467; II: 50, 73.
Cape Suckling, 474.
Cape Trubizin. *See* Edgecumbe.
Capula: town in Michoacán, 72t.
Capulatengo: mining site, 72t.
Capulin: community in Union County, New Mexico, USA, II.70t.

Caracas: city and province in the kingdom of New Granada, now in Venezuela, 11, 113, 115, 140f, 161, 195, 199, 206–7, 239, 274, 279, 285, 289, 307, 328, 344n, 349, 386, 440, 486t, 508, 514, 517, 533, 558, 563, 567n, 571; II: 6–7, 20ff, 31, 198, 232, 241t, 325t, 329, 334t, 338, 357, 367, 374t, 377, 381, 383t, 384, 385n, 387t, 388, 443–44, 454ff, 456n, 457, 460, 466.
Carangas: province in the northern parts of the Bolivian department of Oruro, II: 208, 225.
Caravaya (Carabaya): regionin the kingdom of Quito, now province in the Puno region, southern Peru, II.208.
Carcay Mountains [northeast of Sierra de la Boquilla and Cerro El Grenoso], 424.
Cardonal, Real del: town and municipality in Hidalgo, now Mexico, 133t; II: 72, 78, 179.
Carelmapu: hamlet in Chile, 139.
Caretas: quarry, II.107.
Caripe Mountains: Monagas in eastern Venezuela, II.23.
Carisal: military post, II.458n.
Carlsbad: city in Bohemia, now Czech Republic, II.349.
Carmen Island: Gulf of California, 408.
Carnatic (Carnasic): region in South India between the Eastern and the Western Ghats, II.246n.
Carniola: region that comprised parts of present-day Slovenia, II: 163, 182.
Carolinas Islands: imaginary archipelago, 440.
Caroni: river in today's Venezuela, 237, 514; II: 248, 319.
Carpathian Mountains: Balkan region, 104n.
Cartagena de Indias: [city and province in the kingdom of New Granada, now in Colombia. *See Subject Index*.]
Cartago: Costa Rican city southeast of San José, 158; II: 183, 230n.
Carthage: center of the ancient Carthaginian civilization on the eastern side of the Lake of Tunis, 266, 335.
Casa grande: Mexica ruins, 430, 431.
Casas Grandes: Aztec ruins in Nueva Vizcaya, 51, 64, 130t, 431–32, 437, 441n.
Cascas Mountains: northern kingdom of Quito, now Peru, II.185.
Casiquiare: river in the kingdom of New Granada, now in Venezuela, 514, 520n, 564; II.307.
Castile: historical region of Spain, 98, 173, 188, 328, 356, 410; II: 59, 230, 264, 292, 294, 502.
Castillo del Sisal: colonial castle in the Yucatán, 124t.
Cata: ravine, II: 109–10, 112.
Cataouillou: Catahoula Lake in Louisiana, 422.
Catorce: mining district, 423; description of, II.124.
Catorce Viejo: mining site, II.125.
Cauca: river in the kingdom of New Granada, now in Colombia, 162; II: 182, 228ff, 248.
Caucasus: mountain range in Asia, 104n, 266, 551; II: 121, 260.
Cavite: province in the Philippines on the southern shores of Manila Bay, II.358.
Caya Grande, II.200.
Cayenne: capital of French Guiana, northeast coast of South America, 506n, 521, 523, 525; II.12.
Caylin. *See* Cailín.
Caylloma: region in the kingdom of Quito, now a province in Arequipa, Peru, II: 206n, 208.
Cayman Brac: easternmost of the Cayman Islands, 73t.
Cayman Grande: largest of the Cayman Islands, 73t.

Cayo (Puerto Cayo), in the kingdom of Quito, now in Ecuador, 355.
Cayo de Piedras: island in Cuba, 73.
Cayo Flamingo: Cayo Guillermo in today's Cuba (?), 73t.
Cedar Island: eastern North Carolina, now USA, 127t.
Celaya: town in [central Mexico], 389 [et passim].
Cempoalla [Zempoala, now an important Mesoamerican archaeological site in the Úrsulo Galván municipality, Veracruz, Mexico]: main town of the Totonac, 414.
Ceralvo, Bay of (Isla Cerralvo): now Isla Jacques Cousteau near La Paz, Baja California Sur, Mexico, 443; II.50.
Cerro Blanco: hill and historical landmark in Santiago de Chile, II.72.
Cerro Colorado: borough of the municipality of Tijuana in Baja California, Mexico, II.74.
Cerro Cordo: military post, 88.
Cerro de Buenavista, II.110,
Cerro de Garganta: mountain near Mescala, Mexico, II.133.
Cerro de Guayatlapa, 501.
Cerro de Papantón, II.124.
Cerro de la Brea, II.354.
Cerro de la Cruz: Mexica pyramid, 384.
Cerro de la Cruz del Marqués: peak in the Ajusco Mountain Range located in present-day Cumbres del Ajusco National Park, Mexico City, 39, 178, 320; II.310.
Cerro de Las Cruces: II.310.
Cerro de las Navajas: in Hidalgo, Mexico, 132t, 339n; II.68.
Cerro de Uspalata: mine, II.208.
Cerro del Cubilete: in the Silao municipality in Guanajuato, Mexico, II.81.
Cerro del Ventoso, Durango, Mexico, 320; II.492.
Cerro Duida (Cerro Yennamadi): in Amazonas, Venezuela, 562n.
Cerro Gordo: collection of abandoned mines in the Inyo Mountains near Lone Pine, California, 88; II.414.
Cerro San Cristóbal: hill in northern Santiago, Chile, 80.
Cerro Tacón: mountain in the kingdom of New Granada, now in Colombia, II.230.
Chaboana: village in the Yucatán, 408.
Chachapoyas: city and province, II: 184, 200–201, 233.
Chachiltepec: also Casas Blancas in Guerrero, Mexico, II.74t.
Chacuaco: ancient mine, II.124.
Chaganta: in the kingdom of Quito, now a district in Peru, II.208.
Chagre [Chagres River in central Panama, the largest river in the Panama Canal's watershed]: believed to connect the two oceans, [83], 154ff.
Chalcas: district, 267; II.83.
Chalcha. *See* Hinchinbrook Island.
Chalchicuecan: settlement near Veracruz, Mexico, 410, 414, 446.
Chalchiguitec: municipality in Zacatecas, Mexico, 410.
Chalchihuete: ancient mines, II: [73], 124.
Chalchitepeque, II.72.
Chalchiuhcuecan: ancient name of the beach of Veracruz, [410], 414, [446].
Chalmiagmi: former settlement in Alaska, 473.
Chamacasapa: village in Mexico, II.132.
Chamaya: village in Oaxaca, Mexico, 514n; II.47.
Champotón: river [in Campeche, Mexico] on whose banks a part of the forest of Campeche is located, 407.
Chanate Mountain, 424.
Chapoltepec [Chapultepec near today's Mexico City]: [77, 267, 322–23, 326, 365]; aqueduct carrying potable

water to Mexico City, 332–33; castle built by viceroy Galvez, 352–53; his demise, 352–53; [II: 50, 135, 298, 440, 500].
Charcas [Santa María de las]: market town and mines, 423.
Charo: city in Michoacan, Mexico, 132t, 396.
Chautla: region in Puebla, Mexico, II.147.
Chavani: Russian post, 472.
Chavín (de Huantar): now an archaeological site in Peru, II.200.
Chayamapu archipelago, 165.
Chenchrenae, Gulf of (Kenchreai or Cenchreae), Greece, II.156.
Chiahuitzla: port, 414. [Also mountain near Veracruz, Mexico.]
Chiametlan (Chametla): city in Sinaloa, Mexico, 440.
Chiapas: southern Mexican state bordering Guatemala, 200, 231, 304n, 397, 402, 404, 484, 487–88, 492–93.
Chibuitan: city in Oaxaca, Mexico, II.490t.
Chica, [San Juan de la]: mercury mine, II: 180–81, 314.
Chichi: in former Tejas, no longer in existence, 422.
Chichimequillo (Chichimequillas): village in Querétaro, Mexico, II.108.
Chichíndaro: town in Guanajuato, Mexico, II.74.
Chicle Mountain, Mexico, 77.
Chiconahuatenco: river near Guadalajara (?), 344.
Chiconautla: mountain and village in Mexico, 76, 79t, 80, 123t, 354, 372, 501.
Chihuahua: town, 427 [et passim].
Chigin. *See* Chukchi.
Chiguera ravine, II.203.
Chila: salt mine in the intendancy of Puebla, 386.
Chile: II.208 [et passim. *See also Subject Index*].
Chiloé Island, Chile, 208.
Chilpansingo [Chilpancingo or Chilpanzingo]: municipal town in Mexico, 380.
Chimalapa: river, town and municipality in Oaxaca, Mexico, 21, 82, 147, 150, 180, 499; II: 316ff, 490t, 491, 493t, 508.
Chimaltenango: town in Guatemala, 165.
Chimaltitan: municipality in Jalisco, Mexico, II.72.
Chimborazo: volcano in the kingdom of Quito, now in Ecuador, 101, 104, 111, 181n, 574.
China. *[See Subject Index.]*
Chinameca: village [now municipality in San Miguel, El Salvador], 492.
Chinampas: floating gardens on the lakes of Mexico City, [327], 350–51, [357].
Chinche: city in Pasco in the kingdom of Quito, now in the Province of Daniel Carrión, Peru, 165.
Chinkitané Bay. *See* Sitka Bay.
Chippewah Mountains. *See* Rocky Mountains.
Chiquimula: city in Guatemala, 164n.
Chittagong: port city on the southeastern coast of Bangladesh, II.41.
Chocó: Isthmus of; also province in the kingdom of New Granada, today in Colombia, 82, 84, 146, 162n, 163–64, 170, 179, 430, 508, 552; II: 21, 83, 86–87, 88n, 206, 229ff, 233, 245t, 252, 253t, 268, 320.
Cholollan: historical republic, 142, 244, 390.
Cholula [municipal town of Puebla, Mexico], 412; II.40 [et passim. *See also* Pyramids *in the Subject Index*.]
Chonos, Archipelago of: the Chiloe Indians visited it, [166, 238]; II.465.
Chontale (Chontales), Nicaragua, 478t; II.46.

Chontalpa: area in Tabasco, Mexico, II.72.
Choropampa: region in the kingdom of Quito, now a district in Chota, Peru, II: 203, 206.
Chota: mines, [II: 79, 83, 124], 201–2, [206n, 233].
Choyohuacan, 326.
Chubrusco. *See* Churubusco.
Chucuito: lake and village in the kingdom of Quito, now in Puno Province, Peru, 552; II: 200, 208.
Chucunaque: river in Panama; tributary of the Tuira River in Darién Province, 163.
Chugachi: settlement in Russian America, 473.
Chugatskaia, Gulf of (Chugatsk). *See* Prince William Sound.
Chukchi (Chukchee), 472.
Chukotka. *See* Tchoka.
Chumbivilcas, 262.
Chupiquiyacu: in the kingdom of Quito, now Peru, II.202.
Chuquisaca, kingdom of. *See* Peru.
Churultecal: name Cortés gave to the village of Cholula, II.159.
Cíbola: fabled city, [17], 441.
Cicuic: town in Mexico, II.179n.
Ciénaga Boticaria: swamp near Veracruz, II.400.
Ciénaga de Arjona: swamp near Veracruz, II.400.
Ciénaga del Castillo: Acapulco, II.395.
Cinaloa: province, 429: town, 433.
Citlaltepetl: one of the highest peaks in the Mexican cordillera, 180; meaning of its name, 180n, [42, 93, 113, 114, 131t]. *See [also]* Orizaba Peak.
Ciudad Real: capital of Chiapas, [404], 483–84.
Ciudad Vieja. *See* Almolonga.
Clarke's River: tributary of the Tennessee River, Jackson Purchase region of western Kentucky, now USA, 149n.
Clausthal: city in Lower Saxony, 228; II: 79, 95.
Cloak Bay: south of North Island, Alaska, 238.
Clyde: Forth and Clyde canal, Scotland, 169.
Coadnabaced: name Cortés gave to the city of Cuernavaca, I.380n.
Coahuila: military post, II.458.
Coahuila: province, [18, 19, 61, 172, 183, 202, 203t, 304–5, 307, 308, 311, 312, 316], 418ff, 421n], 423, [480, 481t; II.458].
Coatzacoalcos: major port city in Veracruz on the Coatzacoalcos River, 491, 515, 525; II: 492, 493.
Cochapata (Qochapata): in the kingdom of Quito, now an archaeological site in Cusco, Peru, II: 185–86, 188–87.
Coche: island, II.49.
Cochin (Kochi): city in southwest India, II.446n.
Cochinchina: region encompassing the southern third of present-day Vietnam, 559; II: 34n, 284t.
Cocos Island: in the Pacific Ocean, today administered by Costa Rica, II.54.
Cofre de Perote: one of then highest peaks in the Mexican cordillera, [24, 40, 42, 43n, 45, 93n, 104, 123t, 131t], 180, [188]; description of the mountain, [193], 411; [II: 27, 39].
Cohuixco: region in Mexico, II: 64, 66.
Cojatambo (Cajatambo): in the kingdom of Quito, now province in the Lima region, Peru, II: 200, 206n, 207.
Cojohuacan [Colhuacan, Culhuacan, one of the Nahuatl-speaking pre-Columbian city-states of the Valley of Mexico]: Cortés's favorite place, 343; [267, 268, 322–23, 337–38, 404, 433, 487].
Cojolapan: ancient city in the Zapotecas region in Mexico, II.64n.

Colima: [city and] volcano, 48n; 398 [et passim].
Colipa: village, 409; II: 23, 24, 25, 26, 27t.
Collao: in the kingdom of Quito, now a province in Puno, Peru, 551.
Colombia: country in South America, previously part of the kingdoms of New Granada and Quito, 12, 145t, 161n, 215t, 216t, 460, 486t, 546n.
Colotepeque (Colotepec): settlement near Puerto Escondido, Oaxaca, Mexico, 402, 478t.
Colotlán: now municipality in Jalisco, Mexico, II.5.
Colquisirca (Colquijirca): mining district in Pasco in the kingdom of Quito, now Peru, II.200.
Columbia: river thought to be the Tacoutché-Tessé, 147; how it differs from the Tacoutché-Tessé, 148n; discovered by Quadra, 460. *See [also]* Tacoutché-Tessé.
Comangillas (Comanjilla): village in Guanajuato, Mexico, 390; II: 108, 121.
Comayagua: city in Honduras, 165n, 337n.
Communi: lake, II.208.
Compostela: town, 399.
Conboy, 124t, 125t.
Concepción: city in Chile, II.52.
Concepción de la Vega: in today's Dominican Republic, II.4.
Conchagua: now municipality in the La Unión in El Salvador, 152.
Conchapata: in the kingdom of Quito, now an archaeological site near the city of Ayacucho, Peru, II.200.
Conchucos: in the kingdom of Quito, now district in Pallasca, Peru, II: 183, 200.
Condesuyo: in the kingdom of Quito, now a province in Arequipa, Peru, II.206n.
Condoto: now a municipality in the kingdom of New Granada, now in Chocó, Colombia, II.88.
Coneto: military post, II.458n.
Confederation of the United Mexican States, 161n.
Congo: country in central Africa, 240, 246.
Contoy: small island in Quintana Roo, Mexico, 406.
Cook River. *See* Cook's Inlet.
Cook's Inlet (Cook Inlet), south-central Alaska, USA, 464, 469, 471, 473.
Copala: mines, [20, 305], 399, [428], 433; II: 72, 74, 82t, 83, 166.
Copan: now a Maya archaeological site in western Honduras, 487.
Copiapó: city in northern Chile, II.208.
Copper Island. *See* Mednoi Island.
Copper Mine River: in the North Slave and Kitikmeot regions of the Northwest Territories and Nunavut, Canada, 72.
Copper Mountains, 72.
Copper River. *See* Mednaya River
Coquimbo: port city in Elqui Province, Coquimbo Region of Chile, 191; II: 54–55, 208, 315, 351, 364, 367.
Corazón, 111.
Corbières: mountain region in Languedoc-Roussillon, southeastern France, 169.
Cordillera de Rentema: in the kingdom of Quito, now in Peru; consists of the summits of the Yactán, Naranjos, El Arenal, Picuy, Cumbe, and Los Patos hills; cataract, 436.
Cordilleras of Mexico: 109; description of this mountain range, 71; and its highest peaks, 181; its elevation compared to those of the Alps and the Pyrenees, 182n.
Córdoba: town and district [in Mexio], 417 [et passim].

Corinth, Gulf of: in the Mediterranean, Greece, 156.
Cormolache: mining site in the kingdom of Quito, now in Peru, II: 202, 203.
Coro: settlement near Mérida in Yucátan, Mexico, 206, 406, 442; II: 422, 455.
Coronilla. *See* Puríssima Concepción de Tetela del Río.
Corpus: mining site, II.74.
Corsica: island in the Mediterranean, 315n.
Cosamaloapa River. [*See* Papaloapan.]
Cosamaloapa: town [in the plains of the Sotavento zone in Veracruz], 495.
Coscallan: mining site, II.72.
Cosiquiriachi: military post, II.458; mines, 428, 83, 84t.
Cosoleacaque: village [in Veracruz], 492.
Cotentin Peninsula (Cherbourg Peninsula), in Normandy, France, II: 120, 121.
Cotopaxi: active stratovolcano in the Andes Mountains in the kingdom of Quito, now in Ecuador, 111, 181n, 355, 412n, 466n.
Council Bluffs: city in Pottawattamie County, Iowa, now USA, 55, 56, 72.
Courland: western Latvia, 251.
Cox Canal: near Nootka island, British Columbia, now Canada, 232, 464.
Coxitambo: in the kingdom of Quito, now Cojitambo, archaeological site, mountain and village west of Azogues, Ecuador, II.183n.
Coyame: military post, II.458n.
Coyoacan. See Coyohuacán.
Coyohuacán: municipality in present-day Mexico City, 322, 327, 333, 357; II.6.
Coyotepec: the eastern part of Lake Zumpango, 354.
Coyuca: city in Guerrero, Mexico, 402n.
Cozumel: ancient island populated by Europeans, 406; [II.25].
Crane Mountains. *See* Grulla.
Creuze (Creuse): department in central France, 486t.
Croatia: country in Europe, 544.
Cruz Blanca: hamlet in Mexico, 103.
Cuagolotal, 39.
Cuautitlan: city in the State of Mexico just north of the present-day capital, 371, 376.
Cuautitlan: river, 353, [355, 358–59, 361, 366–67, 369–70, 375–76].
Cuautla de las Amilpas: city in Morelos, Mexico, 488; II.6.
Cuba. *[See Subject Index.]*
Cuchandiro: town in Michoacán, Mexico, 396.
Cucurpe: city in Sonora, Mexico, II.74.
Cuenca: city in the kingdom of Quito, now in southern Ecuador, II.179n.
Cuenca: region in Psain, 486t.
Cuenca: region in the Valley of Mexico, II.183n.
Cuencamé: small city in Durango, Mexico, II.73, 179.
Cuepopan. *See* Tlaquechiahacan.
Cuernava: town in the intendancy of Mexico, [39, 78, 132t, 281, 340], 380; [II: 5, 7, 310, 316n].
Cuernavaca: town in the province of Xochitepec near the entrechment of Xochicalco, [39, 78, 132t, 281], 340, [380; II: 5, 7, 310, 316n].
Cues Plain. *See* Llanos de los Cues.
Cuesta de la Noria: mountain pass in Chihuahua, Mexico, II.121.
Cueva de San Felipe: cave near San Felipe, Guanajuato, Mexico, II.132.
Cuevas Mountain (Cerro de las Cuevas): mountain in Mexico, 395.
Cuinche (Cuincho): minerals baths in Mexico, 396.
Cuitimba: river [in Michoacan] that has disappeared, [392], 394.
Cuitlatecapan: former province in Mexico, II.49.
Cul-de-Sac Plain: Saint-Domingue, Haiti, II.44.
Culebra: part of today's Puerto Rico, 499.

Culhuacan: city ruins, 268, *322*, *323*, 404.
Culiacan: river [city in Sinaloa, Mexico], [181], 429, 432–33, [446, 539].
Cumaná [city in Venezuela]: annual trade balance, II.381.
Cumbre del Peñol: monolith near Guatape in the kingdom of New Granada, now in Antioquia, Colombia, 78t, 79t.
Cundinamarca: ancient Bogotá, 181n, 245, 266, 551.
Cuneza: military post, II.458t.
Cupica, Bay of: canal building project to connect the two oceans there, 161ff; the Andes interrupt it between Cupica and Atrato, 162; [82, 84, 146, 147, 167, 170; II.*320*].
Cupisonora: mining site, II.74.
Curaçao: Dutch Caribbean island, II: 12, 388.
Curimayo plains: gold mining location in the kingdom of Quito, now in Peru, II.206.
Curucupaseo Sinda: mining district, II.74.
Cuyoacan: town in the intendancy of Mexico and convent founded by Cortés, *379*; [II: 497, 503].
Cuyuca beach (Coyuca de Benítez): city and seat of the municipality of Coyuca de Benítez, in the state of Guerrero, southwestern Mexico, II.354.
Cuzco (Cusco): city and intendancy in the kingdom of Quito, now in Peru, 182, 262, 355, 485, 532, 552, 563; II: 36, 200, 250ff, 292, 453.

Dahshur: royal necropolis in the desert on the west bank of the Nile south of Cairo, Egypt, 383, 486.
Damascus: capital of Syria, 348.
Danish Antilles. *See* Antilles, Danish.
Danto cavern: also Pierced Mountain or the Bridge of the Mother of God, II.126.
Danube: river in central Europe, 549.
Darfur: Sudan, 163.
Darién: province in Panama, 167, 179, 191, 552; II: 247ff.
Darién, Gulf of, 72, 141.
Dauphin Lake: near the city of Dauphin in western Manitoba, II.190n.
Davis Strait: northern arm of the Labrador Sea between mid-western Greenland and Nunavut, Canada's Baffin Island, II.51.
Demerara (Demarary): region in Dutch Guiana, II.14.
Denmark: country in Europe, 160, 315; II: 16, 255, 427.
Deptford: southeast London, England, 442.
Derby: England, II.133.
Derbyshire: county in England, II.127.
Desagüe de Martínez, 370.
Desert, The Great [Sahara?], 484.
Dhawalagiri Peak. *See* Mount Dhaulagiri.
Diablerets: village in the canton of Vaud, Switzerland, II.106.
Diego Ramírez Reef or Islands: small group of islands at the southernmost end of Chile, 182.
Diemen's Island: Tasmania, 450.
Diez Brazas, shoals of, 125t.
Dnieper (Dnjepr): river in Russia, 232.
Dolores: Indian village: tableau of births and deaths from 1750 to 1799, II 469t.
Dolores: river, tributary of the Colorado River, in Colorado and Utah, now USA, 91.
Dominica: island in the Caribbean, II: 13, 361.
Dragon's Mouth (Bocas del Dragón): channel in the southeastern Caribbean Sea between Point Peñas in northeastern Venezuela and the northwestern extremity of the island of Trinidad, 464.
Dresden: city in Germany, 20n, 487.

Durango [town and state in Mexico]: position, 54, 426; amount of smeltable iron and nickel found in its vicinity, 426–27; when it was founded, 427. *[For intendancy of Durango, see Subject Index.]*
Durasno [Durazno]: mercury mine, II.180–81.
Dutch Antilles. *See* Antilles, Dutch.

East Indies: sugar exports from, II.16; quantities of gold and silver exported to Europe, II.256–57. *See also* India.
Easter Island: southeastern Pacific Ocean, 566, 558.
Edgecumbe: mountain originally called San Jacinto, 460.
Egypt: country in northern Africa, 25ff.
Ejutla, Partido de, 477t.
El Aguage: in Chihuahua, Mexico, II.74.
El Alacrán: mining site, II.73.
El Carizal: mining site, II.74.
El Cuiche: mountain in Mexico, 392.
El Doctor: mine in the Intendancy of Mexico, [181, 318], 381, [386; II: 71–72, 78, 82–83, 91].
El Fuerte: town [in Sinaloa, Mexico], [57t, 130t], 433.
El Gabilán (Gavilán): mine in Ures, Sonora, Mexico, II.74.
El Guarda: in Mexico State, Mexico, 39.
El Hierro. *See* Ferro Island.
El Jacal: mountains in Hidalgo, Mexico, 132t.
El Oro de Topago: mining site, II.73.
El Paso del Norte: town in the kingdom of New Spain, today in New Mexico, 52n, 480.
El Peñol: hill, 123t.
El Pinal: now Los Amotes, II.72.
El Pópulo: mining site, II.74.
El Realejo (Realexo): port in Nicaragua, 152, 153, 154n, 191.
El Robledal: in the municipality of San Francisco del Rincón, Guanajuato, Mexico, II.178.
El Venado: mining site in San Nicolás de Croix region, II.71.
El Zapote: also Jesús María, town in the municipality of San Martín de Hidalgo, Jalisco, Mexico, II.74, 180n.
Eliso: historic mine, 324n; II.72.
Elyseram: ancient river in India, II.11.
Embarcadero de la Cruz: port area of Tehuantepec, Mexico, 150, 160.
Emoui. *See* Amoy.
Encero farm (near Veracruz), 123; II: 399, 407, 423.
England: country in Europa, 34ff.
Ensenada de Santa Catalina: Catalina Island, California, 194n.
Ensenada de Santa Lucía: port of Acapulco, Mexico, II.350.
Entrada de Eceta (Hezeta): the mouth of the Columbia River in Washington State, now USA, 468.
Equetchecan: Indian village, 408.
Erromanga (Eromanga), New Hebrides, 26.
Erzgebirge: mountain range in Germany, II.80, 158t.
Escapusalco. *See* Ixcapuzalco.
Escaut: former department of France, now part of Belgium and the Netherlands, 315, 544.
Esmeralda Mountains: in the kingdom of Quito, now in Venezuela, 514.
Espartal: swamps outside of Veracruz, II.400.
Esperanza: mine, II: 108, 121, 131; ranch, 527.
Espíritu Santo: archipelago and bay in the New Hebrides, now Vanatu, 406–7, 421n, 439; II: 361, 458, 472t, 480t.
Espíritu Santo: military post, 421n; II.458n.
Essequibo: largest river in Guyana, II: 12, 14.

Estero Aisén. *See* Ayssen Estuary.
Estola, 38–39, 122t.
Estremadura: western Spanish region bordering Portugal, II.562.
Euphrates: river in Mesopotamia, 566.
Europe, 15ff.
Exeter: England, II.105.
Expailly: in Velay, France, II.229.
Ezatlan: town and municipality in Jalisco in central-western Mexico, II: 29, 72.

False-Orizaba: imaginary mountain indicated on Arrowsmith's map, 42.
Farallón del Obispo: island near Isla La Redonda and Playa Hornitos, Guerrero, Mexico, II.350.
Farallones: rocks, 128t.
Father Mountain: small mountain near the Cerro El Potosí, Nuevo León, Mexico, II.224.
Fatu Hiva: southernmost of the Marquesas Islands, French Polynesia, II.361.
Ferro Island: Canary Islands, 27, 29.
Fichtelberg: mountain in the Erzgebirge, Saxony, II.349.
Finland: country in Europe, 209t, 396, 486t.
Flat-Head River. *See* Clarke's River.
Floating gardens. *See* Chinampas.
Flores: farm, 500.
Florida. *[See Subject Index.]*
Florido: river in Chihuahua, Mexico; tributary of the Río Conchos, 88.
Fonseca, Gulf of: part of the Pacific Ocean, bordering El Salvador, Honduras, and Nicaragua, 152.
Fraile Mountain (Frayle): in the Sierra Madre Oriental, Nueva León, Mexico, II.181.
France: country in Europe, 22ff.
Fraser River. *See* Tacoutché-Tessé.
French Antilles. *See* Antilles, French.
Fresnillo: town [in Zacatecas, Mexico], 201 [et passim].
Friedrichswalde: Saxony, Germany, II.119.
Friendly Cove. *See* Nootka, Santa Cruz de.
Fuente de Estola: inn, 112t.
Fuentestiana, II: 133, 200ff.
Fuerte: river in Sinaloa, Mexico, 429, 433.
Funchal: city in Portugal's Madeira region, 194n.

Gaeta: Italy, 412n.
Galapagos archipelago: volcanic archipelago in the Pacific Ocean, 464; II: 51ff, 353.
Galeota Point. *See* Cabo de la Galera.
Galera Point, Trinidad, 27.
Galicia: Spain, 182, 260; II.19, 66n, 351.
Gallega: shoals, 97, 125t, 191, 320.
Galveston: city on the Gulf of Mexico in present-day Texas, now USA, II.457.
Ganges: river in India, 247; II: 34, 257.
Garita de Guadalupe: lodge, 78t, 124t.
Gasave (Guasave): seat of the homonymous municipality in the Mexican state of Sinaloa, II: 128, 180.
Genoa: Italy, 419; II: 26, 119.
Georgetown: now borough in Washington, DC, 73n, 74.
Germany: country in Europe, 18ff.
Ghent: city in Flanders, Belgium, 532.
Giganta [Cerro de la]: mountain in California, 442.
Gigante: mercury mine, [126]; II: [77], 180–81.
Gila: river [in the kingdom of New Spain, now in Mexico]: its junction with the Colorado River, [437], see this word; the Aztecs built their second settllement on the banks of this river, 431.
Giliapa: mine near Veravruz, II.75.
Gilolo: Halmahera, formerly known as Jilolo, Gilolo, or Jailolo, largest of the Maluku Islands, Indonesia, 515.

Gineta: mountain, 402.
Goazacoalcos, 150. *See also* Huasacualco.
Gobi Desert: Mongolia, 174.
Godavari Delta: river delta in India, II.11.
Gojam: kingdom in northwestern Ethiopia, 173n.
Goldcronach (Goldkronach): Fichtelgebirge, Germany, 144n.
Golfo Dulce: Costa Rica, 164n.
Götland: Sweden, 315.
Granada: town in Western Nicaragua, 387.
Grande Pará: Brazil, 165.
Graveneire: Auvergne, France, 66n.
Great Britain: country in Europe, 201ff.
Great Cyclades: island group in the Aegean Sea, 439.
Great Desert: part of the Great Basin between the Sierra Nevada and the Wasatch Range, 484.
Great Island. *See* Prince of Wales Archipelago.
Green Island. *See* Tuk Island.
Greenland: island in the northern Atlantic, 309; II.51.
Greenwich: city in England, 33, 49n, 72, 91, 116, 117; II.366.
Grenoble: city in the French Alpes, II.66n.
Groot Noordhollandsch Kanaal. *See* North Holland Canal.
Gross Kaminsdorf: city in Saxony, Germany, II.66n.
Gruës Mountains: part of the Rocky Mountains in Utah, 149.
Guainopa: military post, II.458n.
Guacarachuco (Huacrachuco): town in central Peru, II.183.
Guachinango: mining site in Mexico, 284; II.72.
Guadalaxara [Guadalajara], town [in the kingdom of New Spain, now in Mexico]: 399; its population, 476; factories in, II: 293, 296. *[See also Subject Index.]*
Guadalcanal: Spain, II: 89, 224.
Guadalquivir: river in Spain, II: 219, 230.
Guadalupe de La Puerta: administrative district in New Spain, II.74t.
Guadalupe de Veta Grande: mining site in New Spain, II.71.
Guadalupe River. *See* Tepeyacac.
Guadiana. *See* Durango.
Guaduas: town in in the kingdom of New Granada, now in Cundinamarca, Colombia, 195; II.6.
Guagua-Pichincha: peak of the Pinchina volcano, II.224n.
Guahan: Guam, island in the Pacific, 358–59.
Guailas (Huaylas): in the kingdom of Quito, now province in the Ancash region, Peru, II.183.
Guajacatan: mining site, II.72.
Gualán: port in Zacapa, Guatemala, 304; II: 17, 22, 363.
Gualgayoc [Peru]: mines, II.201–2; their yield, II: 203, [223, 233].
Gualgayoc (Hualgayoc): province of the Cajamarca Region in Peru, II: 124, 200ff, 206, 207, 223, 233.
Guali: river in the kingdom of New Granada, now in Colombia, II.231.
Guallaga: river [in the kingdom of Quito, now in Peru] that might serve as a trade connection between the two oceans, [82, 146], 165; [II: 200, 233. S*ee also* Huallaga River].
Guallanca (Huallanca or Wallanka): region in the kingdom of Quito, now district of the Bolognesi Province, Ancash region of Peru, II.200.
Guamachuco (Huamachuco): town in the kingdom of Quito, now in northern Peru, 335; II: 200, 244.
Guamalies. *See* Huamalies.
Guamanga (Huamanga): in the intendancy in the kingdom of Quito intendancy, now in Peru, II.200.

Guamoco Mountains: in the kingdom of New Grananda, now in Colombia, II.228.
Guanabacoa: Cuba, II: 104, 120.
Guanacas: territory in the kingdom of New Granada, now in Colombia, 162, 182.
Guanajuato, town: 389; II.98–99; its population, 477. *[See also* Guanaxuato *in Subject Index.]* *See also* Zacatecas.
Guanta: city in the kingdom of New Granada, now in Venezuela, II.200.
Guaraz (Huaraz): town in the kingdom of Quito, now city in the northern Callejón de Huaylas valley, Peru, II: 183, 184.
Guari: mining site in Mexico, II.200.
Guarinó: river in the kingdom of New Granada, now in Colombia, II.231.
Guarisamey: city in Durango, Mexico, 20, 424, 534; II.73, 81, 82t, 83, 84t, 86.
Guarisamey: mine, [20, 424], 534; [II: 73, 81ff, 84t, 86, 489].
Guasacualco: river in [Mexico], 187, 492. [*See also* Coatzacoalcos; Huasacualco.]
Guatimala [Guatemala]: monuments of, 487.
Guatulco (Santa María Huatulco): town in Oaxaca, Mexico, 58, 129t.
Guautitlán (Cuautitlán): river, city and municipality in the State of Mexico, 368, 373t.
Guautla de la Hamilpas: part of Tlaxcala now, 385.
Guaxaca. *See* Oaxaca.
Guayaquil: [port, province, and river, 36 et passim]; annual trade balance of, II.381.
Guaymas: port [and city in Guaymas, Sonora, northwestern Mexico], 430; [II.363].
Guayra, La [port in Caracas, Venezuela], [380]; II: 381, [395, 397].
Guayras. *See* Huayras.
Guazacualcos, Isthmus of: now Coatzacoalcos, 304n.
Guazun: mountains in the kingdom of Quito, now in Ecuador northeast of Azogues, II.183.
Guchilaque heights: in today's Huitzilac region (?), 37, 39, 184, 320, 380; II.310.
Gueguetoque. *See* Huehuetoca.
Guetlachtlan: older name for the intendancy of Veracruz, New Spain, 409.
Guianas: Brazilian, British, Dutch, French; Spanish. 216; II.14t.
Guichichila: mine, 339; [II.72].
Guichichila: village in Guadalajara province, 399.
Guichula. *See* Tabasco.
Guiechapa, village, 58t, 129t.
Guilotitan (Huilotitan): mining site in Mexico, II.72.
Guimilpa: quarry near Huimilpan, Querétaro, Mexico, II.107.
Guinea: country in West Africa, formerly French Guinea, 237, 522, 557, 570; II: 51–52, 251, 256, 354.
Güines canal: project on the island of Cuba, 166; II.466
Güines Valley, 507; II.11.
Guitivis. S*ee* Santa Cruz de Mayo.
Gulf of Murray: Moray Firth in Scotland, 152.
Gunduck (Gandaki): river in India, 182.

Hacienda de Los Morales (near Chapultepec), 565.
Haleb (Aleppo): city in Syria, 348.
Hallaga River (Huallaga): tributary of the Río Marañón in the Amazon Basin, II.184.
Halsbrücke. *[See Subject Index.]*
Hamburg: city in Germany, 544; II.15.

Haro, Strait of: connects the Strait of Georgia to the Strait of Juan de Fuca, British Columbia, Canada, 467.
Harz Mountains: Germany, II: 79, 86, 89, 134, 140, 268, 269t.
Hatun-Potocsi [Cerro de Potosí]. *See* Potosí.
Haucingo: mining site, II.72.
Hautes-Alpes: France, 315, 388, 486t.
Havana. *[See Subject Index.]*
Heceta: original name of the Columbia River, 460n.
Heideberg: small mountain near Zelle, Franconia, II.120.
Hermione, Gulf of (Hermionic Gulf), 391n.
Hiaqui Basin, II.86.
Hiaqui River (Yaqui): in Sonora, northwestern Mexico, 539.
Higane: mining site, II.74.
Highlands of the Mexican Cordillera, 167: the four around Mexico City, 177–78.
Higuera: military post, II.458n.
Himalaya Mountains: their elevation compared to that of the Andes, Alps, and Pyrenees, 183n.
Himmelsfürst, mine in Saxony: compared to that of Valenciana, II.117.
Hinchinbrook Island: Alaska, 474; II.366.
Hindustan: Persian name for India, 16, 86, 182n, 206, 225, 247, 249, 264, 281, 306, 404, 517, 571; II: 31, 259, 284, 447.
Hiva 'Oa: Maquesas Islands, French Polynesia, II.361.
Hocotitlán (Jocotitlán): town in the northwestern part of the State of Mexico, Mexico, 569.
Honda: town in the kingdom of New Granada, now in Colombia, 99n, 155, 163, 412n; II: 6, 23, 309, 460.
Hondo: river in Belize and Mexico alongside the border between the two, 406.
Honduras, Bay of, 142, 405.
Honduras, Coast of, 158n, 407, 164n.
Honduras, Gulf of, 69, 83, 159n; II.22.
Honduras, Province of, II.27.
Hornitos: volcanic mouths near Jorullo, 394.
Hostotipaquillo: mine, II: 72, [82, 92n].
Hraschina: Hungary, 427.
Huacachula: Indian village, 387.
Huailas: one of the twenty provinces of the Ancash Region, Peru, II: 200–201, 206–7, 250.
Huajucingo or Huexotcinco [Huajocongo]: town [Huejotzingo, northwest of the city of Puebla, central Mexico], 387; II.294.
Hualgayoc Mountain: mines of, II: 79, 80, 83, 94, 184n, 204t, 840t, 842t, 849t.
Huallaga. *See* Guallaga.
Huamalies: region in the kingdom of Quito, now provinces in the Huánuco region, Peru, II: 200, 206n.
Huancavelica: city and intendancy in the kingdom of Quito, now in Peru, II: [80, 144, 164, 165, 182–189, 193, 200], 203, [206n, 207, 209n, 225].
Huanuco: river in the kingdom of Quito, now in Peru, 165; II.508.
Huarochiri (Waruchiri): region in the kingdom of Quito, now province in the Lima region of Peru, II.200.
Huasacualco: river that might serve the purpose of connecting the two oceans, 150; how Cortés had already recognized and documented its importance, II.315; its current, 491. [*See also* Guazacualco; Coatzacoalcos.]
Huascazoluya, II.72.
Huatepec. *See* Oatepec.
Huaura: river in the kingdom of Quito, now in the Huaura province, Lima region, Peru, 165.
Huautla de las Amilpas, II.5.

Huaxyacac: one of the capitals of the land of the Zapoteca, 401, [405; II: 40, 64n. *See also* Oaxaca].
Huayna-Potocsi: mountain near Potosí, II.224–25.
Huaytecas: archipelago visited by the Chiloe Indians, II.465.
Hudson Bay, Canada, 183, 190, 365.
Hudson Strait: connects the Atlantic Ocean and Labrador Sea to Hudson Bay, Canada, 485n.
Huehuetoca, city in State of Mexico: elevation 492t; canal 44, 62, 108; II.509; lunar eclipse 30; location, 124t.
Huehuetoca, Desagüe of. *See* Desagüe.
Hueicolhuacan [Culiacán]: town, 433t.
Huejoquilla: mine, II.73.
Huejoquilla (Huejuquilla El Alto): town in Jalisco, central-western Mexico, II.79.
Huejoquillo: military post, II.458n.
Huejotzingo: city in Puebla, Mexico, II.40.
Hueplotitlan: mining site, II.75.
Huexocingo: historic republic. *See* Huajocongo.
Huitepeque (Huitepec): mountain to the west of San Cristóbal de las Casas, Chiapas, Mexico, II.75.
Huitzitzila. *See* Tzintontzan.
Humalies. *See* Tarma.
Hungary: country in Europe, II.15–16, 76–77, 85, 89, 93–84, 97, 102, 106, 130, 180n, 254, 268, 269t, 271.

Iaik: river in Russia. S*ee* Ural River, 484.
Iamiltepec (Jamiltepec), 478t.
Ibagué: city in the kingdom of New Granada, now in Colombia, II.183.
Iberian Peninsula, 260; II.16.
Iceland: island country in Europe, 315, 391, 471, 555.
Ichiaca: former province in what is now Florida, II.49.
Idria (Idrija): known as Idria under Austrian rule, now city in Slovenia, II: 164–66, 182.
Igtschiagik. *See* Igushik.
Iguala: officially Iguala de la Independencia, historic city located near the state capital of Chilpancingo in Guerrero, Mexico, II.206t.
Igushik: river in Alaska, 473.
Île de France: island in the Indian Ocean, now known as Mauritius, 170; II: 11, 14t, 20n.
Illinois River: tributary of the Mississippi River, 150.
Imaglin: Russian island of Big Diomede, part of Chukotka Autonomous Okrug, also known as Imaqliq, Inaliq, Nunarbuk or Ratmanov Island, 472, 472n.
Indehe: mining site, II.73.
India: country in Asia, 86ff.
Ingolstadt: city in Germany, 30, 31, 50, 445.
Inguaran: mine and settlement in Michoacán, Mexico, II.74t.
Iocotan (Jocotán): now municipality in Chiquimula, Guatemala, II.72.
Ipswich: town in Suffolk, England, 442.
Irapuato: town at the foot of the Arandas Hil, in south-central Guanajuato, Mexico, 483, 535, 536; II.5.
Iraputo plains. *See* Irapuato.
Ireland: country in Europe, 22, 158, 267, 396, 546, 553, 554n, 555n; II: 13, 427.
Irkutsk: city in eastern Siberia, 484; II.260, 366n.
Iro: river, 164.
Isabela Islet: largest of the Galápagos Islands, 126t.
Isla de Cubagua: smallest and least populated of the three islands that make up Nueva Esparta, Venezuela, II.49.
Isla de Juan Rodríguez Cabrillo, 128t.

Isla de Sacrificios: island in the Gulf of Mexico near the port of Veracruz, 97, 414n.
Isla de San Benito: island in the Pacific Ocean off the west coast of Baja California, Mexico, 127t.
Isla de San Bernardo: nine coastal coral islands and one artificial island that today belongs to Colombia, 127t.
Isla de San Martín: part of the Leeward Islands, Caribbean Sea, 127t.
Isla de Santa Rosa: second largest of the Channel Islands, California, 128t.
Isla de Ulloa (Prince of Wales Island): one of the islands of the Alexander Archipelago, Alaskan Panhandle, 469.
Isla del Socorro: Revillagigedo Archipelago, Mexico, 129t.
Isla des Coronados. *See* Isla de San Martín.
Isla Guadalupe: volcanic island off the west coast of Baja California, Mexico, 127t.
Isla San Nicolás, 127t.
Isla San Salvador: island of the Bahamas, 127t.
Isla Verde: island in Puerto Rico, 125t; II.320.
Islahuaca: town, 132t.
Islas de las Perlas: group of 200 or more islands and islets, about thirty miles off the Pacific coast of Panama, II.49.
Islote Blanquillas: possibly Blanquilla island, Venezuela, 125t.
Istapa: mining site, II.74.
Istla Valley, 177–78.
Itzacalco: today a borough in Mexico City, 323.
Itzapalapa: one of present-day Mexico City's sixteen boroughs, 343n.
Itztacalco canal, 335, 350–51. .
Iuchipila (Juchipila): town in Zacatecas, Mexico, II.72.
Ixhuacan: village in Veracruz, Mexico, II.28.
Ixmiquilpan: city on the Tula River in Hidalgo, Mexico, II.65.
Ixtaccihuatl. *See* Iztaccihuatl.
Ixtacmaztitlan: town in Puebla in southeastern Mexico, II.75.
Ixtapa: city in Mexico, adjacent to the Pacific Ocean in Guerrero, II.72.
Ixtlán, 477t; II.75.
Izalco Cordillera: stratovolcano in western El Salvador, 164n.
Iztaccihuatl: one of the highest peaks of the Mexican Cordillera, [24, 40, 93, 111–12, 131t, 134], 180, [188, 319–20, 332, 488; II.299].
Iztapalapa: today one of sixteen boroughs in Mexico City, 250, 321–22, 325–26, 329, 333, 357.

Jaén de Bracamoros: province on the banks of the Amazon River, 436, 514n; II: 47, 200.
Jagua, Bay of, Cuba, 73t, 406.
Jalacingo: Veracruz, Mexico, 417.
Jalapita: farm, II.98.
Jalapa. *See* Xalapa.
Jalisco: now a state in Mexico, 212, 304n, 440; II.363.
Jaltipa: village in [Jáltipan, Veracruz, Mexico], 492.
Jamaica [island in the Caribbean]: sugar exports from, II: 10, 13t.
Jamapa: city in Veracruz, Mexico, 415; river in Veracruz, Mexico, 63, 415–16; II.348.
James Peak: mountain in Colorado, 55–56, 71t, 92, 182.
Jamiltepec: now a district in Oaxaca, Mexico, 478t.
Jaruco: town in Mayabeque, Cuba, II.6n.
Jauli: village in Peru, 165.
Java: Indonesian island, 355.
Javita, Isthmus of: Ecudaor (?), II.88n.
Jerez: city in Portugal, II: 122, 178.

Johann-Georgenstadt: Saxony, Germany, II.96.
Jorullo: volcano [cinder cone volcano in Michoacán, Mexico]: its origins, 190; 391 [et passim. *See also* Xorullo].
Juchitlán el Grande: town in Jalisco, Mexico, II.72.
Julimes: military post, II.458n.
Juluapa (Juluapán): Colima, Mexico, II.72.
Jura: mountains and region in Switzerland, 173n, II: 107, 126.
Juruyo. *See* Jorullo.

Kaltschin, Gulf of, 472.
Kaluga: city in western Russia, 315.
Kamchatka Peninsula: Russia, 74, 472n; II: 366.
Kamischezkaja, Gulf of. *See* Bay of Bristol.
Kapalk (Kopal or Qapal): village in Kazakhstan, II.152t.
Kara: river that drains ino the Arctic Kara Sea in northern Siberia, 484.
Karluk: fort, 473.
Kenayskaja, Gulf of: Russian name for the Cook Inlet, Alaska, 473.
Kentucky, now USA: 294, 545, 579; II.37.
Kichtak (Kightag or Kikhtak) Island. *See* Kodiak Island.
Kielwig (Kalvag), Norway, II.120.
Kigiltach: Russian post, 472.
King George Archipelago: now Alexander Archipelago, 458, 474; II.366.
King George's Sound: Australia, 459, 464n; II.366n.
Kodiak Island: south coast of Alaska, 447, 461, 473; II.366n.
Kongsberg: city in Norway, II.92, 93.
Korea: country in Asia, 472n.
Kunlun: one of the longest mountain chains in Asia, 174.
Kuril Islands (Kurile), 472.
Kuymin: village in Russian Alaska, 473.
Kuynegach: village in Russian Alaska, 473.
Kyatkha: Mongolia, II.261t, 365n.

La Albarrada, San Francisco: in the municipality of Temascaltepec in the State of Mexico, II.72.
La Ascención. *See* Columbia River.
La Aurora de Ixtepexi: mining site, II.75.
La Ballena: mining site, II.72.
La Blanca: mining site, II.71.
La Bodega: port in San Francisco, also called Drake's port, 460n, 468n.
La Bouche: department of Allier, France, II.66n.
La Breña: group of cliffs near Durango, 427.
La Cadena: mining site, II.73.
La Carbonera. *See* San Fernando Lake.
La Carlota: municipality in Córdoba, Andalusia, II: 416, 416n.
La Concepción, 481; II: 73, 498.
La Concepción de Haigamé: village in Sonora, Mexico, II.74.
La Cruz del Correo: village in central Mexico, 319.
La Cumbre: village in Oaxaca, Mexico, 319.
La Guairá: capital city of the Venezuelan state of Vargas, II: 403t, 460, 466.
La Laborcilla, 88.
La Lagartija: island in the Gulf of California off Baja California, Mexico, II.73.
La Langosta, Bay of: 380; II.350.
La Ligua: mining district in the Valparaíso region in Chile, II.208.
La Mojonera: village in Mexico, 39.
La Orilla del Mar, II.490t.
La Palma: island of Mallorca, Mediterranean, 31.
La Paz: city in Bolivia, 550, 552; II: 83, 200, 208, 230, 253t.

La Piedad: city in Michoacán, Mexico, 333.
La Plata: capital city of Buenos Aires Province, Argentina, 544; II: 200, 241t.
La Plata, Río de, 24, 465; II: 52, 216n, 457.
La Puna: island off the coast of today's southern Ecuador, 412n.
La Purísima Concepción: village in California, 539t.
La Purísima Concepción de Álamos de Catorce: city in the kingdom of New Spain, now Real de Catorce in San Luis Potosí, Mexico, 423; II: 71, 124.
La Rinconada: town in the Peruvian Andes in the kingdom of Quito near a gold mine, 179; II: 208, 309, 312, 414.
La Roqueta: island in Mexico near Acapulco, Mexico, II.350.
La Sauceda, II: 71, 122, 124.
La Serranilla: submerged reef in the kingdom of New Granada, now in Colombia, II.72.
La Sonora: province, part of the intendancy of Sonora, [63, 305t, 311t, 428], 429.
La Sonora: river, 430.
La Sonora: town, 432.
La Tranca del Conejo, 39.
La Transpana (Transylvania): historical region today in central Romania, 333.
La Ventana. *See* Guadalupe.
Labrador: Canada, 217t, 466, 555.
Ladakh: region in the Indian state of Jammu and Kashmir, 174.
Lagos: town in the intendancy of Guadalaxara, 398–99, [483]; its factories, II: 293, [490t, 492t].
Laguna: island, 496.
Laguna de los Reyes: lake in Durango, Mexico, II.201n.
Laguna de Términos: the largest tidal lagoon located entirely on the Gulf Coast of Mexico, 496.
Laguna de Zitlaltepec: western part of Lake Zumpango, 354.
Laguna Hormiga: intermittent lake near Cerro Pelón and Cerro Las Hormigas, 415; II.400.
Lagunas of the province of Texas, 419, 489.
Lake (Loch) Oich: freshwater loch in the Scottish Highlands, 153.
Lake Chalco. *See* Xochimilco.
Lake Chapala, 388, [390, 398].
Lake Constance: lake bordering Germany, Switzerland and Austria, 187.
Lake Cuiseo (Cuizeo): in Michoacán, Mexico, 395.
Lake Erie: one of the Great Lakes, USA, II: 415, 150.
Lake Geneva: between Switzerland and France, 329n.
Lake León: Nicaragua, 150, 154n, 499.
Lake Manasarovar (Mapam Yumtso): high-altitude freshwater lake in Tibet, 174.
Lake Maracaibo: northwestern Venezuela, 162.
Lake Michigan: one of the Great Lakes, USA, 150.
Lake Mœris: ancient lake northwest of the Faiyum Oasis southwest of Cairo, Egypt, 375n.
Lake Nicaragua, [146, 180, 193, 235, 499; II: 301, 320, 493, 508]; might serve to connect the two oceans, 150ff; its elevation, 499.
Lake Parras: Nueva Vizcaya, 188, 427.
Lake Timpanogos. *See* Lake Utah.
Lake Titicaca: in the kingdom of Quito, now in Peru, II: 208, 250n.
Lake Utah (Lake Timpanogos): 64, 149, 187, 441n. *See also* Teguayo.
Lake Zitlaltepec. *See* Sumpango Lake.
Lamas: city in the kingdom of Quito, now in the San Martín region, northern Peru, II.233.

Lampa: region in the kingdom of Quito, now province in the Puno region, Peru, 262.
Lancaster: city in Pennsylvania, USA, 545; II: 28, 72.
Langara Island. *See* Margarita Island.
Languedoc Canal: now Canal du Midi in the Provence/Languedoc/Camargue region of France, 168, 356.
Lans-le-Bourg: now Lanslebourg-Mont-Cenis, former community in the Savoie department in the Auvergne-Rhône-Alpes region in southeastern France, 379.
Lapland: region in Finland, 315, 471, 476, 506, 511n, 555.
Laredo: city on the banks of the Río Grande, 422.
Las Aguas: mining site near Temascaltepec, II.72.
Las Cruces de Ecatepec: mountain, 372.
Las Grullas: mountains northwest of Santa Fe in New Mexico, now USA, 234.
Las Lajas: river [Río de las Lajas or Devils River, Edwards Plateau in Texas], 187.
Las Mesas: mountain in Veracruz, Mexico, II.73.
Las Playas de Jorullo: farm, 122t, 394.
Las Plomosas: village in Sonora, Mexico, II.72.
Las Ranas, 73t.
Las Salinas: Santo-Domingo, now town in the Barahona province, Dominican Republic, II.71.
Las Vigas (Las Vigas de Ramírez): village in Veracruz, Mexico, 40, 103, 123t; II.39.
Latium (Lazio): region in central western Italy, 200, 265, 326.
Lauricocha: mines, II.201. [*See also* Yauricocha.]
Lavandera: shallows near Veracruz, Mexico, 191; II.321.
Le Havre: port city in France, 168.
Leán: river on the northern Caribbean coast of Honduras, 165n.
Lechaeum (Lechaion): port in ancient Corinthia, 156.
Leghorn (Livorno): port city on the Ligurian Sea, II.393.
Leglelachtok: Russian post, 472.
León: town, 388f.
Lerma: river, 187, [388, 390].
Lerma: village in Mexico, 320, 332, 381, 407.
Levant: area in the Eastern Mediterranean, 348; II: 26, 255, 256–57, 257n, 261t, 284.
Lewis River. *See* Snake River.
Lhasa: capital of Tibet, 174.
Licán: village near Riobamba in the kingdom of Quito, now in Ecuador, II.66n.
Lima [Peru, 3 et passim. *See also Subject Index*].
Limón Mountain: province and mountain range in Costa Rica, II.131.
Linares: city in Tejas, 420, 423; II.179.
Lisbon: capital of Portugal, 30–31, 430, 557; II: 235, 241–42, 267.
Livenza: river in the Italian provinces of Pordenone, Treviso, and Venice, 356.
Liverpool: city in England, 168, 170, 285n.
Livonia: region on the eastern shores of the Baltic Sea, 261, 554n.
Llanitos Mountains: San Carlos District, Panamá Oeste Province, Panama, 388; II.98.
Llano de los Cues: name of the valley where the pyramids of Teotihuacan were found, 340.
Llanos: tropical grassland plain situated to the east of the Andes in the kingdom of New Granada, now in Colombia and Venezuela, 206, 486t, 494–95, 574; II.98.
Llaoin: mining site in Chile, II.208.

Lloró: village on the Río Andagueda in the kingdom of New Granada, now in Colombia, 164; II.229.
Lockwitz Valley: Saxony, Germany, II.119.
Loma de Chiconautla: village near Ecatepec, Mexico, 80.
Loma del Potrero: mountain northwest of Loma Pelona and west of Loma de las Gachupines, 80t.
Lombardy: province in Italy, 185, 254, 485, 511, 456n.
Lomo del Toro: mining region in Zimapán, Hidalgo, Mexico, II: 72, 91, 179, 181.
London: capital of the United Kingdom, 42, 68n, 84ff.
Long-Pendu: lake in Burgundy, France, 56.
Loreto: city on Baja California Peninsula, Mexico, 429, 443ff.
Loreto: military post, 443, 445; II: 448, 458n; mine, II.73–74.
Los Aillones: mining site, II.72.
Los Álamos: town in Sonora, Mexico, 433.
Los Alamos de la Sierra, II.121.
Los Amates: municipality in the Izabal department of today's Guatemala, 39.
Los Angeles: city and port in California, 38; II.363.
Los Baños (Penol or Peñon de Los Baños): Mexico City, 351; II.190.
Los Berrios: small lake near Xalapa, 416.
Los Calzones: mountains, II.181.
Los Carcamos: mining site, II.74.
Los Césares: mythical city in Patagonia, 166.
Los Gavilanes: village in Durango, Mexico, II.73.
Los Guacaros: mountains [in present-day New Mexico], 446.
Los Güines: district in Cuba, 166; II.466.
Los Metates: mountains [in Chihuahua, Mexico], 428.
Los Molinos: likely a village in Chile and a mining site, II.74.
Los Ocotes: in Guanajuato, Mexico, II.72.
Los Órganos: mountains [near Actopán, Hidalgo, Mexico], 424.
Los Otates: city in Actopán, Veracruz, Mexico, 416.
Los Pastos: province in the kingdom of New Granada, now mountainous border between Ecuador and Colombia, 174, 226, 248.
Los Peñoles, II.73.
Los Pozuelos, II.179.
Los Pregones: near Tasco, Mexico, II.180.
Los Sauces: Chilean town and commune in the Malleco Province, Araucanía Region, II.73.
Louisiana, now USA, 15ff.
Lúcuma: river in Panama, 159n.
Lüneburg: city in northern Germany, 486t.
Luzon Island, Philippines, II.356.
Lyon: city in France, 176.
Lys: former department of France, now in Belgium, 315.

Macao: former Portuguese colony in Asia, 457; II: 258, 284t, 364–65.
Mackenzie: river system in Canada, 72, 147, 182; II.190n.
Maconi: mining site, Querétaro, Mexico, II.72.
Macultepec: village and mountain in Tabasco, Mexico, 123t, 416.
Madras: region in India, 348, 555; II: 41, 286, 340.
Magdalena: river in the kingdom of New Granada, now in Colombia, 51, 88, 99n, 155, 161n, 162, 412n; II: 6, 37, 248, 348, 381n, 394n, 457.
Magdalena Island. *See* Fatu Hiva.
Maguarichi: town in Chihuahua, Mexico, II.73.

Malabar: region in South India, 505.
Malacatepec: village in Puebla, Mexico, 205, 205t.
Málaga: city in Spain, 514; II: 8, 376, 393, 405, 407, 415n.
Malinalco: municipality in the Ixtapan region, Mexico State, Mexico, 323.
Malpasso, II.317–18. *See [also]* Passo.
Malpais: area formed by volcanic activity, 392.
Malvinas Islands [Falkland Islands]: they have no stable settlements, [452]; II.52.
Mamanchota. *See* Órganos de Actopán.
Managua: capital of Nicaragua, 154.
Mandanes: village on the upper Missouri River, 56.
Mandinga Bay: Panama on the Caribbean side, 83.
Manila: Philippines, 170ff.
Mannheim: city in Germany, II.20n.
Mansiche (San Salvador de Mansiche): in the kingdom of Quito, now in Trujillo city, Peru, 335.
Mansos: mountains [in New Mexico], 435.
Manzanillo: Pacific Ocean port city in Colima, Mexico, 479.
Mapimi: military post, 428; II.358n.
Mapimi: town [city and now municipal seat of the Mapimí Municipality in the Mexican state of Durango], 428.
Mapimis. *See* Bolsón de Mapimi.
Maquilapa, 484.
Marañón: river in the kingdom of Quito, principal source of the Amazon River northeast of present-day Lima, Peru, 436; II: 183, 201.
Marfil: ravine, valley, and lake in Mato Grosso, Brazil, and in Santa Cruz, Bolivia, 205, 206t, 207, 389; II: 98, 104–6, 107, 110–12, 121, 138.
Margarita Island (Isla de Margarita): in the kingdom of Quito, now part of Venezuela, 214t, 406, 459.
Mariana Islands: Pacific Ocean, 248, 515.
Marías Islands (Islas Marías): archipelago of four islands off the coast of Nayarit, Mexico, 126t.
Mariel: city in Cuba, II.23.
Marienberg: Saxony, Germany, II.96.
Mariquita: city in Spain, II.231.
Marmato Mountain: mining site in the kingdom of New Granada, now in Colombia, II.229–30.
Marqués Bay: on the Pacific Coast in Guerrero, Mexico, II: 350, 363.
Marqués port: part of the port of Acapulco, II.352.
Martinique: island in the Caribbean, 485.
Maryland: now state in the USA, 511, 579.
Mascón (Mashcón): river in Peru, 544.
Matanzas: city and province in Cuba, 531, 602n; II: 10, 20n, 456t.
Matanzillas: mining site, II.71.
Matehuala: city in San Luis Potosí, Mexico, II.71.
Matlatengo: village in Hidalgo, Mexico, 501.
Mauna Loa: volcanoi on Hawai'i, II.361.
Mauritius. *See* Île de France.
Maynas: region in the kingdom of Quito, now province in Loreto, northeastern Peru, 441; II.200.
Mayo: river [in Sonora, Mexico], 429, [433, 539, 552].
Maypures: region in the kingdom of New Granada, now in Colombia, 523.
Mazapil: region in Zacatecas, Mexico, 401; II.71.
Mazatlán: port and town in Sinaloa, Mexico, 19–20, 39, 53, 319; II: 309, 363.
Mecameca (Amecameca): Mexico State, Mexico, 133t, 382, 383; II.299.
Mechigmenskiy, Gulf of: Bering Sea, Russia, 472.
Mechoacan, kingdom of, 390. *See also* Valladolid.

Medehun: possibly one of the Saqqara pyramids in Egypt, 324.
Medellín: city in the kingdom of New Granada, now in Colombia, 416; II: 228n, 499.
Mediterranean, 153, 157, 169, 173n, 393n; II.369.
Mednaya: river in Alaska, also known as Copper River, 474.
Mednoi Island, Alaska, 472n.
Meganos: drifting sand dunes near Veracruz, 494; II.399.
Meganos de Cathalin: on Catalinas Island, II.399.
Meganos del Coyl, II.399.
Melimbo, West Africa: unknown location, 288.
Mellado: mines, 389; II: 81, 104n, 109, 120, 197.
Menchan: island, II.465.
Mendaña Islands: Marquesas Islands, II.361.
Mendieta: vein in the Potosí silver district, now Bolivia, II.223.
Menorca: Mediterranean island, Spain, II.393.
Mercado Mountain (Cerro del Mercado): Durango, Mexico, 179.
Mérida: intendancy of: area, 405; climate, 406; Indians who live there, 407; what it produces, [25, 69, 142, 306, 312–13, 314–15, 405–6], 407, [408; II.450].
Mérida de Yucatán: city, 408.
Mescala (Mezcala): village in Guadalajara, Mexico, 38, 39; II.133.
Mesquic (Mixquic): now in Mexico State, Mexico, 329.
Methona (Methone): ancient city, 391n.
Mexcala: geological formation in Guerrero, Mexico, 122.
Mexicana: river, 420. *See also* Río Mermentas.
Mexico: meaning of this word. 197.
Mexico, Gulf of, 19ff.
Mexico, Valley of, 21ff.
Mexico City. *[See Subject Index.]*
Mextitlan: lake, 188.
Mezquital: city in Durango, Mexico, II: 72, 86.
Miahuatlán (Miahuatlán de Porfirio Díaz): town in Oaxaca, Mexico, 478.
Michuacan. See Mechoacan.
Micuipampa: city in Peru, II: 79–80, 184n, 185n, 200, 202, 207.
Micuipampa: mines, II: [79, 200], 202.
Midi Canal: France, 168.
Miedziana Góra: village in south central Poland, 66n.
Miguitlan. *See* Mitla.
Misantla: river and city in Veracruz, Mexico, 409; II.23–27.
Mississippi River: now USA, 56ff.
Missouri River: now USA, 55n, 56, 65, 90, 149, 182, 234, 240, 296, 309, 432, 435n; II: 307, 451, 457.
Misteca (Mixteca) Alta: region in western Oaxaca and neighboring portions of Puebla in Guerrero, Mexico, 58.
Mitcheldean: village in Herfordshire, UK, II.105.
Mitla: village, now city in the intendancy of Oaxaca, Mexico, 491t. *See also* Mitla, palace of, *in Subject Index*.
Mittelgebirge: lower mountain regions in Germany, 538; II.105.
Mixcoatl (path of the dead): ancient name of the valley where the pyramids of Teotihuacan were found, 340.
Mixteca: mountainous landscape, 402.
Moana-Ro. *See* Mauna Loa.
Mocha: Chilean island off the coast of Arauco province in the Pacific Ocean, II.52.
Moctezuma: river, 187, [62, 177, 360, 362, 378–79, 491].
Moho Tani: uninhabited island of the Marquesas Islands, French Polynesia, II.361.
Mojouera, II.77n.

Moluccas Islands (Maluku Islands): archipelago in eastern Indonesia, 513, 524; II.358.

Mompox: city of the kingdom of Santa Fe, this country's main market for gold and gold panning, [412n]; II: 229, [233].

Monclova: military post, II.423.

Moñino Inlet: today in British Columbia, 469.

Monk: the solitary peak of the Toluca volcano, 24.

Mont Blanc: highest mountain in France, 26n, 112, 173, 319.

Mont Cenis: mountain pass in France, 112, 306, 379; II.311.

Mont d'Or: mountain in Switzerland, 112.

Montagu Island: largest of the South Sandwich Islands in the Weddell Sea off the coast of Antarctica, 466.

Monte de San Nicolás: village in Guanajuato, Mexico, II.70.

Monte Nuovo: cinder cone volcano within the Campi Flegrei caldera, near Naples, southern Italy, 393.

Monte Perdido: third highest mountain in the Pyrenees, Spain, II.185.

Monterrey: presidio and military post, 19, 128, 449, 451; II: 448, 458n.

Monterrey, New California: geographical position, 49; town in the intendancy of San Luis Potosí, 423. [*See also* San Carlos de Monterey; *Subject Index*.]

Montesclaros: town [in northern Minas Gerais, Brazil], 334.

Montevideo: city in Uruguay, 544, 547, 555n; II: 398, 454, 465.

Montpellier: city in France, 250n.

Montradak: Borneo, Indonesia, II.271t.

Montserrat: mountainous Caribbean island, part of the Lesser Antilles, II.13.

Mont-Tonnerre: department in France, II: 163, 183.

Moqui: territory populated by uncivilized Indians, 425; town Father Garcés found there, 437.

Moran: description of its mines, 339; II: [77, 125], 125ff, 128ff, [147, 197].

Moravia: historical region in the Czech Republic, 401; II.15–16.

Moray Firth. *See* Gulf of Murray.

Morocco: country in Northern Africa, II.16.

Morocollo: mountain in Chile, II.208.

Moscow: capital of Russia, 315.

Mosquito Coast: Nicaragua, 151, 170–71; II.441t.

Motage: mining site, 72.

Motagua: river in Guatemala, 165n.

Motepore: mining site, II.74.

Mount Cenis: mountain in the French Savoie Alps, 175.

Mount Dhawalagiri: Nepal, 182.

Mount Edgecumbe: dormant volcano at the southern end of Kruzof Island, Alaska, 473.

Mount Fairweather, British Columbia, 466, 474.

Mount Rainier: active volcano in Washington State, now USA, 148.

Mount Rose: mountain in Nevada, 466n.

Mount Saint Elias: mountain in Alaska on the Yukon border, 180, 180n, 460.

Mount Saint Helens: active stratovolcano in Washington State, now USA, 148.

Mount San Lázaro: mountain in Mexico, 71n.

Mount Timpanogos. *See* Teguayo.

Moyotla: part of the city of Tenochtitlan, II.44.

Mozambique: country in Africa, II: 37, 52, 284.

Muamelula: Oaxaca, Mexico, 478t.

Muapiapam: Oaxaca, Mexico, 478t.

Muerto: desert, 435; swamp, 434.

Müglitz Valley: Saxony, Germany, II.119.

Mulatas Islands, Panama, 159, 159n.

Mulgrave: port on Bering Bay, 116, 466–67, 474.
Multnomah: one of thirty-six counties now in the state of Oregon, now USA, 149n.
Munsfeld (Mansfeld): town in Saxony-Anhalt, Germany, II.107.
Mustung: town in Tibet, I.182.
Mysore: India, 1446n.

Nabajoa: mountains, 425.
Nabajoa: river in southern Utah, 61, 149, 437.
Nacatabori, II.74.
Nacogdoches: Spanish military post closest to the border of Louisiana, 441; [II: 457, 458n].
Nacosari, II.74.
Nacumini, II.74.
Nagasaki: city in Japan, 560n; II: 233, 259.
Nahuelhapi: lake, 139.
Namiquipa: military post, II.458n.
Nantes: city in France, 168.
Nantucket: island off Cape Cod, Massachusetts, now USA, II: 55n, 364.
Napestla: river: perhaps the same as the Arkansas River, 436.
Naples: Italy, 183–84, 284, 313, 331, 346, 393, 412n, 545, 575; II: 98, 409.
Nasas [Nazas]: river, 427.
Natá: Coclé Province, Panama, 154.
Natchez: city in Mississippi, now USA, 422; II.34.
Natchitoches: a county of the USA in the region bordering on the intendancy of San Luis Potosí, [91], 420, [422–23, 436; II.457].
Natividad: mining site, II.72.
Nauhcampatepetl (Cofre de Perote): one of the tallest peaks of Mexican cordillera, 180; meaning of its name, 180n. [*See also* Cofre de Perote.]
Naurouze Pass: southern France, 168.
Nautla: settlement in Veracruz, Mexico, 409; II.24.
Nazarenas: monastery in Celaya, Mexico, 483.
Netschich: Russian post, 472.
Nejapa: municipality in San Salvador, El Salvador, II: 43n, 44, 47.
Nelson River, Manitoba, Canada, 149.
Nespa. *See* Arkansas River.
Ness: river in Scotland that flows from the northern end of Loch Ness northeast to Inverness, 152.
Netzahualcoyotl: city in Mexico, 268, 357.
Nevado de Iztaccíhuatl: dormant volcano in Mexico, 131t.
Nevado de Pelagato: mountain in the kingdom of Quito, now in the Ancash region, Peru, II.183.
Nevado: meaning of this word, 382n.
Nevado de Toluca: mountain in Mexico, 44, 48n, 111, 122t, 188, 319, 395, 402n; II.5.
Nevado of Tolima: mountain in Mexico, II.183.
Nevis: island in the Caribbean, II.13.
New Albion. *See* New California
New Andalusia (Nueva Andalucía): in the kingdom of New Granada, now Venezuela, II.21.
New Biscay, province of [Nueva Vizcaya], 423–24. *See [also]* Durango.
New Brunswick, Canada, 218t.
New California. *See* California, New, *in Subject Index.*
New Cornwall: settlement in the Lunenburg district of Nova Scotia, 232, 458, 470, 473–74; II: 361, 364.
New England: coastal states now in the eastern USA, 224n, 545, 555n; II: 3, 37, 391, 453–54.
New Galicia [Nueva Galicia], kingdom of: area of, [61, 63, 183, 187, 304], 307, [312t, 397, 404, 424, 440t; II: 84t, 183, 440, 450].
New Georgia: largest of the Solomon Islands, 458, 469; II.361n.

New Granada. *See Subject Index.*
New Guinea: islands in the Pacific Ocean, II.362n.
New Hanover: county in North Carolina, now USA, 469, 475.
New Holland, Dutch Brazil, 307, 439, 452, 461, 464n, 556; II.359.
New Iberia: largest city in and the seat of Iberia Parish, Louisiana, now USA, 422.
New Jersey: state in the USA, 209n; II.406n.
New León [Nuevo León], kingdom of, 418, [304n; II.179].
New Mexico, province of: area, 433; climate, 435; rivers in, 435; Indians who live there, 436–37; towns in, 438–39.
New Navarro. *See* Sonora, province.
New Norfolk: town on the Derwent River southeast of Tasmania, Australia, 232, 466, 474.
New Santander: town, 423; province, 418; [region in the Viceroyalty of New Spain, covering the modern Mexican state of Tamaulipas and extending into southern Texas, 312t, 419, 490; II: 448t, 449t].
New Silesia: small province of the Kingdom of Prussia, 173.
New Spain. *See Subject Index.*
New York City: population of, 476.
Newfoundland: region in Canada, 158, 191; II.51.
Nicoya Gulf: Nicaragua, inlet of the Pacific Ocean, 152, 154, 180n, 194n, 266, 499.
Niger: river in West Africa, II.235.
Nile: river in Egypt, 173.
Nirgua: city in Venezuela, 163.
Noanama: village, 163, 164; II.230.
Nochistongo, underground tunnel/canal: history of its construction, 359. [*See also* Huehuetoca.]
Nochistlán: town in Zacatecas, Mexico, 58t, 129t, 478t.
Nombre de Díos [city in Colón Province, Panama], 426–27; [II.392].
Nootka: town, 48ff.
Nootka Bay: position, 49n; Juan Pérez visited it before Cook did, 459–60; he named it port of San Lorenzo, 459; settlements that the Spanish founded there, 461–62; description of the country, 462; ownership disputed by the Spanish and the English, 463.
Nootka Sound: Vancouver Island, British Columbia, Canada, 161, 167.
Nord: department in France, 151n, 315, 544.
Norembega. *See* New England.
Norfolk Bay: southeast of Tasmania, Australia, 232, 450, 460, 561n; II.362.
Noria: ancient mine, II.124.
Norotal: village in Durango, Mexico, II.74.
Norte or Las Juntas: military post, [438]; II.458n.
Nortfolk Sound. *See* Sitka Bay.
North Carolina: state in the USA, 545, 553n.
North Holland Canal: northwest Netherlands, 168, 169.
North River. *See* Río del Norte.
Norton Sound: inlet in the Bering Sea, 472.
Norway: Scandinavian country, 104n, 158n, 463, 512; II: 92–93, 120, 255, 262, 313, 427.
Nova Scotia, Canada, 217t, 218t.
Novita: town in the kingdom of New Granada, now in Chocó, Colombia, 84, 163, 164, 229ff.
Novo-Arkhangelsk: now Sitka near Juneau, Alaska, 366n.
Nublada. *See* San Benedicto.
Nuestra Señora de la Guadalupe: village, 501t.
Nueva Barcelone: 141, 207, 574t; II: 22, 123, 381, 387t.
Nueva Jaén: town, 153n.

Nueva Navarra, 429.
Nueva Veracruz, 63, 95t, 191, 414.
Nuevo Santander. *See* Tamaulipas.
Nugran: Russian post, 472.
Nukan: village in Alaska, 472.
Nuñez Gaona: port, bay, and Spanish colonial fort in present-day Neah Bay, Clallam County, now Washington State, USA, 465.

Oatepec. *See* Ahuatepec.
Oaxaca: city [and state in Mexico], 57ff.
Oaxaca, Valley of: constitutes Cortés's marquisate, 320n, 404–5.
Obejera: mining site, II.70.
Obergebirge: mountain in Saxony, Germany, II.96–97.
Ocosocontla (Ocozocoautla de Espinosa): town and municipality in Chiapas, Mexico, 484.
Ocotelolco: one of the four independent polities of the confederation of Tlaxcallan in present-day Tlaxcala, Mexico, 385.
Ocotepec: small town to the north of Cuernavaca, Mexico, II.72.
Ocotlán: city in Jalisco, Mexico, 103, 386, 478t; II: 47, 311.
Ocotlan [San Juan de]: saline in the intendancy of Puebla. II.157.
Oculma: small town in Mexico, 358, 363.
Ohio: river and state now in the USA, 25, 187, 234, 239n, 240, 421, 546; II: 415, 457.
Ojo Caliente: community in Taos, New Mexico, now USA, 89; II.71t.
Ojo del Água de Matchuala: settlement in San Luis Potosí, Mexico, II.125.
Okhotsk Sea: western Pacific Ocean, 472n.
Old Bahama Channel: strait in the Caribbean between Cuba and the Bahamas, 157, 158n; II.466.
Old California. *[See* California, Old, *in Subject Index.]*
Olinda: Brazil, II: 393, 398.
Olonets (Olonez): town in the Olonetsky District of the Republic of Karelia, Russia, 315, 485t.
Oltenhorn: Switzerland, II.106.
Olumpa: city in Argentina, 427.
Ometepec: city in Guerrero, Mexico, 58, 129t.
Omoa Castle: department of Cortés, Honduras, 499.
Onalaska. *See* Unalaschka.
O-Nateya. *See* Mohotani.
Opelousas: province of Louisiana, border region of Mexico, 421–22.
Oquilcalco, 501.
Øresund (Öresund): body of water in Denmark, 160.
Orinoco: river in South America, 11ff.
Oritucu (Orituco): settlement in Guarico, Venezuela, II.22.
Orizaba: town, 417.
Orizaba Mountain: confusion about this mountain in the maps by Jefferys and Arrowsmith, 42; one of the highest peaks of the Mexican cordillera, 180; description of, 411–12.
Oro: military post, II.458n.
Orsay: port on the Seine in Paris, France, 369n.
Oruro (Uru Uru): city in Bolivia, II: 79, 200, 208, 391.
Ostimury [Hostimuri, former province between Río Fuerte and Río Mayo in Mexico], 429.
O-Taïti. *See* Tahiti.
Ottorocorras. *See* Uttara-Kuru.
Otumba (Otumba de Gómez Farías): town in the northeast of the State of Mexico, Mexico, 205, 321, 338; II.311.
Oude: Awadh region of North India, II.446.
Ounigigah [Unijigah]. *See* River of Peace.
Our Lady of Guadeloupe: thermal springs, 351.

Ovejeras [San Antonio de Ovejera], Guanajuato, Mexico, II.108.
Owyhee Island [Big Island, Hawai'i]: was discovered by the Spanish before Cook, II.461–62.
Oxus: river in central Asia, now Amy Darya, 566.
Oyamel: pine forest, 339; II.126.
Ozark Mountains: mountain range in Arkansas, Oklahoma, and Missouri, now USA, 484.
Ozumbillo: village in Ojo Agua, state of Mexico, Mexico, 501.

Pacaya: active complex volcano in Guatemala, 111n, 180n.
Pachacamac: town in the kingdom of Quito near Lima, now an archaeological site in Peru, 335; II.250.
Pachuca: municipal town in Mexico, 381; description of its mines, II.125.
Pachuca: river, 353.
Pachutla: city in Oaxaca, Mexico, 129t.
Pacific Ocean, 35 et passim.
Pacos: silver mines, 89.
Paducah River: southern branch of Platte River, Nebraska, now USA, 92.
Pájaro: river in the Central Coast region of California, now USA, 125t.
Palcas Mountains, Peru, II.184.
Palenque: its antiquities, 487. [*See also* Culhuacan.]
Palestine: region in the Middle East, 539.
Palizada River, 496
Palmar de Vega, II.70.
Palmas, II.24.
Palmyra: ancient Semitic city in present-day Homs Governorate, Syria, 340.
Palo Blanco: mining site in Sonora, Mexico, II.74.
Palula, 39.
Pampa Fungosa de la Rinconada, II.208.
Pampas del Sacramento, II.200.
Pampatar: city on Isla Margarita, Nueva Esparta, Venezuela, 442.
Pamplona: capital of Navarre, Spain, 162.
Panama, Gulf of, 499, 553; II: 49–50, 53, 83.
Panama, Isthmus of: position, 83; elevation above sea level, 83; prevailing uncertainty about its shape and size, 154; Congress of Panama, 163; causes of the insalubrity of this region, 395.
Panama City, 157, 159ff.
Pansitara: plateau, 544.
Panuco: military post, II.458n.
Panuco: river [in the Valley of Mexico], II: 198, 346–47.
Papagallo valley, 178; II.131.
Papagayo, Gulf of, 150, 152, 154, 499.
Papaloapan: river, 410; its sandbar, 495; its natural harbor, 495. [*See also* Alvarado River.]
Papalotla: river, 353.
Papantla: Indian village in Mexico, 324, 384, 402, 409, 412, 506; II: 23–24, 26, 27t.
Papasquiaro: town, 427.
Paquaro, II.74.
Paraguay, 24, 31, 140, 544.
Paria, 157, 407, 414, 523, 557; II: 23, 248, 250, 253.
Paris, France, 20n, 26ff.
Paropamisus: mountain in the Caucasus, II.235.
Parral: city in present-day Chihuahua, Mexico, 53, 88, 107n, 181, 424; II: 73, 81, 82t, 83, 84t, 92, 164t.
Parras: lake, 188, 427–28.
Parras: town, 428, [438, 571].
Partido, Río: river whose existence is problematic, 151.
Partidos: isolated mines, II.206n.
Pasage: military post, II.458n.
Pasco: city in the kingdom of Quito, now in Peru, 186ff.

Pasco: mine, [186, 79–80, 83, 96n, II: 198], 201–2, [206n, 244t, 246t].

Pascuaro (Patzcuaro): town, [44, 106, 122t, 132t, 134, 175, 344n, 392], 396–97; [II.5].

Paso, river: might be used to connect the two oceans, 318.

Passo del Norte: military post, [183], 438; description of the land and its location, 438.

Pasto, San Juan de: town and province in the kingdom of New Granada, today in Colombia, 344n, 544, 552.

Patagonia: Argentina, 165–66, 237, 239n, 240, 271, 424.

Pataz: city and province in Trujillo, II: 183, 200, 206.

Patorumi: cataract, II.37.

Pátzcuaro: lake, 188, 390.

Payta: port, 54; II.354.

Peak of Tenerife. *See* Teide.

Pecos, river: possibly the Red River of Natchitoches, 436.

Peddapore (Peddapur), India, II.11.

Pelado Mountain, II.493t.

Pelegrino. *See* Scilly Island.

Peloponnese: peninsula in southern Greece, 326.

Penjamo, II.5.

Peñol de los Baños: crag, 76t, 77–78, 338, 363.

Peñol del Márquez: small island, 322.

Peñon Blanco: lake in Durango, Mexico, II: 147, 190.

Pensacola: Florida, now USA, 491; II.379.

Pequeni: village, 156.

Peregrino Valley, 178, 183.

Pernambuco, Brazil, II.393.

Perote: market town, 417.

Persia: region in Asia, 66, 200, 333n, 435, 535, 539; II: 20, 284, 307.

Peru, [11 et passim. *See also Subject Index*.]

Petapa, II: 318–19, 490t.

Petatlan: village [now a city in Guerreo, Mexico, 48n, 126t], 392; [II: 17, 354].

Petén: Mesoamerican region, now Guatemala, 487.

Petit Saint Bernard: mountain pass in the Alps, 112.

Petupa, II.493t.

Philadelphia: [11 et passim]; its population, 476.

Philippines, islands: [39, 235, 466; II: 21, 35, 38, 309, 349, 357, 359, 362, 365], 441t, 444. *[See also Subject Index.]*

Piave River: Venice, Italy, 356.

Picacho de la Variga de Plata, II.125.

Picacho de San Tomás, 134.

Picachos del Mortero, 392.

Picardy, France, 544n, 360.

Pichincha: volcano, 111, 319, 574; II.224n.

Pico del Fraile: highest peak of Nevado de Toluca, 319.

Piedra Blanca, 126t.

Pifo: village in the kingdom of Quito, today in Ecuador, II.38.

Pilaneones, II.200.

Piliza, 173.

Pimería, district of, 430; divided into Alta and Baja, 430.

Pimería Alta: mountains of [now in southern Arizona, USA, 20, 63–64], 181, [429, 430; II.86].

Pimichin: river, II.88n.

Pinahuizapan. *See* Cofre de Perote.

Pinolco: mountain, 500–501.

Piseo: port, II.54.

Pitic: military post, II: 442, 458n.

Pitt's Archipelago, British Columbia, Canada, 469.

Plantanillo: village, II.132.

Platanarito, II.72.

Plateros, II.490.

Platinita: small stream, II.88.

Platte River: Nebraska, now USA, 56, 90ff, 182.

Playa Grande, II.350.

Playa Vicente, 495.
Playas de Jorullo, 48n, 122t, 392ff.
Pochutla (San Pedro Pochutla): city in Oaxaca, Mexico, 58t.
Point Reyes, California. *See* New Albion.
Poland: country in Europe, 18, 261, 544, 578; II: 15–16, 190, 262.
Polochic: river in eastern Guatemala, 165.
Polynesia, 526.
Pomapamba: in the kingdom of Quito, now province in the Ancash region, Peru, II.200.
Pondicherry (Puducherry): French colonial settlement in India until 1954, 227.
Pongo de Manseriche: gorge in the kingdom of Quito, now in northwest Peru, II.184.
Pont d'Austerlitz: bridge across the Seine, Paris, France, 369n.
Pont-Royal: bridge across the Seine, Paris, France, 369n.
Popayán, province of, 164 et passim. *See also Subject Index.*
Popocatépetl: highest peak of theMexican cordillera: 180; meaning of its name, 180n; elevation, 381; eruptions, 488; did Diego Ordaz really visit its crater?, II.299.
Poratich: Hungary, II.180n.
Porche: Coxitambo Mountain, II.183n.
Porco: region in the kingdom of Quito, now district in Peru, II: 99n, 208, 224, 238, 248, 253, 262.
Port Chatham: abandoned town on the Kenai Peninsula, Alaska, 474.
Port Mulgrave: abandoned town in Yorkshire, England, 116n, 467, 474.
Portbelo, city: measures governor Emparán took to improve its climate, II.399.
Portillo: in the Cordilleras, II.493t.
Portsoy, Scotland, II.120.
Posesión, II.72.
Posquelitos, 178.
Potosí: town: temperature there, II.209n.
Potosí: Viceroyalty of Buenos Aires, [181 et passim. *See also Subject Index.*]
Pouilly [Saint-Genis-Pouilly, community in the Ain department in eastern France], 168.
Prince of Wales Archipelago (Alexander Archipelago), Alaska, 458, 460, 473; II.366n.
Prince William Sound. *See* Chugatskaia, Gulf of.
Princess Royal Islands: British Columbia, Canada, 148.
Principe: military post, II.458n.
Provence, France, 442.
Prussia, 208, 209t, 553n, 573; II.409.
Puebla, [40 et passim. *See also Subject Index*].
Puebla: volcano, 23, 48n, 77, 78, 111ff, 131, 133t, 383; II.298. *See also* Popocatépetl.
Puebla de los Angeles, capital of the intendancy of Puebla: population, 386, 476; manufactories, II.294.
Pueblo Nuevo, 39.
Puembo: village, II.38.
Puente Colorada, II.493t.
Puente de Isla, 39.
Puente de Ixtla: farm, 122t.
Puente de Salto: bridge across the stream of the desagüe de Huehuetoca, [80t, 124t], 368.
Puerco: river, 418, 435.
Puerto de Bodega, 455; II.363.
Puerto de Bucareli: port discovered by Quadra, 460.
Puerto de la Navidad: ancient port in modern-day Haiti, 48, 106, 390; II.363.
Puerto de la Paz: also the port of the Marqués del Valle, II.363.
Puerto de los Reyes, 177.
Puerto de Navidad, 48n.
Puerto de San Lorenzo. *See* Santa Cruz de Nootka.

Puerto de Santa Rosa, II.98.
Puerto Escondido, 58; II.363.
Puerto Guatulco, 58t.
Puerto Quemado, 162n, 162n.
Puerto Rico: island, [33 et passim. *See also Subject Index*].
Puerto Real: island, II.357.
Puno, intendancy of, II: 208, 208n.
Punta Careta, 141n.
Punta Charambira, 164.
Punta de Colima, 48n.
Punta de Hicacos, II.350.
Punta de la Bruja, II.350.
Punta de la Desconocida, 124t.
Punta de Lampazos, 538; II.448t.
Punta de Mata-Hambre, 73t.
Punta de Piedras, 406.
Punta del Año Nuevo, 74t, 128t, 457.
Punta Gorda, 125t; II.312, 320.
Punta Grifo, II.350.
Punta Mari Andrea, 125t.
Punta Pariña, II.355.
Punta Pilar, II.350.
Punta San Francisco Solano, 84.
Purificación: town, 399.
Purissíma, famous mine in Catorce: its riches, II.125.
Purissíma Concepción de Alamos de Catorce: mine, 423.
Puy-de-Dôme: dormant volcano in the Auvergne region, France, 105, 173, 177.
Pyrenees: mountain range between France and Spain, 66, 104, 182n, 182t, 388, 520; II: 79, 97, 104n, 119.

Qinling: mountains in China, 174.
Quadra Island: British Columbia, Canada, 451, 465.
Quarta Hoja, 159n.
Quauhnahuac. *See* Cuernevaca.
Quauhquechollan. *See* Huacachula.
Quauhtitlan. *See* Cuautitlan.
Quaxiniquilapa (Cuajinicuilapa): town in the Costa Chica region of Guerrero, Mexico, 39, 402.
Québec, Canada, 463.
Quebrada de la Raspadura: in the kingdom of New Granada, now in Chocó, Colombia, 163; II.87n.
Quebrada de San Pablo: in the kingdom of New Granada, now in Chocó, Colombia, 164.
Quebraloma: mining site, II.230.
Queen Charlotte Islands. See Mendaña Islands.
Queen Charlotte Strait: between Vancouver Island and the Mainland of British Columbia, Canada, 45.
Querétaro: town in the intendancy of Mexico, [19 et passim]; proportion of castes and sexes among its inhabitants, 293; population, 381; its manufactories, II: 291, 295.
Quiabislan. *See* Chiahuitzla.
Quiado: small stream, II.88n.
Quiahutztlán, 385.
Quibdó: city in the kingdom of New Granada, now capital city of Chocó, Colombia, 163, 164; II.230.
Quibdó: river in the kingdom of New Granada, now the Atrato river in Chocó, Colombia, 163, 164; II: 87n, 88n.
Quiechapa (San Pedro Mártir Quiechapa): town and municipality in Oaxaca, Mexico, 478t.
Quilate: forests, II: 24, 25, 26, 27, 461; river in the district of Jacilingo, Jalisco, Mexico, II.24.
Quillota: city in the Aconcagua River valley in central Chile's Valparaíso region, II.208.
Quimixtlan: municipality in Puebla, Mexico, II.28.
Quindiu Mountain (Nevado del Quindio): inactive volcano in the Central Cordillera of the Andes in the kingdom of New Granada, now in Colombia, II.182.

Quispicanchi: one of thirteen provinces in the Cusco Region in the southern highlands of Peru, 262.
Quisuani: settlement in Sonora, Mexico, II.74.
Quito: city in the kingdom of Quito, now in Ecuador, 11ff.
Quivira: fabled town, 431,441n; II.179n.

Racuach (Bacuachito or Bacoachi): mining site in Sonora, Mexico, II.74.
Raijo (Rojo): river in New Mexico, also known as Red River, 91.
Rammelsberg: Harz Mountains, Germany, II.86.
Ramos: in the intendancy of San Luis Potosí, Mexico, 305.
Rapahannoc: river in Virginia, USA, 550.
Raspadura: ravine [in Chocó]: created by a connection between the Atlantic Ocean and the South Sea, 163–64; a small canal dug by a priest between Río San Juan and the River of Quibdó, 163.
Ratan Island (Roatán): in the Caribbean off the northern coast of Honduras, 237n.
Raudal de Maypures: rapids in the kingdom of New Granada, now in Tuparro National Park in Colombia, II.23.
Rawun Rudd (Rawan Hrad): lake in Tibet, 174.
Rayas: mine, 389t.
Real de Ariba: mining site, II.74.
Real de Pozos: mining site, II.180.
Real de Ramos: mining site, 400.
Real de Santa Ana: mining site, II.75.
Real de Todos Santos: mining site, II.73
Real del Cardonal: mining site, now town and one of the eighty-four municipalities of Hidalgo, Mexico, II: 78, 179.
Real del Limón: mining site, II.72.
Real del Monte [town in present-day Hidalgo, Mexico: 132 et passim]; description of its mines, II.215.
Real del Oro: mines, 397.
Rebentón: near Tembladera marsh, which might be Elkhorn Sound, California, II.400t.
Red River of Natchitoches (Cane River): in Louisiana, now USA, 91, 436, 457.
Red Sea: seawater inlet of the Indian Ocean between Africa and Asia, 153, 157; II.257.
Reggio di Calabria: city in southern Italy, 419.
Regla: one of the fifteen boroughs of present-day Havana, Cuba, II: 104, 120.
Reloncaví Gulf (Seno de Reloncaví): body of water in the Los Lagos region of Chile, 139.
Requay: mine, II.207.
Réunion. *See* Bourbon.
Revillagigedo Islands: group of four volcanic islands in the Pacific Ocean, 70, 128t, 469.
Ridge on the road from Cruces to Panama, 155–56.
Rimac: town and river in the kingdom of Quito, now in western Peru and the most important source of potable water for the Lima and Callao metropolitan area, 165, 535. *See also* Lima.
Río Antiguo (River Antigua), Veracruz, Mexico, II.390.
Río Blanco, Cuba, 184n, 511; II.6n.
Río Blanco, [Mexico, 237], 417.
Río Bravo [del Norte]. *See* Río del Norte.
Río Colorado: Texas, 90–91, 146, 149, 421; II.457.
Río de Balzas: old name for the Río Colorado, 429.
Río de Chimalapa: connects two oceans, 150; II.318.

Río de Colorado: its confluence with the Gila, 50; it may help connect the two oceans, 149.
Río de Conchos: Rio Conchos, river in Chihuahua, Mexico, 438.
Rio de Janeiro [Brazil], 476; [II: 41, 51].
Río de la Ascensión, 430.
Río de la Hacha (Riohacha): city in the kingdom of New Granada, now in northern Colombia, II.49.
Río de la Magdalena (Magdalena River): in the kingdom of New Granada, now in Colombia, 88, 155; II: 394, 457.
Río de la Nueces: river in Texas, now USA, 90.
Río de la Plata: river between Argentina and Uruguay, 24, 465; II: 52, 216n, 457.
Río de la Platina, 164.
Río de las Bodegas de Gualán, 165n.
Río de los Brazos de Dios (Brazos River): the eleventh-longest river in the United States, from Blackwater Draw, Curry County, New Mexico, to its mouth at the Gulf of Mexico, 90.
Río de los Nogales (Santa Cruz River?), 423.
Río de Micuipampa (Río Chotano): river in the kingdom of Quito, now in Peru, II.206.
Río de Montezuma: river in Puntarenas province, Costa Rica, 353.
Río de Noanama. *See* Río San Juan.
Río del Norte: might facilitate the trade between the two oceans, 150; description of, 435; its ruin in 1752, 436. [*See also* Río Grande.]
Río del Oro, II.74.
Río Frío: settlement and creek in Texas, now USA, 103, 320.
Río Gila (Gila River): New Mexico and Arizona, now USA, 19 et passim.
Río Grande: river separating Mexico and Texas, 19 et passim.
Río Grande de Motagua (Río Motagua): river in Guatemala, 165n.
Río Mermentas [Mermentau River in southern Louisiana], [309], 420ff.
Río Naipipi (Naixo): small tributary of the Atrato River in the kingdom of New Granada, now Chocó, Colombia, 162, 167n.
Río Negro: largest left tributary of the Amazon River, 11, 214t, 237, 286, 452, 523; II: 88n, 270.
Río Panaloya, 499.
Riobamba: in the kingdom of Quito, now region of the Chimborazo Province, Ecuador, 111, 190, 263; II: 47, 66n.
River of Peace: might be used to help connect the two oceans, 147. [*See also* Columbia River; Ouinigigah River.]
Roanoke: city in North Carolina, now Vierginia, USA, 553, 553n.
Robledo: gorge, narrow pass, 434.
Roca Partida: smallest of the four Revillagigedo Islands in Colima, Mexico, 70, 129t.
Rocky Mountains, 71ff, 181–82, 484.
Rosa de Castilla: Peregrino, II.138.
Rosario: river [in Mexico], 429, 539.
Rosario: town [in Mexico], [20, 57t, 129t, 305, 432], 433; [II: 82, 127, 143t, 197].
Rosario Channel (Gulf of Georgia): now USA, 467.
Rucu-Pichincha: one of the two highest peaks of the Pichincha, an active stratovolcano in the kingdom of Quito, now in Ecuador, whose capital, Quito, wraps around its eastern slopes, II.224.
Ruriquinchay: mining site, II.200.
Russia, Asiatic, 118ff.
Russian America: description of this land, 470ff, [475n; II: 364, 366n].

Sabaleta: river, 162.
Sabatinipa, II.73.

Sabina [Sabine]: river [in Texas and Louisiana, now USA] on whose banks the northeastern Spanish settlements were founded, 442.
Sacabo de Nochistongo. *See* Desagüe and Nochistongo.
Sacapisca, 39.
Sacotecas. *See* Zacatecas.
Sacramento: river in California, now USA, 91.
Sacrificios Islands: in the Gulf of Mexico off the Gulf coastline near the port of Veracruz, Mexico, 97, 414n; II.367.
Sagrario: parish, 292, 325.
Saint Bernard, Petit and Grand: mountain passes in Switzerland, 112, 175, 274, 331n; II.311.
Saint Elias Mountains: elevation of, 113, 180, 460.
Saint Ferréol: reservoir, 328.
Saint George, Gulf of: one of the points where it has been thought possible to connect the two oceans, 166.
Saint Gotthard Pass, II.409.
Saint Joseph, 96.
Saint Kitts (St. Kitts), British Virgin Islands, II.13.
Saint Lucia: British Virgin Islands, II: 397, 408n.
Saint Martin (St. Maarten): island in the Dutch Caribbean, 422.
Saint Michael the Archangel. *See* Archangelsk.
Saint Petersburg, Russia, 275, 330, 468, 474f; II: 268, 308, 366n.
Saint Pierre River, Quebec, Canada, 56t.
Saint Simon: trading post, 474.
Saint Thomas: part of the US Virgin Islands, II:12, 485, 291.
Saint Thomas de Angostura, Guiana, 447.
Saint Vincent: island in the Caribbean, 237; II.13.
Saint-Domingue, [3 et passim. *See also Subject Index*].
Saint-Esprit: archipelago, 27.
Saipan (Saypan): largest of the Northern Mariana Islands, Western Pacific; II.359.
Salamanca: town and plains [in Spain, capital of Salamanca province, part of the Castile and León region], 389.
Saldaña: river in the kingdom of New Granada, now in Colombia, II.183.
Salgado: settlement in Guanajuato, Mexico, II.105.
Salomon Islands: Indian Ocean, II.361.
Saltillo: town [capital and largest city of the northeastern Mexican state of Coahuila], [421n, 422, 426], 428, [538].
Salto de Alvarado: name of a point in Mexico, 342.
Salto del Río de Tula: the end of the Desagüe de Huehuectoca, 368t.
Salvatierra: city in the valley of Huatzindeo in the lowlands of Guanajuato, Mexico, 389, 440, 483; II.5.
Salzburg: Austria, 574; II: 133, 268, 269.
San Agustín de las Cuevas: village in Tlapan, Mexico, 39, 121t, 332–33; II.440.
San Andrés Tenejaca: city in the kingdom of New Granada, now capital city of San Andrés, Colombia, II.493t.
San Andrés Teul, 398.
San Andrés Tuxtla: city and municipality in Veracruz, Mexico, 409; II: 24, 27–28, 493t.
San Ángel: town in Mexico, 332, 562.
San Antonio: city in Texas, now USA. *See also* San Antonio de Bejar.
San Antonio: river in Oaxaca, Mexico, II.75, 86.
San Antonio Abad, 333, 358, 480t.
San Antonio de Alisos, II.74.
San Antonio de Bejar: village, [90], 422–23, [484].
San Antonio de la Cues, 58t.

San Antonio de la Huerta, 74.
San Antonio de la Iguana, II.71.
San Antonio de las Minas, II.70.
San Antonio de las Ventanas, II.73.
San Antonio de los Cues: village, 405t.
San Antonio de los Robles, 51.
San Antonio de Padua: village, 455, [540t; II.486t].
San Antonio de Xacala, II.75.
San Augustin: river, II.230.
San Augustín de las Cuevas: village in Mexico, 562.
San Augustín de Ozumatlan, II.74.
San Benedicto (Isla de los Innocentes): uninhabited, third largest island of the Revillagigedo Islands, 70.
San Bernardino Strait, Philippines, 360.
San Bernardo: lake and plains, 192, 419; II: 114, 120.
San Blas, [8 et passim]; port [in Nayarit, Mexico], 47, 399, 479.
San Blas, Gulf of (Guna Yala): ocean body near San Blas islands, Panama, 83, 159.
San Buenaventura: village, [49, 431, 448, 528], 545, [II.485].
San Buenaventura River, II.76.
San Carlos de Chiloé, 161.
San Carlos de Monterey: capital of New California, II.455.
San Carlos de Perote: fortress, 417.
San Carlos de Vallecillo, II.71.
San Crispín: bastion, 97.
San Cristóbal: lake in Mexico, [109, 177, 328], 353, [355, 359, 362, 370ff; II: 126, 297, 311].
San Cristóbal: mountain and village in Mexico, 75, 79t, 80, 124, 333, 371, 501.
San Demetrio de los Plateros, II.71.
San Diego: military post, II: 448t, 458n.
San Diego: village [in California], 454, 497.
San Dimas, II.73.
San Dionísio: village, 405.
San Eleazario: presidio and military post, II: 436, 458n.
San Esteban, II: 72, 131, 133.
San Felipe: village and plains, 388–89; II: 70, 98, 178, 180.
San Felipe de Bacalar: fort, 406; II.450.
San Felipe Neri: convent, 337n, 342; II.472t.
San Felipe y Santiago: town, 433.
San Fernando: village, 454, [539t; II.485t].
San Fernando Lake, 126
San Francisco: mountain and sandbar, 21, 150.
San Francisco: river in Campeche, Mexico, 406, 408.
San Francisco: village, II.288; northermost settlement of the Spanish [at the time], II.270.
San Francisco Bay, 450.
San Francisco de la Seda: village, II.36.
San Francisco de la Silla, II.73.
San Francisco de Medellín: monastery, II.499.
San Francisco de Pachuca, 481t.
San Francisco del Oro, II.73.
San Francisco Javier de Alisos, II.74.
San Francisco Javier de La Huerta, II.74.
San Francisco los Zacatecas, 482t.
San Francisco Tlaltengo, 355.
San Francisco Xichu, II.72.
San Gabriel: village [in New California], [38, 51, 319, 448], 454, [539t. II: 5, 488t].
San Gerónimo: military post, II.458n.
San Ignacio: mountain, II: 78, 131, 134.
San Jacinto: today [in Humboldt's time] Mount Edgecumbe, discovered by Quadra, 460, [466, 473].
San Jerónimo, II.72.
San Joaquin, II.72.
San Joaquín de los Arrieros, II.73.
San Jorge, Gulf of, 165.
San José: mountain in Peru, Choropampa district, II.206.

San José: village in California: geographical position, 65; details about its mission, II.269.
San José Axusco, 39.
San José de Campeche, Mérida, 482t.
San José de Comangillas: hot springs [in San Jose de Comanjilla, Guanajuato, Mexico], 390.
San José de Copala, II.74.
San José de Gracia: mining site, II.74.
San José de Gracia de Orizaba, 481.
San José de Guichichila: [mining site], 172.
San José de Tayoltita, 173.
San José del Obraje Viejo, 172.
San José del Oro, 172.
San José del Parral: city, 428.
San José Mission, 127t.
San José Tamaulipan, II.71.
San José Tepostitlan, II.72.
San Juan: river in Nicaragua, II.355.
San Juan Bautista: village, 455, [540t].
San Juan Bautista de Río Grande: military post, II.458n.
San Juan Bautista de Panuco, II.71.
San Juan Capistrano: village, [448, 539t; II.485t].
San Juan de la Chica, II: 62, 70, 180, 181.
San Juan de la Cieneguilla, II.73.
San Juan de los Llanos, 388; II.28.
San Juan de Lucanas, II: 146n, 200.
San Juan de Teotihuacán: two Toltec pyramids were found there, 388. *[See also* Pyramids *in Subject Index.]*
San Juan de Ulúa: fort [and castle], [33–34, 150, 193, 395], 414–25; II: [317, 320–21, 348, 410], 459.
San Juan de Ulúa: island and castle in Veracruz, Mexico, 33–34, 150, 193, 365, 414–15; II: 317, 320, 348, 410, 459.
San Juan del Ré: village in Oaxaca, Mexico, II.45.
San Juan del Río: intendancy of and town in Durango, [19, 44, 60, 122t, 132t, 177, 195, 381, 389], 427, [455, 527; II.73].
San Juan Díaz: river, 159n.
San Juan Guetamo: town in the intendancy of Vallaloid, II.178.
San Juan Guetamo Ario, II.74.
San Juan Mountain, 126t.
San Juan Nepomuceno, II.73.
San Juan Sitacora, II.147.
San Juan Tararamco, 396.
San Julián: bay, 166.
San Juaquin: military post, II.458n.
San Lázaro: mountain, plains, 71, 127t, 329, 355, 357; II.480t.
San Lorenzo: name Juan Pérez gave the port of Nootka before Cook). 459–60, [463; II.365]. [*See also* King George's Sound.]
San Lucas: Cape, geographical position, 45ff, [70n, 71n, 74, 126t, 467; II.50]
San Luís de la Paz, 571; II: 70, 180.
San Luis Obispo: village, [51, 448], 454, [539; II.485t].
San Luis Potosí: town, 423 [et passim].
San Luis Rey de Francia: village, 454, [539t].
San Luis, intendancy of: province of San Luis Potosí, 418ff.
San Martín: village, 311.
San Martín Bernalejo, 171.
San Martín de Chacas: city in the kingdom of Quito, now in Peru, II.200.
San Mateo: river and lake, 370, 498.
San Mateo Capulalpa, II.75.
San Mateo del Mar, II.490.
San Mateo el Chico: village, 501.
San Miguel: village, 454.
San Miguel Amatlán, II.75.
San Miguel Chimalapa, II: 490, 493t.
San Miguel Tenango: mine, 388; II.75.
San Miguel de Coneto, II.73.
San Miguel de las Peras, II.75.
San Miguel del Mezquital: city in Zacateca, Mexico, II.73.

San Miguel del Río Blanco, II.72.
San Miguel el Grande: town in the intendancy of Guadalajara, [181], 389, [481; II.178]; manufactories, II.293.
San Nicolás de los Angeles, II.72.
San Nicolás de los Ranchos: village, 114, 122t, 488–89.
San Pablo: isthmus and village, 164, 327, 354; II.87n.
San Pantaleón de la Noria, II.71.
San Pascual: village, 434.
San Pedro: mountain [in intendancy of San Luis Potosí], II: 71, 90.
San Pedro Analco, II.72.
San Pedro Chilchotla, II.28.
San Pedro de Batopilas: town, 428.
San Pedro de los Pozos, II.70.
San Pedro de Tlahua: village built on the dike that separates Lake Chalco from Lake Xochimilco, 354.
San Pedro Island. *See* Mohotani.
San Pedro Lliyapa, 488.
San Pedro Nesicho, II.75.
San Pedro Nolasco, 482; II.491.
San Pedro: river that has disappeared, II.169, 171
San Rafaël: river, 149.
San Rafael de las Flores, II.73.
San Rafael de los Lobos, II: 70, 180.
San Ramón: mining site in Durango, II: 73, 128–29, 139–40, 478t.
San Roque, II.74.
San Roquito, II.103n.
San Saba: river, 418.
San Salvador: country in Central America, 32, 44.
San Salvador Tisayuca, 44.
San Tomás: mining site in the kingdom of Quito, now in Peru, II.200.
San Tomás de Chiconautla: village in Ecatepec de Morelos, Mexico, 354.
San Vicente: province, 72–73, 111n, 180.
San Xavier: river, 149.
San Zimapán, 65n.
Sanagoran: village in the kingdom of Quito, now in Peru, II.200.
Sanchiqueo: village in Michoacán, Mexico, II.74.
Sandwich Islands: former name of the Hawaiian Islands, 45, 307, 439, 456, 463; II: 36, 357–58, 361–62.
Santa Ana: ravine and village, Guanajuato, 46, 205, 206t, 278, 389, 441, 445; II: 70, 183, 231.
Santa Ana: mine, 389.
Santa Barbara: military post and presidio, II.448t, 458n; mines, II: 73, 184ff.
Santa Barbara: village, 454.
Santa Barbara Channel, 449–50, 454, 457.
Santa Barbara de Huancavelica: mountain, II.187.
Santa Buenaventura, 128t.
Santa Catalina: village, 194, 480, 482; II: 48, 73, 75, 200.
Santa Catalina Lachateo, II.75.
Santa Clara, 448–49, 455, 482, 540; II: 74, 486.
Santa Cristina Island. *See* Tahuata.
Santa Cruz: bay in California, 440; village, 455.
Santa Cruz: military post, II.458n.
Santa Cruz de Azulaques, II.72.
Santa Cruz de la Cañada y Taos. *See* Taos.
Santa Cruz de los Flores, II.72.
Santa Cruz de Mayo: port in Mexico, 429; II.363.
Santa Cruz de Mendaña: islands, II.361.
Santa Cruz de Nootka: description of this port, 462–63.
Santa Cruz del Quiche: village, 404.
Santa Eulalia: military post, II.458n.
Santa Fe aqueduct: conveys drinkable water to Mexico City, 332.
Santa Fe: mountain, 56.
Santa Fe [in the Valley of Mexico]: royal gunpowder manufacture, II.298.
Santa Fe [New Mexico, now USA]: 438; about its geographical position, 55.

Santa Fé de Bogotá, Colombia: 99n, 182n, 195, 276, 574; II: 182, 190n, 226t, 228, 250n, 306, 309, 348, 456n.
Santa Fe de Goanajuato. *See* Guanaxuato.
Santa Inés: mountain and village in Mexico, 371, 393–94.
Santa Iñés: hacienda, 79t, 124t, 372
Santa Lucía: village and mountain range in Mexico, 362, 402, 442, 455; II.356.
Santa María: deserted island off the coast of Chile, II.52.
Santa María Chimalapa: town and municipality in Oaxaca, Mexico, II: 490t, 493t.
Santa Maria d'Aorne: port, II: 246, [363].
Santa María de la Mar: village, II.405.
Santa María de las Charcas: market town, 423; [II.71].
Santa María de las Nieves, II.73.
Santa María del Carmen del Sombrero: mining site, II.74.
Santa María del Tule: village, 402.
Santa María Iavecia, II.75.
Santa María, Compuerta de: sluice in the Desague of Huehuetoca, 368t.
Santa Marta: port in Panama, 194; II: 250, 380, 393, 457n.
Santa Rita de Chirangangeo, II.74.
Santa Rosa (Las Adjuntas): mining site, II.74.
Santa Rosa de Cosiguiriachi: town, 427.
Santa Rosa de las Lagunas, II.74.
Santa Rosa de los Osos, 228n.
Santa Rosa Mountain, 388; II: 39, 98, 104, 105, 120, 200.
Santa Rosa Valley, II.182.
Santa Rosa, California, 29, 45, 47.
Santa Teresa: lagoon in New Spain, II.316.
Santander: city in Spain, 125t, 224, 308, 423, 525; river in Spain, 418, 491, 539.
Santiago: mountain, valley, and village, 60, 536, 538; II: 5, 24, 27–28, 71.
Santiago, river: constitutes the border between Mexico and Michoacán, and between the Otomites and Chichimeca, 195, 279; II.179.
Santiago de Buena Esperanza. *See* Purificación.
Santiago de Chucu, II.200.
Santiago de Cuba, II: 9–10, 441t, 456t.
Santiago de las Sabinas, II.71.
Santiago de los Caballeros, II.74.
Santiago de Mapimi, II.73.
Santiago de Tlatelolca: now a suburb of Mexico City, 362.
Santiago de Tuxtla: village, 412.
Santiago Jalisco: Guadalajara, 482t.
Santísima Trinidad, 483.
Santísima Trinidad de Peña Blanca, II.74.
Santíssima Trinidad de Pozole, II.72.
Santixpac (Santispac), Baja California Sur, II.29.
Santo Cristo de las Laxas (Lajas): basilica church in the kingdom of New Granada, now in the southern Colombian Department of Nariño, II.231.
Santo Domingo: island in the Caribbean, II.455. *See also* Saint-Domingue.
Santo Domingo: mining site, II.72.
Santo Domingo de Palenque: city and municipality located in the north of the state of Chiapas, Mexico, 404, 487.
Santo Tomás. *See* Socorro.
Santuario: former settlement on the Papaloapan River, 495.
Saône: river in France, 156, 168.
Saperton (Sapperton): village in Gloucestershire, UK, 360.
Saptin River. *See* Snake River.
Sarabia: river in Mexico, II: 490t, 492n, 493t.

Saracachi (Saracachi): town of Agua Fría in the municipality of Cucurpe, Sonora, Mexico, 74.
Saskatchewan, Canada, 149.
Satlán: town, II.354.
Savoy: region in Central Europe, Western Alps, 173.
Saypan. *See* Saipan.
Sayula: town and municipality in Jalisco, Mexico, 304, 398.
Schemnitz: town in Hungary, 228; II: 79, 93, 109, 127, 130n, 134, 136, 151, 180n.
Schlangenberg: mine in Smeinogorsk, Siberia, II.93.
Schneeberg: mountain in Saxony, Germany, II.93.
Schumagin Islands: group of twenty islands in the Aleutians East Borough south of the mainland of Alaska, 461.
Scilly Island. *See* Society Islands.
Scottish Highlands, 152.
Sea of Cortés. *See* California, Gulf of.
Secho: ravine, II: 108n, 110n.
Segura de la Frontera: now Tepeaca, municipality in Puebla in southeastern Mexico, 387.
Seidewitz Valley: in Saxony, Germany, II.119.
Seine: river and departement in France, 315n, 369.
Selagua: port, 48n, 126t.
Semillete: village near Socorro, New Mexico, no longer in existence, 434.
Senegal: country in West Africa, 384, 505; II.393
Senpualtepec Mountain [Cerro Zempualtepec, Oaxaca, Mexico], 402.
Severn: river in Wales, UK, 148, 360.
Seville: city in Spain, 250, 321, 346, 457n; II: 31, 34, 49, 165n, 300, 369, 405, 411, 415t, 430, 495–96, 498, 500, 505.
Sholey Mountains: New York State, USA, II.66n.
Sianori: mine, II.73.
Sianori: village in Durango, Mexico, II.73.
Siberia, Russia, 148, 233, 472n, 476, 484, 512, 575; II: 16, 33n, 89, 93, 141, 154n, 261, 275, 277, 308, 313, 366n.
Sicily, Italy, 184, 403, 531n; II.396n.
Sierra, provinces of (Argentina): 208.
Sierra de Doña Ana, 435.
Sierra de Guanajuato, II.77.
Sierra de Istepeje, II.48.
Sierra de Acha, 424.
Sierra de los Mimbres, 181.
Sierra de la Summa Paz, 163.
Sierra de los Metates, 428.
Sierra de Pinos, 401, 534; II: 70t, 71, 82t, 90, 92.
Sierra de San Martín, 193, 412.
Sierra de Santa Rosa: description of this mountain chain, [388]; II: 98, [104–5].
Sierra de Tlaxcala, 43, 339n.
Sierra Iztaccihuatl, 488.
Sierra Madre: part of the Mexican cordillera, 181, [191, 426, 430, 500; II.309].
Sierra Negra, II.71.
Sierra Nevada (Iztaccihuatl), 131.
Sierra Nevada de Mérida, 163.
Sierra Nevada: meaning of this name, [24, 40, 62, 112, 131, 134, 163, 320], 382n; [II.299].
Sierra Verde, 149, 181, 187, 426, 435; II.190.
Siguas: in the kingdom of Quito, now district in the province Arequipa, Peru, II.200.
Silao: city in Guanajuato, Mexico, 509, 536; II.492.
Silcai-Yacu (Sidcay): village in the kingdom of Quito, now in Ecuador, II.183n.

Silla de Caracas (Caracas Hills): in the kingdom of New Granada, now in Venezuela, 113.
Silla de Payta: island and hill near Paita in the kingdom of Quito, now in Peru, II.354.
Sillacasa: near Acobamba, Peru (?), II: 185, 187ff.
Simiti Mountains: in the kingdom of New Granada, now region in Colombia, II.228.
Simplon: mountain pass in Switzerland, 112; II: 50, 311.
Sinaloa. *See* Cinaloa.
Sincoque Mountain (Sincoq): in the state of Mexico, Mexico, 177.
Singuilucan: Indian village: tableau of births and deaths from 1750 to 1799, [205–6]; II.466f.
Sinu: river in the kingdom of New Spain): its branches serve the organization of smuggling of gold from Chocó and from Antioquia, [157]; II.233.
Siquani (Sicuani): town in in the kingdom of Quito, now in southern Peru, 263.
Sitiala: near Coyuca de Benítez/Acapulco (?), II.354.
Sitka Bay, Alaska, 474.
Sivirijoa: settlement in Sinaloa, Mexico, II.74.
Sizal (Sisal): seaport town in Hunucmá municipality of Yucatán, Mexico, 408.
Slave Lake: small town in northern Alberta, Canada, 72, 147; II.190.
Slavonia: region in Croatia, 544.
Socabón del Rey: venting drift near Tasco, II.81.
Sochipala: mining site, II: 72, 131.
Society Islands: archipelago in the South Pacific Ocean, 524; II: 358, 361–62.
Soconusco [region in the southwest corner of Chiapas, Mexico, along its border with Guatemala, 36, 142, 180n], 492; [II.49].
Socorro: small volcanic island in the Revillagigedo Islands; also city in New Mexico, 70–71, 129t, 434.
Sola: district, II.45–46.
Soldado point (Soldado Rock): Trinidad and Tobago, 97.
Soledad: village [city in Monterrey, California, now USA], 454, [482n, 540; II: 180, 200, 465].
Sololá: city and department in western Guatemala, 404, 487.
Solomon Islands: South Pacific, 439.
Sombrerete: town, 401; corn is grown in its vicinity, II.124.
Son Mountain: Huayna Potosí in the Cordillera Real, Bolivia, II.224.
Songolica (Zongolica): in the municipality in Veracruz, Mexico, II: 29, 493t.
Sonsonate: city and municipality in El Salvador, 36, 61, 164n, 191; II: 54, 351, 352n, 355.
Sopilote, 39, 77, 126, 132.
Soquitipan (Zoquitipán): settlement in Hidalgo, Mexico, 500.
Sosola (San Jerónimo Sosola): town and municipality in Oaxaca, Mexico, 477.
Soto la Marina: village [now a town in Soto la Marina municipality in Tamaulipas, Mexico, 191–92], 423, 491.
Sotolar Mountain (Cerro El Sotolar): Coahuila, Mexico, II.72.
Sourouma, Japan, II.237.
South Carolina, now USA, 294.
South Pole: Antarctica, magnetic pole, II.398.
South Sea: is it higher than the Atlantic Ocean?, 157.
Spain: country in Europe, 2 et passim.
Spanish Antilles. *See* Antilles, Spanish.
Spanish Lake (Lake Flamand or Lake Tasse) near New Iberia, Louisiana, now USA, 422.

Spanish Peak: in southwestern Huerfano County, Colorado, now USA, 55, 71t, 182, 484.
St. Lawrence: river in Canada, 150.
Steeben: mountains in Bad Steben, Bavaria, Germany, II.144.
Stony Mountains. *See* Rocky Mountains.
Strait of Anian. *See* Anian.
Strait of Juan de Fuca: Salish Sea's outlet to the Pacific Ocean, 45, 47, 74, 59, 447, 465.
Strait of Magellan: navigable sea route in southern Chile separating mainland South America to the north and Tierra del Fuego to the south, 166, 181n; II.397.
Suba Mountain (Cerro de Suba): in the kingdom of New Granada, now in Colombia, 190n.
Subida de Tasco el Viejo (Taxco el Viejo): town in Guerrero, Mexico, II.132.
Suez, Isthmus of, 158, 159n, 162, 168.
Suhl: city in Thuringia, Germany, II.106.
Sultepeque (Sultepec), town and municipality in Mexico State, Mexico, II: 81–82, 95n, 164t, 244t.
Sumatra: Indonesian Island, II.271.
Summer Isles: archipelago in the mouth of Loch Broom, Scottish Highlands, 555.
Sumpango Lake (Zumpango or Tzompango, last of the five interconnected lakes in the Valley of Mexico), 353; is divided into two basins, 354.
Surate (Surat): city in the Indian state of Gujarat, 348.
Suriname: country on the northeastern coast of South America, former Dutch colony, II.12.
Surutato: village in Sinaloa, Mexico, II.74.
Susa Valley: valley in Turin, Piedmont region of northern Italy, 379.
Sutledge River (Sutlej): longest of the five rivers in the Punjab region of northern India and Pakistan, 174.
Swabia: region in southern Germany, 173, 173n.
Sweden: Scandinavian country, 22, 160, 208–9, 315, 388, 400, 544; II: 16, 131, 255, 269, 397n.
Swedish Antilles. *See* Antilles, Swedish.
Swiss Alps, 104n, 110, 182t.
Syracuse: city in Sicily, Italy, 335.
Syria: country in Asia, 435, 533; II.16.
Szkleno. *See* Glashütte.

Tababueto: military post, II.458n.
Tabahueto: town in Durango, Mexico, II.74.
Tabasco: river [now Grijalva River in southeastern Mexico], 492–93, 496.
Tablas: military post, 458n.
Table Mountain: South Africa, 113.
Tacarigua Lake (Lake Valencia), Carabobo and Aragua, Venezuela, 328.
Tachco. *See* Tasco.
Tacoutché-Tessé, river: one of the points that might connect the two oceans, 147; is completely different from the Columbia River, 148.
Tacuba: town in the intendancy of Mexico, [320, 322, 325ff, 341–42, 344, 364], 380, [498. *See also* Tlacopan].
Tacubaya: town in the intendancy of Mexico, [332, 344, 364], 380, [562, 565; II.440].
Tadó: municipality and town in the kingdom of New Granada, now in Chocó, Colombia, 164; II.88.
Tafelberg. *See* Table Mountain.
Tahiti: island in the Pacific, 201, 514, 563; II: 361, 363n, 392n.
Tahuata: part of the Marquesas Islands, French Polynesia, II.361.

Tajo de Ibarra: mine near Zacatecas, Mexico, II.124.
Tajuato: in the kingdom of New Granada, now in Chocó, Colombia, II.88.
Talatlaco. *See* Xalatlaco.
Talea: Villa Talea de Castro, town and municipality in Oaxaca, Mexico, II.75.
Tallenga: Ancash province, Peru, II.207.
Talpan, II: 72, 92n.
Tamana, river, 164, 520n; II.230.
Tamasulapa: river in Guatemala, 402.
Tamaulipas: town [in Tamaulipas, Mexico], II: 247, [213t, 304n, 489, 490, 491].
Tambillo: city in the kingdom of Quito, now in Ecuador, II.200.
Tambo de Calima: in the kingdom of New Granada, now in Colombia, 164.
Tamiagua: [Laguna Tamiahua, in Veracruz, Mexico, 59, 207, 419, 489].
Tamiahua, Barra de, 125t.
Tampico or Pueblo Viejo [city and port in the southeastern part of Tamaulipas, Mexico]: 489; Tampico Lake, 490; climate, 489; population, 490.
Tampico sandbar: Gulf of Mexico, 360, 419, 490.
Tanango (Tenango): Amatzinac River in Morelos and Puebla, Mexico, 319.
Tananitla (Tonanitla): town and municipality in Mexico State, Mexico, 354.
Tancitaro Peak [volcanic mountain in Tancitaro, Michoacán, 68, [132t], 391, [395].
Tanner's Long Peak: mountain in Colorado, now USA, 71t.
Tansitaro. *See* Tancitaro.
Tantoyuca: city in Veracruz, Mexico, 500.
Taos: [town in New Mexico, now USA], [19, 56, 71, 92, 309], 437–38, [446–47]
Tapona: town, II.71.
Tarifa: forest, 82, 150, 317ff.
Tarma: district, intendancy, and province in in the kingdom of Quito, now in Peru, II: 200, 206n.
Tasco: town in the intendancy of Mexico, II.146; mines around it, II.130–31.
Tasco Mountain, 178.
Tase, lake, 422.
Tasis Channel, 462.
Tatarrax: fictive kingdom, 441n.
Tatatila: village and municipality in the mountainous central zone of Veracruz, Mexico, II.28.
Taxco. *See* Tasco.
Tayabamba: city and district in Peru, II.200.
Tayacaxa: mining site, II.206n.
Tchinkegriun: post in Siberia, 472.
Tchinkitané. *See* Sitka Bay.
Tchoka Island (Sakhalin Island), Russia, 472n.
Tecalitan: iron mines, II: 62, 178.
Tecama: village in Veracruz, Mexico, 501.
Tecaxete Mountain, 384.
Teche River: former route of the Mississippi River in Louisiana, now USA, 422.
Tecicapan, II.72.
Tecolutla: river, 412.
Tecomatan: town in Michoacán, Mexico, II.72.
Teguantepec. *See* Tehuantepec.
Teguayo: lake: the Aztecs opened their first trading post there, 437; is possibly the same as Timpanogos, 441n.
Tehuacan de las Granadas: town in the intendancy of Mexico, 387.
Tehuantepec: plains in Oaxaca province, 506; II.318.
Tehuantepec: wind from north-northeast, 1931.
Tehuantepec, Gulf of, 180, 498–99; II: 49, 317, 453.

Tehuantepec, Isthmus of, II.490; one of the points where the two oceans might possibly be connected, 146–47; II.317; the name Cortés gave it, 150.
Tehuantepec, port of: 405, 496ff; inhabitants, 497; climate, 497.
Tehuilotepec: mine in the intendancy of Mexico, II: 78, 132ff.
Tehuilotepec: village in the municipality of Taxco de Alarcön, Guerrero, Mexico, 38, 122t; II: 72, 78, 81, 96, 131ff, 145.
Tehuiloyuca: village in Puebla, Mexico, 76, 79t, 80t, 123t.
Tehuistla (Tehuixtla): city in Morelos, Mexico, II.164t.
Teide: volcano on Tenerife in the Canary Islands, 40, 112–13, 574.
Teignmouth: town in the county of Devon, UK; II.105n.
Teipa: village [in Guerrero, Mexico], 392; [II.17].
Tejuco: village in Brazil, II.89.
Telapón Mountain: volcano in the state of Mexico, Mexico, 320.
Télica Volcano: one of several volcanoes of the Nicaraguan volcanic front, 152.
Temascaltepec: city in the municipality of Temascaltepec, State of Mexico, Mexico, 29, 62; II: 71t, 72, 164, 166.
Temascatio Plains: near San José Temascatio, Guanajuato, Mexico, II.107, 112.
Tembladera: marsh, 415; II.400.
Temetzla: mine, II.75.
Temetzla (Temextla): village in Puebla, Mexico, II.75.
Temihuitlan. *See* Tenochitlan.
Temixtitan: name Cortés gave the capital of Mexico, 321n.
Tenerife: island in the Atlantic, 27, 40, 112–13, 119, 176, 411; II: 184, 395.
Teniztengo: village, 501t.
Tennessee: state in the USA, 545; II: 25, 421.
Tenoriba: mining site in Chihuahua, Mexico, II.74.
Tenoxtitlan. *[See* Tenochtitlan *in the Subject Index.]*
Tenoya: stream in Veracruz, 415n.
Tentila, 478t.
Tentitlan del Camino (Teotitlán), region in Oaxaca, Mexico, 478t.
Tentitlan del Valle (Teotitlán), small village and municipality in the Tlacolula District, Oaxaca, Mexico, 478t.
Teococuilco: town and municipality in Oaxaca, Mexico, 478t.
Teohuacan de las Granadas: in Puebla, Mexico, 387.
Teohuacan de Mizteca. *See* Teohuacan de las Granadas.
Teopectipac: district in Tlaxcala, Mexico, 385.
Teotihuacán: river [in ancient Mesoamerican city in a sub-valley of the Valley of Mexico], 353, 363.
Teotzapotlan: capital of the Zapoteca, 401.
Tepanteria: mining site, II.72.
Tepantitlan (Tepatitlán): city and municipality in Jalisco, Mexico, II.72.
Tepare Valley, 344n.
Tepeaca: mine, II: 164, 493t.
Tepeaca: town in the Cortés's marquisate, 387.
Tepecuacuilco: city in the municipality of Tepecoacuilco de Trujano, Guerrero, Mexico, 38, 122t.
Tepejacac Hill (Tepeyac): elevation in Mexico City, II.109.
Tepetlaque: village in Ecatepec de Morelos, Mexico, 501.
Tepexe de la Seda: in Oaxaca, Mexico, II.38.
Tepeyac: mining site, II.112.
Tepeyacac: river [also city located in the municipality of Zoquitlán, Puebla, Mexico], 353.

Tepic: city [in Nayarit, Mexico], 399.
Teposcolula: district located in the center of the Mixteca Region in Oaxaca, Mexico, 58t, 129t, 478t.
Tequehuen: islands of Chiloe, Chile, II.465.
Tequizquiac: river and municipality in the Zumpango region of the State of Mexico, Mexico, 373.
Tercera Cumbre: mine, II.493.
Términos, Laguna de [largest and one of biologically the richest tidal lagoons on the Gulf Coast of Mexico], 496.
Ternate: military outpost, 430.
Tesechoacan: river and town, 495.
Tetela de Tonatla: mine, II.164.
Tetimpa plain: region near Coapán in central Mexico, 114, 488.
Tetla de Xonotla [settlement in Puebla, Mexico], 388.
Tetlama: city in Morelos, Mexico, 340; II.316n.
Tetlilco: village in the municipality of Zontecomatlan de López y Fuentes, Veracruz, Mexico, II.72.
Tetuachi. *See* Arizpe.
Teutila: San Pedro Teutila, town and municipality in Oaxaca, Mexico, II.26–27.
Texamen, II.73.
Texami: military outpost, II.458n.
Texas: province, 419 [et passim].
Texcoco: city, 76, 123t, 244, 268, 319, 323–23, 326–27, 329, 333n, 337–38, 342n, 343–44, 367, 372, 379.
Texcoco (Texcuco), Lake. *See* Tezcuco, Lake.
Tezcuco: canal project, 372; its dimensions, 372; advantages it would have for trade, 375–76.
Tezcuco: river, 353.
Tezcuco: town, population of, 378; its woolen fabric manufactures, 292. [*See also* Texcoco.]
Tezcuco, Lake: [44, 76, 78, 108, 109], 322; differences of its condition today compared to the time of Cortés; 322; what caused these differences, 326–27; how evaporation has increased due to clearcutting, 328; and also by the Desagüe of Huehuetoca, 328; its size and depth, 328; [332, 338, 351, 353–379, 488; II: 126, 128n, 296, 311, 484.]
Thames: river in England, 148, 360.
Thermal waters: in the valley of Tenochitlán, 351; close to Guanajuato, II.108.
Three Saints Bay: long inlet on the southeast side of Kodiak Island, Alaska, 473.
Thüringerwald: forest in Thuringia, Germany, II.107.
Tiahuanacu plains (Tiwanaku): pre-Columbian polity in western Bolivia, 551.
Tianguetezingo, II.493.
Tibet: country in the Himalayan Mountains, Asia, 173–74, 189, 233, 264–65, 539, 552; II.190.
Tibulca: ancient cavern, 487.
Ticomabacca. *See* Tecomavaca.
Tierra del Fuego: province in Argentina, 182, 236, 462.
Tierra Nueva: northern New Spain territory, 180.
Tierras calientes, 183.
Tigre: river in Costa Rica, 423, 539.
Tihuacan de los Reyes. *See* Teotihuacán.
Tiltil: city and municipality in Chacabuco Province, Santiago Metropolitan Region, Chile, II.208.
Timana: in the kingdom of New Granada, now town and municipality in Huila, Colombia, 162.
Timor: island in the Pacific, 358.
Tinaja Mountain: mountain in New Mexico, now USA, II.123.

Tinta: in the kingdom of Quito, now district in Canchis Province, Peru, 262, 263; II: 206n, 454.
Tisayuca: municipality of Hidalgo, Mexico, 44, 76, 77.
Tivoli: lake in Italy, 350.
Tizapán: town and municipality in Jalisco, Mexico, 323.
Tizatlan: part of the the four independent polities of the confederation of Tlaxcallan in pre-Columbian Mexico; today part of the city of Tlaxcala, Mexico, 385.
Tizicapan, village, 277.
Tlachco. *See* Tasco.
Tlachi, village, II.28.
Tlacochiguaya: town in Oaxaca, Mexico, 402.
Tlacolula (Tlacolula de Matamoros): city and municipality in Oaxaca, Mexico, 500–501; II.28.
Tlacopan. *See* Tacuba.
Tlacotlalpan: town, 417.
Tlahuac, 371.
Tlalmanalco: town and municipality in the far southeastern part of Mexico State, 319.
Tlalnepantla (Tlalnepantla de Baz): city and a municipality in the state of Mexico north of Mexico City, 326.
Tlalpallan. *See* Huehutlapallan.
Tlalpujahua: town and municipality in Michoacán, Mexico, II: 74, 82, 164t, 244t, 253t, 304.
Tlapa (Tlapa de Comonfort): city in the mountain region of Guerrero, Mexico, 385–86; II.164.
Tlapuxahua: mines, 397.
Tlapuxahua [Tlapujagua, town and municipality in the far northeast of Michoacán, Mexico], 397.
Tlaquechiuhcan: district of Tenochtitlan, 327.
Tlascala, town [Tlaxcala, now a small state in central Mexico east of Mexico City]: its manufactures, II.292–93. *[See also Subject Index.]*
Tlatelolco: town founded in 1338, 326; merged with Tenochtitlán, 326.
Tlaxcala Volcano (La Malinche volcano), 42.
Tobolsk: town in Tyumen Oblast, Russia, at the confluence of the Tobol and Irtysh Rivers, 101, 209n, 484–85; II.260.
Tocuistita: mining site, II.74.
Todos los Santos: lake, 139.
Todos los Santos Bay: sheltered bay south of San Diego Bay, 407, 445.
Tolima Peak (Nevado de Tolima): stratovolcano in the kingdom of New Granada, now in Colombia, 181n.
Tollan: country of origin of the Toltec, 232. [*See also* Tula.]
Tollantzinco: Anahuac, 266.
Tolocan. *See* Toluca.
Tololotlan River. *See* Lerma River.
Toltecanila (Toltecamila): in the municipality of Ixcamilpa de Guerrero in Puebla, Mexico, II.75.
Toluca: plains and mountains, 24, 44, 48n, 60, 68, 75, 78, 111, 122t, 131t, 184, 188, 319, 323, 395, 402; II.5.
Toluca: town in the intendancy of Mexico, [44, 62, 122t, 132t, 175, 281, 320], 381, [569; II: 179, 409, 431].
Tomependa, Valley of: Jaén Province, Cajamarca region in Peru, 436, 529; II: 37, 47.
Tomsk: city on the River Tom in Siberia, 484; II.260.
Tonalá: village still part of Jalisco, 304n.
Tongasuca: village, 262; II.453.
Topago: military post, 458n.
Topar: Russian post, 472.
Topia: city in the municipality of Topia, Durango, Mexico, II.73.
Torneo (Torne): river in Sweden on the border with Finland, 242n.

Tortola: island in the British West Indies, II.13.
Tortugas: islands near the Florida Keys in the Gulf of Mexico, II.466.
Tosta: river, 152.
Totamehuacan: now a borough in Puebla, Mexico, 386.
Totolapa: municipality in Chiapas, Mexico, 491.
Totomistla: mining site, II.75.
Totomitxtlahuaca: mining site, II.72.
Totonilco el Grande: basin, baths, and village, 121t, 339; II: 68, 72, 126.
Tototepec: town in Guerrero, Mexico, II.49.
Trancito: mining site, II.74.
Travancore: former kingdom in India, II.446n.
Trinidad: town in Cuba, 72, 73t.
Trinidad: river, 156.
Trinidad: island; support it annually receives from Mexico for administrative expenses, II.441.
Trinity Bay: northeast portion of Galveston Bay, 458.
Tripoli: capital of Libya, II.16.
Trubizin peak. *See* Mount Edgecumbe.
Trujillo: city in Peru, 201, 335; II: 27, 200, 203, 348, 354.
Tschuktschoi-Noss: northeast Cape, Asia, 182.
Tuamini: river in the kingdom of Quito, now in Peru, II.88.
Tubson: military post, II.458n.
Tuguten: Russian post, 472.
Tugulan. *See* Chukchi.
Tuk Island: Green Island, Vancouver, BC, Canada, 474.
Tula: also Tollan, ancient capital of the Toltecs, 232.
Tula: river, 353, 360, 362, 368, 373, 378.
Tulancingo: city in Hidalgo, Mexico, II.75.
Tule (Santa María del Tule): town and a municipality in Oaxaca, Mexico, 402.
Tulha: ruins, capital of Ocosingo, Ocosingo Valley, Chiapas, Mexico, 487.
Tumbez: city in northwestern Peru on the banks of the Tumbes River, 201; II.54.
Tunis: capital of Tunisia, 514, 539; II.16.
Túpica: in the kingdom of New Granasa, now Cúpica, in Chocó, Colombia, 162n, 167n.
Turbaco: in the kingdom of New Granada, now municipality in Bolívar, Colombia, II.23.
Turkey: country in Eurasia, 257; II.20n.
Tuspa Lake: in former province of Tuspa, Jalisco, Mexico, II.132.
Tuspan: port city [also Tuxpan, in Verazcruz, Mexico], 491.
Tustlahuaca (Juxtlahuaca): now district in the Mixteca region, Oaxaca, Mexico, 478.
Tutucuitlalpilco (Totocuitlapilco): village in the municipality of Metepec, 381.
Tuxtla: river, 495; town [capital of Chiapas, Mexico], 484; volcano, 412.
Tuyagualco: former city near Lake Chalco, 565.
Tuyra: river in Panama, 163.
Tyrol: state and region in Austria, 173; II.306.
Tyumen: city on the Tura River in Siberia, Russia, 485.
Tzendales: Maya ruins in Chiapas, Mexico, 404, 487.
Tzintzontzan: capital of the kingdom of Michoacán, 390, 396–97.
Tzompango (Tzompanco). *See* Sumpango.

Uatipan Moutains: near the Sierra Madre, Mexico, 502.
Ubero: Ambergris Caye, island in Belize, 406.
Ucayali: river and city in Peru, 165; II.307.

Ukraine: Eastern Europe, 430; II.138.
Umanak Island: Greenland, 473.
Umea: Westro-Botnia, Sweden, 511n.
Umnak. *See* Umanak Island.
Unalaska Island (Unalaschka): western Aleutians, 447.
Unimak: largest island in the Aleutian chain, Alaska, now USA, 473, 461.
United States of America: population of, 144.
United States of South America, II.455.
Upar Mountains: mountain range in the kingdom of New Granada, now in northeastern Colombia, II.183.
Upper Canada, 218t.
Ural: mountains in Russia, 484; II: 89, 154n, 263, 268n.
Uranienborg: town in Denmark, location of Tycho Brahe's observatory, 30–31.
Uruachi: village in Chihuahua, Mexico, II.73.
Uspanapa [Uxpanapa or Uzpanapa]: river [in Mexico], 492.
Utatlan (Q'umarkaj): now an archaeological site in the southwest of El Quiché, Guatemala, 404, 487.
Utcubamba: province and river in the kingdom of Quito, now in the Amazon region of Peru, II.184.
Uttara-Kuru: ancient kingdom north of India, 430.
Uturicut: [Indian village in the area of the present Gila River Indian reservation], 432.
Utusco: river on the Caxamarca plain in the kingdom of Quito, now in Peru, 544.

Valderas: village in Mexico, 368, 373.
Valdivia: city in southern Chile, 139.
Valencia: city in Columbia, 162–63.
Valencia: county in New Mexico, now USA, 435.
Valenciana: mine in the district of Guanajuato, 389; description of this mine, II.110; comparison with the mine of Himmelsfürst, II.117t.
Valladolid: city. *[See Subject Index.]*
Valladolid: peninsular town in the Yucatán, Mexico, 408 [et passim].
Valladolid de Michoacán: town [also Morelia, in Michoacán, Mexico], 396–97.
Valle San Bartolomé: now Valle de Allende, oldest and now largest city of the municipality of Allende, Chihuahua, Mexico, II.73.
Valles: city in San Luis Potosí, Mexico, 318.
Valparaíso: city in Chile, 164; II.356.
Vancouver Island, Canada, 276, 462, 464, 469.
Var: department in the Provence, France, 486t.
Varinas (Barinas): municipality in Venezuela, II.455–56.
Vega de San Lorenao: San Lorenzo plateau, Tenochtitlán, 544.
Vega de Supia: mining site in New Granada near Marmato, II: 83, 230.
Velay: historical area of France in east Haute-Loire and southeast of the Massif Central, II.229.
Velletri: city south of Rome, Italy, 243, 571.
Venezuela: South America, 141, 145t; II: 21, 385, 455–56.
Venta de Chalco: farm, 121t.
Venta de Chicapa: on the Chicapa River, II.493t.
Venta de Cruces: town on the Chagres River, Panama, 155.
Venta de Soto: farm, 123t.
Venta Vieja: town in the municipality of Eduardo Neri, Guerrero, Mexico, 39.
Ventorillo: dunes in the desert in Veracruz, Mexico, II.399.
Vera Cruz Nueva: city in Mexico, 414.
Veracruz, city: there are three towns with this name. *See* Villarica de la

Veracruz; Veracruz Vieja; Veracruz, port of.
Veracruz, lighthouse of, II.384.
Veracruz, port of. *[See Subject Index.]*
Veracruz Vieja: city, 414.
Veragua: name of five Spanish colonial territorial entities in Central America, 141, 151, 157–58, 179, 335.
Verdozas: mining site in Temascaltepec, Mexico, II.72.
Vermellon (Vermillion) ravine: Clay County, South Dakota, now USA, II.183.
Vermont: northeastern USA, 294.
Vertideros: sluice south of Huehuetoca, 368, 372ff; II.484.
Vesuvius: volcano in southern Italy, 105, 395, 412n, 575.
Vibora shoals: original Spanish name of Pedro Bank, 72.
Viceroys' Staircase: Mexico City, 369.
Victoria: town [old town near Tlacotalpan, Guanajuato, Mexico], 417.
Vieja California. *See* California, Old.
Vienna: capital of Austria, 31, 243, 571, 574; II.416.
Viga canal: Mexico City, 351.
Vigas Mountain: south of Las Vigas, II.92n.
Villa de Armas: ruins in the kingdom of Quito, now in Peru, II.229.
Villa Hermosa: town [in Tabasco, Mexico]: 417, 492–93; population of, 493; its position, 493.
Villa Nueva: town in Zacatecas, Mexico, II.178.
Villa Rica Mountains: near Veracruz, Mexico, 193.
Villalpando: village and mining site near Guanajuato, Mexico, 407.
Villalta: village near Cerro Zempoaltepec north of Zempoaltepetl in Oaxaca, Mexico, 402, 405; II: 75, 84t, 90, 64t.
Villarica de la Veracruz: colony founded by Cortés, II.210.
Vilotepec: [Jilotepec de Molina Enríquez and Jilotepec de Abasolo, town and a municipality in the northwest of the State of Mexico, Mexico, 497.
Virgin Islands: Caribbean Sea, 237n.
Virginia, USA, 511 et passim.
Viroviro (Viro Viro): village in the kingdom of New Granada, now in Chocó, Colombia, II.88.
Visayan Islands (Visayas or Bisayas): central Philippines, 444.
Viscaína: description of this mine, II.128.
Volcán de Agua: stratovolcano in Sacatepéquez and Escuintla, Guatemala, 111, 180n.
Volcán de Fuego (Volcán de Guatemala): active stratovolcano in Guatemala, 111, 180n.
Volcanoes of the Virgins: California, now USA, 59.
Vologda: city in northwestern Russia, 486t.
Volterra: Etruscan city in Tuscany, Italy, 386.
Vuldivui: mining site in Pataz in the kingdom of Quito, now in Peru, II.183.

Wabash River: in Ohio and Indiana, USA, 150.
Wahitaho. *See* Tahuata.
Wartha (Warta): river in western-central Poland, tributary of the Oder, 173.
Washington, city of: its position, 73 [et passim].
West Indies. *See* Antilles.
Wheal Rock: Cornwall, UK, 67.
Wieliczka: Poland, 138; II.190.
Willkapampa (Vilcabamba or Espíritu Pampa): city founded by Manco Inca in 1539, 262.
Wunsiedel: city in the Fichtelgebirge, Germany, II.144.
Württemberg: region in southern Germany, II.16.

Xacala (Jacala): town and municipality in Hidalgo, Mexico, II: 72, 75, 78.
Xalapa: town [in Veracruz, Mexico], 488 [et passim].
Xalpa: town in the municipality of Cochoapa el Grande, Guerrero, Mexico, II.72.
Xaltilolco. *See* Tlatelolco.
Xaltocan: lake [former lake in the northern Valley of Mexico], [321], 354, [362].
Xaltocan: pre-Colombian village and island in the Valley of Mexico in the center of Lake Xaltocan, 45, 79t, 123t, 277, 304.
Xamapa, river: plan to use it to bring water to Veracruz, 416.
Xamiltepec: now Jamiltepec, Oaxaca, Mexico, 58t, 129t.
Xampolan: Indian village [village in the Yucatán region of Mexico], 408.
Xauxa: city in the kingdom of Quito, now Jauja, Peru, II.33n.
Xerecuaro (Jerecuaro): city in the lowlands of Guanajuato state in Mexico, II.74.
Xicayan: village in Oaxaca, Mexico, II.47.
Xico, village: located in the lake of Chaleo, II.97.
Xicochimalco: village in Jalapa, Mexico, also known as Xico, 28.
Xicotlan: saline in the intendancy of Puebla, 386.
Xochicalco: Miacatlán municipality, Morelas State, Mexico, II.28.
Xochicalco: military entrenchment, 340; antique monument, 454.
Xochimilco: lake [in the Valley of Mexico south of Tenochtitlán, 75, 327, 329–30, 350], 353ff, [371, 377, 488].
Xochimilco: part of the city of Tenochtitlán, 327.
Xochitepeque (Xochitepec): municipality in Morelos, Mexico, 340.
Xoconochco. *See* Soconusco.
Xoloc: fort, 326.
Xorullo. *See* Jorullo.
Xuchitepeque: village in Mexico, 39.

Yagualica Mountains (Yahualica): in northeastern Jalisco, Mexico, 502.
Yakutsk: Russian port city in East Siberia, 484.
Yanacanche (Yanacancha): in the province of Tarma, Mexico, II.200.
Yapel (Villa de Cuscus): region in Chile, II.208.
Yaquesila: river [Jaquesila, river in New Mexico, now USA], 437.
Yaqui: river [in Sonora, Mexico, also Sonra River], 429–30.
Yariquisa: cataract in the kingdom of Quito, now in Peru, II.37.
Yatipan: village in Hidalgo, Mexico, 501.
Yauricocha: mine [south of Lima in the kingdom of Quito, now in Peru, II: 79, 197, 200], 201ff, [206n, 207, 233, 244t, 253t].
Ycatoca Mountains: in the kingdom of Quito, now in Peru, II.208.
Yecorato: hamlet in the state of Choix, Sinaloa, II.74.
Yecuatla: now municipality in Veracruz, Mexico, II.24.
Yeniseisk (Yeniseysk): town in Krasnoyarsk Krai, Russia, 484.
Yguala (Iguala): municipality in Guerrero, Mexico, 205–6, 292.
Ynde: military post, II.458n.
Ynguaran: mine, 397.
Yonne: department in France named after the river Yonne, 168.
Yquilao: island on the west coast of Chile, II.465.
Yquique (Iquique): port and town in Chile, II.203.
Yro: mining site, II.88.
Yucatán: province of: original name of New Spain, 142; it became the intendancy of Mérida, 405–6.

Yucatán Peninsula, 25 et passim.
Yucuatl: indigeneous name for Nootka, 462–63, 466. *See* [*also*] Nootka.
Yunguaran: mine, II.178.
Yxtepexi [Yxtepexe]: mines [in Oaxaca, Mexico], 405, [478t; II.91].

Zacatecas, town: aerolite that was found there, 427. *[For intendancy of Zacatecas, see Subject Index.]*
Zacatollan: region between the Río Nexpa and the Río Zacatula, II.66.
Zacatula: river [now Río la Unión near Zacatula], 187.
Zacatula: seaport in the intendancy of Mexico, 380–81.
Zacualtipán: municipality in Hidalgo, Mexico, 501.
Zaguananas River. *See* Colorado River.
Zangling. *See* Qinling.
Zapoteca: region of the Zapotec peoples, primarily in Oaxaca, Kingdom of New Spain, 234, 266, 384, 401ff; II.64.
Zapotilti (now Zapotiltic): town in Jalisco, Mexico, 398.
Zapotlan: village and now municipality in Jalisco, Mexico, 106, 304, 398, 401.
Zataque: former mining site in Mexico, II.74.
Zauchila: pre-Columbian city in Oaxaca, Mexico, 477.
Zautla: now municipality in Puebla, Mexico, II.75.
Zeguengue: former municipality near Popayán, Kingdom of New Granada, II.183.
Zelaya (Celaya): city in Guanajuato, Mexico, 19, 21, 142, 177, 187, 195, 254, 398, 536ff; II.297.
Zimapán: city, 15, 62, 67, 133, 379; II: 78, 81ff, 91–92, 143, 179.
Zimapán: mine in the intendancy of Mexico, 381.
Zimapán Mountains: Hidalgo region, Mexico, II.127.
Zimatlán: now a district west of the Central Valley region in Oaxaca, Mexico, 478t; II: 45, 47.
Zipaquirá: mine, II.190.
Zitara. *See* Quibdó.
Zitara: river, 163–64; II.230.
Zitlaltepec Mountains: in the Zumpango region north of Mexico City, 373.
Zomelahuacan: mine, 417; [II.75].
Zumpango, lake. *See* Sumpango Lake.

Vera M. Kutzinski is the Martha Rivers Ingram Professor of English and professor of comparative literature at Vanderbilt University.

Ottmar Ette is professor of Romance literatures at University of Potsdam, Germany.

J. Ryan Poynter is Associate Vice Provost for Undergraduate Academic Affairs at New York University.

Kenneth Berri is an independent translator.

Giorleny Altamirano Rayo is an independent scholar of political science.

Tobias Kraft is project director of Alexander von Humboldt auf Reisen—Wissenschaft aus der Bewegung at the Berlin-Brandenburg Academy of Sciences in Berlin, Germany.